Handbook of Empirical Economics and Finance

Edited by

Aman Ullah

University of California

Riverside, California, USA

David E. A. Giles

University of Victoria

British Columbia, Canada

CRC Press is an imprint of the
Taylor & Francis Group, an **informa** business
A CHAPMAN & HALL BOOK

Chapman & Hall/CRC
Taylor & Francis Group
6000 Broken Sound Parkway NW, Suite 300
Boca Raton, FL 33487-2742

First issued in paperback 2017

ISBN-13: 978-1-4200-7035-4 (hbk)
ISBN-13: 978-1-138-11366-4 (pbk)

Library of Congress Cataloging-in-Publication Data

Handbook of empirical economics and finance / [edited by] Aman Ullah, David E.A. Giles.
p. cm. -- (Statistics: textbooks and monographs)
Includes bibliographical references and index.
ISBN 978-1-4200-7035-4 (hardcover : alk. paper)
1. Econometrics. 2. Finance--Econometric models. I. Ullah, Aman. II. Giles, David E. A., 1949- III. Title. IV. Series.

HB139.H363 2011
330.01'5195--dc22 2010044029

Visit the Taylor & Francis Web site at
http://www.taylorandfrancis.com

and the CRC Press Web site at
http://www.crcpress.com

Handbook of Empirical Economics and Finance

STATISTICS: Textbooks and Monographs

Recent Titles

The EM Algorithm and Related Statistical Models, *edited by Michiko Watanabe and Kazunori Yamaguchi*

Multivariate Statistical Analysis, Second Edition, Revised and Expanded, *Narayan C. Giri*

Computational Methods in Statistics and Econometrics, *Hisashi Tanizaki*

Applied Sequential Methodologies: Real-World Examples with Data Analysis, *edited by Nitis Mukhopadhyay, Sujay Datta, and Saibal Chattopadhyay*

Handbook of Beta Distribution and Its Applications, *edited by Arjun K. Gupta and Saralees Nadarajah*

Item Response Theory: Parameter Estimation Techniques, Second Edition, *edited by Frank B. Baker and Seock-Ho Kim*

Statistical Methods in Computer Security, *edited by William W. S. Chen*

Elementary Statistical Quality Control, Second Edition, *John T. Burr*

Data Analysis of Asymmetric Structures, *Takayuki Saito and Hiroshi Yadohisa*

Mathematical Statistics with Applications, *Asha Seth Kapadia, Wenyaw Chan, and Lemuel Moyé*

Advances on Models, Characterizations and Applications, *N. Balakrishnan, I. G. Bairamov, and O. L. Gebizlioglu*

Survey Sampling: Theory and Methods, Second Edition, *Arijit Chaudhuri and Horst Stenger*

Statistical Design of Experiments with Engineering Applications, *Kamel Rekab and Muzaffar Shaikh*

Quality by Experimental Design, Third Edition, *Thomas B. Barker*

Handbook of Parallel Computing and Statistics, *Erricos John Kontoghiorghes*

Statistical Inference Based on Divergence Measures, *Leandro Pardo*

A Kalman Filter Primer, *Randy Eubank*

Introductory Statistical Inference, *Nitis Mukhopadhyay*

Handbook of Statistical Distributions with Applications, *K. Krishnamoorthy*

A Course on Queueing Models, *Joti Lal Jain, Sri Gopal Mohanty, and Walter Böhm*

Univariate and Multivariate General Linear Models: Theory and Applications with SAS, Second Edition, *Kevin Kim and Neil Timm*

Randomization Tests, Fourth Edition, *Eugene S. Edgington and Patrick Onghena*

Design and Analysis of Experiments: Classical and Regression Approaches with SAS, *Leonard C. Onyiah*

Analytical Methods for Risk Management: A Systems Engineering Perspective, *Paul R. Garvey*

Confidence Intervals in Generalized Regression Models, *Esa Uusipaikka*

Introduction to Spatial Econometrics, *James LeSage and R. Kelley Pace*

Acceptance Sampling in Quality Control, *Edward G. Schilling and Dean V. Neubauer*

Applied Statistical Inference with MINITAB®, *Sally A. Lesik*

Nonparametric Statistical Inference, Fifth Edition, *Jean Dickinson Gibbons and Subhabrata Chakraborti*

Bayesian Model Selection and Statistical Modeling, *Tomohiro Ando*

Handbook of Empirical Economics and Finance, *Aman Ullah and David E. A. Giles*

Contents

Preface

Econometrics originated as a branch of the classical discipline of mathematical statistics. At the same time it has its foundation in economics where it began as a subject of quantitative economics. While the history of the quantitative analysis of both microeconomic and macroeconomic behavior is long, the formal of the sub-discipline of econometrics per se came with the establishment of the Econometric Society in 1932, at a time when many of the most significant advances in modern statistical inference were made by Jerzy Neyman, Egon Pearson, Sir Ronald Fisher, and their contemporaries. All of this led to dramatic and swift developments in the theoretical foundations of econometrics, followed by commensurate changes that took place in the application of econometric methods over the ensuing decades. From time to time these developments have been documented in various ways, including various "handbooks." Among the other handbooks that have been produced, *The Handbook of Applied Economic Statistics* (1998), edited by Aman Ullah and David. E. A. Giles, and *The Handbook of Applied Econometrics and Statistical Inference* (2002), edited by Aman Ullah, Alan T. K. Wan, and Anoop Chaturvedi (both published by Marcel Dekker), took as their general theme the over-arching importance of the interface between modern econometrics and mathematical statistics.

However, the data that are encountered in economics often have unusual properties and characteristics. These data can be in the form of micro (cross-section), macro (time-series), and panel data (time-series of cross-sections). While cross-section data are more prevalent in the applied areas of microeconomics, such as development and labor economics, time-series data are common in finance and macroeconomics. Panel data have been used extensively in recent years for policy analysis in connection with microeconomic, macroeconomic and financial issues. Associated with each of these types of data are various challenging problems relating to model specification, estimation, and testing. These include, for example, issues relating to simultaneity and endogeneity, weak instruments, average treatment, censoring, functional form, nonstationarity, volatility and correlations, cointegration, varying coefficients, and spatial data correlations, among others. All these complexities have led to several developments in the econometrics methods and applications to deal with the special models arising. In fact many advances have taken place in financial econometrics using time series, in labor economics using cross section, and in policy evaluations using panel data. In the face of all these developments in the economics and financial econometrics, the motivation behind this *Handbook* is to take stock of the subject matter of empirical economics and finance, and where this research field is likely to head in the near future. Given this objective, various econometricians who are acknowledged international experts in their particular fields were commissioned to guide us about the fast, recent growing research in economics and finance. The contributions in this *Handbook* should prove to be useful for researchers, teachers, and graduate students in economics, finance, sociology, psychology, political science, econometrics, statistics, engineering, and the medical sciences.

The *Handbook* contains sixteen chapters that can be divided broadly into the following three parts:

1. Micro (Cross-Section) Models
2. Macro and Financial (Time-Series) Models
3. Panel Data Models

Part I of the *Handbook* consists of chapters dealing with the statistical issues in the analysis of econometric models analysis with the cross-sectional data often arising in microeconomics. The chapter by Cameron and Miller reviews methods to control for regression model error that is correlated within groups or clusters, but is uncorrelated across groups or clusters. The importance of this stems from the fact that failure to control for such clustering can lead to an understatement of standard errors, and hence an overstatement of statistical significance, as emphasized most notably in empirical studies by Moulton and others. These may lead to misleading conclusions in empirical and policy work. Cameron and Miller emphasize OLS estimation with statistical inference based on minimal assumptions regarding the error correlation process, but they also review more efficient feasible GLS estimation, and the adaptation to nonlinear and instrumental variables estimators. Trivedi and Munkin have prepared a chapter on the regression analysis of empirical economic models where the outcome variable is in the form of non-negative count data. Count regressions have been extensively used for analyzing event count data that are common in fertility analysis, health care utilization, accident modeling, insurance, and recreational demand studies, for example. Several special features of count regression models are intimately connected to discreteness and nonlinearity, as in the case of binary outcome models such as the logit and probit models. The present survey goes significantly beyond the previous such surveys, and it concentrates on newer developments, covering both the probability models and the methods of estimating the parameters of these models. It also discusses noteworthy applications or extensions of older topics. Another chapter is by Fagan and Gençay dealing with textual data econometrics. Most of the empirical work in economics and finance is undertaken using categorical or numerical data, although nearly all of the information available to decision-makers is communicated in a linguistic format, either through spoken or written language. While the quantitative tools for analyzing numerical and categorical data are very well developed, tools for the quantitative analysis of textual data are quite new and in an early stage of development. Of course, the problems involved in the analysis of textual data are much greater than those associated with other forms of data. Recently, however, research has shown that even at a coarse level of sophistication, automated textual processing can extract useful knowledge from very large textual databases. This chapter aims to introduce the reader to this new field of textual econometrics, describe the current state-of-the-art, and point interested researchers toward useful public resources.

In the chapter by Golan and Greene an information theoretic estimator is developed for the mixed discrete choice model used in applied microeconomics. They consider an extension of the multinomial model, where parameters are commonly assumed to be a function of the individual's socio-economic characteristics and of an additive term that is multivariate distributed (not necessarily normal) and correlated. This raises a complex problem of determining large number of parameters, and the current solutions are all based on simulated methods. A complementary approach for handling an underdetermined estimation problem is to use an information theoretic estimator, in which (and unlike the class of simulated estimators) the underdetermined problem is converted into a constrained optimization problem where all of the available information enters as constraints and the objective functional is an entropy measure. A friendly guide for applying it is presented. The chapter by Racine looks into the issues that arise when we are dealing with data on economic variables that have nonlinear relationship of some unknown form. Such models are called nonparametric. Within this class of models his contribution emphasizes the case where the regression variables include both continuous and discrete (categorical) data (nominal or ordinal). Recent work that explores the relationship between Bayesian and nonparametric kernel methods is also emphasized. The last two chapters in Part I are devoted to exploring some theoretical contributions. Grendár and Judge introduce fundamental large deviations theory, a subfield of probability theory, where the typical concern is about the asymptotic (large sample) behavior, on a logarithmic scale, of a probability of a given event. The results discussed have implications for the so-called maximum entropy methods, and for the sampling distributions for both nonparametric maximum likelihood and empirical likelihood methods. Finally, Antoine and Renault consider a general framework where weaker patterns of identification may arise in a model. Typically, the data generating process is allowed to depend on the sample size. However, contrary to what is usually done in the literature on weak identification, they suggest not to give up the goal of efficient statistical inference: even fragile information should be processed optimally for the purpose of both efficient estimation and powerful testing. These insights provide a new focus that is especially needed in the studies on weak instruments.

Part II of the *Handbook* contains chapters on time series models extensively used in empirical macroeconomics and finance. The chapter by Fukač and Pagan looks at the development of macro-econometric models over the past sixty years, especially those that have been used for analyzing policy options. They classify them in four generations of models, giving extremely useful details and insights of each generation of models with their designs, the way in which parameters were quantified, and how they were evaluated. Abadir and Talmain explore an issue existing in many macroeconomic and aggregate financial time-series. Specifically, the data follow a nonlinear long-memory process that requires new econometric tools to analyze them. This is because linear ARIMA modeling, often used in standard empirical work, is not consistent with the real world macroeconomic and financial data sets. In view of this Abadir and Talmain have explored econometric aspects of nonlinear modeling guided by economic theory. The chapter by Ludvigson and Ng develops the relationship between bond excess premiums and the macroeconomy by considering factors augmented panel regression of 131 months. Macroeconomic factors are found to have statistically significant predictive power for excess bond returns. Also, they show that forecasts of excess bond returns (or bond risk premia) are countercyclical. This implies that investors are compensated for risks associated with recessions. In another chapter Pesaran explores the predictability of asset returns and the empirical and theoretical basis of the efficient market hypothesis (EMH). He first overviews the statistical properties of asset returns at different frequencies and considers the evidence on return predictability, risk aversion and market efficiency. The chapter then focuses on the theoretical foundation of the EMH, and shows that market efficiency could coexist with heterogeneous beliefs and individual irrationality provided that individual errors are cross-sectionally weakly dependent, but at times of market euphoria or gloom these individual errors are likely to become cross-sectionally strongly dependent, so that the collective outcome could display significant departures from market efficiency. In deviation with the above chapters in this part, which deal with the often used classical point data estimation, Arroyo, González-Rivera and Maté review the statistical literature on the regression analysis and forecasting with the interval-valued and histogram-valued data sets that are increasingly becoming available in economics and finance. Measures of dissimilarities are presented which help us to evaluate forecast errors from different methods. They also provide applications relating to forecasting the daily interval low/high prices of the S&P500 index, and the weekly cross-sectional histogram of the returns to the constituents of the S&P500 index.

Part III of the *Handbook* contains chapters on the types of panel data and spatial models which are increasingly becoming important in analyzing complex economic behavior and policy evaluations. While there has been an extensive growth of the literature in this area in recent years, at least two issues have remained underdeveloped. One of them relates to the econometric issues that arise when analyzing panel models that contain time-series dynamics through the presence of lagged dependent variables. Hsiao, in his chapter, reviews the literature on dynamic panel data models in the presence of unobserved heterogeneity across individuals and over time, from three perspectives: fixed vs. random effects specification; additive vs. multiplicative effects; and the maximum likelihood vs. methods of moments approach. On the other hand, Su and Ullah, in their chapter, explore the often ignored issue of the nonlinear functional form of panel data models by adopting both nonparametric and semiparametric approaches. In their review they focus on the recent developments in the econometrics of conventional panel data models with a one-way error component structure; partially linear panel data models; varying coefficient panel data models; nonparametric panel data models with multi-factor error structure; and nonseparable nonparametric panel data models. Within the framework of panel data or purely cross-sectional data sets we also have the issues that arise when the dependence across cross-sectional units is related to location and distance, as is often found in studies in regional, urban, and agricultural economics. The chapter by Baltagi deals with this area of study and it introduces spatial error component regression models, and the associated methods of estimation and testing. He also discusses some of the issues related to prediction using such models, and studies the performance of various panel unit root tests when spatial correlation is present. Finally, the chapter by Lee and Yu studies the maximum likelihood estimation of spatial dynamic panel data where both the cross-section and time-series observations are large. A new estimation method, based on a particular data transformation approach, is proposed which may eliminate time dummy effects and unstable or explosive components. A bias correction procedure for these estimators is also suggested.

In summary, this *Handbook* brings together both review material and new methodological and applied results which are extremely important to the current and future frontiers in empirical economics and finance. The emphasis is on the inferential issues that arise in the analysis of cross-sectional, time-series, and panel data–based empirical models in economics and finance and in related disciplines. In view of this, the contents and scope of the *Handbook* should have wide appeal. We are very pleased with the final outcome and we owe a great debt to the authors of the various chapters for their marvelous support and cooperation in the preparation of this volume. We are also most grateful to Damaris Carlos and Yun Wang, University of California, Riverside, for the efficient assistance that they provided. Finally, we thank the fine editorial and production staff at Taylor & Francis, especially David Grubbs and Suzanne Lassandro, for their extreme patience, guidance, and expertise.

Aman Ullah

David E. A. Giles

About the Editors

Aman Ullah is a Distinguished Professor and Chair in the Department of Economics at the University of California, Riverside. A Fellow of the *Journal of Econometrics* and the National Academy of Sciences (India), he is the author and coauthor of 8 books and over 125 professional papers in economics, econometrics, and statistics. He is also an Associate Fellow of CIREQ, Montreal, Research Associate of Info-Metrics Institute, Washington, and Senior Fellow of the Rimini Centre for Economic Analysis, Italy. Professor Ullah has been a coeditor of the journal *Econometric Reviews*, and he is currently a member of the editorial boards of *Econometric Reviews, Journal of Nonparametric Statistics, Journal of Quantitative Economics, Macroeconomics and Finance in Emerging Market Economies*, and *Empirical Economics*, among others. Dr. Ullah received the Ph.D. degree (1971) in economics from the Delhi School of Economics, University of Delhi, India.

David E.A. Giles is a Professor in the Department of Economics at the University of Victoria, British Columbia, Canada. He is the author of numerous journal articles and book chapters, and author and coauthor of five books including the book *Seemingly Unrelated Regression Equations Models* (Marcel Dekker, Inc.). He has served as an editor of *New Zealand Economic Papers* as well as an associate editor of *Journal of Econometrics* and *Econometric Theory*. He has been the North American editor of the *Journal of International Trade and Economic Development* since 1996, and he is currently associate editor of *Communications in Statistics* and a member of the editorial boards of the *Journal of Quantitative Economics, Statistical Papers*, and *Economics Research International*, among others. Dr. Giles received the Ph.D. degree (1975) in economics from the University of Canterbury, Christchurch, New Zealand.

List of Contributors

Karim M. Abadir
Business School
Imperial College London
and University of Glasgow
k.m.abadir@imperial.ac.uk

Bertille Antoine
Department of Economics
Simon Fraser University
Burnaby, British Columbia, Canada
bertille_antoine@sfu.ca.

Javier Arroyo
Department of Computer Science
and Artificial Intelligence
Universidad Complutense
de Madrid
Madrid, Spain
javier.arroyo@fdc.ucm.es

Badi H. Baltagi
Department of Economics
and Center for Policy Research
Syracuse University
Syracuse, New York
bbaltagi@maxwell.syr.edu

A. Colin Cameron
Department of Economics
University of California – Davis
Davis, California
accameron@ucdavis.edu

Stephen Fagan
Department of Economics
Simon Fraser University
Burnaby, British Columbia, Canada
sfagan@sfu.ca

Martin Fukač
Reserve Bank of New Zealand
Wellington, New Zealand
martin.fukac@rbnz.govt.nz

Ramazan Gençay
Department of Economics
Simon Fraser University
Burnaby, British Columbia, Canada
gencay@sfu.ca

Amos Golan
Department of Economics
and the Info-Metrics Institute
American University
Washington, DC
agolan@american.edu20016-8029

Gloria González-Rivera
Department of Economics
University of California, Riverside
Riverside, California
gloria.gonzalez@ucr.edu

William H. Greene
Department of Economics
New York University
School of Business
New York, New York
wgreene@stern.nyu.edu

Marian Grendár
Department of Mathematics
FPV UMB,
Bansk a Bystrica, Slovakia
Institute of Mathematics and CS
of Slovak Academy of Sciences
(SAS) and UMB, Bansk a Bystrica
Institute of Measurement Sciences
SAS, Bratislava, Slovakia
marian.grendar@savba.sk

Cheng Hsiao
Department of Economics
University of Southern California
Wang Yanan Institute for Studies in Economics, Xiamen University
Department of Economics and Finance, City University of Hong Kong
chsiao@usc.edu

George Judge
Professor in the Graduate School
University of California
Berkeley, California
gjudge@berkeley.edu

Lung-fei Lee
Department of Economics
The Ohio State University
Columbus, Ohio
lee.1777@osu.edu

Sydney C. Ludvigson
Department of Economics
New York University
New York, New York
sydney.ludvigson@nyu.edu

Carlos Maté
Universidad Pontificia de Comillas
Institute for Research in Technology (IIT)
Advanced Technical Faculty of Engineering (ICAI)
Madrid, Spain
cmate@upcomillas.es

Douglas Miller
Department of Economics
University of California – Davis
Davis, California
dlmiller@ucdavis.edu

Murat K. Munkin
Department of Economics
University of South Florida
Tampa, Florida
mmunkin@coba.usf.edu

Serena Ng
Department of Economics
Columbia University
New York, New York
serena.ng@columbia.edu

Adrian Pagan
School of Economics and Finance
University of Technology Sydney
Sydney, Australia
adrian.pagan@uts.edu.au

M. Hashem Pesaran
Faculty of Economics
University of Cambridge
Cambridge, United Kingdom
mhp1@cam.ac.uk

Jeffrey S. Racine
Department of Economics
McMaster University
Hamilton, Ontario, Canada
racinej@mcmaster.ca

Eric Renault
Department of Economics
University of North Carolina at Chapel Hill
CIRANO and CIREQ
renault@email.unc.edu

Liangjun Su
School of Economics
Singapore Management University
Singapore
ljsu@smu.edu.sg

Gabriel Talmain
Imperial College London
and University of Glasgow
Glasgow, Scotland
g.talmain@lbss.gla.ac.uk

Pravin K. Trivedi
Department of Economics
Indiana University
Bloomington, Indiana
trivedi@indiana.edu

Aman Ullah
Department of Economics
University of California – Riverside
Riverside, California
aman.ullah@ucr.edu

Jihai Yu
Guanghua School of Management
Beijing University
Department of Economics
University of Kentucky
Lexington, Kentucky
jihai.yu@uky.edu

1

Robust Inference with Clustered Data

A. Colin Cameron and Douglas L. Miller

CONTENTS

1.1 Introduction

In this survey we consider regression analysis when observations are grouped in clusters, with independence across clusters but correlation within clusters. We consider this in settings where estimators retain their consistency, but statistical inference based on the usual cross-section assumption of independent observations is no longer appropriate.

Statistical inference must control for clustering, as failure to do so can lead to massively underestimated standard errors and consequent over-rejection using standard hypothesis tests. Moulton (1986, 1990) demonstrated that this problem arises in a much wider range of settings than had been appreciated by microeconometricians. More recently Bertrand, Duflo, and Mullainathan (2004) and Kézdi (2004) emphasized that with state-year panel or repeated cross-section data, clustering can be present even after including state and year effects and valid inference requires controlling for clustering within state. Wooldridge (2003, 2006) provides surveys and a lengthy exposition is given in Chapter 8 of Angrist and Pischke (2009).

A common solution is to use "cluster-robust"standard errors that rely on weak assumptions – errors are independent but not identically distributed across clusters and can have quite general patterns of within-cluster correlation and heteroskedasticity – provided the number of clusters is large. This correction generalizes that of White (1980) for independent heteroskedastic errors. Additionally, more efficient estimation may be possible using alternative estimators, such as feasible Generalized Least Squares (GLS), that explicitly model the error correlation.

The loss of estimator precision due to clustering is presented in Section 1.2, while cluster-robust inference is presented in Section 1.3. The complications of inference, given only a few clusters, and inference when there is clustering in more than one direction, are considered in Sections 1.4 and 1.5. Section 1.6 presents more efficient feasible GLS estimation when structure is placed on the within-cluster error correlation. In Section 1.7 we consider adaptation to nonlinear and instrumental variables estimators. An empirical example in Section 1.8 illustrates many of the methods discussed in this survey.

1.2 Clustering and Its Consequences

Clustering leads to less efficient estimation than if data are independent, and default Ordinary Least Squares (OLS) standard errors need to be adjusted.

1.2.1 Clustered Errors

The linear model with (one-way) clustering is

$$y_{ig} = \mathbf{x}'_{ig}\beta + u_{ig}, \tag{1.1}$$

where i denotes the ith of N individuals in the sample, g denotes the gth of G clusters, $\mathrm{E}[u_{ig} \mid \mathbf{x}_{ig}] = 0$, and error independence across clusters is assumed so that for $i \neq j$

$$\mathrm{E}[u_{ig}u_{jg'} \mid \mathbf{x}_{ig}, \mathbf{x}_{jg'}] = 0, \text{ unless } g = g'. \tag{1.2}$$

Errors for individuals belonging to the same group may be correlated, with quite general heteroskedasticity and correlation. Grouping observations by cluster the model can be written as $\mathbf{y}_g = \mathbf{X}_g\beta + \mathbf{u}_g$, where $\mathbf{y}_g$ and $\mathbf{u}_g$ are $N_g \times 1$ vectors, $\mathbf{X}_g$ is an $N_g \times K$ matrix, and there are N_g observations in cluster g. Further stacking over clusters yields $\mathbf{y} = \mathbf{X}\beta + \mathbf{u}$, where $\mathbf{y}$ and $\mathbf{u}$ are $N \times 1$ vectors, $\mathbf{X}$ is an $N \times K$ matrix, and $N = \sum_g N_g$. The OLS estimator is $\widehat{\beta} = (\mathbf{X}'\mathbf{X})^{-1}\mathbf{X}'\mathbf{y}$. Given error independence across clusters, this estimator has asymptotic variance matrix

$$\mathrm{V}[\widehat{\beta}] = \left(\mathrm{E}[\mathbf{X}'\mathbf{X}]\right)^{-1}\left(\sum_{g=1}^{G}\mathrm{E}[\mathbf{X}'_g\mathbf{u}_g\mathbf{u}'_g\mathbf{X}_g]\right)\left(\mathrm{E}[\mathbf{X}'\mathbf{X}]\right)^{-1}, \tag{1.3}$$

rather than the default OLS variance $\sigma_u^2\left(\mathrm{E}[\mathbf{X}'\mathbf{X}]\right)^{-1}$, where $\sigma_u^2 = \mathrm{V}[u_{ig}]$.

1.2.2 Equicorrelated Errors

One way that within-cluster correlation can arise is in the random effects model where the error $u_{ig} = \alpha_g + \varepsilon_{ig}$, where α_g is a cluster-specific error or common shock that is i.i.d. $(0, \sigma_\alpha^2)$, and ε_{ig} is an idiosyncratic error that is i.i.d. $(0, \sigma_\varepsilon^2)$. Then $\mathrm{Var}[u_{ig}] = \sigma_\alpha^2 + \sigma_\varepsilon^2$ and $\mathrm{Cov}[u_{ig}, u_{jg}] = \sigma_\alpha^2$ for $i \neq j$. It follows that the intraclass correlation of the error $\rho_u = \mathrm{Cor}[u_{ig}, u_{jg}] = \sigma_\alpha^2/(\sigma_\alpha^2 + \sigma_\varepsilon^2)$. The correlation is constant across all pairs of errors in a given cluster. This correlation pattern is suitable when observations can be viewed as exchangeable, with ordering not mattering. Leading examples are individuals or households within a village or other geographic unit (such as state), individuals within a household, and students within a school.

If the primary source of clustering is due to such equicorrelated group-level common shocks, a useful approximation is that for the jth regressor the default OLS variance estimate based on $s^2(\mathbf{X}'\mathbf{X})^{-1}$, where s is the standard error of the regression, should be inflated by

$$\tau_j \simeq 1 + \rho_{x_j}\rho_u(\bar{N}_g - 1), \tag{1.4}$$

where ρ_{x_j} is a measure of the within-cluster correlation of x_j, ρ_u is the within-cluster error correlation, and $\bar{N}_g$ is the average cluster size. This result for equicorrelated errors is exact if clusters are of equal size; see Kloek (1981) for the special case $\rho_{x_j} = 1$, and Scott and Holt (1982) and Greenwald (1983) for the general result. The efficiency loss, relative to independent observations, is increasing in the within-cluster correlation of both the error and the regressor and in the number of observations in each cluster. For clusters of unequal size replace $(\bar{N}_g - 1)$ in formula 1.4 by $((\mathrm{V}[N_g]/\bar{N}_g) + \bar{N}_g - 1)$; see Moulton (1986, p. 387).

To understand the loss of estimator precision given clustering, consider the sample mean when observations are correlated. In this case the entire sample is viewed as a single cluster. Then

$$\mathrm{V}[\bar{y}] = N^{-2}\left\{\sum_{i=1}^{N}\mathrm{V}[y_i] + \sum_{i}\sum_{j\neq i}\mathrm{Cov}[y_i, y_j]\right\}. \tag{1.5}$$

Given equicorrelated errors with $\mathrm{Cov}[y_{ig}, y_{jg}] = \rho\sigma^2$ for $i \neq j$, $\mathrm{V}[\bar{y}] = N^{-2}\{N\sigma^2 + N(N-1)\rho\sigma^2\} = N^{-1}\sigma^2\{1 + \rho(N-1)\}$ compared to $N^{-1}\sigma^2$ in the i.i.d. case. At the extreme $\mathrm{V}[\bar{y}] = \sigma^2$ as $\rho \to 1$ and there is no benefit at all to increasing the sample size beyond $N = 1$.

Similar results are obtained when we generalize to several clusters of equal size (balanced clusters) with regressors that are invariant within cluster, so $y_{ig} = \mathbf{x}_g'\beta + u_{ig}$, where i denotes the ith of N individuals in the sample and g denotes the gth of G clusters, and there are $N_* = N/G$ observations in each cluster. Then OLS estimation of y_{ig} on $\mathbf{x}_g$ is equivalent to OLS estimation in the model $\bar{y}_g = \mathbf{x}_g'\beta + \bar{u}_g$, where $\bar{y}_g$ and $\bar{u}_g$ are the within-cluster averages of the dependent variable and error. If $\bar{u}_g$ is independent and homoskedastic with variance $\sigma^2_{\bar{u}_g}$ then $\mathrm{V}[\hat{\beta}] = \sigma^2_{\bar{u}_g}(\sum_{g=1}^{G}\mathbf{x}_g\mathbf{x}_g')^{-1}$, where the formula for $\sigma^2_{\bar{u}_g}$ varies with the within-cluster correlation of u_{ig}. For equicorrelated errors $\sigma^2_{\bar{u}_g} = N_*^{-1}[1 + \rho_u(N_* - 1)]\sigma_u^2$ compared to $N_*^{-1}\sigma_u^2$ with independent errors, so the true variance of the OLS estimator is $(1 + \rho_u(N_* - 1))$ times the default, as given in formula 1.4 with $\rho_{x_j} = 1$.

In an influential paper Moulton (1990) pointed out that in many settings the adjustment factor τ_j can be large even if ρ_u is small. He considered a log earnings regression using March CPS data ($N = 18,946$), regressors aggregated at the state level ($G = 49$), and errors correlated within state ($\widehat{\rho}_u = 0.032$). The average group size was $18,946/49 = 387$, $\rho_{x_j} = 1$ for a state-level regressor, so $\tau_j \simeq 1 + 1 \times 0.032 \times 386 = 13.3$. The weak correlation of errors within state was still enough to lead to cluster-corrected standard errors being $\sqrt{13.3} = 3.7$ times larger than the (incorrect) default standard errors, and in this example many researchers would not appreciate the need to make this correction.

1.2.3 Panel Data

A second way that clustering can arise is in panel data. We assume that observations are independent across individuals in the panel, but the observations for any given individual are correlated over time. Then each individual is viewed as a cluster. The usual notation is to denote the data as y_{it}, where

i denotes the individual and t the time period. But in our framework (formula 1.1) the data are denoted y_{ig}, where i is the within-cluster subscript (for panel data the time period) and g is the cluster unit (for panel data the individual).

The assumption of equicorrelated errors is unlikely to be suitable for panel data. Instead we expect that the within-cluster (individual) correlation decreases as the time separation increases.

For example, we might consider an AR(1) model with $u_{it} = \rho u_{i,t-1} + \varepsilon_{it}$, where $0 < \rho < 1$ and ε_{it} is i.i.d. $(0, \sigma_\varepsilon^2)$. In terms of the notation in formula 1.1, $u_{ig} = \rho u_{i-1,g} + \varepsilon_{ig}$. Then the within-cluster error correlation $\text{Cor}[u_{ig}, u_{jg}] = \rho^{|i-j|}$, and the consequences of clustering are less extreme than in the case of equicorrelated errors.

To see this, consider the variance of the sample mean $\bar{y}$ when $\text{Cov}[y_i, y_j] = \rho^{|i-j|}\sigma^2$. Then formula 1.5 yields $\text{V}[\bar{y}] = N^{-1}[1 + 2N^{-1}\sum_{s=1}^{N-1} s\rho^s]\sigma_u^2$. For example, if $\rho = 0.5$ and $N = 10$, then $\text{V}[\bar{y}] = 0.26\sigma^2$ compared to $0.55\sigma^2$ for equicorrelation, using $\text{V}[\bar{y}] = N^{-1}\sigma^2\{1 + \rho(N-1)\}$, and $0.1\sigma^2$ when there is no correlation ($\rho = 0.0$). More generally with several clusters of equal size and regressors invariant within cluster, OLS estimation of y_{ig} on $\mathbf{x}_g$ is equivalent to OLS estimation of $\bar{y}_g$ on $\mathbf{x}_g$ (see Subsection 1.2.2), and with an AR(1) error $\text{V}[\widehat{\beta}] = N_*^{-1}[1 + 2N_* \sum_{s=1}^{N_*-1} s\rho^s]\sigma_u^2(\sum_g \mathbf{x}_g\mathbf{x}_g')^{-1}$, less than $N_*^{-1}[1 + \rho_u(N_* - 1)]\sigma_u^2(\sum_g \mathbf{x}_g\mathbf{x}_g')^{-1}$ with an equicorrelated error.

For panel data in practice, while within-cluster correlations for errors are not constant, they do not dampen as quickly as those for an AR(1) model. The variance inflation formula 1.4 can still provide a reasonable guide in panels that are short and have high within-cluster serial correlations of the regressor and of the error.

1.3 Cluster-Robust Inference for OLS

The most common approach in applied econometrics is to continue with OLS, and then obtain correct standard errors that correct for within-cluster correlation.

1.3.1 Cluster-Robust Inference

Cluster-robust estimates for the variance matrix of an estimate are sandwich estimates that are cluster adaptations of methods proposed originally for independent observations by White (1980) for OLS with heteroskedastic errors, and by Huber (1967) and White (1982) for the maximum likelihood estimator.

The cluster-robust estimate of the variance matrix of the OLS estimator, defined in formula 1.3, is the sandwich estimate

$$\widehat{\text{V}}[\widehat{\beta}] = (\mathbf{X}'\mathbf{X})^{-1}\widehat{\mathbf{B}}(\mathbf{X}'\mathbf{X})^{-1}, \tag{1.6}$$

where

$$\widehat{\mathbf{B}} = \left(\sum_{g=1}^{G} \mathbf{X}'_g \widehat{\mathbf{u}}_g \widehat{\mathbf{u}}'_g \mathbf{X}_g \right), \tag{1.7}$$

and $\widehat{\mathbf{u}}_g = \mathbf{y}_g - \mathbf{X}_g \widehat{\beta}$. This provides a consistent estimate of the variance matrix if $G^{-1} \sum_{g=1}^{G} \mathbf{X}'_g \widehat{\mathbf{u}}_g \widehat{\mathbf{u}}'_g \mathbf{X}_g - G^{-1} \sum_{g=1}^{G} \mathrm{E}[\mathbf{X}'_g \mathbf{u}_g \mathbf{u}'_g \mathbf{X}_g] \xrightarrow{p} \mathbf{0}$ as $G \to \infty$.

The estimate of White (1980) for independent heteroskedastic errors is the special case of formula 1.7, where each cluster has only one observation (so $G = N$ and $N_g = 1$ for all g). It relies on the same intuition that $G^{-1} \sum_{g=1}^{G} \mathrm{E}[\mathbf{X}'_g \mathbf{u}_g \mathbf{u}'_g \mathbf{X}_g]$ is a finite-dimensional $(K \times K)$ matrix of averages that can be consistently estimated as $G \to \infty$.

White (1984, pp. 134–142) presented formal theorems that justify use of formula 1.7 for OLS with a multivariate dependent variable, a result directly applicable to balanced clusters. Liang and Zeger (1986) proposed this method for estimation for a range of models much wider than OLS; see Sections 1.6 and 1.7 of their paper for a range of extensions to formula 1.7. Arellano (1987) considered the fixed effects estimator in linear panel models, and Rogers (1993) popularized this method in applied econometrics by incorporating it in Stata. Note that formula 1.7 does not require specification of a model for $\mathrm{E}[\mathbf{u}_g \mathbf{u}'_g]$.

Finite-sample modifications of formula 1.7 are typically used, since without modification the cluster-robust standard errors are biased downwards. Stata uses $\sqrt{c}\widehat{\mathbf{u}}_g$ in formula 1.7 rather than $\widehat{\mathbf{u}}_g$, with

$$c = \frac{G}{G-1} \frac{N-1}{N-K} \simeq \frac{G}{G-1}. \tag{1.8}$$

Some other packages such as SAS use $c = G/(G-1)$. This simpler correction is also used by Stata for extensions to nonlinear models. Cameron, Gelbach, and Miller (2008) review various finite-sample corrections that have been proposed in the literature, for both standard errors and for inference using resultant Wald statistics; see also Section 1.6.

The rank of $\widehat{\mathrm{V}}[\widehat{\beta}]$ in formula 1.7 can be shown to be at most G, so at most G restrictions on the parameters can be tested if cluster-robust standard errors are used. In particular, in models with cluster-specific effects it may not be possible to perform a test of overall significance of the regression, even though it is possible to perform tests on smaller subsets of the regressors.

1.3.2 Specifying the Clusters

It is not always obvious how to define the clusters.

As already noted in Subsection 1.2.2, Moulton (1986, 1990) pointed out for statistical inference on an aggregate-level regressor it may be necessary to cluster at that level. For example, with individual cross-sectional data and a regressor defined at the state level one should cluster at the state level if regression model errors are even very mildly correlated at the state level. In other cases the key regressor may be correlated within group, though not perfectly so, such as individuals within household. Other reasons for clustering include discrete regressors and a clustered sample design.

In some applications there can be nested levels of clustering. For example, for a household-based survey there may be error correlation for individuals within the same household, and for individuals in the same state. In that case cluster-robust standard errors are computed at the most aggregated level of clustering, in this example at the state level. Pepper (2002) provides a detailed example.

Bertrand, Duflo, and Mullainathan (2004) noted that with panel data or repeated cross-section data, and regressors clustered at the state level, many researchers either failed to account for clustering or mistakenly clustered at the state-year level rather than the state level. Let y_{ist} denote the value of the dependent variable for the ith individual in the sth state in the tth year, and let x_{st} denote a state-level policy variable that in practice will be quite highly correlated over time in a given state. The authors considered the difference-in-differences (DiD) model $y_{ist} = \gamma_s + \delta_t + \beta x_{st} + \mathbf{z}'_{ist}\gamma + u_{it}$, though their result is relevant even for OLS regression of y_{ist} on x_{st} alone. The same point applies if data were more simply observed at only the state-year level (i.e., y_{st} rather than y_{ist}).

In general DiD models using state-year data will have high within-cluster correlation of the key policy regressor. Furthermore there may be relatively few clusters; a complication considered in Section 1.4.

1.3.3 Cluster-Specific Fixed Effects

A standard estimation method for clustered data is to additionally incorporate cluster-specific fixed effects as regressors, estimating the model

$$y_{ig} = \alpha_g + \mathbf{x}'_{ig}\beta + u_{ig}. \tag{1.9}$$

This is similar to the equicorrelated error model, except that α_g is treated as a (nuisance) parameter to be estimated. Given N_g finite and $G \rightarrow \infty$ the parameters α_g, $g = 1, \ldots, G$, cannot be consistently estimated. The parameters β can still be consistently estimated, with the important caveat that the coefficients of cluster-invariant regressors (x_g rather than x_{ig}) are not identified. (In microeconometrics applications, fixed effects are typically included to enable consistent estimation of a cluster-varying regressor while controlling for a limited form of endogeneity – the regressor x_{ig} may be correlated with the cluster-invariant component α_g of the error term $\alpha_g + u_{ig}$).

Initial applications obtained default standard errors that assume u_{ig} in formula 1.9 is i.i.d. $(0, \sigma_u^2)$, assuming that cluster-specific fixed effects are sufficient to mop up any within-cluster error correlation. More recently it has become more common to control for possible within-cluster correlation of u_{ig} by using formula 1.7, as suggested by Arellano (1987). Kézdi (2004) demonstrated that cluster-robust estimates can perform well in typical-sized panels, despite the need to first estimate the fixed effects, even when N_g is large relative to G.

It is well-known that there are several alternative ways to obtain the OLS estimator of β in formula 1.9. Less well-known is that these different ways can lead to different cluster-robust estimates of $V[\widehat{\beta}]$. We thank Arindrajit Dube and Jason Lindo for bringing this issue to our attention.

The two main estimation methods we consider are the least squares dummy variables (LSDV) estimator, which obtains the OLS estimator from regression of y_{ig} on $\mathbf{x}_{ig}$ and a set of dummy variables for each cluster, and the mean-differenced estimator, which is the OLS estimator from regression of $(y_{ig} - \bar{y}_g)$ on $(\mathbf{x}_{ig} - \bar{\mathbf{x}}_g)$.

These two methods lead to the same cluster-robust standard errors if we apply formula 1.7 to the respective regressions, or if we multiply this estimate by $G/(G-1)$. Differences arise, however, if we multiply by the small-sample correction c given in formula 1.8. Let K denote the number of regressors including the intercept. Then the LSDV model views the total set of regressors to be G cluster dummies and $(K - 1)$ other regressors, while the mean-differenced model considers there to be only $(K - 1)$ regressors (this model is estimated without an intercept). Then

Model	Finite Sample Adjustment	Balanced Case
LSDV	$c = \frac{G}{G-1}\frac{N-1}{N-G-(k-1)}$	$c \simeq \frac{G}{G-1} \times \frac{N_*}{N_*-1}$
Mean-differenced model	$c = \frac{G}{G-1}\frac{N-1}{N-(k-1)}$	$c \simeq \frac{G}{G-1}.$

In the balanced case $N = N_*G$, leading to the approximation given above if additionally K is small relative to N.

The difference can be very large for small N_*. Thus if $N_* = 2$ (or $N_* = 3$) then the cluster-robust variance matrix obtained using LSDV is essentially 2 times (or 3/2 times) that obtained from estimating the mean-differenced model, and it is the mean-differenced model that gives the correct finite-sample correction.

Note that if instead the error u_{ig} is assumed to be i.i.d. $(0, \sigma_u^2)$, so that default standard errors are used, then it is well-known that the appropriate small-sample correction is $(N - 1)/N - G - (K - 1)$, i.e., we use $s^2(\mathbf{X}'\mathbf{X})^{-1}$, where $s^2 = (N - G - (K - 1))^{-1} \sum_{ig} \widehat{u}_{ig}^2$. In that case LSDV does give the correct adjustment, and estimation of the mean-differenced model will give the wrong finite-sample correction.

An alternative variance estimator after estimation of formula 1.9 is a heteroskedastic-robust estimator, which permits the error u_{ig} in formula 1.9 to be heteroskedastic but uncorrelated across both i and g. Stock and Watson (2008) show that applying the method of White (1980) after mean-differenced estimation of formula 1.9 leads, surprisingly, to inconsistent estimates of $V[\widehat{\beta}]$ if the number of observations N_g in each cluster is small (though it is correct if $N_g = 2$). The bias comes from estimating the cluster-specific means rather than being able to use the true cluster-means. They derive a bias-corrected formula for heteroskedastic-robust standard errors. Alternatively, and more simply, the cluster-robust estimator gives a consistent estimate of $V[\widehat{\beta}]$ even if the errors are only heteroskedastic, though this estimator is more variable than the bias-corrected estimator proposed by Stock and Watson.

1.3.4 Many Observations per Cluster

The preceding analysis assumes the number of observations within each cluster is fixed, while the number of clusters goes to infinity.

This assumption may not be appropriate for clustering in long panels, where the number of time periods goes to infinity. Hansen (2007a) derived asymptotic results for the standard one-way cluster-robust variance matrix estimator for panel data under various assumptions. We consider a balanced panel of N individuals over T periods, so there are NT observations in N clusters with T observations per cluster. When $N \rightarrow \infty$ with T fixed (a short panel), as we have assumed above, the rate of convergence for the OLS estimator $\widehat{\beta}$ is $\sqrt{N}$. When both $N \rightarrow \infty$ and $T \rightarrow \infty$ (a long panel with $N_* \rightarrow \infty$), the rate of convergence of $\widehat{\beta}$ is $\sqrt{N}$ if there is no mixing (his Theorem 2) and $\sqrt{NT}$ if there is mixing (his Theorem 3). By mixing we mean that the correlation becomes damped as observations become further apart in time.

As illustrated in Subsection 1.2.3, if the within-cluster error correlation of the error diminishes as errors are further apart in time, then the data has greater informational content. This is reflected in the rate of convergence increasing from $\sqrt{N}$ (determined by the number of cross-sections) to $\sqrt{NT}$ (determined by the total size of the panel). The latter rate is the rate we expect if errors were independent within cluster.

While the rates of convergence differ in the two cases, Hansen (2007a) obtains the same asymptotic variance for the OLS estimator, so formula 1.7 remains valid.

1.3.5 Survey Design with Clustering and Stratification

Clustering routinely arises in complex survey data. Rather than randomly draw individuals from the population, the survey may be restricted to a randomly selected subset of primary sampling units (such as a geographic area) followed by selection of people within that geographic area. A common approach in microeconometrics is to control for the resultant clustering by computing cluster-robust standard errors that control for clustering at the level of the primary sampling unit, or at a more aggregated level such as state.

The survey methods literature uses methods to control for clustering that predate the references in this paper. The loss of estimator precision due to clustering is called the design effect: "The design effect or Deff is the ratio of the actual variance of a sample to the variance of a simple random sample of the same number of elements"(Kish 1965, p. 258). Kish and Frankel (1974) give the variance inflation formula 1.4 assuming equicorrelated errors in the non-regression case of estimation of the mean. Pfeffermann and Nathan (1981) consider the more general regression case.

The survey methods literature additionally controls for another feature of survey data – stratification. More precise statistical inference is possible after stratification. For the linear regression model, survey methods that do so are well-established and are incorporated in specialized software as well as in some broad-based packages such as Stata.

Bhattacharya (2005) provides a comprehensive treatment in a GMM framework. He finds that accounting for stratification tends to reduce estimated standard errors, and that this effect can be meaningfully large. In his empirical examples, the stratification effect is largest when estimating (unconditional) means and Lorenz shares, and much smaller when estimating conditional means via regression.

The current common approach of microeconometrics studies is to ignore the (beneficial) effects of stratification. In so doing there will be some overestimation of estimator standard errors.

1.4 Inference with Few Clusters

Cluster-robust inference asymptotics are based on $G \to \infty$. Often, however, cluster-robust inference is desired but there are only a few clusters. For example, clustering may be at the regional level but there are few regions (e.g., Canada has only 10 provinces). Then several different finite-sample adjustments have been proposed.

1.4.1 Finite-Sample Adjusted Standard Errors

Finite-sample adjustments replace $\widehat{\mathbf{u}}_g$ in formula 1.7 with a modified residual $\widetilde{\mathbf{u}}_g$. The simplest is $\widetilde{\mathbf{u}}_g = \sqrt{G/(G-1)}\widehat{\mathbf{u}}_g$, or the modification of this given in formula 1.8. Kauermann and Carroll (2001) and Bell and McCaffrey (2002) use $\widetilde{\mathbf{u}}_g^* = [\mathbf{I}_{N_g} - \mathbf{H}_{gg}]^{-1/2}\widehat{\mathbf{u}}_g$, where $\mathbf{H}_{gg} = \mathbf{X}_g(\mathbf{X}'\mathbf{X})^{-1}\mathbf{X}_g'$. This transformed residual leads to $\mathrm{E}[\widehat{\mathrm{V}}[\widehat{\beta}]] = \mathrm{V}[\widehat{\beta}]$ in the special case that $\Omega_g = \mathrm{E}[\mathbf{u}_g\mathbf{u}_g'] = \sigma^2\mathbf{I}$. Bell and McCaffrey (2002) also consider use of $\widetilde{\mathbf{u}}_g^+ = \sqrt{G/(G-1)}[\mathbf{I}_{N_g} - \mathbf{H}_{gg}]^{-1}\widehat{\mathbf{u}}_g$, which can be shown to equal the (clustered) jackknife estimate of the variance of the OLS estimator. These adjustments are analogs of the HC2 and HC3 measures of MacKinnon and White (1985) proposed for heteroskedastic-robust standard errors in the nonclustered case.

Angrist and Lavy (2009) found that using $\widetilde{\mathbf{u}}_g^+$ rather than $\widetilde{\mathbf{u}}_g$ increased cluster-robust standard errors by 10–50% in an application with $G = 30$ to 40.

Kauermann and Carroll (2001), Bell and McCaffrey (2002), Mancl and DeRouen (2001), and McCaffrey, Bell, and Botts (2001) also consider the case where $\Omega_g \neq \sigma^2\mathbf{I}$ is of known functional form, and present extension to generalized linear models.

1.4.2 Finite-Sample Wald Tests

For a two-sided test of $H_0 : \beta_j = \beta_j^0$ against $H_a : \beta_j \neq \beta_j^0$, where β_j is a scalar component of β, the standard procedure is to use Wald test statistic $w = (\widehat{\beta}_j - \beta_j^0)/s_{\widehat{\beta}_j}$, where $s_{\widehat{\beta}_j}$ is the square root of the appropriate diagonal entry in $\widehat{V}[\widehat{\beta}]$. This "t"test statistic is asymptotically normal under H_0 as $G \rightarrow \infty$, and we reject H_0 at significance level 0.05 if $|w| > 1.960$.

With few clusters, however, the asymptotic normal distribution can provide a poor approximation, even if an unbiased variance matrix estimator is used in calculating $s_{\widehat{\beta}_j}$. The situation is a little unusual. In a pure time series or pure cross-section setting with few observations, say $N = 10$, β_j is likely to be very imprecisely estimated so that statistical inference is not worth pursuing. By contrast, in a clustered setting we may have N sufficiently large that β_j is reasonably precisely estimated, but G is so small that the asymptotic normal approximation is a very poor one.

We present two possible approaches: basing inference on the T distribution with degrees of freedom determined by the cluster, and using a cluster bootstrap with asymptotic refinement. Note that feasible GLS based on a correctly specified model of the clustering, see Section 1.6, will not suffer from this problem.

1.4.3 *T* Distribution for Inference

The simplest small-sample correction for the Wald statistic is to use a T distribution, rather than the standard normal. As we outline below in some cases the T_{G-L} distribution might be used, where L is the number of regressors that are invariant within cluster. Some packages for some commands do use the T distribution. For example, Stata uses $G - 1$ degrees of freedom for t-tests and F-tests based on cluster-robust standard errors.

Such adjustments can make quite a difference. For example, with $G = 10$ for a two-sided test at level 0.05 the critical value for T_9 is 2.262 rather than 1.960, and if $w = 1.960$ the p-value based on T_9 is 0.082 rather than 0.05. In Monte Carlo simulations by Cameron, Gelbach, and Miller (2008) this technique works reasonably well. At the minimum one should use the T distribution with $G - 1$ degrees of freedom, say, rather than the standard normal.

Donald and Lang (2007) provide a rationale for using the T_{G-L} distribution. If clusters are balanced and all regressors are invariant within cluster then the OLS estimator in the model $y_{ig} = \mathbf{x}_g'\beta + u_{ig}$ is equivalent to OLS estimation in the grouped model $\bar{y}_g = \mathbf{x}_g'\beta + \bar{u}_g$. If $\bar{u}_g$ is i.i.d. normally distributed then the Wald statistic is T_{G-L} distributed, where $\widehat{V}[\widehat{\beta}] = s^2(\mathbf{X}'\mathbf{X})^{-1}$ and $s^2 = (G - L)^{-1}\sum_g \widehat{\bar{u}_g}^2$. Note that $\bar{u}_g$ is i.i.d. normal in the random effects model if the error components are i.i.d. normal.

Donald and Lang (2007) extend this approach to additionally include regressors $\mathbf{z}_{ig}$ that vary within clusters, and allow for unbalanced clusters. They assume a random effects model with normal i.i.d. errors. Then feasible GLS estimation of β in the model

$$y_{ig} = \mathbf{x}_g'\beta + \mathbf{z}_{ig}'\gamma + \alpha_s + \varepsilon_{is} \tag{1.10}$$

is equivalent to the following two-step procedure. First do OLS estimation in the model $y_{ig} = \delta_g + \mathbf{z}_{ig}'\gamma + \varepsilon_{ig}$, where δ_g is treated as a cluster-specific fixed effect. Then do feasible GLS (FGLS) of $\bar{y}_g - \bar{\mathbf{z}}_g'\widehat{\gamma}$ on $\mathbf{x}_g$. Donald and Lang (2007) give various conditions under which the resulting Wald statistic based on $\widehat{\beta}_j$ is T_{G-L} distributed. These conditions require that if $\mathbf{z}_{ig}$ is a regressor then $\bar{\mathbf{z}}_g$ in the limit is constant over g, unless $N_g \rightarrow \infty$. Usually $L = 2$, as the only regressors that do not vary within clusters are an intercept and a scalar regressor x_g.

Wooldridge (2006) presents an expansive exposition of the Donald and Lang approach. Additionally, Wooldridge proposes an alternative approach based on minimum distance estimation. He assumes that δ_g in $y_{ig} = \delta_g + \mathbf{z}_{ig}'\gamma + \varepsilon_{ig}$ can be adequately explained by $\mathbf{x}_g$ and at the second step uses minimum chi-square methods to estimate β in $\widehat{\delta}_g = \alpha + \mathbf{x}_g'\beta$. This provides estimates of β that are asymptotically normal as $N_g \rightarrow \infty$ (rather than $G \rightarrow \infty$). Wooldridge argues that this leads to less conservative statistical inference. The χ^2 statistic from the minimum distance method can be used as a test of the assumption that the δ_g do not depend in part on cluster-specific random effects. If this test fails, the researcher can then use the Donald and Lang approach, and use a T distribution for inference.

Bester, Conley, and Hansen (2009) give conditions under which the t-test statistic based on formula 1.7 is $\sqrt{G/(G-1)}$ times T_{G-1} distributed. Thus using $\widetilde{\mathbf{u}}_g = \sqrt{G/(G-1)}\widehat{\mathbf{u}}_g$ yields a T_{G-1} distributed statistic. Their result is one that assumes G is fixed while $N_g \rightarrow \infty$; the within group correlation satisfies a mixing condition, as is the case for time series and spatial correlation; and homogeneity assumptions are satisfied including equality of plim $\frac{1}{N_g}\mathbf{X}_g'\mathbf{X}_g$ for all g.

An alternate approach for correct inference with few clusters is presented by Ibragimov and Muller (2010). Their method is best suited for settings where model identification, and central limit theorems, can be applied separately to observations in each cluster. They propose separate estimation of the key parameter within each group. Each group's estimate is then a draw from a normal distribution with mean around the truth, though perhaps with separate variance for each group. The separate estimates are averaged, divided by the sample standard deviation of these estimates, and the test statistic is compared against critical values from a T distribution. This approach has the strength of offering correct inference even with few clusters. A limitation is that it requires identification using only within-group variation, so that the group estimates are independent of one another. For example, if state-year data y_{st} are used and the state is the cluster unit, then the regressors cannot use any regressor z_t such as a time dummy that varies over time but not states.

1.4.4 Cluster Bootstrap with Asymptotic Refinement

A cluster bootstrap with asymptotic refinement can lead to improved finite-sample inference.

For inference based on $G \to \infty$, a two-sided Wald test of nominal size α can be shown to have true size $\alpha + O(G^{-1})$ when the usual asymptotic normal approximation is used. If instead an appropriate bootstrap with asymptotic refinement is used, the true size is $\alpha + O(G^{-3/2})$. This is closer to the desired α for large G, and hopefully also for small G. For a one-sided test or a nonsymmetric two-sided test the rates are instead, respectively, $\alpha + O(G^{-1/2})$ and $\alpha + O(G^{-1})$.

Such asymptotic refinement can be achieved by bootstrapping a statistic that is asymptotically pivotal, meaning the asymptotic distribution does not depend on any unknown parameters. For this reason the Wald t-statistic w is bootstrapped, rather than the estimator $\widehat{\beta}_j$ whose distribution depends on $\mathrm{V}[\widehat{\beta}_j]$ which needs to be estimated. The pairs cluster bootstrap procedure does B iterations where at the bth iteration: (1) form G clusters $\{(\mathbf{y}_1^*, \mathbf{X}_1^*), \ldots, (\mathbf{y}_G^*, \mathbf{X}_G^*)\}$ by resampling with replacement G times from the original sample of clusters; (2) do OLS estimation with this resample and calculate the Wald test statistic $w_b^* = (\widehat{\beta}^*_{j,b} - \widehat{\beta}_j)/s^*_{\widehat{\beta}_{j,b}}$ where $s_{\widehat{\beta}^*_{j,b}}$ is the cluster-robust standard error of $\widehat{\beta}^*_{j,b}$, and $\widehat{\beta}_j$ is the OLS estimate of β_j from the original sample. Then reject H_0 at level α if and only if the original sample Wald statistic w is such that $w < w^*_{[\alpha/2]}$ or $w > w^*_{[1-\alpha/2]}$, where $w^*_{[q]}$ denotes the qth quantile of $w_1^*, \ldots, w_B^*$.

Cameron, Gelbach, and Miller (2008) provide an extensive discussion of this and related bootstraps. If there are regressors that contain few values (such as dummy variables), and if there are few clusters, then it is better to use an alternative design-based bootstrap that additionally conditions on the regressors, such as a cluster Wild bootstrap. Even then bootstrap methods, unlike the method of Donald and Lang, will not be appropriate when there are very few groups, such as $G = 2$.

1.4.5 Few Treated Groups

Even when G is sufficiently large, problems arise if most of the variation in the regressor is concentrated in just a few clusters. This occurs if the key regressor is a cluster-specific binary treatment dummy and there are few treated groups.

Conley and Taber (2010) examine a differences-in-differences (DiD) model in which there are few treated groups and an increasing number of control groups. If there are group-time random effects, then the DiD model is inconsistent because the treated groups random effects are not averaged away. If the random effects are normally distributed, then the model of Donald and Lang (2007) applies and inference can use a T distribution based on the number of treated groups. If the group-time shocks are not random, then the T distribution may be a poor approximation. Conley and Taber (2010) then propose a novel method that uses the distribution of the untreated groups to perform inference on the treatment parameter.

1.5 Multi-Way Clustering

Regression model errors can be clustered in more than one way. For example, they might be correlated across time within a state, and across states within a time period. When the groups are nested (e.g., households within states), one clusters on the more aggregate group; see Subsection 1.3.2. But when they are non-nested, traditional cluster inference can only deal with one of the dimensions.

In some applications it is possible to include sufficient regressors to eliminate error correlation in all but one dimension, and then do cluster-robust inference for that remaining dimension. A leading example is that in a state-year panel of individuals (with dependent variable y_{ist}) there may be clustering both within years and within states. If the within-year clustering is due to shocks that are the same across all individuals in a given year, then including year fixed effects as regressors will absorb within-year clustering and inference then need only control for clustering on state.

When this is not possible, the one-way cluster robust variance can be extended to multi-way clustering.

1.5.1 Multi-Way Cluster-Robust Inference

The cluster-robust estimate of $\mathrm{V}[\widehat{\beta}]$ defined in formulas 1.6 and 1.7 can be generalized to clustering in multiple dimensions. Regular one-way clustering is based on the assumption that $\mathrm{E}[u_i u_j \mid \mathbf{x}_i, \mathbf{x}_j] = 0$, unless observations i and j are in the same cluster. Then formula 1.7 sets $\widehat{\mathbf{B}} = \sum_{i=1}^{N} \sum_{j=1}^{N} \mathbf{x}_i \mathbf{x}'_j \widehat{u}_i \widehat{u}_j \mathbf{1}[i, j \text{ in same cluster}]$, where $\widehat{u}_i = y_i - \mathbf{x}'_i \widehat{\beta}$ and the indicator function $\mathbf{1}[A]$ equals 1 if event A occurs and 0 otherwise. In multi-way clustering, the key assumption is that $\mathrm{E}[u_i u_j | \mathbf{x}_i, \mathbf{x}_j] = 0$, unless observations i and j share any cluster dimension. Then the multi-way cluster robust estimate of $\mathrm{V}[\widehat{\beta}]$ replaces formula 1.7 with $\widehat{\mathbf{B}} = \sum_{i=1}^{N} \sum_{j=1}^{N} \mathbf{x}_i \mathbf{x}'_j \widehat{u}_i \widehat{u}_j \mathbf{1}[i, j \text{ share any cluster}]$.

For two-way clustering this robust variance estimator is easy to implement given software that computes the usual one-way cluster-robust estimate. We obtain three different cluster-robust "variance" matrices for the estimator by one-way clustering in, respectively, the first dimension, the second dimension, and by the intersection of the first and second dimensions. Then add the first two variance matrices and, to account for double counting, subtract the third. Thus,

$$\widehat{\mathrm{V}}_{\text{two-way}}[\widehat{\beta}] = \widehat{\mathrm{V}}_1[\widehat{\beta}] + \widehat{\mathrm{V}}_2[\widehat{\beta}] - \widehat{\mathrm{V}}_{1\cap 2}[\widehat{\beta}], \tag{1.11}$$

where the three component variance estimates are computed using formulas 1.6 and 1.7 for the three different ways of clustering. Similar methods for additional dimensions, such as three-way clustering, are detailed in Cameron, Gelbach, and Miller (2010).

This method relies on asymptotics that are in the number of clusters of the dimension with the fewest number. This method is thus most appropriate when each dimension has many clusters. Theory for two-way cluster robust estimates of the variance matrix is presented in Cameron, Gelbach, and Miller (2006, 2010), Miglioretti and Heagerty (2006), and Thompson (2006). Early empirical applications that independently proposed this method include Acemoglu and Pischke (2003) and Fafchamps and Gubert (2007).

1.5.2 Spatial Correlation

The multi-way robust clustering estimator is closely related to the field of time-series and spatial heteroskedasticity and autocorrelation variance estimation.

In general $\widehat{\mathbf{B}}$ in formula 1.7 has the form $\sum_i \sum_j w(i, j)\mathbf{x}_i\mathbf{x}_j'\widehat{u}_i\widehat{u}_j$. For multi-way clustering the weight $w(i, j) = 1$ for observations who share a cluster, and $w(i, j) = 0$ otherwise. In White and Domowitz (1984), the weight $w(i, j) = 1$ for observations "close" in time to one another, and $w(i, j) = 0$ for other observations. Conley (1999) considers the case where observations have spatial locations, and has weights $w(i, j)$ decaying to 0 as the distance between observations grows.

A distinguishing feature between these papers and multi-way clustering is that White and Domowitz (1984) and Conley (1999) use mixing conditions (to ensure decay of dependence) as observations grow apart in time or distance. These conditions are not applicable to clustering due to common shocks. Instead the multi-way robust estimator relies on independence of observations that do not share any clusters in common.

There are several variations to the cluster-robust and spatial or time-series HAC estimators, some of which can be thought of as hybrids of these concepts.

The spatial estimator of Driscoll and Kraay (1998) treats each time period as a cluster, additionally allows observations in different time periods to be correlated for a finite time difference, and assumes $T \to \infty$. The Driscoll–Kraay estimator can be thought of as using weight $w(i, j) = 1 - D(i, j)/(D_{\max} + 1)$, where $D(i, j)$ is the time distance between observations i and j, and $D_{\max}$ is the maximum time separation allowed to have correlation.

An estimator proposed by Thompson (2006) allows for across-cluster (in his example firm) correlation for observations close in time in addition to within-cluster correlation at any time separation. The Thompson estimator can be thought of as using $w(i, j) = \mathbf{1}[i, j \text{ share a firm, or } D(i, j) \leq D_{\max}]$. It seems that other variations are likely possible.

Foote (2007) contrasts the two-way cluster-robust and these other variance matrix estimators in the context of a macroeconomics example. Petersen (2009) contrasts various methods for panel data on financial firms, where there is concern about both within firm correlation (over time) and across firm correlation due to common shocks.

1.6 Feasible GLS

When clustering is present and a correct model for the error correlation is specified, the feasible GLS estimator is more efficient than OLS. Furthermore, in many situations one can obtain a cluster-robust version of the standard errors for the FGLS estimator, to guard against misspecification of model for the error correlation. Many applied studies nonetheless use the OLS estimator, despite the potential expense of efficiency loss in estimation.

1.6.1 FGLS and Cluster-Robust Inference

Suppose we specify a model for $\Omega_g = \mathrm{E}[\mathbf{u}_g\mathbf{u}_g'|\mathbf{X}_g]$, such as within-cluster equicorrelation. Then the GLS estimator is $(\mathbf{X}'\Omega^{-1}\mathbf{X})^{-1}\mathbf{X}'\Omega^{-1}\mathbf{y}$, where $\Omega = \mathrm{Diag}[\Omega_g]$. Given a consistent estimate $\widehat{\Omega}$ of Ω, the feasible GLS estimator of β is

$$\widehat{\beta}_{\mathrm{FGLS}} = \left(\sum_{g=1}^{G}\mathbf{X}_g'\widehat{\Omega}_g^{-1}\mathbf{X}_g\right)^{-1}\sum_{g=1}^{G}\mathbf{X}_g'\widehat{\Omega}_g^{-1}\mathbf{y}_g. \tag{1.12}$$

The default estimate of the variance matrix of the FGLS estimator, $\left(\mathbf{X}'\widehat{\Omega}^{-1}\mathbf{X}\right)^{-1}$, is correct under the restrictive assumption that $\mathrm{E}[\mathbf{u}_g\mathbf{u}_g'|\mathbf{X}_g] = \Omega_g$.

The cluster-robust estimate of the asymptotic variance matrix of the FGLS estimator is

$$\widehat{\mathrm{V}}[\widehat{\beta}_{\mathrm{FGLS}}] = \left(\mathbf{X}'\widehat{\Omega}^{-1}\mathbf{X}\right)^{-1}\left(\sum_{g=1}^{G}\mathbf{X}_g'\widehat{\Omega}_g^{-1}\widehat{\mathbf{u}}_g\widehat{\mathbf{u}}_g'\widehat{\Omega}_g^{-1}\mathbf{X}_g\right)\left(\mathbf{X}'\widehat{\Omega}^{-1}\mathbf{X}\right)^{-1}, \tag{1.13}$$

where $\widehat{\mathbf{u}}_g = \mathbf{y}_g - \mathbf{X}_g\widehat{\beta}_{\mathrm{FGLS}}$. This estimator requires that $\mathbf{u}_g$ and $\mathbf{u}_h$ are uncorrelated, for $g \neq h$, but permits $\mathrm{E}[\mathbf{u}_g\mathbf{u}_g'|\mathbf{X}_g] \neq \Omega_g$. In that case the FGLS estimator is no longer guaranteed to be more efficient than the OLS estimator, but it would be a poor choice of model for Ω_g that led to FGLS being less efficient.

Not all econometrics packages compute this cluster-robust estimate. In that case one can use a pairs cluster bootstrap (without asymptotic refinement). Specifically B times form G clusters $\{(\mathbf{y}_1^*, \mathbf{X}_1^*), \ldots, (\mathbf{y}_G^*, \mathbf{X}_G^*)\}$ by resampling with replacement G times from the original sample of clusters, each time compute the FGLS estimator, and then compute the variance of the B FGLS estimates $\widehat{\beta}_1, \ldots, \widehat{\beta}_B$ as $\widehat{\mathrm{V}}_{\mathrm{boot}}[\widehat{\beta}] = (B-1)^{-1}\sum_{b=1}^{B}(\widehat{\beta}_b - \overline{\widehat{\beta}})(\widehat{\beta}_b - \overline{\widehat{\beta}})'$. Care is needed, however, if the model includes cluster-specific fixed effects; see, for example, Cameron and Trivedi (2009, p. 421).

1.6.2 Efficiency Gains of Feasible GLS

Given a correct model for the within-cluster correlation of the error, such as equicorrelation, the feasible GLS estimator is more efficient than OLS. The efficiency gains of FGLS need not necessarily be great. For example, if the within-cluster correlation of all regressors is unity (so $\mathbf{x}_{ig} = \mathbf{x}_g$) and $\bar{u}_g$ defined in Subsection 1.2.3 is homoskedastic, then FGLS is equivalent to OLS so there is no gain to FGLS.

For equicorrelated errors and general $\mathbf{X}$, Scott and Holt (1982) provide an upper bound to the maximum proportionate efficiency loss of OLS compared to the variance of the FGLS estimator of $1/[1 + \frac{4(1-\rho_u)[1+(N_{\max}-1)\rho_u}{(N_{\max}\times\rho_u)^2}]$, $N_{\max} = \max\{N_1, \ldots, N_G\}$. This upper bound is increasing in the error correlation ρ_u and the maximum cluster size $N_{\max}$. For low ρ_u the maximal efficiency gain can be low. For example, Scott and Holt (1982) note that for $\rho_u = .05$ and $N_{\max} = 20$ there is at most a 12% efficiency loss of OLS compared to FGLS. But for $\rho_u = 0.2$ and $N_{\max} = 50$ the efficiency loss could be as much as 74%, though this depends on the nature of $\mathbf{X}$.

1.6.3 Random Effects Model

The one-way random effects (RE) model is given by formula 1.1 with $u_{ig} = \alpha_g + \varepsilon_{ig}$, where α_g and ε_{ig} are i.i.d. error components; see Subsection 1.2.2. Some algebra shows that the FGLS estimator in formula 1.12 can be computed by OLS estimation of $(y_{ig} - \widehat{\lambda}_g \bar{y}_i)$ on $(\mathbf{x}_{ig} - \widehat{\lambda}_g \bar{\mathbf{x}}_i)$, where $\widehat{\lambda}_g = 1 - \widehat{\sigma}_\varepsilon / \sqrt{\widehat{\sigma}^2_\varepsilon + N_g \widehat{\sigma}^2_\alpha}$. Applying the cluster-robust variance matrix formula 1.7 for OLS in this transformed model yields formula 1.13 for the FGLS estimator.

The RE model can be extended to multi-way clustering, though FGLS estimation is then more complicated. In the two-way case, $y_{igh} = \mathbf{x}'_{igh}\beta + \alpha_g + \delta_h + \varepsilon_{igh}$. For example, Moulton (1986) considered clustering due to grouping of regressors (schooling, age, and weeks worked) in a log earnings regression. In his model he allowed for a common random shock for each year of schooling, for each year of age, and for each number of weeks worked. Davis (2002) modeled film attendance data clustered by film, theater, and time. Cameron and Golotvina (2005) modeled trade between country pairs. These multi-way papers compute the variance matrix assuming Ω is correctly specified.

1.6.4 Hierarchical Linear Models

The one-way random effects model can be viewed as permitting the intercept to vary randomly across clusters. The hierarchical linear model (HLM) additionally permits the slope coefficients to vary. Specifically

$$y_{ig} = \mathbf{x}'_{ig}\beta_g + u_{ig}, \tag{1.14}$$

where the first component of $\mathbf{x}_{ig}$ is an intercept. A concrete example is to consider data on students within schools. Then y_{ig} is an outcome measure such as test score for the ith student in the gth school. In a two-level model the kth component of β_g is modeled as $\beta_{kg} = \mathbf{w}'_{kg}\gamma_k + v_{kg}$, where $\mathbf{w}_{kg}$ is a vector of school characteristics. Then stacking over all K components of β we have

$$\beta_g = \mathbf{W}_g\gamma + \mathbf{v}_j, \tag{1.15}$$

where $\mathbf{W}_g = \mathrm{Diag}[\mathbf{w}_{kg}]$ and usually the first component of $\mathbf{w}_{kg}$ is an intercept.

The random effects model is the special case $\beta_g = (\beta_{1g}, \beta_{2g})$, where $\beta_{1g} = 1 \times \gamma_1 + v_{1g}$ and $\beta_{kg} = \gamma_k + 0$ for $k > 1$, so v_{1g} is the random effects model's α_g. The HLM model additionally allows for random slopes β_{2g} that may or may not vary with level-two observables $\mathbf{w}_{kg}$. Further levels are possible, such as schools nested in school districts.

The HLM model can be re-expressed as a mixed linear model, since substituting formula 1.15 into formula 1.14 yields

$$y_{ig} = (\mathbf{x}_{ig}'\mathbf{W}_g)\gamma + \mathbf{x}_{ig}'\mathbf{v}_g + u_{ig}. \tag{1.16}$$

The goal is to estimate the regression parameter γ and the variances and covariances of the errors u_{ig} and $\mathbf{v}_g$. Estimation is by maximum likelihood assuming the errors $\mathbf{v}_g$ and u_{ig} are normally distributed. Note that the pooled OLS estimator of γ is consistent but is less efficient.

HLM programs assume that formula 1.15 correctly specifies the within-cluster correlation. One can instead robustify the standard errors by using formulas analogous to formula 1.13, or by the cluster bootstrap.

1.6.5 Serially Correlated Errors Models for Panel Data

If N_g is small, the clusters are balanced, and it is assumed that Ω_g is the same for all g, say $\Omega_g = \Omega$, then the FGLS estimator in formula 1.12 can be used without need to specify a model for Ω. Instead we can let $\widehat{\Omega}$ have ijth entry $G^{-1}\sum_{g=1}^{G} \widehat{u}_{ig}\widehat{u}_{jg}$, where $\widehat{u}_{ig}$ are the residuals from initial OLS estimation.

This procedure was proposed for short panels by Kiefer (1980). It is appropriate in this context under the assumption that variances and autocovariances of the errors are constant across individuals. While this assumption is restrictive, it is less restrictive than, for example, the AR(1) error assumption given in Subsection 1.2.3.

In practice two complications can arise with panel data. First, there are $T(T-1)/2$ off-diagonal elements to estimate and this number can be large relative to the number of observations NT. Second, if an individual-specific fixed effects panel model is estimated, then the fixed effects lead to an incidental parameters bias in estimating the off-diagonal covariances. This is the case for differences-in-differences models, yet FGLS estimation is desirable as it is more efficient than OLS. Hausman and Kuersteiner (2008) present fixes for both complications, including adjustment to Wald test critical values by using a higher-order Edgeworth expansion that takes account of the uncertainty in estimating the within-state covariance of the errors.

A more commonly used model specifies an AR(p) model for the errors. This has the advantage over the preceding method of having many fewer parameters to estimate in Ω, though it is a more restrictive model. Of course, one can robustify using formula 1.13. If fixed effects are present, however, then there is again a bias (of order N_g^{-1}) in estimation of the AR(p) coefficients due to the presence of fixed effects. Hansen (2007b) obtains bias-corrected estimates of the AR(p) coefficients and uses these in FGLS estimation.

Other models for the errors have also been proposed. For example, if clusters are large, we can allow correlation parameters to vary across clusters.

1.7 Nonlinear and Instrumental Variables Estimators

Relatively few econometrics papers consider extension of the complications discussed in this paper to nonlinear models; a notable exception is Wooldridge (2006).

1.7.1 Population-Averaged Models

The simplest approach to clustering in nonlinear models is to estimate the same model as would be estimated in the absence of clustering, but then base inference on cluster-robust standard errors that control for any clustering. This approach requires the assumption that the estimator remains consistent in the presence of clustering.

For commonly used estimators that rely on correct specification of the conditional mean, such as logit, probit, and Poisson, one continues to assume that $\mathrm{E}[y_{ig} \mid \mathbf{x}_{ig}]$ is correctly specified. The model is estimated ignoring any clustering, but then sandwich standard errors that control for clustering are computed. This pooled approach is called a population-averaged approach because rather than introduce a cluster effect α_g and model $\mathrm{E}[y_{ig}|\mathbf{x}_{ig}, \alpha_g]$, see Subsection 1.7.2, we directly model $\mathrm{E}[y_{ig} \mid \mathbf{x}_{ig}] = \mathrm{E}_{\alpha_g}[\,\mathrm{E}[y_{ig} \mid \mathbf{x}_{ig}, \alpha_g]]$ so that α_g has been averaged out.

This essentially extends pooled OLS to, for example, pooled probit. Efficiency gains analogous to feasible GLS are possible for nonlinear models if one additionally specifies a reasonable model for the within-cluster correlation.

The generalized estimating equations (GEE) approach, due to Liang and Zeger (1986), introduces within-cluster correlation into the class of generalized linear models (GLM). A conditional mean function is specified, with $\mathrm{E}[y_{ig} \mid \mathbf{x}_{ig}] = m(\mathbf{x}'_{ig}\beta)$, so that for the gth cluster

$$\mathrm{E}[\mathbf{y}_g|\mathbf{X}_g] = \mathbf{m}_g(\beta), \tag{1.17}$$

where $\mathbf{m}_g(\beta) = [m(\mathbf{x}'_{1g}\beta), \ldots, m(\mathbf{x}'_{N_g g}\beta)]'$ and $\mathbf{X}_g = [\mathbf{x}_{1g}, \ldots, \mathbf{x}_{N_g g}]'$. A model for the variances and covariances is also specified. First given the variance model $\mathrm{V}[y_{ig} \mid \mathbf{x}_{ig}] = \phi h(m(\mathbf{x}'_{ig}\beta)$ where ϕ is an additional scale parameter to estimate, we form $\mathbf{H}_g(\beta) = \mathrm{Diag}[\phi h(m(\mathbf{x}'_{ig}\beta)]$, a diagonal matrix with the variances as entries. Second, a correlation matrix $\mathbf{R}(\alpha)$ is specified with ijth entry $\mathrm{Cor}[y_{ig}, y_{jg} \mid \mathbf{X}_g]$, where α are additional parameters to estimate. Then the within-cluster covariance matrix is

$$\Omega_g = \mathrm{V}[\mathbf{y}_g \mid \mathbf{X}_g] = \mathbf{H}_g(\beta)^{1/2}\mathbf{R}(\alpha)\mathbf{H}_g(\beta)^{1/2}. \tag{1.18}$$

$\mathbf{R}(\alpha) = \mathbf{I}$ if there is no within-cluster correlation, and $\mathbf{R}(\alpha) = \mathbf{R}(\rho)$ has diagonal entries 1 and off diagonal entries ρ in the case of equicorrelation. The resulting GEE estimator $\widehat{\beta}_{\text{GEE}}$ solves

$$\sum_{g=1}^{G} \frac{\partial \mathbf{m}_g'(\beta)}{\partial \beta} \widehat{\Omega}_g^{-1}(\mathbf{y}_g - \mathbf{m}_g(\beta)) = \mathbf{0}, \tag{1.19}$$

where $\widehat{\Omega}_g$ equals Ω_g in formula 1.18 with $\mathbf{R}(\alpha)$ replaced by $\mathbf{R}(\widehat{\alpha})$ where $\widehat{\alpha}$ is consistent for α. The cluster-robust estimate of the asymptotic variance matrix of the GEE estimator is

$$\widehat{V}[\widehat{\beta}_{\text{GEE}}] = \left(\widehat{\mathbf{D}}'\widehat{\Omega}^{-1}\widehat{\mathbf{D}}\right)^{-1} \left(\sum_{g=1}^{G} \mathbf{D}_g'\widehat{\Omega}_g^{-1}\widehat{\mathbf{u}}_g\widehat{\mathbf{u}}_g'\widehat{\Omega}_g^{-1}\mathbf{D}_g\right) \left(\mathbf{D}'\widehat{\Omega}^{-1}\mathbf{D}\right)^{-1}, \tag{1.20}$$

where $\widehat{\mathbf{D}}_g = \partial \mathbf{m}_g'(\beta)/\partial\beta|_{\widehat{\beta}}$, $\widehat{\mathbf{D}} = [\widehat{\mathbf{D}}_1, \ldots, \widehat{\mathbf{D}}_G]'$, $\widehat{\mathbf{u}}_g = \mathbf{y}_g - \mathbf{m}_g(\widehat{\beta})$, and now $\widehat{\Omega}_g = \mathbf{H}_g(\widehat{\beta})^{1/2}\mathbf{R}(\widehat{\alpha})\mathbf{H}_g(\widehat{\beta})^{1/2}$. The asymptotic theory requires that $G \to \infty$.

The result formula 1.20 is a direct analog of the cluster-robust estimate of the variance matrix for FGLS. Consistency of the GEE estimator requires that formula 1.17 holds, i.e., correct specification of the conditional mean (even in the presence of clustering). The variance matrix defined in formula 1.18 permits heteroskedasticity and correlation. It is called a "working" variance matrix as subsequent inference based on formula 1.20 is robust to misspecification of formula 1.18. If formula 1.18 is assumed to be correctly specified then the asymptotic variance matrix is more simply $(\widehat{\mathbf{D}}'\widehat{\Omega}^{-1}\widehat{\mathbf{D}})^{-1}$.

For likelihood-based models outside the GLM class, a common procedure is to perform ML estimation under the assumption of independence over i and g, and then obtain cluster-robust standard errors that control for within-cluster correlation. Let $f(y_{ig} \mid \mathbf{x}_{ig}, \theta)$ denote the density, $\mathbf{s}_{ig}(\theta) = \partial \ln f(y_{ig} \mid \mathbf{x}_{ig}, \theta)/\partial\theta$, and $\mathbf{s}_g(\theta) = \sum_i \mathbf{s}_{ig}(\theta)$. Then the MLE of θ solves $\sum_g \sum_i \mathbf{s}_{ig}(\theta) = \sum_g \mathbf{s}_g(\theta) = \mathbf{0}$. A cluster-robust estimate of the variance matrix is

$$\widehat{V}[\widehat{\beta}_{\text{ML}}] = \left(\sum_g \partial\mathbf{s}_g(\theta)'/\partial\theta|_{\widehat{\theta}}\right)^{-1} \left(\sum_g \mathbf{s}_g(\widehat{\theta})\mathbf{s}_g(\widehat{\theta})'\right) \left(\sum_g \partial\mathbf{s}_g(\theta)/\partial\theta'|_{\widehat{\theta}}\right)^{-1}. \tag{1.21}$$

This method generally requires that $f(y_{ig} \mid \mathbf{x}_{ig}, \theta)$ is correctly specified even in the presence of clustering.

In the case of a (mis)specified density that is in the linear exponential family, as in GLM estimation, the MLE retains its consistency under the weaker assumption that the conditional mean $\mathrm{E}[y_{ig} \mid \mathbf{x}_{ig}, \theta]$ is correctly specified. In that case the GEE estimator defined in formula 1.19 additionally permits incorporation of a model for the correlation induced by the clustering.

1.7.2 Cluster-Specific Effects Models

An alternative approach to controlling for clustering is to introduce a group-specific effect.

For conditional mean models the population-averaged assumption that $E[y_{ig} \mid \mathbf{x}_{ig}] = m(\mathbf{x}_{ig}'\beta)$ is replaced by

$$E[y_{ig} \mid \mathbf{x}_{ig}, \alpha_g] = g(\mathbf{x}_{ig}'\beta + \alpha_g), \tag{1.22}$$

where α_g is not observed. The presence of α_g will induce correlation between y_{ig} and y_{jg}, $i \neq j$. Similarly, for parametric models the density specified for a single observation is $f(y_{ig} \mid \mathbf{x}_{ig}, \beta, \alpha_g)$ rather than the population-averaged $f(y_{ig} \mid \mathbf{x}_{ig}, \beta)$.

In a fixed effects model the α_g are parameters to be estimated. If asymptotics are that N_g is fixed while $G \to \infty$ then there is an incidental parameters problem, as there are N_g parameters $\alpha_1, \ldots, \alpha_G$ to estimate and $G \to \infty$. In general, this contaminates estimation of β so that $\widehat{\beta}$ is a inconsistent. Notable exceptions where it is still possible to consistently estimate β are the linear regression model, the logit model, the Poisson model, and a nonlinear regression model with additive error (so formula 1.22 is replaced by $E[y_{ig} \mid \mathbf{x}_{ig}, \alpha_g] = g(\mathbf{x}_{ig}'\beta) + \alpha_g$). For these models, aside from the logit, one can additionally compute cluster-robust standard errors after fixed effects estimation.

We focus on the more commonly used random effects model that specifies α_g to have density $h(\alpha_g \mid \eta)$ and consider estimation of likelihood-based models. Conditional on α_g, the joint density for the gth cluster is $f(y_{1g}, \ldots, \mid \mathbf{x}_{N_g g}, \beta, \alpha_g) = \prod_{i=1}^{N_g} f(y_{ig} \mid \mathbf{x}_{ig}, \beta, \alpha_g)$. We then integrate out α_g to obtain the likelihood function

$$L(\beta, \eta \mid \mathbf{y}, \mathbf{X}) = \prod_{g=1}^{G} \left\{ \int \left(\prod_{i=1}^{N_g} f(y_{ig} \mid \mathbf{x}_{ig}, \beta, \alpha_g) \right) dh(\alpha_g \mid \eta) \right\}. \tag{1.23}$$

In some special nonlinear models, such as a Poisson model with α_g being gamma distributed, it is possible to obtain a closed-form solution for the integral. More generally this is not the case, but numerical methods work well as formula 1.23 is just a one-dimensional integral. The usual assumption is that α_g is distributed as $\mathcal{N}[0, \sigma_\alpha^2]$. The MLE is very fragile and failure of any assumption in a nonlinear model leads to inconsistent estimation of β.

The population-averaged and random effects models differ for nonlinear models, so that β is not comparable across the models. But the resulting average marginal effects, that integrate out α_g in the case of a random effects model, may be similar. A leading example is the probit model. Then $E[y_{ig} \mid \mathbf{x}_{ig}, \alpha_g] = \Phi(\mathbf{x}_{ig}'\beta + \alpha_g)$, where $\Phi(\cdot)$ is the standard normal c.d.f. Letting $f(\alpha_g)$ denote the $\mathcal{N}[0, \sigma_\alpha^2]$ density for α_g, we obtain $E[y_{ig} \mid \mathbf{x}_{ig}] = \int \Phi(\mathbf{x}_{ig}'\beta + \alpha_g) f(\alpha_g) d\alpha_g = \Phi(\mathbf{x}_{ig}'\beta/\sqrt{1+\sigma_\alpha^2})$; see Wooldridge (2002, p. 470). This differs from $E[y_{ig} \mid \mathbf{x}_{ig}] = \Phi(\mathbf{x}_{ig}'\beta)$ for the pooled or population-averaged probit model. The difference is the scale factor $\sqrt{1+\sigma_\alpha^2}$. However, the marginal effects are similarly rescaled, since $\partial \Pr[y_{ig} = 1 \mid \mathbf{x}_{ig}]/\partial \mathbf{x}_{ig} = \phi(\mathbf{x}_{ig}'\beta/\sqrt{1+\sigma_\alpha^2}) \times \beta/\sqrt{1+\sigma_\alpha^2}$, so in this case PA probit and random effects probit will yield similar estimates of the average marginal effects; see Wooldridge (2002, 2006).

1.7.3 Instrumental Variables

The cluster-robust formula is easily adapted to instrumental variables estimation. It is assumed that there exist instruments $\mathbf{z}_{ig}$ such that $u_{ig} = y_{ig} - \mathbf{x}_{ig}'\beta$ satisfies $\mathrm{E}[u_{ig}|\mathbf{z}_{ig}] = 0$. If there is within-cluster correlation we assume that this condition still holds, but now $\mathrm{Cov}[u_{ig}, u_{jg} \mid \mathbf{z}_{ig}, \mathbf{z}_{jg}] \neq 0$.

Shore-Sheppard (1996) examines the impact of equicorrelated instruments and group-specific shocks to the errors. Her model is similar to that of Moulton, applied to an IV setting. She shows that IV estimation that does not model the correlation will understate the standard errors, and proposes either cluster-robust standard errors or FGLS.

Hoxby and Paserman (1998) examine the validity of overidentification (OID) tests with equicorrelated instruments. They show that not accounting for within-group correlation can lead to mistaken OID tests, and they give a cluster-robust OID test statistic. This is the GMM criterion function with a weighting matrix based on cluster summation.

A recent series of developments in applied econometrics deals with the complication of weak instruments that lead to poor finite-sample performance of inference based on asymptotic theory, even when sample sizes are quite large; see for example the survey by Andrews and Stock (2007), and Cameron and Trivedi (2005, 2009). The literature considers only the nonclustered case, but the problem is clearly relevant also for cluster-robust inference. Most papers consider only i.i.d. errors. An exception is Chernozhukov and Hansen (2008) who suggest a method based on testing the significance of the instruments in the reduced form that is heteroskedastic-robust. Their tests are directly amenable to adjustments that allow for clustering; see Finlay and Magnusson (2009).

1.7.4 GMM

Finally we consider generalized methods of moments (GMM) estimation.

Suppose that we combine moment conditions for the gth cluster, so $\mathrm{E}[\mathbf{h}_g(\mathbf{w}_g, \theta)] = \mathbf{0}$, where $\mathbf{w}_g$ denotes all variables in the cluster. Then the GMM estimator $\widehat{\theta}_{\mathrm{GMM}}$ with weighting matrix $\mathbf{W}$ minimizes $(\sum_g \mathbf{h}_g)'\mathbf{W}(\sum_g \mathbf{h}_g)$, where $\mathbf{h}_g = \mathbf{h}_g(\mathbf{w}_g, \theta)$. Using standard results in, for example, Cameron and Trivedi (2005, p. 175) or Wooldridge (2002, p. 423), the variance matrix estimate is

$$\widehat{\mathrm{V}}[\widehat{\theta}_{\mathrm{GMM}}] = (\widehat{\mathbf{A}}'\mathbf{W}\widehat{\mathbf{A}})^{-1}\widehat{\mathbf{A}}'\mathbf{W}\widehat{\mathbf{B}}\mathbf{W}\widehat{\mathbf{A}}(\widehat{\mathbf{A}}'\mathbf{W}\widehat{\mathbf{A}})^{-1}$$

where $\widehat{\mathbf{A}} = \sum_g \partial\mathbf{h}_g/\partial\theta'|_{\widehat{\theta}}$ and a cluster-robust variance matrix estimate uses $\widehat{\mathbf{B}} = \sum_g \widehat{\mathbf{h}}_g\widehat{\mathbf{h}}_g'$. This assumes independence across clusters and $G \to \infty$. Bhattacharya (2005) considers stratification in addition to clustering for the GMM estimator.

Again a key assumption is that the estimator remains consistent even in the presence of clustering. For GMM this means that we need to assume that the moment condition holds true even when there is within-cluster correlation.

The reasonableness of this assumption will vary with the particular model and application at hand.

1.8 Empirical Example

To illustrate some empirical issues related to clustering, we present an application based on a simplified version of the model in Hersch (1998), who examined the relationship between wages and job injury rates. We thank Joni Hersch for sharing her data with us. Job injury rates are observed only at occupation levels and industry levels, inducing clustering at these levels. In this application we have individual-level data from the Current Population Survey on 5960 male workers working in 362 occupations and 211 industries. For most of our analysis we focus on the occupation injury rate coefficient. Hersch (1998) investigates the surprising negative sign of this coefficient.

In column 1 of Table 1.1, we present results from linear regression of log wages on occupation and industry injury rates, potential experience and its square, years of schooling, and indicator variables for union, nonwhite, and three regions. The first three rows show that standard errors of the OLS estimate increase as we move from default (row 1) to White heteroskedastic-robust (row 2) to cluster-robust with clustering on occupation (row 3). A priori heteroskedastic-robust standard errors may be larger or smaller than the default. The clustered standard errors are expected to be larger. Using formula 1.4 suggests inflation factor $\sqrt{1 + 1 \times 0.169 \times (5960/362 - 1)} = 1.90$, as the within-cluster correlation of model residuals is 0.169, compared to an actual inflation of $0.516/0.188 = 2.74$. The adjustment mentioned after formula 1.4 for unequal group size, which here is substantial, yields a larger inflation factor of 3.77.

Column 2 of Table 1.1 illustrates analysis with few clusters, when analysis is restricted to the 1594 individuals who work in the 10 most common occupations in the dataset. From rows 1 to 3 the standard errors increase, due to fewer observations, and the variance inflation factor is larger due to a larger average group size, as suggested by formula 1.4. Our concern is that with $G = 10$ the usual asymptotic theory requires some adjustment. The Wald two-sided test statistic for a zero coefficient on occupation injury rate is $-2.751/0.994 = 2.77$. Rows 4–6 of column 2 report the associated p-value computed in three ways. First, $p = 0.006$ using standard normal critical values (or the T with $N - K = 1584$ degrees of freedom). Second, $p = 0.022$ using a T distribution based on $G - 1 = 9$ degrees of freedom. Third, when we perform a pairs cluster percentile-T bootstrap, the p-value increases to 0.110. These changes illustrate the importance of adjusting for few clusters in conducting inference. The large increase in p-value with the bootstrap may in part be because the first two p-values are based on cluster-robust standard errors with finite-sample bias; see Subsection 1.4.1. This may also explain why

TABLE 1.1

Occupation Injury Rate and Log Wages: Impacts of Varying Ways of Dealing with Clustering

		1 Main Sample Linear	2 10 Largest Occupations Linear	3 Main Sample Probit
	OLS (or Probit) coefficient on Occupation Injury Rate	−2.158	−2.751	−6.978
1	Default (iid) std. error	0.188	0.308	0.626
2	White-robust std. error	0.243	0.320	1.008
3	Cluster-robust std. error (Clustering on Occupation)	0.516	0.994	1.454
4	P-value based on (3) and Standard Normal		0.006	
5	P-value based on (3) and T(10-1)		0.022	
6	P-value based on Percentile-T Pairs Bootstrap (999 replications)		0.110	
7	Two-way (Occupation and Industry) robust std. error	0.515	0.990	1.516
	Random effects Coefficient on Occupation Injury Rate	−1.652	−2.669	−5.789
8	Default std. error	0.357	1.429	1.106
9	White-robust std. error	0.579	2.058	
10	Cluster-robust std. error (Clustering on Occupation)	0.536	2.148	
	Number of observations (N)	5960	1594	5960
	Number of Clusters (G)	362	10	362
	Within-Cluster correlation of errors (rho)	0.207	0.211	

Note: Coefficients and standard errors multiplied by 100. Regression covariates include Occupation Injury rate, Industry Injury rate, Potential experience, Potential experience squared, Years of schooling, and indicator variables for union, nonwhite, and three regions. Data from Current Population Survey, as described in Hersch (1998). Std. errs. in rows 9 and 10 are from bootstraps with 400 replications. Probit outcome is wages >= $12/hour.

the random effect (RE) model standard errors in rows 8–10 of column 2 exceed the OLS cluster-robust standard error in row 3 of column 2.

We next consider multi-way clustering. Since both occupation-level and industry-level regressors are included, we should compute two-way cluster-robust standard errors. Comparing row 7 of column 1 to row 3, the standard error of the occupation injury rate coefficient changes little from 0.516 to 0.515. But there is a big impact for the coefficient of the industry injury rate. In results, not reported in the table, the standard error of the industry injury rate coefficient increases from 0.563 when we cluster on only occupation to 1.015 when we cluster on both occupation and industry.

If the clustering within occupations is due to common occupation-specific shocks, then a RE model may provide more efficient parameter estimates. From row 8 of column 1 the default RE standard error is 0.357, but if we cluster on occupation this increases to 0.536 (row 10). For these data there is apparently no gain compared to OLS (see row 3).

Finally we consider a nonlinear example, probit regression with the same data and regressors, except the dependent variable is now a binary outcome equal to one if the hourly wage exceeds 12 dollars. The results given in column 3 are qualitatively similar to those in column 1. Cluster-robust standard errors are 2–3 times larger, and two-way cluster robust are slightly larger still. The parameters β of the random effects probit model are rescalings of those of the standard probit model, as explained in Subsection 1.7.2. The RE probit coefficient of -5.789 becomes -5.119 upon rescaling, as $\widehat{\alpha}_g$ has estimated variance 0.279. This is smaller than the standard probit coefficient, though this difference may just reflect noise in estimation.

1.9 Conclusion

Cluster-robust inference is possible in a wide range of settings. The basic methods were proposed in the 1980s, but are still not yet fully incorporated into applied econometrics, especially for estimators other than OLS. Useful references on cluster-robust inference for the practitioner include the surveys by Wooldridge (2003, 2006), the texts by Wooldridge (2002), Cameron and Trivedi (2005) and Angrist and Pischke (2009) and, for implementation in Stata, Nichols and Schaffer (2007) and Cameron and Trivedi (2009).

References

Acemoglu, D., and J.-S. Pischke. 2003. Minimum Wages and On-the-job Training. *Res. Labor Econ.* 22: 159–202.

Andrews, D. W. K., and J. H. Stock. 2007. Inference with Weak Instruments. In *Advances in Economics and Econometrics, Theory and Applications: Ninth World Congress of the Econometric Society*, ed. R. Blundell, W. K. Newey, and T. Persson, Vol. III, Ch. 3. Cambridge, U.K.: Cambridge Univ. Press.

Angrist, J. D., and V. Lavy. 2009. The Effect of High School Matriculation Awards: Evidence from Randomized Trials. *Am. Econ. Rev.* 99: 1384–1414.

Angrist, J. D., and J.-S. Pischke. 2009. *Mostly Harmless Econometrics: An Empiricist's Companion*. Princeton, NJ: Princeton Univ. Press.

Arellano, M. 1987. Computing Robust Standard Errors for Within-Group Estimators. *Oxford Bull. Econ. Stat.* 49: 431–434.

Bell, R. M., and D. F. McCaffrey. 2002. Bias Reduction in Standard Errors for Linear Regression with Multi-Stage Samples. *Surv. Methodol.* 28: 169–179.

Bertrand, M., E. Duflo, and S. Mullainathan. 2004. How Much Should We Trust Differences-in-Differences Estimates?. *Q. J. Econ.* 119: 249–275.

Bester, C. A., T. G. Conley, and C. B. Hansen. 2009. *Inference with Dependent Data Using Cluster Covariance Estimators*. Manuscript, Univ. of Chicago.

Bhattacharya, D. 2005. Asymptotic Inference from Multi-Stage Samples. *J. Econometr.* 126: 145–171.

Cameron, A. C., J. G. Gelbach, and D. L. Miller. 2006. Robust Inference with Multi-Way Clustering. NBER Technical Working Paper 0327.

Cameron, A. C., J. G. Gelbach, and D. L. Miller. 2008. Bootstrap-Based Improvements for Inference with Clustered Errors. *Rev. Econ. Stat.* 90: 414–427.

Cameron, A. C., J. G. Gelbach, and D. L. Miller. 2010. Robust Inference with Multi-Way Clustering. *J. Business and Econ. Stat.*, forthcoming.

Cameron, A. C., and N. Golotvina. 2005. Estimation of Country-Pair Data Models Controlling for Clustered Errors: With International Trade Applications. Working Paper 06-13, U. C. – Davis Department of Economics, Davis, CA.

Cameron, A. C., and P. K. Trivedi. 2005. *Microeconometrics: Methods and Applications.* Cambridge, U.K.: Cambridge Univ. Press.

Cameron, A. C., and P. K. Trivedi. 2009. *Microeconometrics Using Stata.* College Station, TX: Stata Press.

Chernozhukov, V., and C. Hansen. 2008. The Reduced Form: A Simple Approach to Inference with Weak Instruments. *Econ. Lett.* 100: 68–71.

Conley, T. G. 1999. GMM Estimation with Cross Sectional Dependence. *J. Econometr.*, 92, 1–45.

Conley, T. G., and C. Taber. 2010. Inference with 'Difference in Differences' with a Small Number of Policy Changes. *Rev. Econ. Stat.*, forthcoming.

Davis, P. 2002. Estimating Multi-Way Error Components Models with Unbalanced Data Structures. *J. Econometr.* 106: 67–95.

Donald, S. G., and K. Lang. 2007. Inference with Difference-in-Differences and Other Panel Data. *Rev. Econ. Stat.* 89: 221–233.

Driscoll, J. C., and A. C. Kraay. 1998. Consistent Covariance Matrix Estimation with Spatially Dependent Panel Data. *Rev. Econ. Stat.* 80: 549–560.

Fafchamps, M., and F. Gubert. 2007. The Formation of Risk Sharing Networks. *J. Dev. Econ.* 83: 326–350.

Finlay, K., and L. M. Magnusson. 2009. Implementing Weak Instrument Robust Tests for a General Class of Instrumental-Variables Models. *Stata J.* 9: 398–421.

Foote, C. L. 2007. Space and Time in Macroeconomic Panel Data: Young Workers and State-Level Unemployment Revisited. Working Paper 07-10, Federal Reserve Bank of Boston.

Greenwald, B. C. 1983. A General Analysis of Bias in the Estimated Standard Errors of Least Squares Coefficients. *J. Econometr.* 22: 323–338.

Hansen, C. 2007a. Asymptotic Properties of a Robust Variance Matrix Estimator for Panel Data when T is Large. *J. Econometr.* 141: 597–620.

Hansen, C. 2007b. Generalized Least Squares Inference in Panel and Multi-Level Models with Serial Correlation and Fixed Effects. *J. Econometr.* 141: 597–620.

Hausman, J., and G. Kuersteiner. 2008. Difference in Difference Meets Generalized Least Squares: Higher Order Properties of Hypotheses Tests. *J. Econometr.* 144: 371–391.

Hersch, J. 1998. Compensating Wage Differentials for Gender-Specific Job Injury Rates. *Am. Econ. Rev.* 88: 598–607.

Hoxby, C., and M. D. Paserman. 1998. Overidentification Tests with Group Data. Technical Working Paper 0223, New York: National Bureau of Economic Research.

Huber, P. J. 1967. The Behavior of Maximum Likelihood Estimates under Nonstandard Conditions. In *Proceedings of the Fifth Berkeley Symposium*, ed. J. Neyman, 1: 221–233. Berkeley, CA: Univ. of California Press.

Ibragimov, R., and U. K. Muller. 2010. T-Statistic Based Correlation and Heterogeneity Robust Inference. *J. Bus. Econ. Stat.*, forthcoming.

Kauermann, G., and R. J. Carroll. 2001. A Note on the Efficiency of Sandwich Covariance Matrix Estimation. *J. Am. Stat. Assoc.* 96: 1387–1396.

Kézdi, G. 2004. Robust Standard Error Estimation in Fixed-Effects Models. Robust Standard Error Estimation in Fixed-Effects Panel Models. *Hungarian Stat. Rev.* Special Number 9: 95–116.

Kiefer, N. M. 1980. Estimation of Fixed Effect Models for Time Series of Cross-Sections with Arbitrary Intertemporal Covariance. *J. Econometr.* 214: 195–202.

Kish, L. 1965. *Survey Sampling*. New York: John Wiley & Sons.

Kish, L., and M. R. Frankel. 1974. Inference from Complex Surveys with Discussion. *J. Royal Stat. Soc. B Met.* 36: 1–37.

Kloek, T. 1981. OLS Estimation in a Model where a Microvariable is Explained by Aggregates and Contemporaneous Disturbances are Equicorrelated. *Econometrica* 49: 205–07.

Liang, K.-Y., and S. L. Zeger. 1986. Longitudinal Data Analysis Using Generalized Linear Models. *Biometrika* 73: 13–22.

MacKinnon, J. G., and H. White. 1985. Some Heteroskedasticity-Consistent Covariance Matrix Estimators with Improved Finite Sample Properties. *J. Econometr.* 29: 305–325.

Mancl, L. A., and T. A. DeRouen. 2001. A Covariance Estimator for GEE with Improved Finite-Sample Properties. *Biometrics* 57: 126–134.

McCaffrey, D. F., R. M. Bell, and C. H. Botts. 2001. Generalizations of Bias Reduced Linearization. *Proceedings of the Survey Research Methods Section*, Alexandria, VA: American Statistical Association.

Miglioretti, D. L., and P. J. Heagerty. 2006. Marginal Modeling of Nonnested Multilevel Data Using Standard Software. *Am. J. Epidemiol.* 165: 453–463.

Moulton, B. R. 1986. Random Group Effects and the Precision of Regression Estimates. *J. Econometr.* 32: 385–397.

Moulton, B. R. 1990. An Illustration of a Pitfall in Estimating the Effects of Aggregate Variables on Micro Units. *Rev. Econ. Stat.* 72: 334–338.

Nichols, A., and M. E. Schaffer. 2007. Clustered Standard Errors in Stata. Paper presented at United Kingdom Stata Users' Group Meeting.

Pepper, J. V. 2002. Robust Inferences from Random Clustered Samples: An Application Using Data from the Panel Study of Income Dynamics. *Econ. Lett.* 75: 341–345.

Petersen, M. 2009. Estimating Standard Errors in Finance Panel Data Sets: Comparing Approaches. *Rev. Fin. Stud.* 22: 435–480.

Pfeffermann, D., and G. Nathan. 1981. Regression Analysis of Data from a Cluster Sample. *J. Am. Stat. Assoc.* 76: 681–689.

Rogers, W. H. 1993. Regression Standard Errors in Clustered Samples. *Stata Tech. Bull.* 13: 19–23.

Scott, A. J., and D. Holt. 1982. The Effect of Two-Stage Sampling on Ordinary Least Squares Methods. *J. Am. Stat. Assoc.* 77: 848–854.

Shore-Sheppard, L. 1996. The Precision of Instrumental Variables Estimates with Grouped Data. Working Paper 374, Princeton Univ. Industrial Relations Section, Princeton, NJ.

Stock, J. H., and M. W. Watson. 2008. Heteroskedasticity-Robust Standard Errors for Fixed Effects Panel Data Regression. *Econometrica* 76: 155–174.

Thompson, S. 2006. Simple Formulas for Standard Errors That Cluster by Both Firm and Time. SSRN paper. http://ssrn.com/abstract=914002.

White, H. 1980. A Heteroskedasticity-Consistent Covariance Matrix Estimator and a Direct Test for Heteroskedasticity. *Econometrica* 48: 817–838.

White, H. 1982. Maximum Likelihood Estimation of Misspecified Models. *Econometrica* 50: 1–25.

White, H. 1984. *Asymptotic Theory for Econometricians.* San Diego: Academic Press.

White, H., and I. Domowitz. 1984. Nonlinear Regression with Dependent Observations. *Econometrica* 52: 143–162.

Wooldridge, J. M. 2002. *Econometric Analysis of Cross Section and Panel Data.* Cambridge, MA: M.I.T. Press.

Wooldridge, J. M. 2003. Cluster-Sample Methods in Applied Econometrics. *Am. Econ. Rev.* 93: 133–138.

Wooldridge, J. M. 2006. Cluster-Sample Methods in Applied Econometrics: An Extended Analysis. Unpublished manuscript, Michigan State Univ. Department of Economics, East Lansing, MI.

2

Efficient Inference with Poor Instruments: A General Framework

Bertille Antoine and Eric Renault

CONTENTS

2.1 Introduction

The generalized method of moments (GMM) provides a computationally convenient method for inference on the structural parameters of economic models. The method has been applied in many areas of economics but it was in empirical finance that the power of the method was first illustrated. Hansen (1982) introduced GMM and presented its fundamental statistical theory. Hansen and Hodrick (1980) and Hansen and Singleton (1982) showed the potential of the GMM approach to testing economic theories through their empirical analyzes of, respectively, foreign exchange markets and asset pricing. In such contexts, the cornerstone of GMM inference is a set of conditional moment restrictions. More generally, GMM is well suited for the test of an economic theory every time the theory can be encapsulated in the postulated unpredictability of some error term $u(Y_t, \theta)$ given as a known function of p

unknown parameters $\theta \in \Theta \subseteq \mathbb{R}^p$ and a vector of observed random variables Y_t. Then, the testability of the theory of interest is akin to the testability of a set of conditional moment restrictions,

$$E_t[u(Y_{t+1}, \theta)] = 0, \tag{2.1}$$

where the operator $E_t[.]$ denotes the conditional expectation given available information at time t. Moreover, under the null hypothesis that the theory summarized by the restrictions (Equation 2.1) is true, these restrictions are supposed to uniquely identify the true unknown value θ^0 of the parameters. Then, GMM considers a set of H instruments z_t assumed to belong to the available information at time t and to summarize the testable implications of Equation 2.1 by the implied unconditional moment restrictions:

$$E[\phi_t(\theta)] = 0 \quad \text{where} \quad \phi_t(\theta) = z_t \otimes u(Y_{t+1}, \theta). \tag{2.2}$$

The recent literature on weak instruments (see the seminal work by Stock and Wright 2000) has stressed that the standard asymptotic theory of GMM inference may be misleading because of the insufficient correlation between some instruments z_t and some components of the local explanatory variables of $[\partial u(Y_{t+1}, \theta)/\partial\theta]$. In this case, some of the moment conditions (Equation 2.2) are not only zero at θ^0 but rather flat and close to zero in a neighborhood of θ^0.

Many asset pricing applications of GMM focus on the study of a pricing kernel as provided by some financial theory. This pricing kernel is typically either a linear function of the parameters of interest, as in linear-beta pricing models, or a log-linear one as in most of the equilibrium based pricing models where parameters of interest are preference parameters. In all these examples, the weak instruments' problem simply relates to some lack of predictability of some asset returns from some lagged variables.

Since the seminal work of Stock and Wright (2000), it is common to capture the impact of the weakness of instruments by a drifting data generating process (hereafter DGP) such that the informational content of estimating equations $\rho_T(\theta) = E[\phi_t(\theta)]$ about structural parameters of interest is impaired by the fact that $\rho_T(\theta)$ becomes zero for all θ when the sample size goes to infinity. The initial goal of this so-called "weak instruments asymptotics" approach was to devise inference procedures robust to weak identification in the worst case scenario, as made formal by Stock and Wright (2000):

$$\rho_T(\theta) = \frac{\rho_{1T}(\theta)}{\sqrt{T}} + \rho_2(\theta_1) \quad \text{with} \quad \theta = [\theta_1' \; \theta_2']' \quad \text{and} \quad \rho_2(\theta_1) = 0 \Leftrightarrow \theta_1 = \theta_1^0. \tag{2.3}$$

The rationale for Equation 2.3 is the following. While some components θ_1 of θ would be identified in a standard way if the other components θ_2 were known, the latter ones are so weakly identified that for sample sizes typically available in practice, no significant increase of accuracy of estimators can be noticed when the sample size increases: the typical root-T consistency is completely erased by the DGP drifting at the same rate through the term $\rho_{1T}(\theta)/\sqrt{T}$. It is then clear that this drifting rate is a worst case scenario, sensible when robustness to weak identification is the main concern, as it is the case for popular micro-econometric applications: for instance the study of Angrist and Krueger (1991) on returns to education.

The purpose of this chapter is somewhat different: taking for granted that some instruments may be poor, we nevertheless do not give up the efficiency goal of statistical inference. Even fragile information must be processed optimally, for the purpose of both efficient estimation and powerful testing. This point of view leads us to a couple of modifications with respect to the traditional weak instruments asymptotics.

First, we consider that the worst case scenario is a possibility but not the general rule. Typically, we revisit the drifting DGP (Equation 2.3) with a more general framework like:

$$\rho_T(\theta) = \frac{\rho_{1T}(\theta)}{T^{\lambda}} + \rho_2(\theta_1) \quad \text{with} \quad 0 \leq \lambda \leq 1/2.$$

The case $\lambda = 1/2$ has been the main focus of interest of the weak instruments literature so far because it accommodates the observed lack of consistency of some GMM estimators (typically estimators of θ_2 in the framework of Equation 2.3) and the implied lack of asymptotic normality of the consistent estimators (estimators of θ_1 in the framework of Equation 2.3). We rather set the focus on an intermediate case, $0 < \lambda < 1/2$, which has been dubbed nearly weak identification by Hahn and Kuersteiner (2002) in the linear case and Caner (2010) for nonlinear GMM. Standard (strong) identification would take $\lambda = 0$. Note also that nearly weak identification is implicitly studied by several authors who introduce infinitely many instruments: the large number of instruments partially compensates for the genuine weakness of each of them individually (see Han and Phillips 2006; Hansen, Hausman, and Newey 2008; Newey and Windmeijer 2009).

However, following our former work in Antoine and Renault (2009, 2010a), our main contribution is above all to consider that several patterns of identification may show up simultaneously. This point of view appears especially relevant for the asset pricing applications described above. Nobody would pretend that the constant instrument is weak. Therefore, the moment condition, $E[u(Y_{t+1}, \theta)] = 0$, should not display any drifting feature (as it actually corresponds to $\lambda = 0$). Even more interestingly, Epstein and Zin (1991) stress that the pricing equation for the market return is poorly informative about the difference between the risk aversion coefficient and the inverse of the elasticity of substitution. Individual asset returns should be more informative.

This paves the way for two additional extensions in the framework (Equation 2.3). First, one may consider, depending on the moment conditions, different values of the parameter λ of drifting DGP. Large values of λ would be assigned to components $[z_{it} \times u_j(Y_{t+1}, \theta)]$ for which either the pricing of asset j or the lagged value of return i are especially poorly informative. Second, there is no such thing as a parameter θ_2 always poorly identified or parameter θ_1 which would be strongly identified if the other parameters θ_2 were known. Instead, one must define directions in the parameter space (like the difference between risk aversion and inverse of elasticity of substitution) that may be poorly identified by some particular moment restrictions.

This heterogeneity of identification patterns clearly paves the way for the device of optimal strategies for inferential use of fragile (or poor) information. In this chapter, we focus on a case where asymptotic efficiency of estimators is well-defined through the variance of asymptotically normal distributions. The price to pay for this maintained tool is to assume that the set of moment conditions that are not genuinely weak ($\lambda < 1/2$) is sufficient to identify the true unknown value θ^0 of the parameters. In this case, normality must be reconsidered at heterogeneous rates smaller than the standard root-T in different directions of the parameter space (depending on the strength of identification about these directions). At least, non-normal asymptotic distributions introduced by situations of partial identification as in Phillips (1989) and Choi and Phillips (1992) are avoided in our setting. It seems to us that, by considering the large sample sizes typically available in financial econometrics, working with the maintained assumption of asymptotic normality of estimators is reasonable; hence, the study of efficiency put forward in this chapter. However, there is no doubt that some instruments are poorer and that some directions of the parameter space are less strongly identified. Last but not least: even though we are less obsessed by robustness to weak identification in the worst case scenario, we do not want to require from the practitioner a prior knowledge of the identification schemes. Efficient inference procedures must be feasible without requiring any prior knowledge neither of the different rates λ of nearly weak identification, nor of the heterogeneity of identification patterns in different directions in the parameter space.

To delimit the focus of this chapter, we put an emphasis on efficient inference. There are actually already a number of surveys that cover the earlier literature on inference robust to weak instruments. For example, Stock, Wright, and Yogo (2002) set the emphasis on procedures available for detecting and handling weak instruments in the linear instrumental variables model. More recently, Andrews and Stock (2007) wrote an excellent review, discussing many issues involved in testing and building confidence sets robust to the weak instrumental variables problem. Smith (2007) revisited this review, with a special focus on empirical likelihood-based approaches. This chapter is organized as follows. Section 2.2 introduces framework and identification procedure with poor instruments; the consistency of all GMM estimators is deduced from an empirical process approach. Section 2.3 is concerned with asymptotic theory and inference. Section 2.4 compares our approach to others: we specifically discuss the linear instrumental variables regression model, the (non)equivalence between efficient two-step GMM and continuously updated GMM and the GMM-score test of Kleibergen (2005). Section 2.5 concludes. All the proofs are gathered in the appendix.

2.2 Identification with Poor Instruments

2.2.1 Framework

We consider the true unknown value θ^0 of the parameter $\theta \in \Theta \subset \mathbb{R}^p$ defined as the solution of the moment conditions $E[\phi_t(\theta)] = 0$ for some known function $\phi_t(.)$ of size K. Since the seminal work of Stock and Wright (2000), the weakness of the moment conditions (or instrumental variables) is usually captured through a drifting DGP such that the informational content of the estimating equations shrinks toward zero (for all θ) while the sample size T grows to infinity.

More precisely, the population moment conditions obtained from a set of *poor* instruments are modeled as a function $\rho_T(\theta)$ that depends on the sample size T and becomes zero when it goes to infinity. The statistical information about the estimating equations $\rho_T(\theta)$ is given by the sample mean $\bar{\phi}_T(\theta) = (1/T)\sum_{t=1}^{T} \phi_t(\theta)$ and the asymptotic behavior of the empirical process $\sqrt{T}[\bar{\phi}_T(\theta) - \rho_T(\theta)]$.

Assumption 2.1 *(Functional CLT)*
(i) There exists a sequence of deterministic functions ρ_T such that the empirical process $\sqrt{T}\left[\bar{\phi}_T(\theta) - \rho_T(\theta)\right]$, for $\theta \in \Theta$, weakly converges (for the sup-norm on Θ) toward a Gaussian process on Θ with mean zero and covariance $S(\theta)$.
(ii) There exists a sequence A_T of deterministic nonsingular matrices of size K and a bounded deterministic function c such that

$$\lim_{T\to\infty} \sup_{\theta\in\Theta} \|c(\theta) - A_T \rho_T(\theta)\| = 0.$$

The rate of convergence of coefficients of the matrix A_T toward infinity characterizes the degree of global identification weakness. Note that we may not be able to replace $\rho_T(\theta)$ by the function $A_T^{-1}c(\theta)$ in the convergence of the empirical process since

$$\sqrt{T}\left[\rho_T(\theta) - A_T^{-1}c(\theta)\right] = \left(\frac{A_T}{\sqrt{T}}\right)^{-1} [A_T \rho_T(\theta) - c(\theta)],$$

may not converge toward zero. While genuine weak identification like Stock and Wright (2000) means that $A_T = \sqrt{T} Id_K$ (with Id_K identity matrix of size K), we rather consider nearly weak identification where some rows of the matrix A_T may go to infinity strictly slower than $\sqrt{T}$. Standard GMM asymptotic theory based on strong identification would assume $A_T = Id_K$ and $\rho_T(\theta) = c(\theta)$ for all T. In this case, it would be sufficient to assume asymptotic normality of $\sqrt{T}\bar{\phi}_T(\theta^0)$ at the true value θ^0 of the parameters (while $\rho_T(\theta^0) = c(\theta^0) = 0$). By contrast, as already pointed out by Stock and

Wright (2000), the asymptotic theory with (nearly) weak identification is more involved since it assumes a functional central limit theorem uniform on Θ. However, this uniformity is not required in the linear case,[1] as now illustrated.

Example 2.1 (Linear IV regression)
We consider a structural linear equation: $y_t = x_t'\theta + u_t$ *for* $t = 1, \cdots, T$*, where the* p *explanatory variables* x_t *may be endogenous. The true unknown value* θ^0 *of the structural parameters is defined through* $K \geq p$ *instrumental variables* z_t *uncorrelated with* $(y_t - x_t'\theta^0)$*. In other words, the estimating equations for standard IV estimation are*

$$\bar{\phi}_T(\hat{\theta}_T) = \frac{1}{T} Z'(y - X\hat{\theta}_T) = 0, \tag{2.4}$$

where X *(respectively* Z*) is the* (T, p) *(respectively* (T, K)*) matrix which contains the available observations of the* p *explanatory variables (respectively the* K *instrumental variables) and* $\hat{\theta}_T$ *denotes the standard IV estimator of* θ*. Inference with poor instruments typically means that the required rank condition is not fulfilled, even asymptotically:*

$$\texttt{Plim}\left[\frac{Z'X}{T}\right] \quad \textit{may not be of full rank.}$$

Weak identification means that only $\texttt{Plim}[\frac{Z'X}{\sqrt{T}}]$ *has full rank, while intermediate cases with nearly weak identification have been studied by Hahn and Kuersteiner (2002). The following assumption conveniently nests all the above cases.*

Assumption L1 *There exists a sequence* A_T *of deterministic nonsingular matrices of size* K *such that* $\texttt{Plim}[A_T \frac{Z'X}{T}] = \Pi$ *is full column rank.*

While standard strong identification asymptotics assume that the largest absolute value of all coefficients of the matrix A_T*,* $\| A_T \|$*, is of order* $\mathcal{O}(1)$*, weak identification means that* $\| A_T \|$ *grows at rate* $\sqrt{T}$*. The following assumption focuses on nearly weak identification, which ensures consistent IV estimation under standard regularity conditions as explained below.*

Assumption L2 *The largest absolute value of all coefficients of the matrix* A_T *is* $o(\sqrt{T})$*.*

To deduce the consistency of the estimator $\hat{\theta}_T$*, we rewrite Equation (2.4) as follows and pre-multiply it by* A_T*:*

$$\frac{Z'X}{T}(\hat{\theta}_T - \theta^0) + \frac{Z'u}{T} = 0 \Rightarrow A_T \frac{Z'X}{T}(\hat{\theta}_T - \theta^0) + A_T \frac{Z'u}{T} = 0. \tag{2.5}$$

After assuming a central limit theorem for $(Z'u/\sqrt{T})$ *and after considering (for simplicity) that the unknown parameter vector* θ *evolves in a bounded subset of* $\mathbb{R}^p$*,*

[1] Note also that uniformity is not required in the linear-in-variable case.

we get

$$\Pi(\hat{\theta}_T - \theta^0) = o_P(1).$$

Then, the consistency of $\hat{\theta}_T$ directly follows from the full column rank assumption on Π. Note that uniformity with respect to θ does not play any role in the required central limit theorem since we have

$$\sqrt{T}[\bar{\phi}_T(\theta) - \rho_T(\theta)] = \frac{Z'u}{\sqrt{T}} + \sqrt{T}\left[\frac{Z'X}{T} - E[z_t x_t']\right](\theta^0 - \theta)$$

with

$$\rho_T(\theta) = E[z_t x_t'](\theta^0 - \theta).$$

Linearity of the moment conditions with respect to unknown parameters allows us to factorize them out and uniformity is not an issue.

It is worth noting that in the linear example, the central limit theorem has been used to prove consistency of the IV estimator and not to derive its asymptotic normal distribution. This nonstandard proof of consistency will be generalized for the nonlinear case in the next subsection, precisely thanks to the uniformity of the central limit theorem over the parameter space. As far as asymptotic normality of the estimator is concerned, the key issue is to take advantage of the asymptotic normality of $\sqrt{T}\bar{\phi}_T(\theta^0)$ at the true value θ^0 of the parameters (while $\rho_T(\theta^0) = c(\theta^0) = 0$). The linear example again shows that, in general, doing so involves additional assumptions about the structure of the matrix A_T. More precisely, we want to stress that when several degrees of identification (weak, nearly weak, strong) are considered simultaneously, the above assumptions are not sufficient to derive a meaningful asymptotic distributional theory. In our setting, it means that the matrix A_T is not simply a scalar matrix $\lambda_T A$ with the scalar sequence λ_T possibly going to infinity but not faster than $\sqrt{T}$. This setting is in contrast with most of the literature on weak instruments (see Kleibergen 2005; Caner 2010 among others).

Example 2.1 (Linear IV regression – continued)
To derive the asymptotic distribution of the estimator $\hat{\theta}_T$, pre-multiplying the estimating equations by the matrix A_T may not work. However, for any sequence of deterministic nonsingular matrices $\tilde{A}_T$ of size p, we have

$$\frac{Z'X}{T}(\hat{\theta}_T - \theta^0) + \frac{Z'u}{T} = 0 \Rightarrow \frac{Z'X}{T}\tilde{A}_T\sqrt{T}\tilde{A}_T^{-1}(\hat{\theta}_T - \theta^0) = -\frac{Z'u}{\sqrt{T}}. \quad (2.6)$$

If $[\frac{Z'X}{T}\tilde{A}_T]$ converges toward a well-defined matrix with full column rank, a central limit theorem for $(Z'u/\sqrt{T})$ ensures the asymptotic normality of $\sqrt{T}\tilde{A}_T^{-1}(\hat{\theta}_T - \theta^0)$. In general, this condition cannot be deduced from Assumption L1 unless the matrix A_T appropriately commutes with $[\frac{Z'X}{T}]$. Clearly, this is not an issue if A_T is simply a scalar matrix $\lambda_T Id_K$. In case of nearly weak identification ($\lambda_T = o(\sqrt{T})$), it delivers *asymptotic normality of the estimator at slow rate $\sqrt{T}/\lambda_T$ while, in case of genuine weak identification ($\lambda_T = \sqrt{T}$), consistency is not ensured and asymptotic Cauchy distributions show up.*

In the general case, the key issue is to justify the existence of a sequence of deterministic nonsingular matrices $\tilde{A}_T$ of size p such that $[\frac{Z'X}{T}\tilde{A}_T]$ converges toward a well-defined matrix with full column rank. In the just-identified case ($K = p$), it follows directly from Assumption L1 with $\tilde{A}_T = \Pi^{-1}A_T$:

$$\texttt{Plim}\left[\frac{Z'X}{T}\Pi^{-1}A_T\right] = \texttt{Plim}\left[\frac{Z'X}{T}\left(A_T\frac{Z'X}{T}\right)^{-1}A_T\right] = Id_p.$$

In the overidentified case ($K > p$), it is rather the structure of the matrix A_T (and not only its norm, or largest coefficient) that is relevant. Of course, by Equation 2.5, we know that

$$\frac{Z'X}{T}\sqrt{T}\left(\hat{\theta}_T - \theta^0\right) = -\frac{Z'u}{\sqrt{T}}$$

is asymptotically normal. However, in case of lack of strong identification, ($Z'X/T$) is not asymptotically full rank and some linear combinations of $\sqrt{T}(\hat{\theta}_T - \theta^0)$ may blow up. To provide a meaningful asymptotic theory for the IV estimator $\hat{\theta}_T$, the following condition is required. In the general case, we explain why such a sequence $\tilde{A}_T$ always exists and how to construct it (see Theorem 2.3).

Assumption L3 *There exists a sequence $\tilde{A}_T$ of deterministic nonsingular matrices of size p such that $\texttt{Plim}[\frac{Z'X}{T}\tilde{A}_T]$ is full column rank.*

It is then straightforward to deduce that $\sqrt{T}\tilde{A}_T^{-1}(\hat{\theta}_T - \theta^0)$ is asymptotically normal. Hansen, Hausman, and Newey (2008) provide a set of assumptions to derive similar results in the case of many weak instruments asymptotics. In their setting, considering a number of instruments growing to infinity can be seen as a way to ensure Assumption L2, even though weak identification (or $\|A_T\|$ of order $\sqrt{T}$) is assumed for any given finite set of instruments.

The above example shows that, in case of (nearly) weak identification, a relevant asymptotic distributional theory is not directly about the common sequence $\sqrt{T}(\hat{\theta}_T - \theta^0)$ but rather about a well-suited reparametrization $\tilde{A}_T^{-1}\sqrt{T}(\hat{\theta}_T - \theta^0)$. Moreover, lack of strong identification means that the matrix of reparametrization $\tilde{A}_T$ also involves a rescaling (going to infinity with the sample size) in order to characterize slower rates of convergence. For sake of structural interpretation, it is worth disentangling the two issues: first, the rotation in the parameter space, which is assumed well-defined at the limit (when $T \to \infty$); second, the rescaling. The convenient mathematical tool is the singular value decomposition of the matrix A_T (see Horn and Johnson 1985, pp.414–416, 425). We know that the nonsingular matrix A_T can always be written as: $A_T = M_T\Lambda_T N_T'$ with M_T, N_T, and Λ_T three square matrices of size K, M_T, and N_T orthogonal and Λ_T diagonal with nonzero entries. In our context of rates of convergence, we want to see the singular values of the matrix A_T (that is the diagonal coefficients of Λ_T) as positive and, without loss of generality, ranked in increasing order. If we consider Assumption 2.1(ii) again, N_T' can intuitively be seen as selecting appropriate linear combinations of the moment conditions and Λ_T as rescaling appropriately these combinations. On the other hand, M_T is related to selecting linear combinations of the deterministic vector c.

Without loss of generality, we always consider the singular value decomposition $A_T = M_T \Lambda_T N_T'$ such that the diagonal matrix sequence Λ_T has positive diagonal coefficients bounded away from zero and the two sequences of orthogonal matrices M_T and N_T have well-defined limits[2] when $T \to \infty$, M and N, respectively, both orthogonal matrices.

2.2.2 Consistency

In this subsection, we set up a framework where consistency of a GMM estimator is warranted in spite of lack of strong identification. The key is to ensure that a sufficient subset of the moment conditions is not impaired by genuine weak identification: in other words, the corresponding rates of convergence of the singular values of A_T are slower than $\sqrt{T}$. As explained above, specific rates of convergence are actually assigned to appropriate linear combinations of the moment conditions:

$$d(\theta) = M^{-1}c(\theta) = \lim_T \left[\Lambda_T N_T' \rho_T(\theta)\right].$$

Our maintained identification assumption follows:

Assumption 2.2 *(Identification)*
(i) The sequence of nonsingular matrices A_T writes $A_T = M_T \Lambda_T N_T'$ with $\lim_T[M_T] = M$, $\lim_T[N_T] = N$, M, and N orthogonal matrices.

(ii) The sequence of matrices Λ_T is partitioned as $\Lambda_T = \begin{bmatrix} \tilde{\Lambda}_T & 0 \\ 0 & \breve{\Lambda}_T \end{bmatrix}$, such that $\tilde{\Lambda}_T$ and $\breve{\Lambda}_T$ are two diagonal matrices, respectively, of size $\tilde{K}$ and $(K - \tilde{K})$, with[3] $\|\tilde{\Lambda}_T\| = o(\sqrt{T})$, $\|\breve{\Lambda}_T\| = \mathcal{O}(\sqrt{T})$ and $\breve{\Lambda}_T^{-1} = o(\|\tilde{\Lambda}_T\|^{-1})$.

(iii) The vector d of moment conditions, with $d(\theta) = M^{-1}c(\theta) = \lim_T[\Lambda_T N_T' \rho_T(\theta)]$, is partitioned accordingly as $d = [\tilde{d}' \ \breve{d}'\,]'$ such that θ^0 is a well-separated zero of the vectorial function $\tilde{d}$ of size $\tilde{K} \leq p$:

$$\forall \epsilon > 0 \quad \inf_{\|\theta - \theta^0\| > \epsilon} \|\tilde{d}(\theta)\| > 0.$$

(iv) The first $\tilde{K}$ elements of $N_T \rho_T(\theta^0)$ are identically equal to zero for any T.

---

[2] It is well known that the group of real orthogonal matrices is compact (see Horn and Johnson 1985, p. 71). Hence, one can always define M and N for convergent subsequences, respectively M_{T_n} and N_{T_l}. To simplify the notations, we only refer to sequences and not subsequences.

[3] $\|M\|$ denotes the largest element (in absolute value) of any matrix M.

As announced, the above identification assumption ensures that the first $\tilde{K}$ moment conditions are only possibly nearly weak (and not genuinely weak), $\|\tilde{\Lambda}_T\| = o(\sqrt{T})$, and sufficient to identify the true unknown value θ^0:

$$\tilde{d}(\theta) = 0 \Leftrightarrow \theta = \theta^0.$$

The additional moment restrictions, as long as they are strictly weaker ($\breve{\Lambda}_T^{-1} = o(\|\tilde{\Lambda}_T\|^{-1})$, may be arbitrarily weak and even misspecified, since we do not assume $\breve{d}(\theta^0) = 0$. It is worth noting that the above identification concept is nonstandard, since all singular values of the matrix A_T may go to infinity. In such a case, we have

$$\texttt{Plim}\left[\bar{\phi}_T(\theta)\right] = 0 \;\; \forall\, \theta \in \Theta. \tag{2.7}$$

This explains why the following consistency result of a GMM estimator cannot be proved in a standard way. The key argument is actually tightly related to the uniform functional central limit theorem of Assumption 2.1.

Theorem 2.1 *(Consistency of $\hat{\theta}_T$)*
We define a GMM-estimator:

$$\hat{\theta}_T = \arg\min_{\theta\in\Theta} \left[\bar{\phi}_T'(\theta)\Omega_T\bar{\phi}_T(\theta)\right] \tag{2.8}$$

with Ω_T a sequence of symmetric positive definite random matrices of size K which converges in probability toward a positive definite matrix Ω.

Under the Assumptions 2.1 and 2.2, any GMM estimator like Equation 2.8 is weakly consistent.

We now explain why the consistency result cannot be deduced from a standard argument based on a simple rescaling of the moment conditions to avoid asymptotic degeneracy of Equation 2.7. The GMM estimator (Equation 2.8) can be rewritten as

$$\hat{\theta}_T = \arg\min_{\theta\in\Theta} \left\{[\Lambda_T N_T'\bar{\phi}_T(\theta)]' W_T [\Lambda_T N_T'\bar{\phi}_T(\theta)]\right\}$$

with a weighting matrix sequence, $W_T = \Lambda_T^{-1} N_T' \Omega_T N_T \Lambda_T^{-1}$, and rescaled moment conditions $[\Lambda_T N_T'\bar{\phi}_T(\theta)]$ such that

$$\texttt{Plim}[\Lambda_T N_T'\bar{\phi}_T(\theta)] = \lim_T[\Lambda_T N_T'\rho_T(\theta)] = d(\theta) \neq 0 \text{ for } \theta \neq \theta^0.$$

However, when all singular values of A_T go to infinity, the weighting matrix sequence W_T is such that

$$\texttt{Plim}\,[W_T] = \lim_T \left[\Lambda_T^{-1} N' \Omega N \Lambda_T^{-1}\right] = 0.$$

In addition, the limit of the GMM estimator in Theorem 2.1 is solely determined by the strongest moment conditions that identify θ^0. There is actually no need to assume that the last $(K - \tilde{K})$ coefficients in $[\Lambda_T N_T'\rho_T(\theta^0)]$, or even their limits $\check{d}(\theta^0)$, are equal to zero. In other words, the additional estimating equations $\check{d}(\theta) = 0$ may be biased and this has no consequence on the limit value of the GMM estimator insofar as the additional moment restrictions are strictly weaker than the initial ones, $\check{\Lambda}_T^{-1} = o(\|\tilde{\Lambda}_T\|^{-1})$. They may even be genuinely weak with $\|\check{\Lambda}_T\| = \sqrt{T}$. This result has important consequences on the power of the overidentification test defined in the next section.

2.3 Asymptotic Distribution and Inference

2.3.1 Efficient Estimation

In our setting, rates of convergence slower than square-root T are produced because some coefficients of A_T may go to infinity while the asymptotically identifying equations are given by $\rho_T(\theta) \overset{a}{\sim} A_T^{-1}c(\theta)$. Since we do not want to introduce other causes for slower rates of convergence (like singularity of the Jacobian matrix of the moment conditions, as done in Sargan 1983), first-order local identification is maintained.

Assumption 2.3 *(Local identification)*

(i) $\theta \to c(\theta)$, $\theta \to d(\theta)$ *and* $\theta \to \rho_T(\theta)$ *are continuously differentiable on the interior of* Θ.

(ii) θ^0 *belongs to the interior of* Θ.

(iii) *The* $(\tilde{K}, p)$*-matrix* $[\partial\tilde{d}(\theta^0)/\partial\theta']$ *has full column rank* p.

(iv) $\Lambda_T N_T'[\partial\rho_T(\theta)/\partial\theta']$ *converges uniformly on the interior of* Θ *toward* $M^{-1}[\partial c(\theta)/\partial\theta'] = \partial d(\theta)/\partial\theta'$.

(v) *The last* $(K - \tilde{K})$ *elements of* $N_T\rho_T(\theta^0)$ *are either identically equal to zero for any* T, *or genuinely weak with the corresponding element of* $\check{\Lambda}_T$ *equal to* $\sqrt{T}$.

Assumption 2.3(iv) states that rates of convergence are maintained after differentiation with respect to the parameters. Contrary to the linear case, this does not follow automatically in the general case. Then, we are able to show that the structural parameters are identified at the slowest rate available from the set of identifying equations. Assumption 2.3(v) ensures that the additional moment restrictions (the ones not required for identification) are either well-specified or genuinely weak: this ensures that these conditions do not deteriorate the rate of convergence of the GMM estimator (see Theorem 2.2). Intuitively, a GMM estimator is always a linear combination of the moment conditions. Hence, if some moments are misspecified and do not *disappear* as fast as $\sqrt{T}$, they can only deteriorate the rate of convergence of the estimator.

Theorem 2.2 *(Rate of convergence)*
Under Assumptions 2.1 to 2.3, any GMM estimator $\hat{\theta}_T$ like Equation 2.8 is such that

$$\|\hat{\theta}_T - \theta^0\| = \mathcal{O}_p(\|\tilde{\Lambda}_T\|/\sqrt{T}).$$

The above result is quite poor, since it assigns the slowest possible rate to all components of the structural parameters. We now show how to identify faster directions in the parameter space. The first step consists in defining a matrix $\tilde{A}_T$ similar to the one introduced in the linear example. The following result justifies its existence: in the appendix, we also explain in details how to construct it.

Theorem 2.3 *Under Assumptions 2.1 to 2.3, there exists a sequence $\tilde{A}_T$ of deterministic nonsingular matrices of size p such that the smallest eigenvalue of $\tilde{A}_T'\tilde{A}_T$ is bounded away from zero and*

$$\lim_T \left[\Lambda_T^{-1}M^{-1}\frac{\partial c(\theta^0)}{\partial\theta'}\tilde{A}_T\right] \quad \textit{exists and is full column rank with } \|\tilde{A}_T\| = \mathcal{O}(\|\tilde{\Lambda}_T\|).$$

Following the approach put forward in the linear example, Theorem 2.3 is used to derive the asymptotic theory of the estimator $\hat{\theta}_T$. Since,

$$\frac{\partial\bar{\phi}_T(\theta^0)}{\partial\theta'}\sqrt{T}(\hat{\theta}_T - \theta^0) = \frac{\partial\bar{\phi}_T(\theta^0)}{\partial\theta'}\tilde{A}_T\sqrt{T}\tilde{A}_T^{-1}(\hat{\theta}_T - \theta^0),$$

a meaningful asymptotic distributional theory is not directly about the common sequence $\sqrt{T}(\hat{\theta}_T - \theta^0)$, but rather about a well-suited reparametrization $\tilde{A}_T^{-1}\sqrt{T}(\hat{\theta}_T - \theta^0)$. Similar to the structure of A_T, $\tilde{A}_T$ involves a reparametrization and a rescaling. In others words, specific rates of convergence are actually assigned to appropriate linear combinations of the structural parameters.

Assumption 2.4 *(Regularity)*

$$(i)\ \sqrt{T}\left[\frac{\partial\bar{\phi}_T(\theta^0)}{\partial\theta'} - A_T^{-1}\frac{\partial c(\theta^0)}{\partial\theta'}\right] = \mathcal{O}_P(1)$$

$$(ii)\ \sqrt{T}\frac{\partial}{\partial\theta}\left[\frac{\partial\bar{\phi}_T(\theta)}{\partial\theta'}\right]_{k.} - \frac{\partial}{\partial\theta}\left[A_T^{-1}\frac{\partial c(\theta)}{\partial\theta'}\right]_{k.} = \mathcal{O}_P(1) \textit{ and } \frac{\partial}{\partial\theta}\left[\frac{\partial c(\theta)}{\partial\theta'}\right]_{k.} = \mathcal{O}_P(1)$$

for any $1 \le k \le K$, uniformly on the interior of Θ with $[M]_{k.}$ the kth row the matrix M.

With additional regularity Assumption 2.4(i), Corollary 2.1 extends Theorem 2.3 to rather consider the empirical counterparts of the moment conditions: it is the nonlinear analog of Assumption L3.

Corollary 2.1 *(Nonlinear extension of L3)*
Under Assumptions 2.1–2.3 and 2.4(i), we have

$$\Gamma(\theta^0) \equiv \texttt{Plim}\left[\frac{\partial\bar{\phi}_T(\theta^0)}{\partial\theta'}\tilde{A}_T\right] \text{ exists and is full column rank.}$$

In order to derive a standard asymptotic theory for the GMM estimator $\hat{\theta}_T$, we need to impose an assumption on the homogeneity of identification.

Assumption 2.5 *(Homogenous identification)*

$$\frac{\sqrt{T}}{\underline{\tilde{\Lambda}}_T} = o\left(\frac{\sqrt{T}}{\|\tilde{\Lambda}_T\|}\right)^2$$

where $\|M\|$ $\underline{M}$ *denote respectively the largest and the smallest absolute values of all nonzero coefficients of the matrix M.*

Intuitively, assumption 2.5 ensures that second-order terms in Taylor expansions remain negligible in front of the first-order central limit theorem terms. Note that a sufficient condition for homogenous identification is dubbed nearly-strong and writes: $\|\tilde{\Lambda}_T\|^2 = o(\sqrt{T})$. It corresponds to the above homogenous identification condition when some moment conditions are strong, that is $\underline{\tilde{\Lambda}}_T = 1$. Then we want to ensure that the slowest possible rate of convergence of parameter estimators is strictly faster than $T^{1/4}$. This nearly-strong condition is actually quite standard in semiparametric econometrics to control for the impact of infinite dimensional nuisance parameters (see Andrews' (1994) MINPIN estimators and Newey's (1994) linearization assumption).

The asymptotic distribution of the rescaled estimated parameters $\sqrt{T}\tilde{A}_T^{-1}(\hat{\theta}_T - \theta^0)$ can now be characterized by seemingly standard GMM formulas:

Theorem 2.4 *(Asymptotic distribution of* $\hat{\theta}_T$*)*
Under Assumptions 2.1–2.5, any GMM estimator $\hat{\theta}_T$ *like Equation 2.8 is such that* $\sqrt{T}\tilde{A}_T^{-1}(\hat{\theta}_T - \theta^0)$ *is asymptotically normal with mean zero and variance* $\Sigma(\theta^0)$ *given by*

$$\Sigma(\theta^0) = \left[\Gamma'(\theta^0)\Omega\Gamma(\theta^0)\right]^{-1}\Gamma'(\theta^0)\Omega S(\theta^0)\Omega\Gamma(\theta^0)\left[\Gamma'(\theta^0)\Omega\Gamma(\theta^0)\right]^{-1},$$

where $S(\theta^0)$ *is the asymptotic variance of* $\sqrt{T}\bar{\phi}_T(\theta^0)$.

Theorem 2.4 paves the way for a concept of efficient estimation in presence of poor instruments. By a common argument, the unique limit weighting matrix Ω minimizing the above covariance matrix is clearly $\Omega = [S(\theta^0)]^{-1}$.

Theorem 2.5 *(Efficient GMM estimator)*
Under Assumptions 2.1–2.5, any GMM estimator $\hat{\theta}_T$ *like Equation 2.8 with a weighting matrix* $\Omega_T = S_T^{-1}$, *where* S_T *denotes a consistent estimator of* $S(\theta^0)$,

is such that $\sqrt{T}\tilde{A}_T^{-1}(\hat{\theta}_T - \theta^0)$ *is asymptotically normal with mean zero and variance* $[\Gamma'(\theta^0)S^{-1}(\theta^0)\Gamma(\theta^0)]^{-1}$.

In our framework, the terminology "efficient GMM" and "standard formulas" for asymptotic covariance matrices must be carefully qualified. On the one hand, it is true that for all practical purposes, Theorem 2.5 states that, for T large enough, $\sqrt{T}\tilde{A}_T^{-1}(\hat{\theta}_T - \theta^0)$ can be seen as a Gaussian vector with mean zero and variance consistently estimated by

$$\tilde{A}_T^{-1}\left[\frac{\partial\bar{\phi}_T'(\hat{\theta}_T)}{\partial\theta}S_T^{-1}\frac{\partial\bar{\phi}_T(\hat{\theta}_T)}{\partial\theta'}\right]^{-1}\tilde{A}_T^{-1\prime}, \tag{2.9}$$

since $\Gamma(\theta^0) = \texttt{Plim}[\frac{\partial\bar{\phi}_T(\theta^0)}{\partial\theta'}\tilde{A}_T]$. However, it is incorrect to deduce from Equation (2.9) that, after simplifications on both sides by $\tilde{A}_T^{-1}$, $\sqrt{T}(\hat{\theta}_T - \theta^0)$ can be seen (for T large enough) as a Gaussian vector with mean zero and variance consistently estimated by

$$\left[\frac{\partial\bar{\phi}_T'(\hat{\theta}_T)}{\partial\theta}S_T^{-1}\frac{\partial\bar{\phi}_T(\hat{\theta}_T)}{\partial\theta'}\right]^{-1}. \tag{2.10}$$

This is wrong since the matrix $[\frac{\partial\bar{\phi}_T'(\hat{\theta}_T)}{\partial\theta}S_T^{-1}\frac{\partial\bar{\phi}_T(\hat{\theta}_T)}{\partial\theta'}]$ is asymptotically singular. In this sense, a truly standard GMM theory does not apply and at least some components of $\sqrt{T}(\hat{\theta}_T - \theta^0)$ must blow up. Quite surprisingly, it turns out that the spurious feeling that Equation 2.10 estimates the asymptotic variance (as usual) is tremendously useful for inference as explained in Subsection 2.3.2. Intuitively, it explains why standard inference procedures work, albeit for nonstandard reasons. As a consequence, for all practical purposes related to inference about the structural parameters θ, the knowledge of the matrices A_T and $\tilde{A}_T$ is not required.

However, the fact that the correct understanding of the "efficient GMM" covariance matrix as estimated by Equation 2.9 involves the sequence of matrices $\tilde{A}_T$ is important for two reasons.

First, it is worth reminding that the construction of the matrix $\tilde{A}_T$ only involves the first $\tilde{K}$ components of the rescaled estimating equations $[N_T'\rho_T(\theta)]$. This is implicit in the rate of convergence of $\|\tilde{A}_T\|$ put forward in Theorem 2.3 and quite clear in its proof. In other words, when the total number of moment conditions K is strictly larger than $\tilde{K}$, the last $(K - \tilde{K})$ rows of the matrix $\Gamma(\theta^0) = \texttt{Plim}[\frac{\partial\bar{\phi}_T(\theta^0)}{\partial\theta'}\tilde{A}_T]$ are equal to zero. Irrespective of the weighting matrix's choice for GMM estimation, the associated estimator does not depend asymptotically on these last moment conditions. Therefore, there is an obvious waste of information: the so-called efficient GMM estimator of Theorem 2.5 does not make use of all the available information. Moment conditions based on poorer instruments (redundant for the purpose of identification) should actually be used for improved accuracy of the estimator, as explicitly shown in Antoine and Renault (2010a).

Second, the interpretation of the matrix $\tilde{A}_T$ in terms of reparametrization is underpinned by the proof of Theorem 2.3 which shows that

$$\tilde{A}_T = [\, \lambda_{1T} R_1 \vdots \lambda_{2T} R_2 \vdots \cdots \vdots \lambda_{LT} R_L \,] = R\Delta_T \text{ with } R = [\, R_1 \vdots R_2 \vdots \cdots \vdots R_L \,].$$

R is a nonsingular matrix of size p with each submatrix R_i of size (p, s_i); Δ_T is a diagonal matrix with L diagonal blocks equal to $\lambda_{iT} Id_{s_i}$. It is worth reinterpreting Theorem 2.5 in terms of the asymptotic distribution of the estimator of a new parameter vector[4]:

$$\eta = R^{-1}\theta = [\eta_1' \; \eta_2' \; \cdots \; \eta_L']'.$$

Theorem 2.5 states that $(R^{-1}\hat{\theta}_T)$ is a consistent asymptotically normal estimator of the true unknown value $\eta^0 = R^{-1}\theta^0$, while each subvector η_i of size s_i is attached to a specific (slower) rate of convergence $\sqrt{T}/\lambda_{iT}$. It is clear in the appendix that this reparametrization is performed according to the directions which span the range of the Jacobian matrix of the rescaled "efficient" moment conditions $\tilde{d}(\theta)$, that is according to the columns of the matrix R. Even though the knowledge of the matrix R (and corresponding rates λ_{iT}) is immaterial for the practical implementation of inference procedures on structural parameters (as shown in Section 2.3.2), it may matter for a fair assessment of the accuracy of this inference. As an illustration, Subsection 2.4.3 studies the power of score-type tests against sequences of local alternatives in different directions.

In the context of the consumption-based capital asset pricing model (CCAPM) discussed in Stock and Wright (2000) and Antoine and Renault (2009), there are two structural parameters: θ_1, the subjective discount factor and θ_2, the coefficient of relative risk aversion of a representative investor. Antoine and Renault (2009) provide compelling evidence that a first parameter η_1, estimated at fast rate $\sqrt{T}$, is very close to θ_1 (the estimation results show that $\eta_1 = 0.999\theta_1 - 0.007\theta_2$), while any other direction in the parameter space, like for instance the risk aversion parameter θ_2, is estimated at a much slower rate. In other words, all parameters are consistently estimated as shown in Stock and Wright's (2000) empirical results (and contrary to their theoretical framework), but the directions with $\sqrt{T}$-consistent estimation are now inferred from data instead of being considered as a prior specification.

The practical way to consistently estimate the matrix R from the sample counterpart of the Jacobian matrix of the moment conditions is extensively discussed in Antoine and Renault (2010a). Of course, since this Jacobian matrix involves in general the unknown structural parameters θ, there is little hope to consistently estimate R at a rate faster than the slowest one, namely $\sqrt{T}/\|\tilde{\Lambda}_T\|$. Interestingly enough, this slower rate does not impair the faster rates involved in Theorem 2.5. When R is replaced by its consistent estimator

---

[4] Note that the structural parameter θ is such that $\theta = \sum_{i=1}^{L} R_i \eta_i$.

$\hat{R}$, in the context of Theorem 2.5,

$$\sqrt{T}\Delta_T^{-1}(\hat{R}^{-1}\hat{\theta}_T - \hat{R}^{-1}\theta^0)$$

is still asymptotically normal with mean zero and variance $[\Gamma'(\theta^0)S^{-1}(\theta^0) \times \Gamma(\theta^0)]^{-1}$. The key intuition comes from the following decomposition:

$$\hat{R}^{-1}\hat{\theta}_T - \hat{R}^{-1}\theta^0 = R^{-1}(\hat{\theta}_T - \theta^0) + (\hat{R}^{-1} - R^{-1})(\hat{\theta}_T - \theta^0).$$

The potentially slow rates of convergence in the second term of the right-hand side do not deteriorate the fast rates in the relevant directions of $R^{-1}(\hat{\theta}_T - \theta^0)$: these slow rates show up as $T/\|\tilde{\Lambda}_T\|^2$ at worst, which is still faster than the fastest rate $\sqrt{T}/\lambda_{1T}$ by our nearly strong identification Assumption 2.5.

2.3.2 Inference

As discussed in the previous section, inference procedures are actually more involved than one may believe at first sight from the apparent similarity with standard GMM formulas. Nonetheless, the seemingly standard "efficient" asymptotic distribution theory of Theorem 2.5 paves the way for two usual results: the overidentification test and the Wald test.

Theorem 2.6 *(J test)*
Let S_T^{-1} *be a consistent estimator of* $\lim_T[\mathtt{Var}(\sqrt{T}\bar{\phi}_T(\theta^0))]^{-1}$.
Under Assumptions 2.1–2.5, for any GMM estimator like Equation (2.8), we have

$$T\bar{\phi}_T'(\hat{\theta}_T)S_T^{-1}\bar{\phi}_T(\hat{\theta}_T) \xrightarrow{d} \chi^2(K-p).$$

As already announced, Theorem 2.1 has important consequences for the practice of GMM inference. We expect the above overidentification test to have little power to detect the misspecification of moment conditions when this misspecification corresponds to a subset of moment conditions of heterogeneous strengths. The proofs of Theorems 2.1 and 2.3 actually show that

$$T\bar{\phi}_T(\hat{\theta}_T)S_T^{-1}\bar{\phi}_T(\hat{\theta}_T) = \mathcal{O}_P\left(\frac{T}{\|\breve{\Lambda}_T\|^2}\right).$$

In other words, the standard J-test statistic for overidentification will not diverge as fast as the standard rate T of divergence and will even not diverge at all if the misspecified moment restrictions are genuinely weak ($\|\breve{\Lambda}_T\| = \sqrt{T}$).

Second, we are interested in testing the null hypothesis, $H_0 : g(\theta) = 0$, where the function $g : \Theta \to \mathbb{R}^q$ is continuously differentiable on the interior of Θ. We focus on Wald testing since it avoids estimation under the null which may affect the reparametrization[5] previously defined. The following example illustrates how the standard delta-theorem is affected in our framework.

Example 2.2

Consider the null hypothesis $H_0 : g(\theta) = 0$ with g a vector of size q such that

$$\left[\frac{\partial g_j(\theta^0)}{\partial \theta}\right] \notin \texttt{col}\left[\frac{\partial \tilde{d}_1'(\theta^0)}{\partial \theta}\right] \quad \forall\, j = 1, \cdots, q$$

and a diagonal matrix Λ_T,

$$\Lambda_T = \begin{bmatrix} \lambda_{1T} Id_{K_1} & O \\ O & \lambda_{2T} Id_{K-K_1} \end{bmatrix} \text{with } \lambda_{1T} = o(\lambda_{2T}),\ \lambda_{2T} \to \infty, \text{ and } \lambda_{2T} = o(\sqrt{T}).$$

Applying the standard argument to derive the Wald test, we have that, under the null,

$$\left[\frac{\sqrt{T}}{\lambda_{2T}} g(\hat{\theta}_T)\right] \overset{a}{\sim} \left[\frac{\partial g(\theta^0)}{\partial \theta'} \frac{\sqrt{T}}{\lambda_{2T}} (\hat{\theta}_T - \theta^0)\right].$$

In other words, for T large enough, $\left[\frac{\sqrt{T}}{\lambda_{2T}} g(\hat{\theta}_T)\right]$ can be seen as a normal random variable with mean 0 and variance

$$\frac{\partial g(\theta^0)}{\partial \theta'} \left[\frac{\partial \bar{\phi}_T'(\theta^0)}{\partial \theta} [S(\theta^0)]^{-1} \frac{\partial \bar{\phi}_T(\theta^0)}{\partial \theta'}\right]^{-1} \frac{\partial g'(\theta^0)}{\partial \theta}.$$

Suppose now that there exists a nonzero vector α such that

$$\left[\frac{\partial g'(\theta^0)}{\partial \theta} \alpha\right] \in \texttt{col}\left[\frac{\partial \tilde{d}_1'(\theta^0)}{\partial \theta}\right].$$

Then, under the null, $\left[\frac{\sqrt{T}}{\lambda_{1T}} \alpha' g(\hat{\theta}_T)\right]$ is asymptotically normal and thus

$$\frac{\sqrt{T}}{\lambda_{2T}} \alpha' g(\hat{\theta}_T) = \frac{\lambda_{1T}}{\lambda_{2T}} \frac{\sqrt{T}}{\lambda_{1T}} \alpha' g(\hat{\theta}_T) \overset{P}{\to} 0.$$

This means that even when a full rank assumption is maintained for the constraints to be tested, $\left[\frac{\sqrt{T}}{\lambda_{2T}} g(\hat{\theta}_T)\right]$ does not behave asymptotically like a normal with a nonsingular variance matrix. This explains why deriving the asymptotic distributional theory for the Wald test statistic is nonstandard.

Surprisingly enough, the above asymptotic singularity issue is immaterial and the standard Wald-type inference holds without additional regularity

[5] Typically, with additional information, the linear combinations of θ estimated respectively at specific rates of convergence may be defined differently. Caner (2010) derives the standard asymptotic equivalence results for the trinity of tests because he only considers testing when all parameters converge at the same nearly weak rate.

assumption as stated in Theorem 2.7. The intuition is the following. Consider a fictitious situation where the range of $[\partial \tilde{d}_1'(\theta^0)/\partial\theta]$ is known. Then, one can always define a nonsingular matrix H of size q and the associated vector h, $h(\theta) = Hg(\theta)$, in order to avoid the asymptotic singularity issue portrayed in Example 2.2. More precisely, with a (simplified) matrix A_T as in the above example, we consider

$$\begin{aligned} &\text{for } j = 1, \cdots, q_1 : \left[\partial h_j(\theta^0)/\partial\theta\right] \in \texttt{col}\left[\partial \tilde{d}_1'(\theta^0)/\partial\theta\right]; \\ &\text{for } j = q_1+1, \cdots, q : \left[\partial h_j(\theta^0)/\partial\theta\right] \notin \texttt{col}\left[\partial \tilde{d}_1'(\theta^0)/\partial\theta\right] \\ &\text{and no linear combinations of } \left[\partial h_j(\theta^0)/\partial\theta\right] \text{ does.} \end{aligned}$$

Note that the new restrictions $h(\theta) = 0$ should be interpreted as a nonlinear transformations of the initial ones $g(\theta) = 0$ (since the matrix H depends on θ). It turns out that, for all practical purposes, by treating H as known, the Wald-type test statistics written with $h(.)$ or $g(.)$ are numerically equal; see the proof of Theorem 2.7 in the appendix.

Theorem 2.7 *(Wald test)*
Under Assumptions 2.1–2.5, the Wald test statistic ξ_T, for testing $H_0 : g(\theta) = 0$ with g twice continuously differentiable,

$$\xi_T = Tg'(\hat{\theta}_T)\left\{\frac{\partial g(\hat{\theta}_T)}{\partial\theta'}\left[\frac{\partial\bar{\phi}_T'(\hat{\theta}_T)}{\partial\theta}S_T^{-1}\frac{\partial\bar{\phi}_T(\hat{\theta}_T)}{\partial\theta'}\right]^{-1}\frac{\partial g'(\hat{\theta}_T)}{\partial\theta}\right\}^{-1} g(\hat{\theta}_T)$$

is asymptotically distributed as a chi-square with q degrees of freedom under the null.

In our framework, the standard result holds with respect to the size of the Wald test. Of course, the power of the test heavily depends on the strength of identification of the various constraints to test as extensively discussed in Antoine and Renault (2010a). See also the discussion in Subsection 2.4.3.

2.4 Comparisons with Other Approaches

2.4.1 Linear IV Model

Following the discussion in Examples 2.1 and 2.1, several matrices Π_T may be considered in the linear model with poor instruments. We now show that this choice is not innocuous.

(i) Staiger and Stock (1997) consider a framework with the same genuine weak identification pattern for all the parameters: $\Pi_T = C/\sqrt{T}$. To maintain Assumption L2, we can consider it as the limit case of: $\Pi_T = C/T^{\lambda}$, for $0 < \lambda < 1/2$ and C full column rank. Then $A_T = T^{\lambda} Id_K$ fulfills Assumption L1. Similarly, $\tilde{A}_T = T^{\lambda} Id_p$ fulfills

Assumption L3. Note that in this simple example, $\|A_T\|$ and $\|\tilde{A}_T\|$ grow at the same rate, which corresponds to the unique degree of nearly weak identification.

(ii) Stock and Wright (2000) reinterpret the above framework to accommodate simultaneously strong and weak identification patterns. This distinction is done at the parameter level and the structural parameter θ is (a priori) partitioned: $\theta = [\theta_1' \vdots \theta_2']'$ with θ_1 of dimension p_1 strongly identified and θ_2 of dimension $p_2 = p - p_1$ weakly identified. Following their approach, while maintaining Assumption L2, we consider the matrix

$$\Pi_T = \begin{bmatrix} \pi_{11} & \pi_{12}/T^{\lambda} \\ \pi_{21} & \pi_{22}/T^{\lambda} \end{bmatrix} = \Pi D_T^{-1},$$

with $0 < \lambda < 1/2$ while $\lambda = 1/2$ in Stock and Wright (2000); $\Pi = \begin{bmatrix} \pi_{11} & \pi_{12} \\ \pi_{21} & \pi_{22} \end{bmatrix}$ and D_T a (p, p)-diagonal matrix (with 1 as the first p_1 coefficients and T^{λ} as the remaining ones). $\tilde{A}_T = D_T$ directly fulfills Assumption L3. Note that the degree of identification of each parameter has to be known (assumed) a priori in Stock and Wright's (2000) specification.

(iii) Antoine and Renault (2009) choose to distinguish between strong and nearly weak identification at the instrument level (see in particular their Subsection 2.3.2). They suppose that the set of K instruments can be partitioned between K_1 strong ones and $(K - K_1)$ nearly weak ones, so that

$$\Pi_T = \begin{bmatrix} \pi_{11} & \pi_{12} \\ \pi_{21}/T^{\lambda} & \pi_{22}/T^{\lambda} \end{bmatrix} = \Lambda_T^{-1}\Pi,$$

with Λ_T a (K, K)-diagonal matrix (with 1 as the first K_1 coefficients and T^{λ} as the K_2 remaining ones). The limit case with $\lambda = 1/2$ is the framework of Hahn, Ham, and Moon (2009).

Interestingly enough, the above approaches (ii) and (iii) lead to the same concentration matrix, a well-known measure of the strength of the instruments. As a consequence, one concludes that both approaches capture similar patterns of weak identification. In Examples 2.1 and 2.1, the concentration matrix and its determinant are respectively equal to

$$\mu = \Sigma_V^{-1/2'}\Pi_T' Z' Z \Pi_T \Sigma_V^{1/2}$$

and

$$\det(\mu) = \frac{1}{T^{2\lambda}} \det(Z'Z) \det(\Sigma_V^{-1}) \det(\Pi)^2$$

with $\Sigma_V \equiv \mathtt{Var}[V]$. With standard weak asymptotics ($T^{\lambda} = \sqrt{T}$), the concentration matrix has a finite limit (see also Andrews and

Stock 2007). Nearly weak asymptotics allow an infinite limit for the determinant of the concentration matrix, but at a rate smaller than $\det[Z'Z] = \mathcal{O}(T)$. In this respect, there is no difference between the two approaches, only the rate of convergence to zero of respectively a row or a column of the matrix Π_T matters.

(iv) Phillips (1989) introduces partial identification where Π_T matrices that may not be of full rank are considered. Generalization to asymptotic rank condition failures (at rate T^{λ}) comes at the price of having to specify which row (or column) asymptotically goes to zero. At least, Antoine and Renault's (2009) approach (iii) works with "estimable functions" of the structural parameters, or functions that can be identified and square-root T consistently estimated. By contrast, the approach (ii) implies directly a partition of the structural parameters between strongly and weakly identified ones.

(v) Antoine and Renault (2010a) generalize the above approach (iii) to accommodate matrices of reduced form like $\Pi_T = \Lambda_T^{-1}\Pi$ with Λ_T a (K, K)-diagonal matrix such that $\|\Lambda_T\| = o(\sqrt{T})$. Then $A_T = \Lambda_T E_{zz}^{-1}$ fulfills Assumption L1. By contrast with the former examples, the case where instruments may not be mutually orthogonal and may display different levels of strength leads to a nondiagonal matrix A_T. However, in this case, it is easy to imagine a standardization of instruments such that A_T eventually becomes diagonal (i.e., $A_T = \Lambda_T$). Then, a sequence of matrices $\tilde{A}_T$ fulfilling Assumption L3 can be built according to the general result provided in Theorem 2.3. The detailed construction provided in the appendix shows that we can actually choose $\tilde{A}_T = R\tilde{\Lambda}_T$ with R nonsingular (p, p)-matrix whose columns provide a basis for the orthogonal of the null space of Π while $\tilde{\Lambda}_T$ is a diagonal (p, p)-matrix such that $\|\tilde{\Lambda}_T\| \leq \|\Lambda_T\|$. In other words, all parameters are estimated with a rate of convergence at least equal to $\sqrt{T}/\|\tilde{\Lambda}_T\|$ irrespective of the slowest rate $\sqrt{T}/\|\Lambda_T\|$. The key is that some instruments (among the weakest) may be irrelevant, depending on the range of Π'. This analysis actually provides primitive conditions for the high-level Assumption 2 in Hansen, Hausman, and Newey (2008) where they assume that $\Upsilon = \Pi_T' z_t$ (where z_t denotes the tth observation of the K instruments) can be rewriten as $\Upsilon = S_T \tilde{z}_T$ for some p-dimensional vector $\tilde{z}_T$. This transformation exactly corresponds to our transformation of A_T into $\tilde{A}_T$ which is made explicit in the above detailed discussion. As also done in Antoine and Renault (2009, 2010a), Hansen, Hausman, and Newey (2008) take advantage of the matrix S_T to characterize how some linear combinations of the parameters may be identified at different rates.

2.4.2 Continuously Updated GMM

We now show that the nearly strong identification Assumption 2.5 is exactly needed to ensure that any direction in the parameter space is equivalently estimated by efficient two-step GMM and continuously updated GMM. This will also explain the equivalence between GMM score test and Kleibergen's modified score test discussed in the next section. Hansen, Heaton, and Yaron (1996) define the continuously updated GMM estimator $\hat{\theta}_T^{CU}$ as follow:

Definition 2.1 *Let $S_T(\theta)$ be a family of nonsingular random matrices such that*[6]

(i) $S_T(\theta^0)$ *is a (unfeasible) consistent estimator of* $S \equiv \lim_T[\mathtt{Var}(\sqrt{T}\bar{\phi}_T(\theta^0))]$.

(ii) $\|S_T^{-1}(\theta^0)\| = \mathcal{O}_P(1)$.

(iii) $\sup_{\theta\in\Theta}\|S_T(\theta)\| = \mathcal{O}_P(1)$.

(iv) $\sup_{\|\theta-\theta^0\|<\delta_T}\|S_T^{-1}(\theta) - S^{-1}\| = o_p(1)$ *for some real sequence* δ_T.

The continuously updated GMM estimator $\hat{\theta}_T^{CU}$ of θ^0 is then defined as

$$\hat{\theta}_T^{CU} = \arg\min_{\theta\in\Theta}\left[\bar{\phi}_T'(\theta)S_T^{-1}(\theta)\bar{\phi}_T(\theta)\right]. \tag{2.11}$$

PROPOSITION 2.1 *(Equivalence between CU-GMM and efficient 2S-GMM) Under Assumptions 2.1–2.5, any direction in the parameter space is equivalently estimated by efficient two-step GMM and continuously updated GMM. That is,*

$$\sqrt{T}\tilde{A}_T^{-1}(\hat{\theta}_T^{CU} - \hat{\theta}_T) = o_p(1).$$

In the special case where the same degree of global identification weakness λ_T is assumed for all coefficients of $\tilde{A}_T$, CU-GMM and efficient 2S-GMM are equivalent without the homogenous identification Assumption 2.5 (insofar as $\lambda_T = o(\sqrt{T})$).

Several comments are in order.

First, since nondegenerate asymptotic normality is obtained for $\sqrt{T}\tilde{A}_T^{-1}(\hat{\theta}_T - \theta^0)$ (and not for $\sqrt{T}(\hat{\theta}_T - \theta^0)$), the relevant (nontrivial) equivalence result between two-step efficient GMM and continuously updated GMM relates to the suitably *rescaled* difference $\sqrt{T}\tilde{A}_T^{-1}(\hat{\theta}_T - \hat{\theta}_T^{CU})$.

Second, the case with nearly weak (and not homogenous) identification ($\|\tilde{\Lambda}_T\|^2/\sqrt{T} = o(1)$) breaks down the standard theory of efficient GMM: the proof shows that there is no reason to believe that continuously updated GMM may be an answer. Two-step GMM and continuously updated GMM, albeit no longer equivalent, are both perturbed by higher-order terms with ambiguous effects on asymptotic distributions. The intuition given by higher-order asymptotics in standard identification settings cannot be extended to the case of nearly weak identification. While the latter approach shows that continuously updated GMM is, in general, higher-order efficient (see Newey

[6] The following regularity assumptions are standard when defining the continuously updated GMM estimator. See Pakes and Pollard (1989, pp. 1044–1046).

and Smith 2004; Antoine, Bonnal, and Renault 2007), there is no clear ranking of asymptotic performances under weak identification.

Third, it is important to keep in mind that all these difficulties are due to the fact that we consider realistic circumstances where several degrees of global identification weakness are simultaneously involved. Standard results (equivalence, or rankings between different approaches) carry on when the same rate λ_T is assumed for all coefficients of $\tilde{A}_T$.

2.4.3 GMM Score-Type Testing

As already explained, when the same degree of global identification weakness λ_T is assumed for all coefficients of the matrix Λ_T, standard procedures and results hold. One of the contribution of this paper is to characterize the heterogeneity of the informational content of moment conditions along different directions in the parameter space. We now illustrate how the power of tests is affected. More precisely, we are interested in testing the null hypothesis: $\underline{H}_0 : \theta = \theta_0$. To simplify the exposition, we focus here on a diagonal matrix A_T:

$$A_T = \begin{bmatrix} Id_{K_1} & O \\ O & \lambda_T Id_{K-K_1} \end{bmatrix} \quad \text{with} \quad \lambda_T \to \infty \quad \text{and} \quad \lambda_T = o(\sqrt{T}).$$

Assumption 2.3 is modified accordingly:

(simplified) Assumptions 2.3

$$\begin{bmatrix} Id_{K_1} & O \\ O & \lambda_T Id_{K-K_1} \end{bmatrix} \frac{\partial \bar{\phi}_T(\theta_0)}{\partial \theta'} = \begin{bmatrix} Id_{K_1} & O \\ O & \lambda_T Id_{K-K_1} \end{bmatrix} \begin{pmatrix} \frac{\partial \bar{\phi}_{1T}(\theta^0)}{\partial \theta'} \\ \frac{\partial \bar{\phi}_{2T}(\theta^0)}{\partial \theta'} \end{pmatrix}$$

$$\to \frac{\partial \tilde{d}(\theta^0)}{\partial \theta'} \equiv \begin{pmatrix} \frac{\partial \tilde{d}_1(\theta^0)}{\partial \theta'} \\ \frac{\partial \tilde{d}_2(\theta^0)}{\partial \theta'} \end{pmatrix}$$

with the (K, p)*-matrix* $[\partial \tilde{d}(\theta^0)/\partial \theta']$ *full column rank.*

The following (simple) example illustrates our focus of interest.

Example 2.3

Consider the functions ϕ_{1t} *and* ϕ_{2t} *defined as*

$$\phi_{1t}(\theta) = Y_{1t} - g(\theta) \quad \text{and} \quad \phi_{2t}(\theta) = -Z_t \otimes (Y_{2t} - X_{2t}\theta),$$

and associated moment conditions

$$E[Y_{1t}] = g(\theta^0) \quad \text{and} \quad E\left[Z_t \otimes (Y_{2t} - X_{2t}\theta^0)\right] = 0.$$

The instruments Z_t introduced in ϕ_{2t} are only nearly weak instruments since

$$E[Z_t \otimes X_{2t}] = \frac{1}{\lambda_T}\frac{\partial \tilde{d}_2(\theta^0)}{\partial \theta'} \quad \textit{with} \quad \lambda_T \xrightarrow{T} \infty, \quad \textit{and} \quad \frac{\lambda_T}{\sqrt{T}} \xrightarrow{T} 0.$$

Then the associated Jacobian matrices are

$$\texttt{Plim}\left[\frac{\partial \bar{\phi}_{1T}(\theta^0)}{\partial \theta'}\right] = \frac{\partial g(\theta^0)}{\partial \theta'} = \frac{\partial \tilde{d}_1(\theta^0)}{\partial \theta'}$$

$$\texttt{Plim}\left[\lambda_T \frac{\partial \bar{\phi}_{2T}(\theta^0)}{\partial \theta'}\right] = \texttt{Plim}\left[\lambda_T \frac{1}{T}\sum_{t=1}^{T}(Z_t \otimes X_{2t})\right]$$
$$= \lim_T [\lambda_T E\,(Z_t \otimes X_{2t})] = \frac{\partial \tilde{d}_2(\theta^0)}{\partial \theta'},$$

and we assume that $\left[\frac{\partial \tilde{d}_1'(\theta^0)}{\partial \theta} \vdots \frac{\partial \tilde{d}_2'(\theta^0)}{\partial \theta}\right]'$ *has full column rank.*

The GMM score-type testing approach wonders whether the test value θ_0 is close to fulfill the first-order conditions of the (efficient) two-step GMM minimization, that is whether the score vector is close to zero. The score vector is defined at the test value θ_0 as

$$V_T(\theta_0) = \frac{\partial \bar{\phi}_T'(\theta_0)}{\partial \theta} S_T^{-1}(\theta_0)\bar{\phi}_T(\theta_0).$$

The GMM score test statistic (Newey and West 1987) is then a suitable norm of $V_T(\theta_0)$:

$$\xi_T^{NW} = T V_T'(\theta_0)\left[\frac{\partial \bar{\phi}_T'(\theta_0)}{\partial \theta} S_T^{-1}(\theta_0)\frac{\partial \bar{\phi}_T(\theta_0)}{\partial \theta'}\right]^{-1} V_T(\theta_0).$$

Kleibergen's (2005) approach rather considers the first-order conditions of the CU-GMM minimization. The corresponding score vector is defined at the test value θ_0 as

$$V_T^{CU}(\theta_0) = \frac{\partial \phi_T^{CU\prime}(\theta_0)}{\partial \theta} S_T^{-1}(\theta_0)\bar{\phi}_T(\theta_0),$$

where each row of $\left[\frac{\partial \phi_T^{CU}(\theta_0)}{\partial \theta'}\right]$ is the residual of the long-term affine regression of $\left[\frac{\partial \bar{\phi}_T(\theta_0)}{\partial \theta'}\right]_{[i.]}$ on $\bar{\phi}_T(\theta_0)$:

$$\left[\frac{\partial \phi_T^{CU}(\theta_0)}{\partial \theta'}\right]'_{[i.]} = \left[\frac{\partial \bar{\phi}_T(\theta_0)}{\partial \theta'}\right]'_{[i.]}$$
$$-\mathrm{Cov}_{as}\left(\sqrt{T}\left[\frac{\partial \bar{\phi}_T(\theta_0)}{\partial \theta'}\right]'_{[i.]}, \sqrt{T}\bar{\phi}_T(\theta_0)\right)\mathrm{Var}_{as}\left(\sqrt{T}\bar{\phi}_T(\theta_0)\right)^{-1}\bar{\phi}_T(\theta_0) \tag{2.12}$$

where $\text{Var}_{as}(\sqrt{T}\bar{\phi}_T(\theta_0)) = S^0$ is the long-term covariance matrix of the moment conditions $\phi_t(\theta_0)$ and $\text{Cov}_{as}\left(\sqrt{T}\left[\frac{\partial\bar{\phi}_T(\theta_0)}{\partial\theta}\right]'_{[i.]'}, \sqrt{T}\bar{\phi}_T(\theta_0)\right)$ is the long-term covariance between $\left[\frac{\partial\phi_t(\theta_0)}{\partial\theta}\right]_{[i.]}$ and $\phi_t(\theta_0)$ (which is assumed well-defined).

This characterization of the score of continuously updated GMM in terms of residual of an affine regression is extensively discussed in Antoine, Bonnal, and Renault (2007) through their Euclidean empirical likelihood approach. It explains the better finite sample performance of CU-GMM since the regression allows to remove the perverse correlation between the Jacobian matrix and the moment conditions. In finite sample, this perverse correlation implies that the first order conditions of standard (two-step) efficient GMM are biased. As clearly explained by Kleibergen (2005), this perverse correlation is even more detrimental with genuinely weak instruments since it does not even vanish asymptotically. This is the reason why Kleibergen (2005) puts forward a modified version of the Newey-West (1987) score test statistic:

$$\xi_T^K = T V_T^{CU\prime}(\theta_0)\left[\frac{\partial\phi_T^{CU\prime}(\theta_0)}{\partial\theta} S_T^{-1}(\theta_0)\frac{\partial\phi_T^{CU}(\theta_0)}{\partial\theta'}\right]^{-1} V_T^{CU}(\theta_0).$$

In contrast with Kleibergen (2005), we show that with nearly weak instruments, the aforementioned correlation does not matter asymptotically and that the standard GMM score test statistic ξ_T^{NW} works. It is actually asymptotically equivalent to the modified Kleibergen's score test statistic under the null:

PROPOSITION 2.2 *(Equivalence under the null)*
Under the null $\underline{H}_0 : \theta = \theta_0$, we have: $\texttt{Plim}\left[\xi_T^{NW} - \xi_T^K\right] = 0$. *Both ξ_T^{NW} and ξ_T^K converge in distribution toward a chi-square with p degrees of freedom.*

The following example illustrates how a proper characterization of the heterogeneity of the informational content of moment conditions matters when considering power of tests under sequences of local alternatives.

Example 2.3 (continued)
Consider a sequence of local alternatives defined by a given deterministic sequence $(\gamma_T)_{T\geq 0}$ in $\mathbb{R}^p$, going to zero when T goes to infinity, and such that the true unknown value θ_0 is defined as: $\theta_T = \theta_0 + \gamma_T$. For T large enough, $g(\theta_T)$ can be seen as $g(\theta_0) + [\partial g(\theta_0)/\partial\theta']\gamma_T$. Therefore, the strongly identified moment restrictions $E[Y_{1t} - g(\theta_T)] = 0$ are informative with respect to the violation of the null $(\theta_T \neq \theta_0)$ if and only if: $[\partial g(\theta_0)/\partial\theta']\gamma_T \neq 0$.

As a consequence, we expect GMM-based tests of $\underline{H}_0 : \theta = \theta_0$ to have power against sequences of local alternatives converging at standard rate $\sqrt{T}$, $\theta_T = \theta_0 + \gamma/\sqrt{T}$, if and only if $[\partial g(\theta_0)/\partial\theta']\gamma \neq 0$, or, when γ does not belong to the null space of $[\partial g(\theta_0)/\partial\theta'] = [\partial\tilde{d}_1(\theta_0)/\partial\theta']$. By contrast, if $[\partial\tilde{d}_1(\theta_0)/\partial\theta']\gamma = 0$, violations of the null can only be built from the other identifying conditions:

$$E[Z_t \otimes Y_t] = \frac{\partial\tilde{d}_2(\theta_0)}{\partial\theta'}\frac{\theta_T}{\lambda_T}.$$

We show that the sequences of local alternatives relevant to characterize nontrivial power are necessarily such that $\theta_T = \theta_0 + \lambda_T \frac{\gamma}{\sqrt{T}}$.

In other words, the degree of weakness of the moment conditions λ_T *downplays the standard rate* $[\gamma/\sqrt{T}]$ *of sequences of local alternatives against which the tests have nontrivial local power. Under such a sequence of local alternatives,*

$$E\left[Z_t \otimes Y_t\right] = \frac{\partial \tilde{d}_2(\theta_0)}{\partial \theta'}\left[\frac{\theta_0}{\lambda_T} + \frac{\gamma}{\sqrt{T}}\right]$$

differs from its value under the null by the standard scale $1/\sqrt{T}$.

PROPOSITION 2.3 *(Local power of GMM score tests)*
(i) With a (drifted) true unknown value, $\theta_T = \theta_0 + \gamma/\sqrt{T}$, *for some* $\gamma \in \mathbb{R}^p$, *we have* $\texttt{Plim}[\xi_T^{NW} - \xi_T^K] = 0$, *and both* ξ_T^{NW} *and* ξ_T^K *converge in distribution toward a noncentral chi-square with p degrees of freedom and noncentrality parameter*

$$\mu = \left(\gamma' \frac{\partial \tilde{d}_1'(\theta_0)}{\partial \theta} \vdots 0\right)[S(\theta^0)]^{-1}\begin{pmatrix} \frac{\partial \tilde{d}_1(\theta_0)}{\partial \theta'}\gamma \\ 0 \end{pmatrix}.$$

(ii) In case of nearly strong identification ($\lambda_T^2 = o(\sqrt{T})$), *with a (drifted) true unknown value* $\theta_T = \theta_0 + \gamma/\lambda_T$, *for some* $\gamma \in \mathbb{R}^p$ *such that* $\frac{\partial \tilde{d}_1(\theta_0)}{\partial \theta'}\gamma = 0$, *we have* $\texttt{Plim}[\xi_T^{NW} - \xi_T^K] = 0$, *and both* ξ_T^{NW} *and* ξ_T^K *converge in distribution toward a noncentral chi-square with p degrees of freedom and noncentrality parameter*

$$\mu = \left(0 \vdots \gamma' \frac{\partial \tilde{d}_2'(\theta_0)}{\partial \theta}\right)[S(\theta^0)]^{-1}\begin{pmatrix} 0 \\ \frac{\partial \tilde{d}_2(\theta_0)}{\partial \theta'}\gamma \end{pmatrix}.$$

Two additional conclusions follow from Proposition 2.3:

(i) First, if $\frac{\partial \tilde{d}_1(\theta_0)}{\partial \theta'}\gamma \neq 0$, the two GMM score tests behave more or less as usual against sequences of local alternatives in the direction γ. They are asymptotically equivalent and both consistent against sequences converging slower than $\sqrt{T}$. They both follow asymptotically a noncentral chi-square against sequences with the usual rate $\sqrt{T}$.

(ii) Second, if $\frac{\partial \tilde{d}_1(\theta_0)}{\partial \theta'}\gamma = 0$, the two GMM score tests have no power against sequences of local alternatives $\theta_T = \theta_0 + \gamma/\sqrt{T}$. They may have power against sequences $\theta_T = \theta_0 + \gamma\lambda_T/\sqrt{T}$ (or slower); their behavior is pretty much the standard one, but only in the homogenous identification case where $\lambda_T^2 = o(\sqrt{T})$.

We now explain why nonstandard asymptotic behavior of both score tests may arise when we consider sequences of local alternatives in the weak directions ($\theta_T = \theta_0 + \gamma\lambda_T/\sqrt{T}$ with $\frac{\partial \tilde{d}_1(\theta_0)}{\partial \theta'}\gamma = 0$) with severe nearly weak identification issues. Recall that the genuine weak identification usually considered in the literature ($\lambda_T = \sqrt{T}$) is a limit case, since we always maintain the nearly weak identification condition $\lambda_T = o(\sqrt{T})$. Under such a sequence of local alternatives, while $\sqrt{T}\phi_T(\theta_T)$ is asymptotically normal with zero mean, the key to get a standard noncentral chi-square for the asymptotic distribution of a score test statistic is to ensure that $\sqrt{T}\bar{\phi}_T(\theta_0)$ is asymptotically normal with nonzero mean if and only if γ is not zero. This result should follow from the Taylor approximation around θ_0 with θ_T^* between θ_0 and θ_T:

$$\sqrt{T}\bar{\phi}_T(\theta_T) \approx \sqrt{T}\bar{\phi}_T(\theta_0) + \sqrt{T}\frac{\partial\bar{\phi}_T(\theta_T^*)}{\partial\theta'}(\theta_T - \theta_0) \approx \sqrt{T}\bar{\phi}_T(\theta_0) + \begin{pmatrix} 0 \\ \frac{\partial \bar{d}_2(\theta_0)}{\partial\theta'}\gamma \end{pmatrix}.$$

This approximation is justified by (simplified) Assumption 2.3 as long as

$$\frac{\partial \bar{d}_1(\theta_0)}{\partial\theta'}\gamma = 0 \Rightarrow \texttt{Plim}\left[\lambda_T\frac{\partial\bar{\phi}_{1T}(\theta_T^*)}{\partial\theta'}\gamma\right] = 0. \tag{2.13}$$

This is not an issue if, as in Kleibergen (2005), the same degree of global identification weakness[7] is assumed for all coefficients of the matrix A_T. In other words, we can easily state that in the interesting case with mixture of strong and nearly weak identification (or nonempty subsets of components ϕ_1 and ϕ_2), Equation 2.13 should follow from

$$\frac{\partial \bar{d}_1(\theta_0)}{\partial\theta'}\gamma = 0 \Rightarrow \texttt{Plim}\left[\lambda_T\frac{\partial\bar{\phi}_{1T}(\theta_T)}{\partial\theta'}\gamma\right] = 0. \tag{2.14}$$

Fragile identification may be wasted by Kleibergen's modification precisely because it comes with another piece of information which is stronger. To see this, the key is the aforementioned lack of logical implication from Equation 2.14 to Equation 2.13. As a result, the modified score statistic and the original one may have quite different asymptotic behaviors. It is quite evident from Equation 2.12 that, when $\sqrt{T}\bar{\phi}_T(\theta_0)$ is not $\mathcal{O}_P(1)$, the modified score statistic may have an arbitrarily nasty asymptotic behavior. However, it is worth noting that if we maintain Assumption 2.1 of a functional central limit theorem,

$$\sqrt{T}\bar{\phi}_T(\theta_0) - \sqrt{T}\rho_T(\theta_0) = \mathcal{O}_P(1),$$

a sufficient condition to ensure that Kleibergen's modified test statistic is well-behaved under the sequence of local alternatives $\theta_T = \theta_0 + \gamma\frac{\lambda_T}{\sqrt{T}}$ is

$$\sqrt{T}\rho_T\left(\theta_T - \gamma\frac{\lambda_T}{\sqrt{T}}\right) = \mathcal{O}(1).$$

The proof of Theorem 2.2 in the appendix allows us to think that this condition is plausible, since it precisely states that the rate of convergence of any GMM

[7] Smith (2007) already pointed out that the standard equivalence between tests holds when only one rate of convergence is considered.

estimator $\hat{\theta}_T$ ($\|\hat{\theta}_T - \theta_T\| = \mathcal{O}_P(\lambda_T/\sqrt{T})$) precisely comes from the fact (see Lemma 2.1 in the appendix) that:

$$\sqrt{T}\rho_T(\hat{\theta}_T) = \mathcal{O}_P(1).$$

To put it differently, Kleibergen's modified score test is well-behaved under a given sequence of local alternatives insofar as this sequence behaves as well as any GMM estimator. Such a result is not surprising. The novel feature introduced by nearly weak instruments asymptotics is that the rate of sequences of local alternatives must be assessed not only in the parameter space ($\|\theta_T - \theta_0\| = \mathcal{O}(\lambda_T/\sqrt{T})$) but also in the moments space ($\sqrt{T}\rho_T(\theta_0) = \mathcal{O}(1)$).

2.5 Conclusion

To conclude, we have proposed a general framework where weaker patterns of identification may arise without giving up the efficiency goal of statistical inference. We actually believe that even fragile information should be processed optimally for the purpose of both efficient estimation and powerful testing.

Our main contribution has been to consider that several patterns of identification may arise simultaneously. This heterogeneity of identification schemes paved the way for the device of optimal strategies for inferential use of information of poor quality. More precisely, we focus on a case where asymptotic efficiency of estimators is well-defined through the variance of asymptotically normal distributions. The price to pay for this maintained tool was to assume that the set of moment conditions that are not genuinely weak was sufficient to identify the true unknown value of the parameters. In this case, normality was characterized at heterogeneous rates smaller than the standard root-T in different directions of the parameter space. Finally, we were able to show that in such a case standard efficient estimation procedures still hold and are even feasible without requiring the prior knowledge of the identification schemes.

As emphasized in the survey of Andrews and Stock (2007), there are three main topics related to inference with weak identification: hypothesis tests and confidence intervals that are robust to weak instruments; point estimation; and pretesting for weak instruments. Andrews and Stock (2007) have focused on the first topic *"for which a solution is closer at hand than it is for estimation."* Our paper focuses on point estimation as well as power. This can only be done because we consider that the worst case scenario of genuine weak identification is not always warranted. As far as testing for strong/weak identification is concerned, the framework put forward in the present chapter allows us in a companion paper (Antoine and Renault 2010b) to add to the available literature that includes Hahn and Hausman (2002) and Hahn, Ham, and Moon (2009) among others.

Appendix

Notations

- For any vector v with element $(v_i)_{1\leq i\leq H}$, we define: $\|v\|^2 = \sum_{i=1}^{H} v_i^2$.
- For any matrix M with elements m_{ij} that is not a vector, we define: $\|M\| = \max_{i,j} |m_{ij}|$.
- Id_l denotes the identity matrix of size l.
- $[M]_{k.}$ denotes the kth row of the matrix M.
- $\texttt{col}[M]$ denotes the subspace generated by the columns of the matrix M.
- $\texttt{col}[M]^{\perp}$ denotes the subspace orthogonal to the one generated by $\texttt{col}[M]$.

We start with a preliminary result useful for the proofs of consistency and rates of convergence.

Lemma 2.1 *(i) Under Assumptions 2.1 and 2.2,*

$$\|\tilde{\rho}_T(\tilde{\theta}_T)\| = \mathcal{O}_P\left(\frac{1}{\sqrt{T}}\right) \quad \textit{with} \quad \tilde{\rho}_T(\theta) = [Id_{\tilde{K}} \vdots O_{\tilde{K},K-\tilde{K}}]N_T'\rho_T(\theta)$$

where $\tilde{\theta}_T$ is the GMM-estimator deduced from the partial set of moment conditions as follows:

$$\tilde{\theta}_T = \arg\min_{\theta\in\Theta} \tilde{Q}_T(\theta) = \arg\min_{\theta\in\Theta} \left[\tilde{\phi}_T'(\theta)\tilde{\Omega}_T\tilde{\phi}_T(\theta)\right] \quad \textit{with}$$

$$\tilde{\phi}_T(\theta) = [Id_{\tilde{K}} \vdots O_{\tilde{K},K-\tilde{K}}]N_T'\bar{\phi}_T(\theta)$$

where $\tilde{\Omega}_T$ is a sequence of symmetric positive definite random matrices of size $\tilde{K}$ converging toward a positive definite matrix $\tilde{\Omega}$.
(ii) Under Assumptions 2.1, 2.2, and 2.3(v),

$$\|\rho_T(\hat{\theta}_T)\| = \mathcal{O}_P\left(\frac{1}{\sqrt{T}}\right),$$

where $\hat{\theta}_T$ is the GMM-estimator defined in (2.8).

Proof of Lemma 2.1 First, we prove (ii); second, we show how (i) directly follows.

From Assumption 2.1(i), the objective function is written as follows:

$$TQ_T(\theta) \equiv T\phi_T'(\theta)\Omega_T\phi_T(\theta) = \left[\Psi_T(\theta) + \sqrt{T}\rho_T(\theta)\right]'\Omega_T\left[\Psi_T(\theta) + \sqrt{T}\rho_T(\theta)\right],$$

where the empirical process $(\Psi_T(\theta))_{\theta\in\Theta}$, is asymptotically Gaussian. Since $\hat{\theta}_T$ is the minimizer of Q_T, we have

$$\begin{aligned} Q_T(\hat{\theta}_T) \leq Q_T(\theta^0) \Leftrightarrow\ & T\rho_T'(\hat{\theta}_T)\Omega_T\rho_T(\hat{\theta}_T) + 2\sqrt{T}\rho_T'(\hat{\theta}_T)\Omega_T\Psi_T(\hat{\theta}_T) \\ & +\Psi_T'(\hat{\theta}_T)\Omega_T\Psi_T(\hat{\theta}_T) \\ & \leq T\rho_T'(\theta^0)\Omega_T\rho_T(\theta^0) + 2\sqrt{T}\rho_T'(\theta^0)\Omega_T\Psi_T(\theta^0) \\ & +\Psi_T'(\theta^0)\Omega_T\Psi_T(\theta^0). \end{aligned} \quad (2.15)$$

Following the notations introduced in Assumption 2.2, we define: $N_T\rho_T(\theta) = [\tilde{\rho}_T(\theta)'\ \breve{\rho}_T(\theta)']'$.

From Assumption 2.2(iv), we have: $\tilde{\rho}_T(\theta^0) = 0$ for any T. From Assumptions 2.2(ii) and (iii), we have: $\|\breve{\Lambda}_T\breve{\rho}_T(\theta)\| = \mathcal{O}_P(1)$. Following Assumption 2.3(v), we distinguish two cases[8]:
(a) the additional restrictions are well-specified, $\breve{\rho}_T(\theta^0) = 0$, and we have

$$(2.15) \Rightarrow T\rho_T'(\hat{\theta}_T)\Omega_T\rho_T(\hat{\theta}_T) + 2\sqrt{T}\rho_T'(\hat{\theta}_T)\Omega_T\Psi_T(\hat{\theta}_T) + h_T \leq 0, \quad (2.16)$$

with $h_T = \Psi_T'(\hat{\theta}_T)\Omega_T\Psi_T(\hat{\theta}_T) - \Psi_T'(\theta^0)\Omega_T\Psi_T(\theta^0)$.
(b) the additional restrictions are not well-specified, but genuinely weak, $\breve{\Lambda}_T = \sqrt{T} Id_{K-\tilde{K}}$ which implies $\|\sqrt{T}\breve{\rho}_T(\theta)\| = \mathcal{O}_P(1)$, and we have

$$(2.15) \Rightarrow T\rho_T'(\hat{\theta}_T)\Omega_T\rho_T(\hat{\theta}_T) + 2\sqrt{T}\rho_T'(\hat{\theta}_T)\Omega_T\Psi_T(\hat{\theta}_T) + h_T + \epsilon_T \leq 0, \quad (2.17)$$

with $\epsilon_T = \mathcal{O}_P(1)$.

Hence, we can always write:

$$T\rho_T'(\hat{\theta}_T)\Omega_T\rho_T(\hat{\theta}_T) + 2\sqrt{T}\rho_T'(\hat{\theta}_T)\Omega_T\Psi_T(\hat{\theta}_T) + h_T + \nu_T \leq 0, \quad (2.18)$$

with h_T defined above and $\nu_T = \mathcal{O}_P(1)$: actually, $\nu_T = 0$ in case (a) and $\nu_T = \epsilon_T$ in case (b).

Then, after defining μ_T as the smallest eigenvalue of Ω_T, it follows that

$$T\mu_T\|\rho_T(\hat{\theta}_T)\|^2 - 2\sqrt{T}\|\rho_T(\hat{\theta}_T)\| \times \|\Omega_T\Psi_T(\hat{\theta}_T)\| + h_T + \nu_T \leq 0.$$

In other words, $x_T \equiv \|\sqrt{T}\rho_T(\hat{\theta}_T)\|$ solves the inequality:

$$x_T^2 - \frac{2\|\Omega_T\Psi_T(\hat{\theta}_T)\|}{\mu_T}x_T + \frac{h_T + \nu_T}{\mu_T} \leq 0.$$

Therefore, we must have $\Delta_T \geq 0$

$$\text{with} \quad \Delta_T = \frac{\|\Omega_T\Psi_T(\hat{\theta}_T)\|^2}{\mu_T^2} - \frac{h_T + \nu_T}{\mu_T},$$

[8] Note that a combination of these two cases also works similarly: by combination, we have in mind that some components of $\breve{\rho}_T(\theta^0)$ are well-specified whereas some others are not well-specified but genuinely weak.

$$\text{and} \quad \frac{\|\Omega_T \Psi_T(\hat{\theta}_T)\|}{\mu_T} - \sqrt{\Delta_T} \le x_T \le \frac{\|\Omega_T \Psi_T(\hat{\theta}_T)\|}{\mu_T} + \sqrt{\Delta_T}.$$

We want to show that $x_T = \mathcal{O}_P(1)$, that is

$$\frac{\|\Omega_T \Psi_T(\hat{\theta}_T)\|}{\mu_T} = \mathcal{O}_P(1) \quad \text{and} \quad \Delta_T = \mathcal{O}_P(1),$$

which amounts to show that

$$\frac{\|\Omega_T \Psi_T(\hat{\theta}_T)\|}{\mu_T} = \mathcal{O}_P(1) \quad \text{and} \quad \frac{\|\Omega_T \Psi_T(\theta^0)\|}{\mu_T} = \mathcal{O}_P(1).$$

Denote by det(M) the determinant of any square matrix M. Since $\det(\Omega_T) \xrightarrow{P} \det(\Omega) > 0$, no subsequence of μ_T can converge in probability toward zero and thus we can assume (for T sufficiently large) that μ_T remains lower bounded away from zero with asymptotic probability one. Therefore, we just have to show that

$$\|\Omega_T \Psi_T(\hat{\theta}_T)\| = \mathcal{O}_P(1) \quad \text{and} \quad \|\Omega_T \Psi_T(\theta^0)\| = \mathcal{O}_P(1).$$

Denote by $\mathtt{tr}(M)$ the trace of any square matrix M. Since $\mathtt{tr}(\Omega_T) \xrightarrow{P} \mathtt{tr}(\Omega)$ and the sequence $\mathtt{tr}(\Omega_T)$ is upper bounded in probability, so are all the eigenvalues of Ω_T. Therefore the required boundedness in probability follows from the functional CLT in Assumption 2.1(i) which ensures

$$\sup_{\theta \in \Theta} \|\Psi_T(\theta)\| = \mathcal{O}_P(1).$$

This completes the proof of (ii).
(i) easily follows after realizing that Assumption 2.3(v) is irrelevant since dealing with the additional moment restrictions and that an inequality similar to (2.18) can be obtained as follows:

$$\tilde{Q}_T(\tilde{\theta}_T) \le \tilde{Q}_T(\theta^0) \Leftrightarrow T\tilde{\rho}'_T(\tilde{\theta}_T)\tilde{\Omega}_T\tilde{\rho}_T(\tilde{\theta}_T) + 2\sqrt{T}\tilde{\rho}'_T(\tilde{\theta}_T)\tilde{\Omega}_T\tilde{\Psi}_T(\tilde{\theta}_T) + \tilde{h}_T \le 0,$$

with $\tilde{h}_T = \tilde{\Psi}'_T(\tilde{\theta}_T)\tilde{\Omega}_T\tilde{\Psi}_T(\tilde{\theta}_T) - \tilde{\Psi}'_T(\theta^0)\tilde{\Omega}_T\tilde{\Psi}_T(\theta^0)$.

Proof of Theorem 2.1 *(Consistency)*
Consider the GMM-estimators $\hat{\theta}_T$ defined in (2.8) and $\tilde{\theta}_T$ deduced from the partial set of moment conditions as follows:

$$\tilde{\theta}_T = \arg\min_{\theta \in \Theta} \tilde{Q}_T(\theta) = \arg\min_{\theta \in \Theta} \left[\tilde{\phi}'_T(\theta)\tilde{\Omega}_T\tilde{\phi}_T(\theta)\right] \text{ with}$$
$$\tilde{\phi}_T(\theta) = [Id_{\tilde{K}} \vdots O_{\tilde{K},K-\tilde{K}}]N'_T\bar{\phi}_T(\theta),$$

where $\tilde{\Omega}_T$ is a sequence of symmetric positive definite random matrices of size $\tilde{K}$ converging toward a positive definite matrix $\tilde{\Omega}$. The proof of consistency of $\hat{\theta}_T$ is divided into two steps: (1) we show that $\tilde{\theta}_T$ is a consistent estimator of θ^0; (2) we show that $\text{Plim}[\hat{\theta}_T] = \text{Plim}[\tilde{\theta}_T]$.

(1) The weak consistency of $\tilde{\theta}_T$ follows from a contradiction argument. If $\tilde{\theta}_T$ were not consistent, there would exist some positive ϵ such that

$$P\left[\|\tilde{\theta}_T - \theta^0\| > \epsilon\right]$$

does not converge to zero. Then we can define a subsequence $(\tilde{\theta}_{T_n})_{n\in\mathbb{N}}$ such that, for some positive η:

$$P\left[\|\tilde{\theta}_{T_n} - \theta^0\| > \epsilon\right] \geq \eta \quad \text{for } n \in \mathbb{N}$$

From Assumption 2.2(ii), we have

$$\alpha \equiv \inf_{\|\theta-\theta^0\|>\epsilon} \|\tilde{d}(\theta)\| > 0.$$

Note that since c is bounded and the orthogonal matrix M_T is norm-preserving, $[\tilde{\Lambda}_T \vdots O_{\tilde{K},K-\tilde{K}}]N_T'\rho_T(\theta)$ converges to $\tilde{d}(\theta)$ uniformly on Θ by Assumption 2.1. Then, by Assumption 2.2(ii), we have

$$\inf_{\|\theta-\theta^0\|>\epsilon} \|[\tilde{\Lambda}_T \vdots O_{\tilde{K},K-\tilde{K}}]N_T'\rho_T(\theta)\| \geq \frac{\alpha}{2} \quad \text{for all } T \text{ sufficiently large.}$$

That is, for all T sufficiently large, we have

$$\inf_{\|\theta-\theta^0\|>\epsilon} \|\tilde{\Lambda}_T\tilde{\rho}_T(\theta)\| \geq \frac{\alpha}{2} \quad \text{where} \quad \tilde{\rho}_T(\theta) = [Id_{\tilde{K}} \vdots O_{\tilde{K},K-\tilde{K}}]N_T'\rho_T(\theta).$$

Since $\|\tilde{\Lambda}_T\|/\sqrt{T} = o(1)$ by Assumption 2.2(ii) and $\sqrt{T}\tilde{\rho}(\tilde{\theta}_T) = \mathcal{O}_P(1)$ by Lemma 2.1, we get a contradiction when considering a subsequence T_n. We conclude that $\tilde{\theta}_T$ is a consistent estimator of θ^0.

(2) We now show that $\theta^0 = \mathtt{Plim}[\hat{\theta}_T]$, by showing that it is true for any subsequence. If we could find a subsequence which does not converge toward θ^0, we could find a sub-subsequence with a limit in probability $\theta^1 \neq \theta^0$ (by assumption Θ is compact). To avoid cumbersome notations with sub-subsequences, it is sufficient to show that: $\mathtt{Plim}[\hat{\theta}_T] = \theta^1 \Rightarrow \theta^1 = \theta^0$. Consider the criterion function: $Q_T(\theta) = \bar{\phi}_T'(\theta)\Omega_T\bar{\phi}_T(\theta)$. We show that

(i) $Q_T(\tilde{\theta}_T) = \mathcal{O}_P(1/\|\breve{\Lambda}_T\|^2)$

(ii) $\theta^1 \neq \theta^0 \Rightarrow \|\breve{\Lambda}_T\|^2 Q_T(\hat{\theta}_T) \xrightarrow{T} \infty$.

This would lead to a contradiction with the definition of GMM estimators: $Q_T(\hat{\theta}_T) \leq Q_T(\tilde{\theta}_T)$.To show (i) and (ii), we assume without loss of generality that the weighting matrices Ω_T, $\tilde{\Omega}_T$, Ω and $\tilde{\Omega}$ are all identity matrices; otherwise, this property would come with a convenient rescaling of the moment conditions.

$$T Q_T(\theta) = T\bar{\phi}'_T(\theta)\bar{\phi}_T(\theta)$$

$$= \|\sqrt{T}[Id_{\breve{K}} \vdots O_{\breve{K},K-\breve{K}}]\bar{\phi}_T(\theta)\|^2 + \|\sqrt{T}[O_{K-\breve{K},\breve{K}} \vdots Id_{K-\breve{K}}]\bar{\phi}_T(\theta)\|^2$$

$$= \|\sqrt{T}[Id_{\breve{K}} \vdots O_{\breve{K},K-\breve{K}}]\bar{\phi}_T(\theta)\|^2 + \|[O_{K-\breve{K},\breve{K}} \vdots Id_{K-\breve{K}}](\Psi(\theta) - \sqrt{T}\rho_T(\theta))\|^2$$

From Lemma 2.1: $\|\sqrt{T}[Id_{\breve{K}} \vdots O_{\breve{K},K-\breve{K}}]\bar{\phi}_T(\tilde{\theta}_T)\| = \mathcal{O}_P(1)$.

From Assumption 2.2(ii): $[O_{K-\breve{K},\breve{K}} \vdots \breve{\Lambda}_T]N'_T\rho_T(\theta) \to \breve{d}(\theta)$ uniformly. Thus: $Q_T(\theta) = \mathcal{O}_P(1/(\|\breve{\Lambda}_T\|^2))$.

$$\|\breve{\Lambda}_T\|^2 Q_T(\hat{\theta}_T)$$

$$\geq \left\| \frac{\|\breve{\Lambda}_T\|}{\sqrt{T}}[Id_{\breve{K}} \vdots O_{\breve{K},K-\breve{K}}]\Psi(\hat{\theta}_T) + \|\breve{\Lambda}_T\|[Id_{\breve{K}} \vdots O_{\breve{K},K-\breve{K}}]\rho_T(\hat{\theta}_T)\right\|^2$$

$$\geq \left[\|\breve{\Lambda}_T\|\|[Id_{\breve{K}} \vdots O_{\breve{K},K-\breve{K}}]\rho_T(\hat{\theta}_T)\| - \frac{\|\breve{\Lambda}_T\|}{\sqrt{T}}\|[Id_{\breve{K}} \vdots O_{\breve{K},K-\breve{K}}]\Psi(\hat{\theta}_T)\|\right]^2$$

From Assumption 2.2(ii): $\frac{\|\breve{\Lambda}_T\|}{\sqrt{T}}\|[Id_{\breve{K}} \vdots O_{\breve{K},K-\breve{K}}]\Psi(\hat{\theta}_T)\| = \mathcal{O}_P\left(\|\Psi(\hat{\theta}_T)\|\right) = \mathcal{O}_P(1)$, while

$$\|\breve{\Lambda}_T\|\|[Id_{\breve{K}} \vdots O_{\breve{K},K-\breve{K}}]\rho_T(\hat{\theta}_T)\| \geq \frac{\|\breve{\Lambda}_T\|\|[\tilde{\Lambda}_T \vdots O_{\breve{K},K-\breve{K}}]\rho_T(\hat{\theta}_T)\|}{\|\tilde{\Lambda}_T\|},$$

with $\|[\tilde{\Lambda}_T \vdots O_{\breve{K},K-\breve{K}}]N'_T\rho_T(\hat{\theta}_T)\| \to \|\breve{d}(\theta^1)\| \neq 0$. Thus,

$$\frac{\|\breve{\Lambda}_T\|\|[\tilde{\Lambda}_T \vdots O_{\breve{K},K-\breve{K}}]\rho_T(\hat{\theta}_T)\|}{\|\tilde{\Lambda}_T\|} \to +\infty,$$

and we get the announced result.

Proof of Theorem 2.2 *(Rate of convergence)*
From Lemma 2.1, $\|\rho_T(\hat{\theta}_T)\| = \mathcal{O}_P(1/\sqrt{T})$. We know that N_T is bounded. Hence, we have: $\|N'_T\rho_T(\hat{\theta}_T)\| = \mathcal{O}_P(1/\sqrt{T})$. Recall also that from Assumption 2.2(iii), the first $\tilde{K}$ elements of $N_T\rho_T(\theta^0)$ are identically zero. The mean-value theorem, for some $\tilde{\theta}_T$ between $\hat{\theta}_T$ and θ^0 (component by component), yields to

$$\left\|[\tilde{\Lambda}_T\ 0]N'_T\frac{\partial\rho_T(\tilde{\theta}_T)}{\partial\theta'}\left(\hat{\theta}_T - \theta^0\right)\right\| = \mathcal{O}_P\left(\frac{\|\tilde{\Lambda}_T\|}{\sqrt{T}}\right).$$

Note that (by a common abuse of notation) we omit to stress that $\tilde{\theta}_T$ actually depends on the component of ρ_T. Define now z_T as follows:

$$z_T \equiv \frac{\partial \tilde{d}(\theta^0)}{\partial \theta'}(\hat{\theta}_T - \theta^0). \tag{2.19}$$

Since $[\partial \tilde{d}(\theta^0)/\partial \theta']$ is full column rank by Assumption 2.3(iii), we have

$$(\hat{\theta}_T - \theta^0) = \left[\frac{\partial \tilde{d}'(\theta^0)}{\partial \theta}\frac{\partial \tilde{d}(\theta^0)}{\partial \theta'}\right]^{-1}\frac{\partial \tilde{d}'(\theta^0)}{\partial \theta} z_T.$$

Hence, we only need to prove that $\|z_T\| = \mathcal{O}_P(\|\tilde{\Lambda}_T\|/\sqrt{T})$. By definition of z_T, we have

$$z_T = \left[\frac{\partial \tilde{d}(\theta^0)}{\partial \theta'} - [\tilde{\Lambda}_T \; 0]N_T'\frac{\partial \rho_T(\tilde{\theta}_T)}{\partial \theta'}\right](\hat{\theta}_T - \theta^0) + w_T \quad \text{with}$$

$$\|w_T\| = \mathcal{O}_P\left(\frac{\|\tilde{\Lambda}_T\|}{\sqrt{T}}\right). \tag{2.20}$$

However, since $\tilde{\theta}_T \xrightarrow{P} \theta^0$ and $[\tilde{\Lambda}_T \; 0]N_T'[\partial \rho_T(\theta)/\partial \theta']$ converges uniformly on the interior of Θ toward $[\partial \tilde{d}(\theta)/\partial \theta']$ by Assumption 2.3(iv), we have

$$\begin{aligned}&\left[\frac{\partial \tilde{d}(\theta^0)}{\partial \theta'} - [\tilde{\Lambda}_T \; 0]N_T'\frac{\partial \rho_T(\tilde{\theta}_T)}{\partial \theta'}\right](\hat{\theta}_T - \theta^0),\\ &= \left[\frac{\partial \tilde{d}(\theta^0)}{\partial \theta'} - [\tilde{\Lambda}_T \; 0]N_T'\frac{\partial \rho_T(\tilde{\theta}_T)}{\partial \theta'}\right]\left[\frac{\partial \tilde{d}'(\theta^0)}{\partial \theta}\frac{\partial \tilde{d}(\theta^0)}{\partial \theta'}\right]^{-1}\frac{\partial \tilde{d}'(\theta^0)}{\partial \theta} z_T,\\ &= D_T z_T,\end{aligned}$$

for some matrix D_T such that $\|D_T\| \xrightarrow{P} 0$. Therefore: $\|z_T\| \leq \epsilon_T \|z_T\| + \|w_T\|$ with $\epsilon_T \to 0$. Hence, $\|z_T\| = \mathcal{O}_P(\|\tilde{\Lambda}_T\|/\sqrt{T})$ and we get: $\|\hat{\theta}_T - \theta^0\| = \mathcal{O}_P(\|\tilde{\Lambda}_T\|/\sqrt{T})$.

Proof of Theorem 2.3

Without loss of generality, we write the diagonal matrix Λ_T as:

$$\Lambda_T = \left(\begin{array}{ccc|ccc} \lambda_{1T} Id_{K_1} & & & & & \\ & \ddots & & & & \\ & & \lambda_{LT} Id_{K_L} & & & \\ \hline & & & \lambda_{L+1,T} Id_{K_{L+1}} & & \\ & & & & \ddots & \\ & & & & & \lambda_{\overline{L},T} Id_{K_{\overline{L}}} \end{array}\right) = \left(\begin{array}{c|c} \tilde{\Lambda}_T & O \\ \hline O & \check{\Lambda}_T \end{array}\right)$$

with $\overline{L} \leq K$, $\sum_{l=1}^{\overline{L}} K_l = K$ and $\lambda_{lT} = o(\lambda_{l+1,T})$. For convenience, we also rewrite the (p, K)-matrix $\left[\frac{\partial c'(\theta^0)}{\partial \theta} M^{-1'}\right]$ by stacking horizontally $\overline{L}$ blocks of size (p, K_l) denoted J_l, $(l = 1, \cdots, \overline{L})$ as follows:

$$\frac{\partial c'(\theta^0)}{\partial \theta} M^{-1\prime} = (J_1 \cdots J_L | J_{L+1} \cdots J_{\overline{L}}) = \left(\frac{\partial \tilde{d}'(\theta^0)}{\partial \theta'} \;\middle|\; \frac{\partial \check{d}'(\theta^0)}{\partial \theta'} \right),$$

$$\text{with } J_1' \equiv \begin{pmatrix} \left[M^{-1}\frac{\partial c(\theta^0)}{\partial \theta'}\right]_{[1.]} \\ \vdots \\ \left[M^{-1}\frac{\partial c(\theta^0)}{\partial \theta'}\right]_{[K_1]} \end{pmatrix} \text{ and } J_l' \equiv \begin{pmatrix} \left[M^{-1}\frac{\partial c(\theta^0)}{\partial \theta'}\right]_{[K_1+\cdots+K_{l-1}+1.]} \\ \vdots \\ \left[M^{-1}\frac{\partial c(\theta^0)}{\partial \theta'}\right]_{[K_1+\cdots+K_l.]} \end{pmatrix}$$

$$\text{for } l = 2, \cdots, \overline{L}.$$

Recall also that by Assumption 2.3(iii), the columns of $\frac{\partial \tilde{d}'(\theta^0)}{\partial \theta}$ span the whole space $\mathbb{R}^p$. We now introduce the square matrix of size p, $R = [R_1 \; R_2 \; \cdots \; R_L]$ which spans $\mathbb{R}^p$. The idea is that each (p, s_l)-block R_l defined through $\texttt{col}[J_l]$ collects the directions associated with the specific rate $\lambda_{lT}, l = 1, \cdots, L$ and $\sum_{l=1}^{L} s_l = p$. Then, the matrix $\tilde{A}_T$ is built as

$$\tilde{A}_T = \left[\lambda_{1T} R_1 \vdots \lambda_{2T} R_2 \vdots \cdots \vdots \lambda_{LT} R_L \right].$$

By convention, $\lambda_{lT} = o(\lambda_{l+1,T})$ for any $1 \leq l \leq L-1$. We now explain how to construct the matrix R. The idea is to separate the parameter space into L subspaces. More specifically:

- R_L is defined such that $J_i' R_L = 0$ for $1 \leq i < L$ and $\texttt{rk}[R_L] = \texttt{rk}[J_L]$. In other words, R_L spans $\texttt{col}[J_1 \; J_2 \; \cdots \; J_{L-1}]^{\perp}$.
- R_{L-1} is defined such that $J_i' R_{L-1} = 0$ for $1 \leq i < L-1$ and $\texttt{rk}[R_{L-1} \; R_L] = \texttt{rk}[J_{L-1} \; J_L]$.
- And so on, until R_2 is defined such that $J_1' R_2 = 0$ and $\texttt{rk}[R_2 \; \cdots \; R_L] = \texttt{rk}[J_2 \; \cdots \; J_L]$.
- Finally, R_1 is defined such that $R = [R_1 \; R_2 \; \cdots \; R_L]$ is full rank.

We now check that $\lim_T [\Lambda_T^{-1} M^{-1} \frac{\partial c(\theta^0)}{\partial \theta'} \tilde{A}_T]$ exists and is full column rank. First, recall that we have

$$\lim_T \left(\Lambda_T^{-1} M^{-1} \frac{\partial c(\theta^0)}{\partial \theta'} \tilde{A}_T \right) = \lim_T \left(\begin{bmatrix} \tilde{\Lambda}_T^{-1} & 0 \\ 0 & \check{\Lambda}_T^{-1} \end{bmatrix} \begin{bmatrix} \frac{\partial \tilde{d}(\theta^0)}{\partial \theta'} \\ \frac{\partial \check{d}(\theta^0)}{\partial \theta'} \end{bmatrix} \tilde{A}_T \right)$$

$$= \lim_T \begin{pmatrix} \tilde{\Lambda}_T^{-1} \frac{\partial \tilde{d}(\theta^0)}{\partial \theta'} \tilde{A}_T \\ 0 \end{pmatrix},$$

since $\check{\Lambda}_T^{-1} = o(\|\tilde{\Lambda}_T^{-1}\|)$ and $\|\tilde{A}_T\| = \mathcal{O}(\tilde{\Lambda}_T)$.

We now show that $[\tilde{\Lambda}_T^{-1} \frac{\partial \tilde{d}(\theta^0)}{\partial \theta'} \tilde{A}_T]$ converges to a block diagonal matrix of rank p.

$$\tilde{\Lambda}_T^{-1} \frac{\partial \tilde{d}(\theta^0)}{\partial \theta'} \tilde{A}_T = \begin{pmatrix} \lambda_{1T}^{-1} Id_{K_1} & & \\ & \ddots & \\ & & \lambda_{LT}^{-1} Id_{K_L} \end{pmatrix}$$
$$\times \left[\lambda_{1T} \frac{\partial \tilde{d}(\theta^0)}{\partial \theta'} R_1 \vdots \lambda_{2T} \frac{\partial \tilde{d}(\theta^0)}{\partial \theta'} R_2 \vdots \cdots \vdots \lambda_{LT} \frac{\partial \tilde{d}(\theta^0)}{\partial \theta'} R_L \right].$$

- The L diagonal blocks are equal to $J_l' R_l$; these (K_l, s_l)-blocks are full column rank s_l by construction of the matrices R_l with $\sum_{l=1}^{L} s_l = p$.
- The lower triangular blocks converge to zero since $\lambda_{jT} = o(\lambda_{lT})$ for any $1 \leq j < l \leq L$.
- The upper triangular blocks converge to zero by construction of the matrices R_l since $J_l' R_i = 0$ for any $1 \leq l < i \leq L$.

Proof of Corollary 2.1 *(Extended Theorem 2.3)*
From Assumption 2.4(i):

$$\frac{\partial \bar{\phi}_T(\theta^0)}{\partial \theta'} - A_T^{-1} \frac{\partial c(\theta^0)}{\partial \theta'} = \mathcal{O}_P\left(\frac{1}{\sqrt{T}}\right)$$
$$\Leftrightarrow \frac{\partial \bar{\phi}_T(\theta^0)}{\partial \theta'} - N_T^{-1} \Lambda_T^{-1} M_T^{-1} \frac{\partial c(\theta^0)}{\partial \theta'} = \mathcal{O}_P\left(\frac{1}{\sqrt{T}}\right)$$
$$\Rightarrow \frac{\partial \bar{\phi}_T(\theta^0)}{\partial \theta'} \tilde{A}_T - N_T^{-1} \Lambda_T^{-1} M_T^{-1} \frac{\partial c(\theta^0)}{\partial \theta'} \tilde{A}_T = \mathcal{O}_P\left(\frac{\|\tilde{\Lambda}_T\|}{\sqrt{T}}\right)$$
$$\Rightarrow \frac{\partial \bar{\phi}_T(\theta^0)}{\partial \theta'} \tilde{A}_T - N^{-1} H = \mathcal{O}_P\left(\frac{\|\tilde{\Lambda}_T\|}{\sqrt{T}}\right),$$

with H full column rank matrix from Theorem 2.3.

Proof of Theorem 2.4 *(Asymptotic distribution)*
Mean-value expansion of the moment conditions around θ^0 for $\tilde{\theta}_T$ between $\hat{\theta}_T$ and θ^0,

$$\bar{\phi}_T(\hat{\theta}_T) = \bar{\phi}_T(\theta^0) + \frac{\partial \bar{\phi}_T(\tilde{\theta}_T)}{\partial \theta'} (\hat{\theta}_T - \theta^0), \tag{2.21}$$

combined with the first-order conditions,

$$\frac{\partial \bar{\phi}_T'(\hat{\theta}_T)}{\partial \theta} \Omega_T \bar{\phi}_T(\hat{\theta}_T) = 0$$

yields to

$$\frac{\partial \bar{\phi}_T'(\hat{\theta}_T)}{\partial \theta} \Omega_T \bar{\phi}_T(\theta^0) + \frac{\partial \bar{\phi}_T'(\hat{\theta}_T)}{\partial \theta} \Omega_T \frac{\partial \bar{\phi}_T(\tilde{\theta}_T)}{\partial \theta'} (\hat{\theta}_T - \theta^0) = 0$$

$$\Leftrightarrow \tilde{A}_T' \frac{\partial \bar{\phi}_T'(\hat{\theta}_T)}{\partial \theta} \Omega_T \sqrt{T} \bar{\phi}_T(\theta^0)$$

$$+ \tilde{A}_T' \frac{\partial \bar{\phi}_T'(\hat{\theta}_T)}{\partial \theta} \Omega_T \frac{\partial \bar{\phi}_T(\tilde{\theta}_T)}{\partial \theta'} \tilde{A}_T \tilde{A}_T^{-1} (\hat{\theta}_T - \theta^0) = 0$$

$$\Leftrightarrow \tilde{A}_T^{-1} \sqrt{T} (\hat{\theta}_T - \theta^0) = -\left[\tilde{A}_T' \frac{\partial \bar{\phi}_T'(\hat{\theta}_T)}{\partial \theta} \Omega_T \frac{\partial \bar{\phi}_T(\tilde{\theta}_T)}{\partial \theta'} \tilde{A}_T \right]^{-1}$$

$$\times \tilde{A}_T' \frac{\partial \bar{\phi}_T'(\hat{\theta}_T)}{\partial \theta} \Omega_T \sqrt{T} \bar{\phi}_T(\theta^0)$$

$$\Rightarrow \tilde{A}_T^{-1} \sqrt{T} (\hat{\theta}_T - \theta^0) = \mathcal{O}_P(1). \tag{2.22}$$

We then get the expected result after justifying the invertibility of $\left[\tilde{A}_T' \frac{\partial \bar{\phi}_T'(\hat{\theta}_T)}{\partial \theta} \Omega_T \frac{\partial \bar{\phi}_T(\tilde{\theta}_T)}{\partial \theta'} \tilde{A}_T \right]$ for T large enough.

Lemma 2.2 *(Extension of Corollary 2.1)*
Under Assumptions 2.1–2.5, for any consistent estimator θ_T *s.t.* $\|\theta_T - \theta^0\| = \mathcal{O}_P(\|\tilde{\Lambda}_T\|/\sqrt{T})$*,*

$$\texttt{Plim} \left[\frac{\partial \bar{\phi}_T(\theta_T)}{\partial \theta'} \tilde{A}_T \right] \quad \textit{exists and is full column rank.}$$

Proof Mean-value expansion of the kth row of $\partial[\bar{\phi}_T(\theta_T)/\partial\theta']$ around θ^0 for $\tilde{\theta}_T$ between $\hat{\theta}_T$ and θ^0:

$$\left[\frac{\partial \bar{\phi}_T(\theta_T)}{\partial \theta'} \tilde{A}_T \right]_{k.} = \left[\frac{\partial \bar{\phi}_T(\theta^0)}{\partial \theta'} \tilde{A}_T \right]_{k.} + (\theta_T - \theta^0)' \frac{\partial}{\partial \theta} \left[\frac{\partial \bar{\phi}_T(\tilde{\theta}_T)}{\partial \theta'} \tilde{A}_T \right]_{k.}'$$

$$\Leftrightarrow \left[\frac{\partial \bar{\phi}_T(\theta_T)}{\partial \theta'} \tilde{A}_T \right]_{k.} - \left[\frac{\partial \bar{\phi}_T(\theta^0)}{\partial \theta'} \tilde{A}_T \right]_{k.} = \frac{\sqrt{T}}{\|\tilde{\Lambda}_T\|} (\theta_T - \theta^0)$$

$$\times \frac{\|\tilde{\Lambda}_T\|}{\sqrt{T}} \frac{\partial}{\partial \theta} \left[\frac{\partial \bar{\phi}_T(\tilde{\theta}_T)}{\partial \theta'} \tilde{A}_T \right]_{k.}.$$

From Assumption 2.4(ii), the Hessian term is such that

$$\begin{aligned}
\frac{\partial}{\partial\theta}\left[\frac{\partial\bar{\phi}_T(\tilde{\theta}_T)}{\partial\theta'}\right]_{k.} &= \frac{\partial}{\partial\theta}\left[A_T^{-1}\frac{\partial c(\tilde{\theta}_T)}{\partial\theta'}\right]_{k.} + \mathcal{O}_P\left(\frac{1}{\sqrt{T}}\right)\\
&= \frac{\partial}{\partial\theta}\left[N_T^{-1}\Lambda_T^{-1}M_T^{-1}\frac{\partial c(\tilde{\theta}_T)}{\partial\theta'}\right]_{k.} + \mathcal{O}_P\left(\frac{1}{\sqrt{T}}\right)\\
&= \mathcal{O}_P\left(\frac{1}{\lambda_{lT}}\right) + \mathcal{O}_P\left(\frac{1}{\sqrt{T}}\right) \quad \text{from Assumption 2.4(ii)}\\
&= \mathcal{O}_P\left(\frac{1}{\lambda_{lT}}\right), \qquad (2.23)
\end{aligned}$$

for any k such that $K_1 + \cdots + K_{l-1} < k \leq K_1 + \cdots + K_l$ and $l = 1, \cdots, L$.

Recall that $\tilde{A}_T = [\lambda_{1T}R_1 \vdots \cdots \vdots \lambda_{LT}R_L]$. To get the final result, we distinguish two cases to show that the RHS of the following equation is $o_p(1)$.

$$\left(\left[\frac{\partial\bar{\phi}_T(\theta_T)}{\partial\theta'}\right]_{k.} - \left[\frac{\partial\bar{\phi}_T(\theta^0)}{\partial\theta'}\right]_{k.}\right)\lambda_{iT}R_i$$

$$= \frac{\sqrt{T}}{\|\tilde{\Lambda}_T\|}(\theta_T - \theta^0) \times \frac{\|\tilde{\Lambda}_T\|}{\sqrt{T}}\frac{\partial}{\partial\theta}\left[\frac{\partial\bar{\phi}_T(\tilde{\theta}_T)}{\partial\theta'}\right]_{k.}\lambda_{iT}R_i.$$

- For $1 \leq i \leq l$, $\lambda_{iT} = o(\lambda_{lT})$ and the result directly follows from equation (2.23).
- For $i > l$, $\lambda_{lT} = o(\lambda_{iT})$ and the result follows from nearly-strong identification Assumption 2.5.

Note that when the same degree of global identification weakness is assumed, the asymptotic theory is available under Assumptions 2.1–2.4, since Lemma 2.2 holds without the nearly-strong identification Assumption 2.5.

Proof of Theorem 2.6 *(J test)*
Plugging (2.22) into (2.21), we get

$$\sqrt{T}\bar{\phi}_T(\hat{\theta}_T) = \sqrt{T}\bar{\phi}_T(\theta^0) - \frac{\partial\bar{\phi}_T(\tilde{\theta}_T)}{\partial\theta'}\tilde{A}_T\left[\tilde{A}_T'\frac{\partial\bar{\phi}_T'(\hat{\theta}_T)}{\partial\theta}\Omega_T\frac{\partial\bar{\phi}_T(\tilde{\theta}_T)}{\partial\theta'}\tilde{A}_T\right]^{-1}$$

$$\tilde{A}_T'\frac{\partial\bar{\phi}_T'(\hat{\theta}_T)}{\partial\theta}\Omega_T\sqrt{T}\bar{\phi}_T(\theta^0)$$

$$\Rightarrow TQ_T(\hat{\theta}_T) = \left[\sqrt{T}\bar{\phi}_T(\theta^0)\right]'\Omega_T^{\prime 1/2}[Id_K - P_X]\Omega_T^{1/2}\left[\sqrt{T}\bar{\phi}_T(\theta^0)\right] + o_P(1),$$

with $\Omega_T = \Omega_T^{\prime 1/2}\Omega_T^{1/2}$ and $P_X = X(X'X)^{-1}X'$ for $X = \Omega_T^{1/2}\frac{\partial\bar{\phi}_T(\hat{\theta}_T)}{\partial\theta'}\tilde{A}_T$.

Proof of Theorem 2.7 *(Wald test)*
To simplify the exposition, the proof is performed with Λ_T as defined in Example 2.2. In step 1, we define an algebraically equivalent formulation of $H_0 : g(\theta) = 0$ as $H_0 : h(\theta) = 0$ such that its first components are identified at the

fast rate λ_{1T}, while the remaining ones are identified at the slow rate λ_{2T} without any linear combinations of the latter being identified at the fast rate. In step 2, we show that the Wald test statistic $\xi_T^W(h)$ on $H_0 : h(\theta) = 0$ asymptotically converges to the proper chi-square distribution with q degrees of freedom and that it is numerically equal to the Wald test statistic $\xi_T^W(g)$ on $H_0 : g(\theta) = 0$.

Step 1: The space of fast directions to be tested is

$$I^0(g) = \left[\text{col}\frac{\partial g'(\theta^0)}{\partial \theta}\right] \cap \left[\text{col}\frac{\partial \tilde{d}_1'(\theta^0)}{\partial \theta}\right].$$

Denote $n^0(g)$ the dimension of $I^0(g)$. Then, among the q restrictions to be tested, $n^0(g)$ are identified at the fast rate and the $(q - n^0(g))$ remaining ones are identified at the slow rate.
Define q vectors of $\mathbb{R}^q$ denoted as ϵ_j $(j = 1, \cdots, q)$ such that $\left[(\partial g'(\theta^0)/\partial\theta)\epsilon_j\right]_{j=1}^{q_1}$ is a basis of $I^0(g)$ and $\left[(\partial g'(\theta^0)/\partial\theta)\epsilon_j\right]_{j=q_1+1}^{q}$ is a basis of

$$[I^0(g)]^{\perp} \cap \left[\text{col}\left(\frac{\partial g'(\theta^0)}{\partial \theta}\right)\right].$$

We can then define a new formulation of the null hypothesis $H_0 : g(\theta) = 0$ as, $H_0 : h(\theta) = 0$ where $h(\theta) = Hg(\theta)$ with H invertible matrix such that $H' = [\epsilon_1 \ \cdots \ \epsilon_q]$. The two formulations are algebraically equivalent since $h(\theta) = 0 \iff g(\theta) = 0$. Moreover,

$$\texttt{Plim}\left[D_T^{-1}\frac{\partial h(\theta^0)}{\partial \theta'}\tilde{A}_T\right] = B^0 \quad \text{with} \quad D_T = \begin{bmatrix} \lambda_{1T} Id_{n^0(g)} & 0 \\ 0 & \lambda_{2T} Id_{q-n^0(g)} \end{bmatrix},$$

and B^0 a full column rank (q, p)-matrix.
Step 2: we show that the two induced Wald test statistics $\xi_T^W(g)$ and $\xi_T^W(h)$ are equal.

$$\begin{aligned}
\xi_T^W(g) &= Tg'(\hat{\theta}_T)\left\{\frac{\partial g(\hat{\theta}_T)}{\partial \theta'}\left[\frac{\partial \bar{\phi}_T'(\hat{\theta}_T)}{\partial \theta}S_T^{-1}\frac{\partial \bar{\phi}_T(\hat{\theta}_T)}{\partial \theta'}\right]^{-1}\frac{\partial g'(\hat{\theta}_T)}{\partial \theta}\right\}^{-1} g(\hat{\theta}_T) \\
&= TH'g'(\hat{\theta}_T)\left\{H\frac{\partial g(\hat{\theta}_T)}{\partial \theta'}\left[\frac{\partial \bar{\phi}_T'(\hat{\theta}_T)}{\partial \theta}S_T^{-1}\frac{\partial \bar{\phi}_T(\hat{\theta}_T)}{\partial \theta'}\right]^{-1}\frac{\partial g'(\hat{\theta}_T)}{\partial \theta}H'\right\}^{-1} Hg(\hat{\theta}_T) \\
&= \xi_T^W(h).
\end{aligned}$$

Then, we show $\xi_T^W(h)$ is asymptotically distributed as a chi-square with q degrees of freedom. First, a preliminary result naturally extends the above convergence toward B^0 when θ^0 is replaced by a λ_{2T}-consistent estimator θ_T^*:

$$\texttt{Plim}\left[D_T^{-1}\frac{\partial h(\theta_T^*)}{\partial \eta'}\tilde{A}_T\right] = B^0.$$

The proof is very similar to Lemma 2.2 and is not reproduced here. The Wald test statistic $\xi_T^W(h)$ now writes:

$$\xi_T^W(h) = \left[\sqrt{T} D_T^{-1} h(\hat{\theta}_T)\right]' \left\{ D_T^{-1} \frac{\partial h(\hat{\theta}_T)}{\partial \theta'} \tilde{A}_T^{-1} \left[\tilde{A}_T' \frac{\partial \bar{\phi}_T'(\hat{\theta}_T)}{\partial \theta} S_T^{-1} \frac{\partial \bar{\phi}_T(\hat{\theta}_T)}{\partial \theta'} \tilde{A}_T \right]^{-1} \right.$$

$$\left. \times \tilde{A}_T \frac{\partial h'(\hat{\theta}_T)}{\partial \theta} D_T^{-1} \right\}^{-1} \times \left[\sqrt{T} D_T^{-1} h(\hat{\theta}_T)\right].$$

From Lemma 2.2,

$$\left[\tilde{A}_T' \frac{\partial \bar{\phi}_T'(\hat{\theta}_T)}{\partial \theta} S_T^{-1} \frac{\partial \bar{\phi}_T(\hat{\theta}_T)}{\partial \theta'} \tilde{A}_T \right] \xrightarrow{P} \Sigma \text{ nonsingular matrix.}$$

Now, from the mean-value theorem under H_0 we deduce

$$\sqrt{T} D_T^{-1} h(\hat{\theta}_T) = \sqrt{T} D_T^{-1} \frac{\partial h(\theta_T^*)}{\partial \theta'} (\hat{\theta}_T - \theta^0) = \left[D_T^{-1} \frac{\partial h(\theta_T^*)}{\partial \theta'} \tilde{A}_T \right] \sqrt{T} \tilde{A}_T^{-1} (\hat{\theta}_T - \theta^0)$$

with

$$\left[D_T^{-1} \frac{\partial h(\theta_T^*)}{\partial \theta'} \tilde{A}_T \right] \xrightarrow{P} B^0 \quad \text{and} \quad \sqrt{T} \tilde{A}_T^{-1} \left(\hat{\theta}_T - \theta^0\right) \xrightarrow{d} \mathcal{N}(0, \Sigma^{-1}).$$

Finally we get

$$\xi_T^W(h) = \left[\sqrt{T} \tilde{A}_T^{-1} (\hat{\theta}_T - \theta^0)\right]' B_0' (B_0 \Sigma B_0')^{-1} B_0 \left[\sqrt{T} \tilde{A}_T^{-1} (\hat{\theta}_T - \theta^0)\right] + o_P(1).$$

Following the proof of Theorem 2.6 we get the expected result.

Proof of Proposition 2.1 *(Equivalence between CU-GMM and 2S-GMM)* FOC of the CU-GMM optimization problem can be written as follows (see Antoine, Bonnal, and Renault 2007):

$$\sqrt{T} \frac{\partial \bar{\phi}_T'(\hat{\theta}_T^{CU})}{\partial \theta} S_T^{-1}(\hat{\theta}_T^{CU}) \sqrt{T} \bar{\phi}_T(\hat{\theta}_T^{CU})$$

$$- P \sqrt{T} \frac{\partial \bar{\phi}_T'(\hat{\theta}_T^{CU})}{\partial \theta} S_T^{-1}(\hat{\theta}_T^{CU}) \sqrt{T} \bar{\phi}_T(\hat{\theta}_T^{CU}) = 0,$$

where P is the projection matrix onto the moment conditions. Recall that

$$P \sqrt{T} \frac{\partial \bar{\phi}_T^{(j)}(\hat{\theta}_T^{CU})}{\partial \theta} = \text{Cov}\left(\frac{\partial \bar{\phi}_T^{(j)}(\hat{\theta}_T^{CU})}{\partial \theta}, \bar{\phi}_T(\hat{\theta}_T^{CU}) \right) S_T^{-1}(\hat{\theta}_T^{CU}) \sqrt{T} \bar{\phi}_T(\hat{\theta}_T^{CU}).$$

With a slight abuse of notation, we define conveniently the matrix of size (p, K^2) built by stacking horizontally the K matrices of size (p, K),

$\mathrm{Cov} \times (\partial\bar{\phi}_{j,T}(\hat{\theta}_T^{CU})/\partial\theta, \bar{\phi}_T(\hat{\theta}_T^{CU}))$, as

$$\mathrm{Cov}\left(\frac{\partial\bar{\phi}'_T(\hat{\theta}_T^{CU})}{\partial\theta}, \bar{\phi}_T(\hat{\theta}_T^{CU})\right)$$

$$\equiv \left[\mathrm{Cov}\left(\frac{\partial\bar{\phi}_T^{(1)}(\hat{\theta}_T^{CU})}{\partial\theta}, \bar{\phi}_T(\hat{\theta}_T^{CU})\right) \dots \mathrm{Cov}\left(\frac{\partial\bar{\phi}_T^{(j)}(\hat{\theta}_T^{CU})}{\partial\theta}, \bar{\phi}_T(\hat{\theta}_T^{CU})\right) \dots \right.$$

$$\left. \times \mathrm{Cov}\left(\frac{\partial\bar{\phi}_T^{(K)}(\hat{\theta}_T^{CU})}{\partial\theta}, \bar{\phi}_T(\hat{\theta}_T^{CU})\right)\right].$$

Then, we can write:

$$P\sqrt{T}\frac{\partial\bar{\phi}_T(\hat{\theta}_T^{CU})}{\partial\theta} = \mathrm{Cov}\left(\frac{\partial\bar{\phi}'_T(\hat{\theta}_T^{CU})}{\partial\theta}, \bar{\phi}_T(\hat{\theta}_T^{CU})\right)$$

$$\times\left(Id_K \otimes [S_T^{-1}(\hat{\theta}_T^{CU})\sqrt{T}\bar{\phi}_T(\hat{\theta}_T^{CU})]\right) \equiv H_T,$$

where $H_T = \mathcal{O}_P(1)$. Next, pre-multiply the above FOC by $\tilde{A}'_T/\sqrt{T}$ to get

$$\tilde{A}'_T\frac{\partial\bar{\phi}'_T(\hat{\theta}_T^{CU})}{\partial\theta}S_T^{-1}(\hat{\theta}_T^{CU})\sqrt{T}\bar{\phi}_T(\hat{\theta}_T^{CU}) - \frac{\tilde{A}'_T}{\sqrt{T}}H_T S_T^{-1}(\hat{\theta}_T^{CU})\sqrt{T}\bar{\phi}_T(\hat{\theta}_T^{CU}) = 0.$$

To get the equivalence between both estimators, we now show that the second element of the LHS is equal to $o_P(1)$.

From Assumption 2.1, we have $\sqrt{T}\bar{\phi}_T(\hat{\theta}_T^{CU}) = \sqrt{T}\rho_T(\hat{\theta}_T^{CU}) + \Psi_T(\hat{\theta}_T^{CU})$ with Ψ_T a Gaussian process. Hence, we have

$$\frac{\tilde{A}'_T}{\sqrt{T}}H_T S_T^{-1}(\hat{\theta}_T^{CU})\sqrt{T}\bar{\phi}_T(\hat{\theta}_T^{CU}) = \frac{\tilde{A}'_T}{\sqrt{T}}H_T S_T^{-1}(\hat{\theta}_T^{CU})\Psi_T(\hat{\theta}_T^{CU})$$

$$-\frac{\tilde{A}'_T}{\sqrt{T}}H_T S_T^{-1}(\hat{\theta}_T^{CU})\sqrt{T}\rho_T(\hat{\theta}_T^{CU}).$$

The first term of the RHS is obviously $o_p(1)$ since $\|\tilde{A}_T\| = o(\sqrt{T})$. The same remains to be shown for the second term,

$$\frac{\tilde{A}'_T}{\sqrt{T}}H_T\sqrt{T}S_T^{-1}(\hat{\theta}_T^{CU})\rho_T(\hat{\theta}_T^{CU}).$$

A result and proof similar to Lemma 2.1 for $\hat{\theta}_T^{CU}$ yield to: $\|\sqrt{T}\rho_T(\hat{\theta}_T^{CU})\| = \mathcal{O}_P(1)$.

Also, we already know that $H_T = \mathcal{O}_P(1)$ and $\|\tilde{A}_T\| = o(\sqrt{T})$. So, we combine these results to get

$$\frac{\tilde{A}'_T}{\sqrt{T}}H_T\sqrt{T}S_T^{-1}(\hat{\theta}_T^{CU})\rho_T(\hat{\theta}_T^{CU}) = o_p(1).$$

We conclude that both estimators are always defined by the same set of equations. To deduce that they are equivalent, we need Assumption 2.5 in order to get the same asymptotic theory. When the same degree of global identification weakness is assumed, the asymptotic theory holds without Assumption 2.5.

References

Andrews, D. W. K. 1994. Asymptotics for Semiparametric Econometric Models via Stochastic Equicontinuity, *Econometrica* 62: 43–72.

Andrews, D. W. K., and J.H. Stock. 2007. *Inference with Weak Instruments*. Econometric Society Monograph Series, vol. 3, ch. 8 in Advances in Economics and Econometrics, Theory and Applications: Ninth World Congress of the Econometric Society, Cambridge, U.K.: Cambridge University Press.

Angrist, J. D., and A. B. Krueger. 1991. Does Compulsory School Attendance Affect Schooling and Earnings? *Quarterly Journal of Economics* 106: 979–1014.

Antoine, B., H. Bonnal, and E. Renault. 2007. On the Efficient Use of the Informational Content of Estimating Equations: Implied Probabilities and Euclidean Empirical Likelihood. *Journal of Econometrics* 138(2): 461–487.

Antoine, B., and E. Renault. 2009. Efficient GMM with Nearly-Weak Instruments. *Econometric Journal* 12: 135–171.

———. 2010a. Efficient Minimum Distance Estimation with Multiple Rates of Convergence. *Journal of Econometrics*, forthcoming.

———. 2010b. Specification Tests for Strong Identification. *Working Paper*, UNC-CH.

Caner, M. 2010. Testing, Estimation in GMM and CUE with Nearly-Weak Identification. *Econometric Reviews*, 29(3): 330–363.

Choi, I., and P. C. B. Phillips. 1992. Asymptotic and Finite Sample Distribution Theory for IV Estimators and Tests in Partially Identified Structural Equations. *Journal of Econometrics* 51: 113–150.

Epstein, L. G., and S. E. Zin. 1991. Substitution, Risk Aversion, and the Temporal Behavior of Consumption and Asset Returns: An Empirical Analysis. *Journal of Political Economy* 99(2): 263–286.

Hahn, J., J. Ham, and H. R. Moon. 2009. The Hausman Test and Weak Instruments. *Working Paper*, USC.

Hahn, J., and J. Hausman. 2002. A new Specification Test for the Validity of Instrumental Variables. *Econometrica* 70(1): 163–190.

Hahn, J., and G. Kuersteiner. 2002. Discontinuities of Weak Instruments Limiting Distributions. *Economics Letters* 75: 325–331.

Han, C., and P. C. B. Phillips. 2006. GMM with Many Moment Conditions. *Econometrica* 74(1): 147–192.

Hansen, C., J. Hausman, and W. Newey. 2008. Estimation with Many Instrumental Variables. *Journal of Business and Economic Statistics* 26: 398–422.

Hansen, L. P. 1982. Large Sample Properties of Generalized Method of Moments Estimators. *Econometrica* 50(4): 1029–1054.

Hansen, L. P., J. Heaton, and A. Yaron. 1996. Finite Sample Properties of Some Alternative GMM Estimators. *Journal of Business and Economic Statistics* 14: 262–280.

Hansen, L. P., and R. J. Hodrick. 1980. Forward Exchange Rates as Optimal Predictors of Future Spot Rates: An Econometric Analysis. *Journal of Political Economy* 88: 829–853.

Hansen, L. P., and K. J. Singleton. 1982. Generalized Instrumental Variables Estimation of Nonlinear Rational Expectations Models. *Econometrica* 50: 1269–1286.

Horn, R. A., and C. R. Johnson. 1985. *Matrix Analysis*. Cambridge, U.K.: Cambridge University Press.

Kleibergen, F. 2005. Testing Parameters in GMM without Assuming that They Are Identified. *Econometrica* 73: 1103–1123.

Newey, W. K. 1994. The Asymptotic Variance of Semiparametric Estimators. *Econometrica* 62: 1349–1382.

Newey, W. K., and R. J. Smith. 2004. Higher-Order Properties of GMM and Generalized Empirical Likelihood Estimators. *Econometrica* 72: 219–255.

Newey, W. K., and K. D. West. 1987. Hypothesis Testing with Efficient Method of Moments Estimation. *International Economic Review* 28: 777–787.

Newey, W. K., and F. Windmeijer. 2009. GMM with Many Weak Moment Conditions. *Econometrica* 77: 687–719.

Pakes, A., and D. Pollard. 1989. Simulation and the Asymptotics of Optimization Estimators. *Econometrica* 57(5): 1027–1057.

Phillips, P. C. B. 1989. Partially Identified Econometric Models. *Econometric Theory* 5: 181–240.

Sargan, J. D. 1983. Identification and Lack of Identification. *Econometrica* 51(6): 1605–1634.

Smith, R. J. 2007. *Weak Instruments and Empirical Likelihood: A discussion of the Papers by D. W. K. Andrews and J. H. Stock and Y. Kitamura*. Econometric Society Monograph Series, vol. 3, ch. 8 in Advances in Economics and Econometrics, Theory and Applications: Ninth World Congress of the Econometric Society, pp. 238–260, Cambridge, U.K.: Cambridge University Press.

Staiger, D., and J. Stock. 1997. Instrumental Variables Regression with Weak instruments. *Econometrica* 65: 557–586.

Stock, J. H., and J. H. Wright. 2000. GMM with Weak Identification. *Econometrica* 68(5): 1055–1096.

Stock, J. H., J. H. Wright, and M. Yogo. 2002. A Survey of Weak Instruments and Weak Identification in Generalized Method of Moments. *Journal of Business and Economic Statistics* 20: 518–529.

3

An Information Theoretic Estimator for the Mixed Discrete Choice Model

Amos Golan and William H. Greene

CONTENTS

3.1 Introduction

There is much work in the social sciences on discrete choice models. Among those, the multinomial logit is the most common model used for analyzing survey data when the number of choices is greater than 2. In many cases, however, the underlying assumptions leading to the traditional maximum likelihood estimator (MLE) for the logit model are inconsistent with the perceived process that generated the observed data. One of these cases is the random parameter (RP) logit model (also known as "mixed logit" – see Revelt and Train 1998) that can be viewed as a variant of the multinomial choice model. In this chapter we formulate an Information-Theoretic (IT) estimator for the RP mixed logit model. Our estimator is easy to use and is computationally much less demanding than its competitors — the simulated likelihood class of estimators.

The objective of this work is to develop an estimation approach that is not simulation based, does not build on an underlying normal structure and is computationally efficient. This method is simple to use and apply and it works well for all sample sizes, though it is especially useful for smaller or ill-behaved data. The random parameters logit model is presented in Section 3.2. We discuss the information theoretic model and the motivation for constructing it in Section 3.3. In Section 3.4 we extend and generalize our basic model. In Section 3.5 we provide the necessary statistics for diagnostics and inference. We note, however, that the emphasis in this chapter is not on providing the large sample properties of our estimator, but rather to present the reader with convincing arguments that this model works well (relative to its competitors) for finite data and to provide the user with the correct set of tools to apply it. In Section 3.6 we provide simulated examples and contrast our IT estimator with competing simulated methods. We conclude in Section 3.7.

3.2 The Random Parameters Logit

The RP model is somewhat similar to the random coefficients model for linear regressions. (See, for example, Bhat 1996; Jain, Vilcassim, and Chintagunta 1994; Revelt and Train 1998; Train, Revelt, and Ruud, 1996; and Berry, Levinsohn, and Pakes 1995.) The core model formulation is a multinomial logit model, for individuals $i = 1, \ldots, N$ in choice setting t. Let y_{it} be the observed choice ($t = j$) of individual i and neglecting for the moment the error components aspect of the model, we begin with the basic form of the multinomial logit model, with alternative specific constants α_{ji} and a K-dimensional attributes vector $\mathbf{x}_{jit}$,

$$\text{Prob}(y_{it} = j) = \frac{\exp(\alpha_{ji} + \boldsymbol{\beta}_i' \mathbf{x}_{jit})}{\sum_{q=1}^{J_i} \exp(\alpha_{qi} + \boldsymbol{\beta}_i' \mathbf{x}_{qit})}. \tag{3.1}$$

The RP model emerges as the form of the individual specific parameter vector, $\boldsymbol{\beta}_i$ is developed. In the most familiar, simplest version of the model

$$\beta_{ki} = \beta_k + \sigma_k v_{ki},$$

and

$$\alpha_{ji} = \alpha j + \sigma_j v_{ji},$$

where β_k is the population mean, v_{ki} is the individual specific heterogeneity, with mean zero and standard deviation one, and σ_k is the standard deviation of the distribution of β_{ik}'s around β_k. The choice specific constants, α_{ji} and the elements of $\boldsymbol{\beta}_i$ are distributed randomly across individuals with fixed means. A refinement of the model is to allow the means of the parameter distributions to be heterogeneous with observed data, $\mathbf{z}_i$, (which does not include a constant term). This would be a set of choice invariant characteristics that produce individual heterogeneity in the means of the randomly distributed coefficients, so that

$$\beta_{ki} = \beta_k + \boldsymbol{\delta}_k' \mathbf{z}_i + \sigma_k v_{ki},$$

and likewise for the constants. The basic model for heterogeneity is not limited to the normal distribution. We consider several alternatives below. One important variation is the lognormal model,

$$\beta_{ki} = \exp(\phi_k + \boldsymbol{\delta}_k' \mathbf{z}_i + \sigma_k v_{ki}).$$

The v's are individual unobserved random disturbances, the source of the heterogeneity. Thus, as stated above, in the population, if the random terms are normally distributed, then

$$\alpha_{ji} \text{ or } \beta_{ki} \sim \text{ Normal or Lognormal } [\phi_{j \text{ or } k} + \boldsymbol{\delta}_{j \text{ or } k}' \mathbf{z}_i, \sigma_{j \text{ or } k}^2].$$

Other distributions may be specified in a similar fashion.

For the full vector of K random coefficients in the model, we may write the set of random parameters as

$$\boldsymbol{\phi}_i = \boldsymbol{\phi} + \boldsymbol{\Delta} z_i + \boldsymbol{\Gamma} \mathbf{v}_i. \tag{3.2}$$

where $\boldsymbol{\Gamma}$ is a diagonal matrix which contains σ_k on its diagonal. For convenience at this point, we will simply gather the parameters, choice specific or not, under the subscript 'k.' (The notation is a bit more cumbersome for the lognormally distributed parameters.)

We can go a step farther and allow the random parameters to be correlated. All that is needed to obtain this additional generality is to allow $\boldsymbol{\Gamma}$ to be a lower triangular matrix with nonzero elements below the main diagonal. Then, the full covariance matrix of the random coefficients is $\boldsymbol{\Sigma} = \boldsymbol{\Gamma}\boldsymbol{\Gamma}'$. The standard case of uncorrelated coefficients has $\boldsymbol{\Gamma} = \text{diag}(\sigma_1, \sigma_2, \ldots, \sigma_K)$. If the coefficients are freely correlated, $\boldsymbol{\Gamma}$ is an unrestricted lower triangular matrix and $\boldsymbol{\Sigma}$ will have nonzero off diagonal elements. It is convenient to aggregate this one step farther. We may gather the entire parameter vector for the model in this formulation simply by specifying that for the nonrandom parameters in the model, the corresponding rows in $\boldsymbol{\Delta}$ and $\boldsymbol{\Gamma}$ are zero. We also define the data and parameter vector so that any choice specific aspects are handled by appropriate placements of zeros in the applicable parameter vector. This is the approach we take in Section 3.3.

An additional extension of the model allows the distribution of the random parameters to be heteroscedastic. As stated above, the variance of v_{ik} is taken to be a constant. The model is made heteroscedastic by assuming, instead, that

$$\text{Var}[v_{ik}] = \sigma jk^2 [\exp(\boldsymbol{\omega}_k \text{ } '\mathbf{hr}_i)]^2$$

where $\mathbf{hr}_i$ is a vector of covariates, and $\boldsymbol{\omega}_k$ is the associated set of parameters. A convenient way to parameterize this is to write the full model (Equation 3.2) as

$$\boldsymbol{\phi}_i = \boldsymbol{\phi} + \boldsymbol{\Delta} \mathbf{z}_i + \boldsymbol{\Gamma} \boldsymbol{\Omega}_i \text{ } \mathbf{v}_i \tag{3.3}$$

where $\boldsymbol{\Omega}_i$ is a diagonal matrix of individual specific standard deviation terms: $\omega_{ik} = \exp(\boldsymbol{\omega}_k' \mathbf{hr}_i)$.

The list of variations above produces an extremely flexible, general model. Typically, depending on the problem at hand, we use only some of these variations, though in principle, all could appear in the model at once. The probabilities defined above (Equation 3.1) are conditioned on the random terms, $\mathbf{v}_i$. The unconditional probabilities are obtained by integrating v_{ik} out of the conditional probabilities: $P_j = E_{\mathbf{v}}[P(j|\mathbf{v}_i)]$. This is a multiple integral which does not exist in closed form. Therefore, in these types of problems, the integral is approximated by sampling R draws from the assumed populations and averaging. The parameters are estimated by maximizing the simulated log-likelihood,

$$\log L_s = \sum_{i=1}^{N} \log \frac{1}{R} \sum_{r=1}^{R} \prod_{t=1}^{T_i} \sum_{j=1}^{J_{it}} d_{ijt} \frac{\exp[\alpha_{ji} + \boldsymbol{\beta}_{ir}' \mathbf{x}_{jit}]}{\sum_{q=1}^{J_{it}} \exp[\alpha_{qi} + \boldsymbol{\beta}_{ir}' \mathbf{x}_{qit}]}, \tag{3.4}$$

with respect to $(\boldsymbol{\beta}, \boldsymbol{\Delta}, \boldsymbol{\Gamma}, \boldsymbol{\Omega})$, where

$d_{ijt} = 1$ if individual i makes choice j in period t, and zero otherwise,
$R =$ the number of replications,
$\boldsymbol{\beta}_{ir} = \boldsymbol{\beta} + \boldsymbol{\Delta}\mathbf{z}_i + \boldsymbol{\Gamma}\boldsymbol{\Omega}_i\mathbf{v}_{ir} =$ the rth draw on $\boldsymbol{\beta}_i$,
$\mathbf{v}_{ir} =$ the rth multivariate draw for individual i.

The heteroscedasticity is induced first by multiplying $\mathbf{v}_{ir}$ by $\boldsymbol{\Omega}_i$, then the correlation is induced by multiplying $\boldsymbol{\Omega}_i\mathbf{v}_{ir}$ by $\boldsymbol{\Gamma}$. See Bhat (1996), Revelt and Train (1998), Train (2003), Greene (2008), Hensher and Greene (2003), and Hensher, Greene, and Rose (2006) for further formulations, discussions and examples.

3.3 The Basic Information Theoretic Model

Like the basic logit models, the basic mixed logit model discussed above (Equation 3.1) is based on the utility functions of the individuals. However, in the mixed logit (or RP) models in Equation 3.1, there are many more parameters to estimate than there are data points in the sample. In fact, the construction of the simulated likelihood (Equation 3.4) is based on a set of restricting assumptions. Without these assumptions (on the parameters and on the underlying error structure), the number of unknowns is larger than the number of data points regardless of the sample size leading to an underdetermined problem. Rather than using a structural approach to overcome the identification problem, we resort here to the basics of information theory (IT) and the method of Maximum Entropy (ME) (see Shannon 1948; Jaynes 1957a, 1957b). Under that approach, we can maximize the total entropy of the system subject to the observed data. All the observed and known information enters as constraints within that optimization. Once the optimization is done, the problem is converted to its concentrated form (profile likelihood), allowing us to identify the natural set of parameters of that model. We now formulate our IT model.

The model we develop here is a direct extension of the IT, generalized maximum entropy (GME) multinomial choice model of Golan, Judge, and Miller (1996) and Golan, Judge, and Perloff (1996). To simplify notations, in the formulation below we include all unknown signal parameters (the constants and choice specific covariates) within $\boldsymbol{\beta}$ so that the covariates $\mathbf{X}$ also include the choice specific constants. Specifically, and as we discussed in Section 3.2, we gather the entire parameter vector for the model by specifying that for the nonrandom parameters in the model, the corresponding rows in $\boldsymbol{\Delta}$ and $\boldsymbol{\Gamma}$ are zero. Further, we also define the data and parameter vector so that any choice specific aspects are handled by appropriate placements of zeros in the applicable parameter vector. This is the approach we take below.

Instead of considering a specific (and usually unknown) $F(\cdot)$, or a likelihood function, we express the observed data and their relationship to the unobserved probabilities, P, as

$$y_{ij} = F(\mathbf{x}'_{ji}\boldsymbol{\beta}_j) + \varepsilon_{ij} = p_{ij} + \varepsilon_{ij},\, i = 1, \ldots, N, j = 1, \ldots, J,$$

where p_{ij} are the unknown multinomial probabilities and ε_{ij} are additive noise components for each individual. Since the observed Y's are either zero or one, the noise components are naturally contained in $[-1, 1]$ for each individual. Rather than choosing a specific $F(\cdot)$, we connect the observables and unobservables via the cross moments:

$$\sum_i y_{ij}x_{ijk} = \sum_i x_{ijk}p_{ij} + \sum_i x_{ijk}\varepsilon_{ij} \tag{3.5}$$

where there are $(N \times (J - 1))$ unknown probabilities, but only $(K \times J)$ data points or moments. We call these moments "stochastic moments" as the last term is different from the traditional (pure) moment representation of $\sum_i y_{ij}x_{ijk} = \sum_i x_{ijk}p_{ij}$.

Next, we reformulate the model to be consistent with the mixed logit data generation process. Let each p_{ij} be expressed as the expected value of an M-dimensional discrete random variable $\mathbf{s}$ (or an equally spaced support) with underlying probabilities π_{ij}. Thus, $p_{ij} \equiv \sum_m^M s_m\pi_{ijm}$, $s_m \in [0, 1]$ and $m = 1, 2, \ldots, M$ with $M \geq 2$ and where $\sum_m^M \pi_{ijm} = 1$. (We consider an extension to a continuous version of the model in Section 3.4.) To formulate this model within the IT-GME approach, we need to attach each one of the unobserved disturbances ε_{ij} to a proper probability distribution. To do so, let ε_{ij} be the expected value of an H-dimensional support space (random variable) $\mathbf{u}$ with corresponding H-dimensional vector of weights, $\mathbf{w}$. Specifically, let $\mathbf{u} = (-1/\sqrt{N}, \ldots, 0, \ldots 1/\sqrt{N})'$, so $\varepsilon_{ij} \equiv \sum_{h=1}^H u_h w_{ijh}$ (or $\varepsilon_i = E[u_i]$) with $\sum_h w_{ijh} = 1$ for each ε_{ij}.

Thus, the H-dimensional vector of weights (proper probabilities) $\mathbf{w}$ converts the errors from the $[-1, 1]$ space into a set of $N \times H$ proper probability distributions within **u**. We now reformulate Equation 3.5 as

$$\sum_i y_{ij}x_{ijk} = \sum_i x_{ijk}p_{ij} + \sum_i x_{ijk}\varepsilon_{ij} = \sum_{i,m} x_{ijk}s_m\pi_{ijm} + \sum_{i,h} x_{ijk}u_h w_{ijh}. \tag{3.6}$$

As we discussed previously, rather than using a simulated likelihood approach, our objective is to estimate, with minimal assumptions, the two sets of unknown $\boldsymbol{\pi}$ and $\mathbf{w}$ simultaneously. Since the problem is inherently underdetermined, we resort to the Maximum Entropy method (Jaynes 1957a, 1957b, 1978; Golan, Judge, and Miller, 1996; Golan, Judge, and Perloff, 1996). Under that approach, one uses an information criterion, called entropy (Shannon 1948), to choose one of the infinitely many probability distributions consistent with the observed data (Equation 3.6). Let $H(\boldsymbol{\pi}, \mathbf{w})$ be the joint entropies of $\boldsymbol{\pi}$ and $\mathbf{w}$, defined below. (See Golan, 2008, for a recent review and formulations of that class of estimators.) Then, the full set of unknown $\{\boldsymbol{\pi}, \mathbf{w}\}$ is estimated by maximizing $H(\boldsymbol{\pi}, \mathbf{w})$ subject to the observed stochastic moments (Equation 3.6) and the requirement that $\{\boldsymbol{\pi}\}$, $\{\mathbf{w}\}$ and $\{\mathrm{P}\}$ are proper probabilities. Specifically,

$$\underset{\boldsymbol{\pi},\mathbf{w}}{\text{Max}}\left\{H(\boldsymbol{\pi},\mathbf{w}) = -\sum_{ijm} \pi_{ijm}\log\pi_{ijm} - \sum_{ijh} w_{ijh}\log w_{ijh}\right\} \tag{3.7}$$

subject to

$$\begin{aligned}\sum_i y_{ij}x_{ijk} &= \sum_i x_{ijk}p_{ij} + \sum_i x_{ijk}\varepsilon_{ij} \\ &= \sum_{i,m} x_{ijk}s_m\pi_{ijm} + \sum_{i,h} x_{ijk}u_h w_{ijh}\end{aligned} \tag{3.8}$$

$$\sum_m \pi_{ijm} = 1, \ \sum_h w_{ijh} = 1 \tag{3.9a}$$

$$\sum_{j,m} s_m\pi_{ijm} = 1 \tag{3.9b}$$

with $\mathbf{s} \in [0, 1]$ and $\mathbf{u} \in (-1, 1)$.

Forming the Lagrangean and solving yields the IT estimators for $\boldsymbol{\pi}$

$$\hat{\pi}_{ijm} = \frac{\exp\left[s_m\left(-\sum_k \hat{\lambda}_{kj}x_{ijk} - \hat{\rho}_i\right)\right]}{\sum_{m=1}^{M}\exp\left[s_m\left(-\sum_k \hat{\lambda}_{kj}x_{ijk} - \hat{\rho}_i\right)\right]} \equiv \frac{\exp\left[s_m\left(-\sum_k \hat{\lambda}_{kj}x_{ijk} - \hat{\rho}_i\right)\right]}{\Omega_{ij}(\hat{\lambda}, \hat{\rho})}, \tag{3.10}$$

and for $\mathbf{w}$

$$\hat{w}_{ijh} = \frac{\exp\left(-u_h\sum_k x_{ijk}\hat{\lambda}_{jk}\right)}{\sum_h \exp\left(-u_h\sum_k x_{ijk}\hat{\lambda}_{jk}\right)} \equiv \frac{\exp\left(-u_h\sum_k x_{ijk}\hat{\lambda}_{jk}\right)}{\Psi_{ij}(\hat{\lambda})} \tag{3.11}$$

where $\boldsymbol{\lambda}$ is the set of $K \times (J-1)$ Lagrange multiplier (estimated coefficients) associated with (Equation 3.8) and $\boldsymbol{\rho}$ is the N-dimensional vector of Lagrange multipliers associated with Equation 3.9a). Finally, $\hat{p}_{ij} = \sum_m s_m \hat{\pi}_{ijm}$ and $\hat{\varepsilon}_{ij} = \sum_h u_h \hat{w}_{ijh}$. These $\boldsymbol{\lambda}$'s are the $\boldsymbol{\alpha}$'s and $\boldsymbol{\beta}$'s defined and discussed in Section 3.1: $\boldsymbol{\lambda}' = (\boldsymbol{\alpha}', \boldsymbol{\beta}')$. We now can construct the concentrated entropy (profile likelihood) model which is just the dual version of the above constrained optimization model. This allows us to concentrate the model on the lower dimensional, real parameters of interest ($\boldsymbol{\lambda}$ and $\boldsymbol{\rho}$). That is, we move from the {P, W} space to the $\{\boldsymbol{\lambda}, \boldsymbol{\rho}\}$ space.

The concentrated entropy (likelihood) model is

$$\underset{\lambda,\rho}{\text{Min}} \left\{ -\sum_{ijk} y_{ij} x_{ijk} \lambda_{kj} + \sum_{i} \rho_i + \sum_{ij} \ln \Omega_{ij}(\lambda, \rho) + \sum_{ij} \ln \Psi_{ij}(\lambda) \right\}. \tag{3.12}$$

Solving with respect to $\boldsymbol{\lambda}$ and $\boldsymbol{\rho}$, we use Equation 3.10 and Equation 3.11 to get $\hat{\boldsymbol{\pi}}$ and $\hat{\mathbf{w}}$ that are then transformed to $\hat{\mathbf{p}}$ and $\hat{\boldsymbol{\varepsilon}}$.

Returning to the mixed logit (Mlogit) model discussed earlier, the set of parameters $\boldsymbol{\lambda}$ and $\boldsymbol{\rho}$ are the parameters in the individual utility functions (Equation 3.2 or 3.3) and represent both the population means and the random (individual) parameters. But unlike the simulated likelihood approach, no simulations are done here. Under this general criterion function, the objective is to minimize the joint entropy distance between the data and the state of complete ignorance (the uniform distribution or the uninformed empirical distribution). It is a dual-loss criterion that assigns equal weights to prediction (P) and precision (W). It is a shrinkage estimator that simultaneously shrinks the data and the noise to the center of their pre-specified supports. Further, looking at the basic primal (constrained) model, it is clear that the estimated parameters reflect not only the unknown parameters of the distribution, but also the amount of information in each one of the stochastic moments (Equation 3.8). Thus, λ_{kj} reflects the informational contribution of moment kj. It is the reduction in entropy (increase in information) as a result of incorporating that moment in the estimation. The $\boldsymbol{\rho}$'s reflect the individual effects.

As common to these class of models, the analyst is not (usually) interested in the parameters, but rather in the marginal effects. In the model developed here, the marginal effects (for the continuous covariates) are

$$\frac{\partial p_{ij}}{\partial x_{ijk}} = \sum_m s_m \frac{\partial \pi_{ijm}}{\partial x_{ijk}}$$

with

$$\frac{\partial \pi_{ijm}}{\partial x_{ijk}} = \pi_{ijm} \left(s_m \lambda_{kj} - \sum_m \pi_{ijm} s_m \lambda_{kj} \right)$$

and finally

$$\frac{\partial p_{ij}}{\partial x_{ijk}} = \sum_m s_m \left[\pi_{ijm} \left(s_m \lambda_{kj} - \sum_m \pi_{ijm} s_m \lambda_{kj} \right) \right].$$

3.4 Extensions and Discussion

So far in our basic model (Equation 3.12) we used discrete probability distributions (or similarly discrete spaces) and uniform (uninformed) priors. We now extend our basic model to allow for continuous spaces and for nonuniform priors. We concentrate here on the noise distributions.

3.4.1 Triangular Priors

Under the model formulated above, we maximize the joint entropies subject to our constraints. This model can be reconstructed as a minimization of the entropy distance between the (yet) unknown posteriors and some priors (subject to the same constraints). This class of methods is also known as "cross entropy" models (e.g., Kullback 1959; Golan, Judge, and Miller, 1996). Let, w^0_{ijh} be a set of prior (proper) probability distributions on $\mathbf{u}$. The normalization factors (partition functions) for the errors are now

$$\Psi_{ij} = \sum_h w^0_{ijh} \exp\left(u_h \sum_k x_{ijk}\lambda_{jk}\right)$$

and the concentrated IT criterion (Equation 3.12) becomes

$$\operatorname*{Max}_{\lambda,\rho}\left\{\sum_{ijk} y_{ij}x_{ijk}\lambda_{kj} - \sum_i \rho_i - \sum_{ij} \ln\Omega_{ij}(\lambda,\rho) - \sum_{ij} \ln\Psi_{ij}(\lambda)\right\}.$$

The estimated $\mathbf{w}$'s are:

$$\tilde{w}_{ijh} = \frac{w^0_{ijh}\exp\left(u_h \sum_k x_{ijk}\lambda_{jk}\right)}{\sum_h w^0_{ijh}\exp\left(u_h \sum_k x_{ijk}\lambda_{jk}\right)} \equiv \frac{w^0_{ijh}\exp\left(u_h \sum_k x_{ijk}\tilde{\lambda}_{jk}\right)}{\Psi_{ij}(\tilde{\lambda})}$$

and $\tilde{\varepsilon}_{ij} = \sum_h u_h \tilde{w}_{ijh}$. If the priors are all uniform ($w^0_{ijh} = 1/H$ for all i and j) this estimator is similar to Equation 3.12. In our model, the most reasonable prior is the triangular prior with higher weights on the center (zero) of the support $\mathbf{u}$. For example, if $H = 3$ one can specify $w^0_{ij1} = 0.25$, $w^0_{ij2} = 0.5$ and $w^0_{ij3} = 0.25$ or for $H = 5$, $\mathbf{w}^0 = (0.05, 0.1, 0.7, 0.1, 0.05)'$ or any other triangular prior the user believes to be consistent with the data generating process. Note that like the uniform prior, the a priori mean (for each ε_{ij}) is zero. Similarly, if such information exists, one can incorporate the priors for the signal. However, unlike the noise priors just formulated, we cannot provide here a natural source for such priors.

3.4.2 Bernoulli

A special case of our basic model is the Bernoulli priors. Assuming equal weights on the two support bounds, and letting $\eta_{ij} = \sum_k x_{ijk}\lambda_{jk}$ and u_1 is the support bound such that $\mathbf{u} \in [-u_1, u_1]$, then the errors' partition function is

$$\Psi(\lambda) = \prod_{ij} \frac{1}{2}\left(e^{\sum_k x_{ijk}\lambda_{jk}u_1} + e^{-\sum_k x_{ijk}\lambda_{jk}u_1}\right)$$
$$= \prod_{ij} \frac{1}{2}\left(e^{\eta_{ij}u_1} + e^{-\eta_{ij}u_1}\right) = \prod_{ij} \cosh(\eta_{ij}u_1).$$

Then Equation 3.12 becomes

$$\underset{\lambda,\rho}{\text{Max}}\left\{\sum_{ijk} y_{ij}x_{ijk}\lambda_{kj} - \sum_i \rho_i - \sum_{ij} \ln\Omega_{ij}(\lambda,\rho) - \sum_{ij} \ln\Psi_{ij}(\lambda)\right\}$$

where

$$\sum_{ij} \ln\Psi_{ij}(\lambda) = \sum_{ij} \ln\left[\frac{1}{2}\left(e^{\eta_{ij}u_1} + e^{-\eta_{ij}u_1}\right)\right] = \sum_{ij} \ln\cosh(\eta_{ij}u_1).$$

Next, consider the case of Bernoulli model for the signal $\boldsymbol{\pi}$. Recall that $s_m \in [0, 1]$ and let the priors weights be q_1 and q_2 on zero (s_1) and one (s_2), respectively. The signal partition function is

$$\Omega(\lambda,\rho) = \prod_{ij}\left(q_1 e^{s_1\left(\sum_k x_{ijk}\lambda_{jk}+\rho_i\right)} + q_2 e^{s_2\left(\sum_k x_{ijk}\lambda_{jk}+\rho_i\right)}\right)$$
$$= \prod_{ij}\left(q_1 + q_2 e^{\sum_k x_{ijk}\lambda_{jk}+\rho_i}\right) = \prod_{ij}\left(q_1 + q_2 e^{\eta_{ij}+\rho_i}\right)$$

and Equation 3.12 is now

$$\underset{\lambda,\rho}{\text{Max}}\left\{\sum_{ijk} y_{ij}x_{ijk}\lambda_{kj} - \sum_i \rho_i - \sum_{ij} \ln\Omega_{ij}(\lambda,\rho) - \sum_{ij} \ln\Psi_{ij}(\lambda)\right\}$$

where

$$\sum_{ij} \ln\Omega_{ij}(\lambda,\rho) = \sum_{ij} \ln\left[q_1 + q_2 e^{\eta_{ij}+\rho_i}\right].$$

Traditionally, one would expect to set uniform priors ($q_1 = q_2 = 0.5$).

3.4.3 Continuous Uniform

Using the same notations as above and recalling that $\mathbf{u} \in [-u_1, u_1]$, the errors' partition functions for continuous uniform priors are

$$\Psi_{ij}(\lambda) = \frac{e^{\eta_{ij}u_1} - e^{-\eta_{ij}u_1}}{2u_1\eta_{ij}} = \frac{\sinh(u_1\eta_{ij})}{u_1\eta_{ij}}.$$

The right-hand side term of Equation 3.12 becomes

$$\sum_{ij} \ln \Psi_{ij}(\lambda) = \sum_{ij} \left[\ln \left(\frac{1}{2} \left(e^{\eta_{ij} u_1} + e^{-\eta_{ij} u_1} \right) \right) - \ln \left(\eta_{ij} u_1 \right) \right]$$
$$= \sum_{ij} \left[\ln \left(\sinh \left(\eta_{ij} u_1 \right) \right) - \ln \left(\eta_{ij} u_1 \right) \right].$$

Similarly, and in general notations, for any uniform prior $[a, b]$, the signal partition function for each i and j is

$$\Omega_{ij}(\lambda, \rho) = \frac{e^{a(-\eta_{ij} - \rho_i)} - e^{b(-\eta_{ij} - \rho_i)}}{(b - a)\eta_{ij}}.$$

This reduces to

$$\Omega_{ij}(\lambda, \rho) = \frac{1 - e^{-\eta_{ij} - \rho_i}}{\eta_{ij}}$$

for the base case $[a, b] = [0, 1]$ which is the natural support for the signal in our model. The basic model is then

$$\underset{\lambda,\rho}{\text{Min}} \left\{ \sum_{ijk} y_{ij} x_{ijk} \lambda_{kj} - \sum_i \rho_i - \sum_{ij} \left[\ln \left(1 - e^{-\eta_{ij} - \rho_i} \right) - \ln \left(\eta_{ij} \right) \right] \right.$$
$$\left. - \sum_{ij} \left[\ln \left(\sinh \left(\eta_{ij} u_1 \right) \right) - \ln \left(2\eta_{ij} u_1 \right) \right] \right\}$$
$$= \underset{\lambda,\rho}{\text{Min}} \left\{ \sum_{ijk} y_{ij} x_{ijk} \lambda_{kj} - \sum_i \rho_i - \sum_{ij} \ln \Omega_{ij}(\lambda, \rho) - \sum_{ij} \ln \Psi_{ij}(\lambda) \right\}.$$

Finally, the estimator for P (the individuals' choices) is

$$\hat{p}_{ij} = \frac{1}{(b - a)} \left\{ \frac{a e^{a(-\hat{\eta}_{ij} - \hat{\rho}_i)} - b e^{b(-\hat{\eta}_{ij} - \hat{\rho}_i)}}{\hat{\eta}_{ij}} + \frac{e^{a(-\hat{\eta}_{ij} - \hat{\rho}_{ii})} - e^{b(-\hat{\eta}_{ij} - \hat{\rho}_i)}}{\hat{\eta}_{ij}^2} \right\}$$

for any $[a, b]$ and

$$\hat{p}_{ij} = \frac{-e^{-\hat{\eta}_{ij} - \hat{\rho}_i}}{\hat{\eta}_{ij}} + \frac{1 - e^{-\hat{\eta}_{ij} - \hat{\rho}_i}}{\hat{\eta}_{ij}^2}$$

for our problem of $[a, b] = [0, 1]$.

In this section we provided further detailed derivations and background for our proposed IT estimator. We concentrated here on prior distributions that seem to be consistent with the data generating process. Nonetheless, in some very special cases, the researcher may be interested in specifying other structures that we did not discuss here. Examples include normally distributed errors or possibly truncated normal with truncation points at −1 and 1. These imply normally distributed $\mathbf{w}_i$s within their supports. Though, mathematically, we can provide these derivations, we do not do it here as it does not seem to be in full agreement with our proposed model.

3.5 Inference and Diagnostics

In this section we provide some basic statistics that allow the user to evaluate the results. We do not develop here large sample properties of our estimator. There are two basic reasons for that. First, and most important, using the error supports $\mathbf{v}$ as formulated above, it is trivial to show that this model converges to the ML Logit. (See Golan, Judge, and Perloff, 1996, for the proof of the simpler IT-GME model.) Therefore, basic statistics developed for the ML logit are easily modified for our model. The second reason is simply that our objective here is to provide the user with the necessary tools for diagnostics and inference when analyzing finite samples.

Following Golan, Judge, and Miller (1996) and Golan (2008) we start by defining the information measures, or normalized entropies

$$S_1(\hat{\boldsymbol{\pi}}) \equiv \frac{-\sum_{ijm} \hat{\pi}_{ijm} \ln \hat{\pi}_{ijm}}{(N \times J)\ln(M)}$$

and

$$S_2(\hat{\boldsymbol{\pi}}_{ij}) \equiv \frac{-\sum_{m} \hat{\pi}_{ijm} \ln \hat{\pi}_{ijm}}{\ln(M)},$$

where both sets of measures are between zero and one, with one reflecting uniformity (complete ignorance: $\boldsymbol{\lambda} = \mathbf{0}$) of the estimates, and zero reflecting perfect knowledge. The first measure reflects the (signal) information in the whole system, while the second one reflects the information in each i and j. Similar information measures of the form $I(\hat{\boldsymbol{\pi}}) = 1 - S_j(\hat{\boldsymbol{\pi}})$ are also used (e.g., Soofi, 1994).

Following the traditional derivation of the (empirical) likelihood ratio test (within the likelihood literature), the empirical likelihood literature (Owen 1988, 1990, 2001; Qin and Lawless 1994), and the IT literature, we can construct an entropy ratio test. (For additional background on IT see also Mittelhammer, Judge, and Miller, 2000.) Let ℓ_Ω be the unconstrained entropy model Equation 3.12, and ℓ_ω be the constrained one where, say $\boldsymbol{\gamma}' = (\boldsymbol{\lambda}', \boldsymbol{\rho}') = \mathbf{0}$, or similarly $\boldsymbol{\beta} = \boldsymbol{\alpha} = \mathbf{0}$ (in Section 3.2). Then, the entropy ratio statistic is $2(\ell_\omega - \ell_\Omega)$. The value of the unconstrained problem ℓ_Ω is just the value of $\text{Max}\{H(\boldsymbol{\pi}, \mathbf{w})\}$, or similarly the maximal value of Equation 3.12, while $\ell_\omega = (N \times J)\ln(M)$ for uniform $\boldsymbol{\pi}$'s. Thus, the entropy-ratio statistic is just

$$W(IT) = 2(\ell_\omega - \ell_\Omega) = 2(N \times J)\ln(M)[1 - S_1(\hat{\boldsymbol{\pi}})].$$

Under the null hypothesis, W(IT) converges in distribution to $\chi^2_{(n)}$ where "n" reflects the number constraints (or hypotheses). Finally, we can derive the Pseudo-R^2 (McFadden 1974) which gives the proportion of variation in the data that is explained by the model (a measure of model fit):

$$\text{Pseudo-}R^2 \equiv 1 - \frac{\ell_\Omega}{\ell_\omega} = 1 - S_1(\hat{\pi}).$$

To make it somewhat clearer, the relationship of the entropy criterion and the χ^2 statistic can be easily shown. Consider, for example the cross entropy criterion discussed in Section 3.4. This criterion reflects the entropy distance between two proper distributions such as a prior and post-data (posterior) distributions. Let $I(\pi||\pi^0)$ be the entropy distance between some distribution $\boldsymbol{\pi}$ and its prior π^0. Now, with a slight abuse of notations, to simplify the explanation, let $\{\boldsymbol{\pi}\}$ be of dimension M. Let the null hypothesis be $H_0 : \pi = \pi^0$. Then,

$$\chi^2_{(M-1)} = \sum_m \frac{1}{\pi^0_m}(\pi_m - \pi^0_m)^2.$$

Looking at the entropy distance (cross entropy) measure$I\left(\pi||\pi^0\right)$ and formulating a second order approximation yields

$$I(\boldsymbol{\pi}||\boldsymbol{\pi}^0) \equiv \sum_m \pi_m \log(\pi_m/\pi^0_m) \cong \frac{1}{2}\sum_m \frac{1}{\pi^0_m}(\pi_m - \pi^0_m)^2$$

which is just the entropy (log-likelihood) ratio statistic of this estimator. Since 2 times the log-likelihood ratio statistic corresponds approximately to χ^2, the relationship is clear. Finally, though we used here a certain prior π^0, the derivation holds for all priors, including the uniform (uninformed) priors (e.g., $\pi_m = 1/M$) used in Section 3.3.

In conclusion, we stress the following: Under our IT-GME approach, one investigates how "far" the data pull the estimates away from a state of complete ignorance (uniform distribution). Thus, a high value of χ^2 implies the data tell us something about the estimates, or similarly, there is valuable information in the data. If, however, one introduces some priors (Section 3.4), the question becomes how far the data take us from our initial (a priori) beliefs — the priors. A high value of χ^2 implies that our prior beliefs are rejected by the data. For more discussion and background on goodness of fit statistics for multinomial type problems see Greene (2008). Further discussion of diagnostics and testing for ME-ML model (under zero moment conditions) appears in Soofi (1994). He provides measures related to the normalized entropy measures discussed above and provides a detailed formulation of decomposition of these information concepts. For detailed derivations of statistics for a whole class of IT models, including discrete choice models, see Golan (2008) as well as Good (1963). All of these statistics can be used in the model developed here.

3.6 Simulated Examples

Sections 3.3 and 3.4 have developed our proposed IT model and some extensions. We also discussed some of the motivations for using our proposed model, namely that it is semiparametric, and that it is not dependent on simulated likelihood approaches. It remains to investigate and contrast the IT model with its competitors. We provide a number of simulated examples for different sample sizes and different level of randomness. Among the appeals of the Mixed Logit, (RP) models is its ability to predict the individual choices. The results below include the in-sample and out-of-sample prediction tables for the IT models as well.

The out-of-sample predictions for the simulated logit is trivial and is easily done using NLOGIT (discussed below). For the IT estimator, the out-of-sample prediction involves estimating the $\boldsymbol{\rho}$'s as well. Using the first sample and the estimated $\boldsymbol{\rho}$'s from the IT model (as the dependent variables), we run a Least Squares model and then use these estimates to predict the out-of-sample $\boldsymbol{\rho}$'s. We then use these predicted $\boldsymbol{\rho}$'s and the estimated $\boldsymbol{\lambda}$'s from the first sample to predict out-of-sample.

3.6.1 The Data Generating Process

The simulated model is a five-choice setting with three independent variables. The utility functions are based on random parameters on the attributes, and five nonrandom choice specific intercepts (the last of which is constrained to equal zero). The random errors in the utility functions (for each individual) are iid extreme value in accordance with the multinomial logit specification. Specifically, $\mathbf{x}_1$ is a randomly assigned discrete (integer) uniform in [1, 5], $\mathbf{x}_2$ is from the uniform (0, 1) population and $\mathbf{x}_3$ is normal (0, 1). The values for the $\boldsymbol{\beta}$'s are: $\beta_{1i} = 0.3 + 0.2u_1$, $\beta_{2i} = -0.3 + 0.1u_2$, and $\beta_{3i} = 0.0 + 0.4u_3$, where u_1, u_2 and u_3 are iid normal (0, 1). The values for the choice specific intercept ($\boldsymbol{\alpha}$) are 0.4, 0.6, −0.5, 0.7 and 0.0 respectively for choices $j = 1, \ldots, 5$. In the second set of experiments, $\boldsymbol{\alpha}$'s are also random. Specifically, $\alpha_{ij} = \alpha_j + 0.5u_{ij}$, where u_j is iid normal(0,1) and $j = 1, 2, \ldots, 5$.

3.6.2 The Simulated Results

Using the software NLOGIT (Nlogit) for the MLogit model, we created 100 samples for the simulated log-likelihood model. We used GAMS for the IT-GME models – the estimator in NLOGIT was developed during this writing. For a fair comparison of the two different estimators, we use the correct model for the simulated likelihood (Case A) and a model where all parameters are taken to be random (Case B). In both cases we used the correct likelihood. For the IT estimator, we take all parameters to be random and there is no need for incorporating distributional assumptions. This means that if the IT dominates when it's not the correct model, it is more robust for the underlying

TABLE 3.1

In and Out-of-Sample Predictions for Simulated Experiments. All Values Are the Percent of Correctly Predicted

	N = 100 In/Out	N = 200 In/Out	N = 500 In/Out	N = 1000 In/Out	N = 1500 In/Out	N = 3000 In
Case 1: Random $\boldsymbol{\beta}$						
MLogit - A	29/28	34/38.5	34.4/33.6	35.5/33.3	34.6/34.0	33.8
MLogit - B	29/28	32.5/28.5	31.4/26.8	29.9/28.9	28.5/29	29.4
IT-GME*	41/23	35/34	33.6/35.6	36.4/34.6	34.4/33.9	34.8
Case 2: Random $\boldsymbol{\beta}$ and $\boldsymbol{\alpha}$						
MLogit	31/22	31/27	34.2/26.8	32/28.9	30.3/31.9	31
IT-GME*	45/29	40.5/29.5	38.4/32.4	37/34.2	37.1/34.9	36.3

Note: A: The correct model.
B: The incorrect model (both $\boldsymbol{\beta}$ and $\boldsymbol{\alpha}$ random).
*All IT-GME models are for both $\boldsymbol{\beta}$ and $\boldsymbol{\alpha}$ random.

structure of the parameters. The results are presented in Table 3.1. We note a number of observations regarding these experiments. First, the IT-GME model converges far faster than the simulated likelihood approach–since no simulation is needed, all expressions are in closed form. Second, in the first set of experiments (only the $\boldsymbol{\beta}$'s are random) and using the correct simulated likelihood model (Case 1A), both models provide very similar (on average) predictions, though the IT model is slightly superior. In the more realistic case, when the user does not know the exact model and uses RP for all parameters (Case 1B), the IT method is always superior. Third, for the more complicated data (generated with RP for both $\boldsymbol{\beta}$'s and $\boldsymbol{\alpha}$'s) – Case 2 – the IT estimator dominates for all sample sizes.

In summary, though the IT estimator seems to dominate for all samples and structures presented, it is clear that its relative advantage increases as the sample size decreases and as the complexity (number of random parameters) increases. From the analyst's point of view, it seems that for data with many choices and with much uncertainty about the underlying structure of the model, the IT is an attractive method to use. For the less complicated models and relatively large data sets, the simulated likelihood methods are proper (but are computationally more demanding and are based on a stricter set of assumptions).

3.7 Concluding Remarks

In this chapter we formulate and discuss an IT estimator for the mixed discrete choice model. This model is semiparametric and performs well relative to the class of simulated likelihood methods. Further, the IT estimator is computationally more efficient and is easy to use. This chapter is written in a way that makes it possible for the potential user to easily use this estimator. A detailed formulation of different potential priors and frameworks, consistent with the way we visualize the data generating process, is provided as well. We also provide the concentrated model that can be easily coded in some software.

References

Berry, S., J. Levinsohn and A. Pakes. 1995. Automobile Prices in Market Equilibrium. *Econometrica* 63(4): 841–890.

Bhat, C. 1996. *Accommodating Variations in Responsiveness to Level-of-Service Measures in Travel Mode Choice Modeling*. Department of Civil Engineering, University of Massachusetts, Amherst.

Golan, A. 2008. Information and Entropy Econometrics – A Review and Synthesis. *Foundations and Trends® in Econometrics* 2(1–2): 1–145.

Golan, A., G. Judge, and D. Miller. 1996. *Maximum Entropy Econometrics: Robust Estimation with Limited Data*. New York: John Wiley & Sons.

Golan, A., G. Judge, and J. Perloff. 1996. A Generalized Maximum Entropy Approach to Recovering Information from Multinomial Response Data. *Journal of the American Statistical Association* 91: 841–853.

Good, I.J. 1963. Maximum Entropy for Hypothesis Formulation, Especially for Multidimensional Contingency Tables. *Annals of Mathematical Statistics* 34: 911–934.

Greene, W.H. 2008. *Econometric Analysis*, 6th ed. Upper Saddle River, NJ: Prentice Hall.

Hensher D.A., and W.H., Greene. 2003. The Mixed Logit Model: The State of Practice. *Transportation* 30(2): 133–176.

Hensher, D.A., J. M. Rose, and W.H. Greene. 2006. *Applied Choice Analysis*. Cambridge, U.K.: Cambridge University Press.

Jain, D., N. Vilcassim, and P. Chintagunta. 1994. A Random-Coefficients Logit Brand Choice Model Applied to Panel Data. *Journal of Business and Economic Statistic* 12(3): 317–328.

Jaynes, E.T. 1957a. Information Theory and Statistical Mechanics. *Physics Review* 106: 620–630.

Jaynes, E.T. 1957b. Information Theory and Statistical Mechanics II. *Physics Review* 108: 171–190.

Jaynes, E.T. 1978. Where Do We Stand on Maximum Entropy. In *The Maximum Entropy Formalis*, eds. R.D. Levine and M. Tribus, pp. 15–118. Cambridge, MA: MIT Press.

Kullback, S. 1959. *Information Theory and Statistics*. New York: John Wiley & Sons.

McFadden, D. 1974. Conditional Logit Analysis of Qualitative Choice Behavior. In *Frontiers of Econometrics*, ed. P. Zarembka. New York: Academic Press, pp. 105–142.

Mittelhammer, R.C., G. Judge, and D. M. Miller. 2000. *Econometric Foundations*. Cambridge, U.K.: Cambridge University Press.

Owen, A. 1988. Empirical Likelihood Ratio Confidence Intervals for a Single Functional. *Biometrika* 75(2): 237–249.

Owen, A. 1990. Empirical Likelihood Ratio Confidence Regions. *The Annals of Statistics* 18(1): 90–120.

Owen, A. 2001. *Empirical Likelihood*. Boca Raton, FL: Chapman & Hall/CRC.

Qin, J., and J. Lawless. 1994. Empirical Likelihood and General Estimating Equations. *The Annals of Statistics* 22: 300–325.

Revelt, D., and K. Train. 1998. Mixed Logit with Repeated Choices of Appliance Efficiency Levels. *Review of Economics and Statistics* LXXX (4): 647–657.

Shannon, C.E. 1948. A Mathematical Theory of Communication. *Bell System Technical Journal* 27: 379–423.

Soofi, E.S. 1994. Capturing the Intangible Concept of Information. *Journal of the American Statistical Association* 89(428): 1243–1254.

Train, K.E. 2003. *Discrete Choice Methods with Simulation*. New York: Cambridge University Press.

Train, K., D. Revelt, and P. Ruud. 1996. *Mixed Logit Estimation Routine for Cross-Sectional Data.* UC Berkeley, http://elsa.berkeley.edu/Software/abstracts/train0196.html.

4

Recent Developments in Cross Section and Panel Count Models

Pravin K. Trivedi and Murat K. Munkin

CONTENTS

4.1 Introduction

Count data regression is now a well-established tool in econometrics. If the outcome variable is measured as a nonnegative count, $y, y \in \mathbb{N}_0 = \{0, 1, 2, \ldots\}$, and the object of interest is the marginal impact of a change in the variable x on the regression function $\mathrm{E}[y|x]$, then a count regression is a relevant tool of analysis. Because the response variable is discrete, its distribution places probability mass at nonnegative integer values only. Fully parametric formulations of count models accommodate this property of the distribution. Some semiparametric regression models only accommodate $y \geq 0$, but not discreteness. Given the discrete nature of the outcome variable, a linear regression is usually not the most efficient method of analyzing such data. The standard count model is a nonlinear regression.

Several special features of count regression models are intimately connected to discreteness and nonlinearity. As in the case of binary outcome models like the logit and probit, the use of count data regression models is very widespread in empirical economics and other social sciences. Count regressions have been extensively used for analyzing event count data that are common in fertility analysis, health care utilization, accident modeling, insurance, recreational demand studies, analysis of patent data.

Cameron and Trivedi (1998), henceforth referred to as CT (1998), and Winkelmann (2005) provided monograph length surveys of econometric count data methods. More recently, Greene (2007b) has also provided a selective survey of newer developments. The present survey also concentrates on newer developments, covering both the probability models and the methods of estimating the parameters of these models, as well as noteworthy applications or extensions of older topics. We cover specification and estimation issues at greater length than testing.

Given the length restrictions that apply to this article, we will cover cross-section and panel count regression but not time series count data models. The reader interested in time series of counts is referred to two recent survey papers; see Jung, Kukuk, and Liesenfeld (2006), and Davis, Dunsmuir, and Streett (2003). A related topic covers hidden Markov models (multivariate time series models for discrete data) that have been found very useful in modeling discrete time series data; see MacDonald and Zucchini (1997). This topic is also not covered even though it has connections with several themes that we do cover.

The natural stochastic model for counts is derived from the Poisson point process for the occurrence of the event of interest, which leads to Poisson distribution for the number of occurrences of the event, with probability mass function

$$\Pr[Y = y] = \frac{e^{-\mu}\mu^{y}}{y!}, \qquad y = 0, 1, 2, \ldots, \tag{4.1}$$

where μ is the intensity or rate parameter. The first two moments of this distribution, denoted $\mathcal{P}[\mu]$, are $\mathrm{E}[Y] = \mu$, and $\mathrm{V}[Y] = \mu$, demonstrating the well-known equidispersion property of the Poisson distribution. The Poisson regression follows from the parameterization $\mu = \mu(\mathbf{x})$, where $\mathbf{x}$ is a K-dimensional vector of exogenous regressors. The usual specification of the conditional mean is

$$\mathrm{E}[y|\mathbf{x}] = \exp(\mathbf{x}'\beta). \tag{4.2}$$

Standard estimation methods are fully parametric Poisson maximum likelihood, or "semiparametric" methods such as nonlinear least squares, or moment-based estimation, based on the moment condition $\mathrm{E}[y-\exp(\mathbf{x}'\beta)|\mathbf{x}] = \mathbf{0}$, possibly further augmented by the equidispersion restriction used to generate a weight function.

Even when the analysis is restricted to cross-section data with strictly exogenous regressors, the basic Poisson regression comes up short in empirical work in several respects. The mean-variance equality restriction is inconsistent with the presence of significant unobserved heterogeneity in cross-section data. This feature manifests itself in many different ways. For example, Poisson model often under-predicts the probability of zero counts, in a data situation often referred to as the *excess zeros* problem. A closely related deficiency of the Poisson is that in contrast to the equidispersion property, data more usually tend to be overdispersed, i.e., (conditional) variance usually exceeds the (conditional) mean. Overdispersion can result from many different sources (see CT, 1998, 97–106). Overdispersion can also lead to the problem of excess zeros (or *zero inflation*) in which there is a much larger probability mass at the zero value than is consistent with the Poisson distribution. The literature on new functional forms to handle overdispersion is already large and continues to grow. Despite the existence of a plethora of models for overdispersed data, a small class of models, including especially the negative binomial regression (NBR), the two-part model (TPM), and the zero-inflated Poisson (ZIP) and zero-inflated negative binomial (ZINB), has come to dominate the applied literature. In what follows we refer to this as the set of basic or benchmark parametric count regression models, previously comprehensively surveyed in CT (1998, 2005).

Beyond the cross-section count regression econometricians are also interested in applying count models to time series, panel data, as well as multivariate models. These types of data generally involve patterns of dependence more general than those for cross-section analysis. For example, serial dependence of outcomes is likely in time series and panel data, and a variety of dependence structures can arise for multivariate count data. Such data provide considerable opportunity for developing new models and methods.

Many of the newer developments surveyed here arise from relaxing the strong assumptions underlying the benchmark models. These new developments include the following:

- A richer class of models of unobserved heterogeneity some of which permit nonseparable heterogeneity
- A richer parameterization of regression functions
- Relaxing the assumption of conditional independence of $y_i|\mathbf{x}_i$ ($i = 1, \ldots, N$)
- Relaxing the assumption that the regressors $\mathbf{x}_i$ are exogenous
- Allowing for self-selection in the samples
- Extending the standard count regression to the multivariate case
- Using simulation-based estimation to handle the additional complications due to more flexible functional form assumptions

The remainder of the chapter is arranged as follows. Section 4.2 concentrates on extensions of the standard model involving newer functional forms. Section 4.3 deals with issues of cross-sectional dependence in count data. Section 4.4 deals with the twin interconnected issues of count models with endogenous regressors and/or self-selection. Sections 4.4 and 4.5 cover panel data and multivariate count models, respectively. The final Section 4.6 covers computational matters.

4.2 Beyond the Benchmark Models

One classic and long-established extension of the Poisson regression is the negative binomial (NB) regression. The NB distribution can be derived as a Poisson-Gamma mixture. Given the Poisson distribution $f(y|x, \nu) = \exp(-\mu\nu)(\mu\nu)^y/y!$ with the mean $\mathrm{E}[y|\mathbf{x}, \nu] = \mu(\mathbf{x})\nu$, $\nu > 0$, where the random variable ν, representing multiplicative unobserved heterogeneity, a latent variable, has Gamma density $g(\nu) = \nu^{\alpha-1}\exp(-\nu)/\Gamma(\alpha)$, with $\mathrm{E}[\nu] = 1$, and variance $\alpha (\alpha > 0)$. The resulting mixture distribution is the NB:

$$f(y|\mu(\mathbf{x})) = \int_0^\infty f(y|\mu(\mathbf{x}), \nu)g(\nu)d\nu = \frac{\mu(\mathbf{x})^y\Gamma(y+\alpha)}{y!\Gamma(\alpha)}\left(\frac{1}{\mu(\mathbf{x})+\alpha}\right)^{y+\alpha}, \tag{4.3}$$

which has mean $\mu(\mathbf{x})$ and variance $\mu(\mathbf{x})[\mathbf{1} + \alpha\mu(\mathbf{x})] > \mathrm{E}[y|\mathbf{x}]$, thus accommodating the commonly observed overdispersion. The gamma heterogeneity assumption is very convenient, but the same approach can be used with other mixing distributions.

This leading example imposes a particular mathematical structure on the model. Specifically, the latent variable reflecting unobserved heterogeneity is separable from the main object of identification, the conditional mean. This is a feature of many established mixture models. Modern approaches, however, deal with more flexible models where the latent variables are nonseparable. In such models unobserved heterogeneity impacts the entire disribution of the outcome of interest. Quantile regression and finite mixtures are two examples of such nonseparable models.

There are a number of distinctive ways of allowing for unobserved heterogeneity. It may be treated as an additive or a multiplicative random effect (uncorrelated with included regressors) or a fixed effect (potentially correlated with included regressors). Within the class of random effects models, heterogeneity distributions may be treated as continuous or discrete. Examples include a random intercept in cross-section and panel count models, fixed effects in panel models of counts. Second, both intercept and slope parameters may be specified to vary randomly and parametrically, as in finite mixture count models. Third, heterogeneity may be modeled in terms of both observed and unobserved variables using mixed models, hierarchical models and/or models of clustering. The approach one adopts and the manner in which it is combined with other assumptions has important implications for computation. The second and third approaches are reflected in many recent developments.

4.2.1 Parametric Mixtures

The family of random effects count models is extensive. In Table 4.1 we show some leading examples that have featured in empirical work. By far the most popular is the negative binomial specification with either a linear variance function (NB1) or a quadratic variance function (NB2). Both these functional forms capture extra-Poisson probability mass at zero and in the right tail, as would other mixtures, e.g., Poisson lognormal. But the continuing popularity of the NB family rests on computational convenience, even though (as we discuss later in this chapter) computational advances have made other models empirically accessible. When the right tail of the distribution is particularly heavy, the Poisson-inverse Gaussian mixture (P-IG) with a cubic variance function is attractive, but again this consideration must be balanced against additional computational complexity (see Guo and Trivedi 2002).

The foregoing models are examples of continuous mixture models based on a continuous distribution of heterogeneity. Mixture models that also allow for finite probability point mass, such as the hurdle ("two part") model and zero inflated models shown in Table 4.1, that appeared in the literature more than a decade ago (see Gurmu and Trivedi 1996) have an important advantage – they

TABLE 4.1

Selected Mixture Models

	Distribution	$f(y) = \Pr[Y = y]$	Mean; Variance
1	Poisson	$e^{-\mu}\mu^y/y!$	$\mu(\mathbf{x}); \mu(\mathbf{x})$
2	NB1	As in NB2 below with α^{-1} replaced by $\alpha^{-1}\mu$	$\mu(\mathbf{x}); (1+\alpha)\,\mu(\mathbf{x})$
3	NB2	$\frac{\Gamma(\alpha^{-1}+y)}{\Gamma(\alpha^{-1})\Gamma(y+1)}\left(\frac{\alpha^{-1}}{\alpha^{-1}+\mu}\right)^{\frac{1}{\alpha}}\left(\frac{\mu}{\mu+\alpha^{-1}}\right)^{y}$	$\mu(\mathbf{x}); (1+\alpha\mu(\mathbf{x}))\,\mu(\mathbf{x})$
4	P-IG	$\Pr(Y=0)\times\frac{\mu^k}{\Gamma(k+1)}(1+2\eta)^{-k/2}$	$\mu(\mathbf{x})\,;\,\mu\,(\mathbf{x})+\mu\,(\mathbf{x})^3/\tau$
	$k \geq 1$	$\times\sum_{i=0}^{k-1}\frac{\Gamma(k+i)}{\Gamma(k-i)\Gamma(i+1)}\left(\left(\frac{\eta}{2\mu(\mathbf{x})}\right)^{i}(1+2\eta)^{-i/2}\right.$,	
		where $\Pr(Y=0)=\exp\left[\frac{\mu}{\eta}(1-\sqrt{1+2\eta})\right]$, $(\eta=\mu^2/\tau)$	
5	Hurdle	$\begin{cases} f_1(0) & \text{if } y=0,\\ \frac{1-f_1(0)}{1-f_2(0)}f_2(y) & \text{if } y\geq 1.\end{cases}$	$\Pr[y>0\vert\mathbf{x}]\mathrm{E}_{y>0}[y\vert y>0,\mathbf{x}]$;
			$\Pr[y>0\vert\mathbf{x}]\mathrm{V}_{y>0}[y\vert y>0,\mathbf{x}]$ $+\Pr[y=0\vert\mathbf{x}]\mathrm{E}_{y>0}[y\vert y>0\vert\mathbf{x}]$
6	Zero-inflated	$\begin{cases} f_1(0)+(1-f_1(0))f_2(0) & \text{if } y=0,\\ (1-f_1(0))f_2(y) & \text{if } y\geq 1.\end{cases}$	$(1-f_1(0))(\mu(\mathbf{x})+f_1(0)\mu^2(\mathbf{x}))$
7	Finite mixture	$\sum_{j=1}^{m}\pi_j f_j(y\vert\theta_j)$	$\Sigma_{i=1}^2\pi_i\mu_i(\mathbf{x}); \Sigma_{i=1}^2\pi_i[\mu_i(\mathbf{x})+\mu_i^2(\mathbf{x})]$
8	PPp	$h_2(y\vert\mu,\mathbf{a})=\frac{e^{-\mu}\mu^y}{y!}\frac{(1+a_1y+a_2y^2)^2}{\eta_2(\mathbf{a},\mu)}$ where	
		$\eta_2(\mathbf{a},\mu)=1+2a_1m_1+(a_1^2+2a_2)m_2+2a_1a_2m_3+a_2^2m_4$	Complicated

relax the restrictions on both the conditional mean and variance functions. There are numerous ways of attaining such an objective using latent variables, latent classes, and a combination of these. This point is well established in the literature on generalized linear models. Skrondal and Rabe-Hesketh (2004) is a recent survey.

4.2.1.1 Hurdle and Zero-Inflated Models

Hurdle and zero-inflated models are motivated by the presence of "excess zeros" in the data. The hurdle model or two-part model (TPM) relaxes the assumption that the zeros and the positives come from the same data-generating process. Suppressing regressors for notational simplicity, the zeros are determined by the density $f_1(\cdot)$, so that $\Pr[y=0] = f_1(0)$ and $\Pr[y>0] = 1-f_1(0)$. The positive counts are generated by the truncated density $f_2(y|y>0) = f_2(y)/(1-f_2(0))$, that is multiplied by $\Pr[y>0]$ to ensure a proper distribution. Thus, $f(y) = f_1(0)$ if $y=0$ and $f(y) = [1-f_1(0)]f_2(y)/[1-f_2(0)]$ if $y \geq 1$. This generates the standard model only if $f_1(\cdot) = f_2(\cdot)$.

Like the hurdle model, zero-inflated model supplements a count density $f_2(\cdot)$ with a binary process with density $f_1(\cdot)$. If the binary process takes value 0, with probability $f_1(0)$, then $y=0$. If the binary process takes value 1, with probability $f_1(1)$, then y takes count values $0, 1, 2, \ldots$ from the count density $f_2(\cdot)$. This lets zero counts occur in two ways: either as a realization of the binary process or a count process. The zero-inflated model has density $f(y) = f_1(0) + [1-f_1(0)]f_2(0)$ if $y=0$, and $f(y) = [1-f_1(0)]f_2(y)$ if $y \geq 1$. As in the case of the hurdle model the probability $f_1(0)$ may be parameterized through a binomial model like the logit or probit, and the set of variables in the $f_1(\cdot)$ density may differ from those in the $f_2(\cdot)$ density.

4.2.1.1.1 Model Comparison in Hurdle and ZIP Models Zero-inflated variants of the Poisson (ZIP) and the negative binomial (ZINB) are especially popular. For the empirical researcher this generates an embarrassment of riches. The challenge comes from having to evaluate the goodness of fit of these models and selecting the "best" model according to some criterion, such as the AIC or BIC. It is especially helpful to have software that can simultaneously display the relevant information for making an informed choice. Care must be exercised in model selection because even when the models under comparison have similar overall fit, e.g., log-likelihood, they may have substantially different implications regarding the marginal effect parameters, i.e., $\partial \mathrm{E}[y|\mathbf{x}]/\partial x_j$. A practitioner needs suitable software for model interpretation and comparison.

A starting point in model selection is provided by a comparison of fitted probabilities of different models and the empirical frequency distribution of counts. Lack of fit at specific frequencies may be noticeable even in an informal comparison. Implementing a formal goodness-of-fit model comparison is easier when the rival models are nested, in which case we can apply a likelihood ratio test. However, some empirically interesting pairs of models are not nested, e.g., Poisson and ZIP, and negative binomial and ZINB. In these cases the so-called Vuong test (Vuong 1989), essentially a generalization of the likelihood ratio test, may be used to test the null hypothesis of equality of two distributions, say f and g. For example, consider the log of the ratio of fitted probabilities of Poisson and ZIP models, denoted $r_i = \ln\{\widehat{\Pr}_P(y_i|\mathbf{x}_i)/\widehat{\Pr}_{ZIP}(y_i|\mathbf{x}_i)\}$. Let $\bar{r} = N^{-1}\sum r_i$ and s_r denotes the standard deviation of r_i; then the test statistic $T_{vuong} = \bar{r}/(s_r/\sqrt{N})$ has asymptotic standard normal distribution. So the test can be based on the critical values of the standard normal. A large value of T_{vuong} in this case implies a departure from the null in the direction of Poisson, and a large negative value in the direction of ZIP. For other empirically interesting model pairs, e.g., ZIP and ZINB, the same approach can be applied, although it is less common for standard software to make these statistics available also. In such cases model selection information criteria such as the AIC and BIC are commonly used.

Two recent software developments have been very helpful in this regard. First, these models are easily estimated and compared in many widely used microeconometrics packages such as Stata and Limdep; see, for example, CT (2009) and Long and Freese (2006) for coverage of options available in Stata. For example, Stata provides goodness-of-fit and model comparison statistics in a convenient tabular form for the Poisson, NB2, ZIP, and ZINB. Using packaged commands it has become easy to compare the fitted and empirical frequency distribution of counts in a variety of parametric models. Second, mere examination of estimated coefficients and their statistical significance provides an incomplete picture of the properties of the model. In empirical work, a key parameter of interest is the average marginal effect (AME), $N^{-1}\sum_{i=1}^{N}\partial \mathrm{E}[y_i|\mathbf{x}_i]/\partial x_{j,i}$, or the marginal effect evaluated at a "representative" value of $\mathbf{x}$ (MER). Again, software developments have made estimation of these parameters very accessible.

4.2.1.2 Finite Mixture Specification

An idea that is not "recent" in principle, but has found much traction in recent empirical work of discrete or mixtures of count distributions. Unlike the NB model, which has a continuous mixture representation, the finite mixture approach instead assumes a discrete representation of unobserved heterogeneity. It encompasses both intercept and slope heterogeneity and hence the full distribution of outcomes. This generates a class of flexible parametric models called finite mixture models (FMM) – a subclass of latent class models; see Deb (2007), CT (2005, Chapter 20.4.3).

A FMM specifies that the density of y is a linear combination of m different densities, where the jth density is $f_j(y|\beta_j)$, $j = 1, 2, \ldots, m$. An m-component finite mixture is defined by

$$f(y|\beta, \pi) = \sum\nolimits_{j=1}^{m} \pi_j f_j(y|\beta_j), \quad 0 < \pi_j < 1, \sum\nolimits_{j=1}^{m} \pi_j = 1. \tag{4.4}$$

A simple example is a two-component ($m = 2$) Poisson mixture of $\mathcal{P}[\mu_1]$ and $\mathcal{P}[\mu_2]$. This may reflect the possibility that the sampled population contains two "types" of cases, whose y outcomes are characterized by distributions $f_1(y|\beta_1)$ and $f_2(y|\beta_2)$ that are assumed to have different moments. The mixing fraction π_1 is in general an unknown parameter. In a more general formulation it too can be parameterized in terms of observed variable(s) $\mathbf{z}$.

The FMM specification is attractive for empirical work in cross-section analysis because it is flexible. Mixture components may come from different parametric families, although commonly they are specified to come from the same family. The mixture components permit differences in conditional moments of the components, and hence in the marginal effects. In an actual empirical setting, the latent classes often have a convenient interpretation in terms of the differences between the underlying subpopulations.

Application of FMM to panel data is straightforward if the panel data can be treated as pooled cross section. However, when the T-dimension of a panel is high in the relevant sense, a model with fixed mixing probabilities may be tenuous as transitions between latent classes may occur over time. Endogenous switching models allow the transition probability between latent classes to be correlated with outcomes and hidden Markov models allow the transition probabilities to depend upon past states; see Fruhwirth-Schnatter (2006) and MacDonald and Zucchini (1997).

There are a number of applications of the FMM framework for cross-section data. Deb and Trivedi (1997) use Medical Expenditure Panel Survey data to study the demand for care by the elderly using models of two- and three-component mixtures of several count distributions. Deb and Trivedi (2002) re-examine the Rand Health Insurance Experiment (RHIE) pooled cross-section data and show that FMM fit the data better than the hurdle (two-part) model. Of course, this conclusion, though not surprising, is specific to their data set. Lourenco and Ferreira (2005) apply the finite mixture model to model doctor visits to public health centers in Portugal using truncated-at-zero samples. Bohning and Kuhnert (2006) study the relationship between mixtures of truncated count distributions and truncated mixture distributions and give conditions for their equivalence.

Despite its attractions, the FMM class has potential limitations. First, maximum likelihood (ML) estimation is not straightforward because, in general, the log-likelihood function may have multiple maxima. The difficulties are greater if the mixture components are not well separated. Choosing a suitable optimization algorithm is important. Second, it is easy to overparameterize mixture models. When the number of components is small, say 2, and the means of the component distribution are far apart, discrimination between the components is easier. However, as additional components are added, there is a tendency to "split the difference" and unambiguous identification of all components becomes difficult because of the increasing overlap in the distributions. In particular, the presence of outliers may give rise to components that account for a small proportion (small values of π_j) of the observations. That is, identification of individual components may be fragile. CT (2009, Chapter 17) give examples using Stata's FMM estimation (Deb, 2007) command and suggest practical ways of detecting estimation problems.

Recent biometric literature offers promise of more robust estimation of finite mixtures via alternative to maximum likelihood. Lu, Hui, and Lee (2003), following Karlis and Xekalaki (1998), use minimum Hellinger distance estimation (MHDE) for finite mixtures of Poisson regressions; Xiang et al. (2008) use MHDE for estimating a k-component Poisson regression with random effects. The attraction of MHDE relative to MLE is that it is expected to be more robust to the presence of outliers and when mixture components are not well separated, and/or when the model fit is poor.

4.2.1.3 Hierarchical Models

While cross-section and panel data are by far the most common in empirical econometrics, sometimes other data structures are also available. For example, sample survey data may be collected using a multi-level design; an example is state-level data further broken down by counties, or province-level data clustered by communes (see Chang and Trivedi [2003]). When multi-level covariate information is available, hierarchical modeling becomes feasible. Such models have been widely applied to the generalized linear mixed model (GLMM) class of which Poisson regression is a member. For example, Wang, Yau, and Lee (2002) consider a hierarchical Poisson mixture regression to account for the inherent correlation of outcomes of patients clustered within hospitals. In their set-up data are in m clusters, with each cluster having n_j ($j = 1, \ldots, m$) observations, let $n = \sum n_j$. For example, the following Poisson-lognormal mixture can be interpreted as a one-level hierarchical model.

$$\begin{aligned} y_{ij} &\sim \mathcal{P}(\mu_{ij}), \qquad i = 1, \ldots, n_j; \, j = 1, \ldots, m \\ \log \mu_{ij} &= \mathbf{x}'_{ij}\beta + \varepsilon_{ij}, \, \varepsilon_{ij} \sim \mathcal{N}(0, \sigma^2). \end{aligned} \tag{4.5}$$

An example of a two-level model, also known as a hierarchical Poisson mixture, that incorporates covariate information at both levels is as follows:

$$y_{ij} \sim \mathcal{P}(\mu_{ij}), \qquad i = 1, \ldots, n_j; \, j = 1, \ldots, m \tag{4.6}$$

$$\log \mu_{ij} = \mathbf{x}'_{ij}\beta_j + \varepsilon_{ij}, \quad \varepsilon_{ij} \sim \mathcal{N}(0, \sigma_\varepsilon^2)$$

$$\beta_{kj} = \mathbf{w}'_{kj}\gamma + v_{kj}; \quad v_{kj} \sim \mathcal{N}(0, \sigma_v^2), \; k = 1, \ldots K; \, j = 1, \ldots, m. \tag{4.7}$$

In this case coefficients vary by clusters, and cluster-specific variables $\mathbf{w}_{kj}$ enter at the second level to determine the first-level parameters β_j, whose elements are β_{kj}. The parameter vector γ, also called hyperparameter, is the target of statistical inference. Both classical (Wang, Yau, and Lee 2002) and Bayesian analyses can be applied.

4.2.2 Quantile Regression for Counts

Quantile regression (QR) is usually applied to continuous response data; see Koenker (2005) for a thorough treatment of properties of QR. QR is consistent under weak stochastic assumptions and is equivariant to monotone transformations. A major attraction of QR is that it potentially allows for response heterogeneity at different conditional quantiles of the variables of interest. If the method could be extended to counts, then one could go beyond the standard and somewhat restrictive models of unobserved heterogeneity based on strong distributional assumptions. Also QR facilitates a richer interpretation of the data because it permits the study of the impact of regressors on both the location and scale parameters of the model, while at the same time avoiding strong distributional assumptions about data. Moreover, advances made in quantile regression such as handling endogenous regressors can be exploited for count data. The problem, however, is that the quantiles of discrete variables are not unique since the c.d.f. is discontinuous with discrete jumps between flat sections. By convention the lower boundary of the interval defines the quantile in such a case. However, recent theoretical advances have extended QR to a special case of count regression; see Machado and Santos Silva (2005), Miranda (2006, 2008), Winkelmann (2006).

The key step in the quantile count regression (QCR) model of Machado and Santos Silva (2005) involves replacing the discrete count outcome y with a continuous variable $z = h(y)$, where $h(\cdot)$ is a smooth continuous transformation. The standard linear QR methods are then applied to z. The particular continuation transformation used is $z = y + u$, where $u \sim \mathcal{U}[0, 1]$ is a pseudo-random draw from the uniform distribution on $(0, 1)$. This step is called "jittering" the count. Point and interval estimates are then retransformed to the original y-scale using functions that preserve the quantile properties.

Let $Q_q(y|\mathbf{x})$ and $Q_q(z|\mathbf{x})$ denote the qth quantiles of the conditional distributions of y and z, respectively. The conditional quantile for $Q_q(z|\mathbf{x})$ is specified to be

$$Q_q(z|\mathbf{x}) = q + \exp(\mathbf{x}'\beta_q). \tag{4.8}$$

The additional term q appears in the equation because $Q_q(z|\mathbf{x})$ is bounded from below by q, due to the jittering operation.

To be able to estimate a quantile model in the usual linear form $\mathbf{x}'\beta$, a log transformation is applied so that $\ln(z - q)$ is modelled, with the adjustment that if $z - q < 0$ then we use $\ln(\varepsilon)$ where ε is a small positive number. The transformation is justified by the equivariance property of the quantiles and the property that quantiles above the censoring point are not affected by censoring from below. Post-estimation transformation of the z-quantiles back to y-quantiles uses the ceiling function, with

$$Q_q(y|\mathbf{x}) = \lceil Q_q(z|\mathbf{x}) - 1 \rceil, \tag{4.9}$$

where the symbol $\lceil r \rceil$ in the right-hand side of Equation 4.9 denotes the smallest integer greater than or equal to r.

To reduce the effect of noise due to jittering, the model is estimated multiple times using independent draws from $\mathcal{U}(0, 1)$ distribution, and the multiple estimated coefficients and confidence interval endpoints are averaged.

Hence the estimates of the quantiles of y counts are based on $\widehat{Q}_q(y|\mathbf{x}) = \lceil Q_q(z|\mathbf{x}) - 1\rceil = \lceil q + \exp(\mathbf{x}'\bar{\bar{\beta}}_q) - 1\rceil$, where $\bar{\bar{\beta}}$ denotes the average over the jittered replications.

Miranda (2008) applies the QCR to analysis of Mexican fertility data. Miranda (2006) describes Stata's add-on `qcount` command for implementing QCR. CT (2009, Chapter 7.5) discuss an empirical illustration in detail, with special focus on marginal effects. The specific issue of how to choose the quantiles is discussed by Winkelmann (2006), the usual practice being to select a few values such as q equal to 25, .50, and .75. This practice has to be modified to take account of the zeros problem because it is not unusual to have (say) 35% zeros in a sample, in which case q must be greater than .35.

4.3 Adjusting for Cross-Sectional Dependence

The assumption of cross-sectionally independent observations was common in the econometric count data literature during and before the 1990s. Recent theoretical and empirical work pays greater attention to the possibility of cross-sectional dependence. Two sources of dependence in cross-sectional data are stratified survey sampling design and, in geographical data, dependence due spatially correlated unobserved variables.

Contrary to a common assumption in cross-section regression, count data used in empirical studies are more likely to come from complex surveys derived from stratified sampling. Data from the stratified random survey samples, also known as complex surveys, are usually dependent. This may be due to use of survey design involving interviews with multiple households in the same street or block that may be regarded as natural clusters, where by cluster is meant a set whose elements are subject to common shocks. Such a sampling scheme is likely to generate correlation within cluster due to variation induced by common unobserved cluster-specific factors. Cross-sectional dependence between outcomes invalidates the use of variance formulae based on assumption of simple random samples.

Cross-sectional dependence also arises when the count outcomes have a spatial dimension, as when the data are drawn from geographical regions. In such cases the outcomes of units that are spatially contiguous may display dependence that must be controlled for in regression analysis.

There are two broad approaches for controlling for dependence within cluster, the key distinction being between random and fixed cluster effects analogous to panel data analysis.

4.3.1 Random Effects Cluster Poisson Regression

To clarify this point additional notation is required. Consider a sample with total N observations, which are distributed in C clusters with each cluster having $N_c(c = 1, \ldots, C)$ observations and $\sum_{c=1}^{C} N_c = N$. If the number of observations per cluster varies, the data correspond to an unbalanced panel. Intra-cluster correlation refers to correlation between $y_{i,c}$ and $y_{j,c}$, $i \neq j$, $c = 1, \ldots, C$. A common assumption is of nonzero intra-cluster correlation and zero between-cluster correlation, i.e., corr $[y_{i,c}, y_{j,c'}, i \neq j, c \neq c'] = 0$. Additional complications arise according to the assumptions regarding N_c and C, i.e., whether there are many small clusters or few large clusters. The notation for handling clusters is similar to that for panel data, and a number of important results we cover also parallel similar ones in the panel data literature.

For specificity, we consider the Poisson regression for clustered data. A popular assumption states that the cluster-specific effects enter the model through the intercept term alone. The clustered count data, denoted $y_{ij}, i = 1, \ldots, C$, $j = 1, \ldots, N_j$, are Poisson distributed with $\mathrm{E}[y_{ij}|\mathbf{x}_{ij}, \alpha_i] = \exp(\alpha_i + \alpha + \mathbf{x}_{ij}'\beta)$, where $\mathbf{x}_{ij}$ are linearly independent covariates; the term α_i is the deviation of cluster-specific intercept from the population-averaged fixed intercept α. This model is referred to as cluster-specific intercept Poisson regression. A number of results are available corresponding to different assumptions about α_i. Demidenko (2007) presents several results for the case in which $\alpha_i (i = 1, \ldots, C)$ are i.i.d., C approaches ∞, and $N_c < \infty$. In the econometrics literature this corresponds to the cluster-specific random effects (CSRE) Poisson regression.

1. Under the assumptions stated above, the standard M-estimators for the Poisson regression applied to pooled clustered data are consistent but not efficient.
2. In the case where C is relatively small, e.g., $C < \min N_i$, a separate dummy variable corresponding to each cluster-specific intercept can be introduced in the conditional mean function and standard maximum likelihood procedure can be applied to the resulting model.
3. Under a specific assumption about the conditional covariance structure for the data, a generalized estimating equations (GEE), or (in econometrics terminology) nonlinear generalized least squares, procedure may be applied for efficiency gain over the simple Poisson. This requires a working matrix as an estimator of the unknown true variance matrix. Under the assumption that the cluster-random effects are equicorrelated, the working matrix can be parameterized in terms of a single correlation parameter.
4. Under a strong parametric assumption about the distribution of α_i, maximum likelihood can be applied. Computationally this is more demanding and may require simulation-based estimation.
5. Under somewhat special assumptions in which all clusters have the same number of observations, and the covariates are identical across clusters the methods mentioned above are all equivalent.

4.3.2 Cluster-Robust Variance Estimation

An approach that does not require a distributional assumption for the random component is to simply use robust variance estimation, i.e., "cluster robust" standard errors obtained by adapting the so-called Eicker–White robust variance estimator to handle clustered data. Specifically, if f_i $(i = 1, \ldots, n)$ denotes the density for the ith observation, θ denotes the vector of unknown parameters, then the cluster-robust variance estimator evaluated at the maximum likelihood estimate $\widehat{\theta}_{MLE}$ is given by

$$V_C = \left[\sum_{j=1}^{C}\sum_{i=1}^{N_j}\frac{\partial^2 \ln f_{ij}}{\partial\theta\partial\theta'}\right]^{-1}\left[\sum_{j=1}^{C}\sum_{i=1}^{N_j}\sum_{k=1}^{N_j}\frac{\partial \ln f_{ij}}{\partial\theta}\frac{\partial \ln f_{kj}}{\partial\theta'}\right] \times \left[\sum_{j=1}^{C}\sum_{i=1}^{N_j}\frac{\partial^2 \ln f_{ij}}{\partial\theta\partial\theta'}\right]^{-1}\Bigg|_{\widehat{\theta}_{MLE}}. \tag{4.10}$$

If, instead of ML estimation, another consistent M-estimator is used, e.g., that defined by the nonlinear estimating equations $\sum_{c=1}^{C}\sum_{j=1}^{N_c}\mathbf{h}(y_{jc}, \mathbf{x}_{jc}, \theta) = \mathbf{0}$, the above formula is adjusted by replacing the score function $\partial \ln f_{ij}/\partial\theta$ by $\mathbf{h}_{i,j}(\widehat{\theta})$; see CT (2005, Chapter 24.5.6). Observe how within each cluster we do not use the likelihood score for each observation as in the case of independent observations; instead we replace it by the sum of likelihood scores over all cluster elements. The usual regularity conditions for the validity of the "sandwich" variance formula are required.

4.3.3 Cluster-Specific Fixed Effects

The Poisson fixed effects cluster model specifies

$$\begin{aligned} y_{ij} &\sim \mathcal{P}[\mu_{ij}], \quad i = 1, \ldots, C, \; j = 1, \ldots, N_i, \\ \mu_{ij} &= \alpha_i \exp(\mathbf{x}'_{ij}\beta), \end{aligned} \tag{4.11}$$

where $\mathbf{x}_{ij}$ excludes an intercept and any cluster-invariant regressors. The difference from the standard Poisson model is that the usual conditional mean $\exp(\mathbf{x}'_{ij}\beta)$ is scaled multiplicatively by the cluster-specific fixed effect (FE) α_i. Because of its similarity to the panel Poisson model, we defer a longer treatment of estimation to Section 4.5, but simply note that one can use either the conditional maximum likelihood approach in which inference is carried out conditionally on sufficient statistics for the fixed effects (i.e., the parameters α_i are eliminated), or we can introduce cluster-specific dummy variables and apply the standard ML estimation.

4.3.4 Spatial Dependence

We now consider models in which outcomes are counts with a spatial distribution; hence it becomes necessary to adjust for spatial correlation between neighboring counts. Such spatial dependence is characterized by some underlying data generating process. Griffith and Haining (2006) survey the early literature on spatial Poisson regression and a number of its modern extensions.

Besag (1974) defined a general class of "auto-models" suitable for different types of spatially correlated data. In the special case, dubbed the "auto-Poisson" model, $\mathcal{P}[Y(i) = y(i)|\{Y(j)\}, j \in \mathsf{N}(i)]$ denotes the conditional probability that the random variable $Y(i)$, defined at location i, realizes value $y(i)$, given the values of Y at the sites in the neighborhood of i, denoted $\mathsf{N}(i)$. If the $\{Y(i)\}$ have an auto-Poisson distribution with intensity parameter $\mu(i)$, then

$$\mathcal{P}[Y(i) = y(i)|\{Y(j)\}, j \in \mathsf{N}(i)] = \frac{e^{-\mu(i)}\mu(i)^{y(i)}}{y(i)!}$$

$$\log \mu(i) = \alpha(i) + \sum_{j \in \mathsf{N}(i)} \beta(i, j) y(j)) \tag{4.12}$$

where the parameter $\alpha(i)$ is an area-specific effect, and $\beta(i, j) = \beta(j, i)$. The standard set-up specified $\beta(i, j) = \gamma w(i, j)$, where γ is a spatial autoregressive parameter, and $w(i, j)$ (= 0 or 1) represents the neighborhood structure. Let $\mathsf{N}(i)$ denote the set of neighbors of area i; then $w(i, j) = 1$ if i and j are neighbors $[j \in \mathsf{N}(i)]$, and otherwise $w(i, j) = 0$. In addition, $w(i, i) = 0$ must be assumed. The difficulty with this auto-Poisson model is under the restriction $\sum_i \mathcal{P}[Y(i) = y(i)|\{Y(j)\}, j \in \mathsf{N}(i)] = 1$, implies $\gamma \leq 0$, which implies dependence property with negative spatial correlation. The spatial count literature has evolved in different directions to overcome this difficulty; see Kaiser and Cressie (1997) and Griffith and Haining (2006). One line of development uses spatial weighting dependent on assumptions about the spatial dependence structure.

A different approach for modeling either positive or negative spatial autocorrelation in a Poisson model is based on a mixture specification. The conditional mean function is defined as

$$\ln[\mu(i)] = \alpha(i) + S(i) \tag{4.13}$$

where $S(i)$ is a random effect defined by a conditional autoregressive model with a general covariance structure. For example, an N-dimensional multivariate normal specification with covariance matrix $\sigma^2(\mathbf{I}_N - \tau\mathbf{W})^{-1}$, where $\mathbf{W} = [\mathbf{w}(i, j)]$ is the spatial weighting matrix, can capture negative or positive dependence between outcomes through the spatial autoregressive parameter τ. The resulting model is essentially a Poisson mixture model with a particular dependence structure. Estimation of this mixture specification by maximum likelihood would typically require simulation because its likelihood will not be expressed in a closed form. Griffith and Haining (2006) suggest Bayesian Markov chain Monte Carlo (MCMC); the use of this method is illustrated in Subsection 4.6.3. They compare the performance of several alternative estimators for the Poisson regression applied to georeferenced data.

Spatial dependence modeled using a mixture specification induces overdispersion. If the conditional mean is correctly specified, the resulting model can be consistently estimated using the Poisson model under pseudo-likelihood. Robust variance estimator should be used to adjust for the effects of overdispersion. However, an advantage of estimating the full specification is that one obtains information about the structure of spatial dependence.

4.4 Endogeneity and Self-Selection

Endogenous regressors, both categorical and continuous, arise naturally in many count regression models. A well-known example from health economics involves models of counts of health services, e.g., doctor visits, with one of the regressors being the health insurance status of the individual. Assumption that choice of health insurance and the count outcome equation are conditionally uncorrelated is unrealistic when data are observational and insurance status is not exogenously assigned, rather it is self-selected. The case of endogenous dummy variables occurs commonly in empirical work and thus will get special attention in this section.

Important earlier analyses of count models with endogenous regressors include Mullahy (1997) and Windmeier and Santos Silva (1997) who proposed moment-based estimators within a GMM framework, and Terza (1998) who provided a full-information parametric analysis as well as a "semiparametric" sequential two-step estimator. They are motivated by a desire for more robust estimators. Discreteness of count outcomes, and often also that of the endogenous regressor, is typically ignored.

4.4.1 Moment-Based Estimation

Consider the exponential mean model $\mathrm{E}[y_i|\mathbf{x}_i] = \exp(\mathbf{x}_i'\beta)$, where at least one component of $\mathbf{x}$ is endogenous. To introduce endogeneity into this model the first step is to introduce another source of randomness in the specification. This is done using unobserved heterogeneity, either additively or multiplicatively. For example, Mullahy specified the moment as $\mathrm{E}[y_i|\mathbf{x}_i, \nu_i] = \exp(\mathbf{x}_i'\beta)\nu_i$, where $\mathrm{E}[\nu_i] = 1$, with $(\mathbf{x}_i, \nu_i)$ being jointly dependent. Instead of introducing further parametric assumptions about the dependence structure of $(\mathbf{x}_i, \nu_i)$, the GMM approach postulates the availability of a vector of instruments, $\mathbf{z}$, $\dim(\mathbf{z}) \geq \dim(\mathbf{x})$, that satisfy the moment condition

$$\mathrm{E}[y_i \exp(-\mathbf{x}_i'\beta) - v_i|\mathbf{z}_i] = \mathbf{0}. \tag{4.14}$$

By assumption, $\mathbf{z}$ is orthogonal to ν_i, and may have common elements with $\mathbf{x}$, but has some distinct elements that are excluded from $\mathbf{x}$. The instruments are assumed to be valid and relevant, in the sense that there is nontrivial correlation between $\mathbf{z}$ and $\mathbf{x}$. Note that this moment condition is different from that for the case in which ν_i enters the conditional mean additively – IVs that are orthogonal under multiplicative error are not so in general under additive errors.

A specific feature of the count model is discreteness and heteroskedasticity. GMM estimation of this model usually ignores the first feature; however, two-step efficient GMM estimators can accommodate heteroskedasticity. This topic has been surveyed in CT (2005) who also provide a discussion of the practical aspects of implementing efficient GMM estimation, but this discussion is broader in scope and not just for count regressions.

4.4.2 Control Function Approach

Since endogenous regressors cause significant complication in estimation of nonlinear models, one strategy is to first test for the presence of endogenous regressors, and then to use a suitable new estimator only if the null hypothesis of zero correlation between the regressors and the equation error is rejected. When the estimator is defined by a moment condition, the equation error is also implicitly defined by it. The error term, say u, is explicitly defined in linear two-stage least squares. There, and in the related literature on Durbin–Wu–Hausman tests of endogeneity (see Davidson and MacKinnon 2004, Chapter 8.7), the following test procedure is recommended. Suppose the regression of interest with dependent variable y_1 has a scalar right-hand side variables y_2 and exogenous variables $\mathbf{X}$. Let $\mathbf{W} = [\mathbf{Z}\ \mathbf{X}]$ denote the set of instrumental variables. Let $\mathbf{P}_W y_2$ be the linear projection of y_2 on $\mathbf{W}$. Then a test of endogeneity of y_2 is a test of $H_0 : \delta = 0$ in the OLS regression $y_1 = \mathbf{X}\beta_1 + y_2\beta_2 + \mathbf{P}_W y_2 \delta + u$. Because of the least squares identity $y_2 \equiv P_W y_2 + \widehat{v}_2$, this procedure is equivalent to testing $H_0 : \delta^* = 0$ in the regression

$$y_1 = \mathbf{X}\beta_1 + (\beta_2 + \delta^*) y_2 - \delta^* \widehat{v} + u,$$

in which the right-hand side is augmented by the reduced form residual $\widehat{v}$. If the null hypothesis is rejected, this OLS regression is equivalent to the standard two-stage least squares. In essence, adding the variable $\widehat{v}_2$ controls for the endogeneity of y_2; once it is included the standard OLS estimator yields consistent point estimates. Hence we refer to this approach as the control function approach.

An interesting question is whether this approach can be extended to standard count regression models. Differences from the two-stage least squares case are due to the exponential conditional mean function and multiplicative error term. Terza (1998) considered a bivariate model in which the counted variable y depends on exogenous variables $\mathbf{x}$ and an endogenous treatment dummy variable $d (\equiv y_2)$. He provided both maximum likelihood and a two-step ("semiparametric") estimator for this model. His two-step estimator can be interpreted as a control function estimator.

Hardin, Schmiediche, and Carroll (2003) propose a method for estimating the Poisson regression with endogenous regressors that can also be interpreted as a control function type approach. Their method is intended to apply to models in the linear exponential family, Poisson regression being a special case. A linear-reduced form regression is estimated for the endogenous variable. As in two-stage least squares, valid instruments, $\mathbf{x}_2$, are assumed to be available. The linear-reduced form regression, estimated by OLS, generates predicted values for the endogenous variable y_2, denoted $\widehat{y}_2$. The original Poisson regression is estimated after replacing the endogenous variable by its predicted value. The variances are obtained using a bootstrap to allow for the fact that $\widehat{y}_2$ is a generated regressor subject to sampling variability.

Formally, the Poisson regression is estimated given the conditional mean function

$$E(y|\widehat{y}_2, x_1) = \mu = \exp(\beta_1\widehat{y}_2 + \mathbf{x}_1'\beta_2), \tag{4.15}$$

where $\widehat{y}_2 = \mathbf{x}_1'\widehat{\gamma}_1 + \mathbf{x}_2'\widehat{\gamma}_2$. This approach does not combine testing for endogeneity with estimation; it instead estimates the model assuming endogeneity. Despite its similarity with moment-based methods, the basis of the moment condition being used is not clear.

Finally we consider another somewhat ad hoc fitted-value method that resembles two-stage least squares that has been used in the context of Poisson regression with one endogenous dummy variable. This set-up is common in empirical work. Consider the overdispersed Poisson model,

$$\begin{aligned} y_i &\sim \mathcal{P}[\mu_i\eta_i], i = 1, \ldots, N \\ \mathrm{E}[y_i|\mu_i, \eta_i] &= \mu_i\eta_i \\ &= \exp(\mathbf{x}_i'\beta + \gamma d_i + \varepsilon_i), \end{aligned} \tag{4.16}$$

where $\eta_i = \exp(\varepsilon_i)$, d_i is the endogenous dummy variable, ε_i is unobserved heterogeneity uncorrelated with $\mathbf{x}_i'$.

Suppose $\mathbf{z}_i$ be a set of valid instruments, $\dim(\mathbf{z}_i) > \dim(\mathbf{x}_i)$. Assume $\mathrm{E}[y_i - \mu_i|\mathbf{z}_i] = \mathbf{0}$, but $\mathrm{E}[y_i - \mu_i|\mathbf{x}_i] \neq \mathbf{0}$. Consider the following two-step estimator: (1) generate a fitted value $\widehat{d}_i(\mathbf{z}_i)$ from a "reduced form" of d_i, using instruments $\mathbf{z}_i$; (2) replace d_i by $\widehat{d}_i(\mathbf{z}_i)$ and estimate the new Poisson regression by MLE. Though appealing in its logic, it is not clear that the two-step estimator is consistent. Let $d_i = \widehat{d}_i(\mathbf{z}_i) + \widehat{v}_i$, where $\widehat{d}_i(\mathbf{z}_i)$ is a predicted ("fitted") value of d_i from linear regression on $\mathbf{z}_i$. Hence,

$$\begin{aligned} \mathrm{E}(y_i|\widehat{d}_i, \widehat{v}_i) &= \exp(\mathbf{x}_i'\beta + \gamma\widehat{d}_i(\mathbf{z}_i) + \varepsilon_i + \gamma\widehat{v}_i) \\ &= \exp(\mathbf{x}_i'\beta + \gamma\widehat{d}_i(\mathbf{z}_i)) \exp(\varepsilon_i + \gamma\widehat{v}_i). \end{aligned} \tag{4.17}$$

Consistency requires $\mathrm{E}[\exp(\varepsilon_i + \gamma\widehat{v}_i)|\mathbf{x}_i, \widehat{d}_i] = \mathbf{0}$, but it is not obvious that this condition can be satisfied regardless of the functional form used to generate $\widehat{d}_i(\mathbf{z}_i)$.

4.4.3 Latent Factor Models

An alternative to the above moment-based approaches is a pseudo-FIML approach of Deb and Trivedi (2006a) who consider models with count outcome and endogenous treatment dummies. The model is used to study the impact of health insurance status on utilization of care. Endogeneity in these models arises from the presence of common latent factors that impact both the choice of treatments a (interpreted as treatment variables) and the intensity of utilization (interpreted as an outcome variable). The specification is consistent with selection on unobserved (latent) heterogeneity. In this model the endogenous variables in the count outcome equations are categorical, but the approach can be extended to the case of continuous variables.

The model includes a set of J dichotomous treatment variables that correspond to insurance plan dummies. These are endogenously determined by mixed multinomial logit structure (MMNL)

$$\Pr(\mathbf{d}_i|\mathbf{z}_i, \mathbf{l}_i) = \frac{\exp(\mathbf{z}_i'\alpha_j + \delta_j l_{ij})}{1 + \sum_{k=1}^{J}\exp(\mathbf{z}_i'\alpha_k + \delta_k l_{ik})}. \tag{4.18}$$

where d_j is observed treatment dummies, $\mathbf{d}_i = [d_{i1}, d_{i2}, \ldots, d_{iJ}]$, $j = 0, 1, 2, \ldots, J$, $\mathbf{z}_i$ is exogenous covariates, $\mathbf{l}_i = [l_{i1}, l_{i2}, \ldots, l_{iJ}]$, and l_{ij} are latent or unobserved factors.

The expected outcome equation for the counted outcomes is

$$\mathrm{E}(y_i|\mathbf{d}_i, \mathbf{x}_i, \mathbf{l}_i) = \exp\left(\mathbf{x}_i'\beta + \sum\nolimits_{j=1}^{J}\gamma_j d_{ij} + \sum\nolimits_{j=1}^{J}\lambda_j l_{ij}\right), \tag{4.19}$$

where $\mathbf{x}_i$ is a set of exogenous covariates. When the factor loading parameter $\lambda_j > 0$, treatment and outcome are positively correlated through unobserved characteristics, i.e., there is positive selection. Deb and Trivedi (2006a) assume that the distribution of y_i is negative binomial

$$f(y_i|\mathbf{d}_i, \mathbf{x}_i, \mathbf{l}_i) = \frac{\Gamma(y_i + \psi)}{\Gamma(\psi)\Gamma(y_i + 1)}\left(\frac{\psi}{\mu_i + \psi}\right)^{\psi}\left(\frac{\mu_i}{\mu_i + \psi}\right)^{y_i}, \tag{4.20}$$

where $\mu_i = \mathrm{E}(y_i|\mathbf{d}_i, \mathbf{x}_i, \mathbf{l}_i) = \exp(\mathbf{x}_i'\beta + \mathbf{d}_i'\gamma + \mathbf{l}_i'\lambda)$ and $\psi \equiv 1/\alpha$ $(\alpha > 0)$ is the overdispersion parameter.

The parameters in the MMNL are only identified up to a scale. Hence a scale normalization for the latent factors is required; accordingly, they set $\delta_j = 1$ for each j. Although the model is identified through nonlinearity when $\mathbf{z}_i = \mathbf{x}_i$, they include some variables in $\mathbf{z}_i$ that are not included $\mathbf{x}_i$.

Joint distribution of treatment and outcome variables is

$$\begin{aligned}\Pr(y_i, \mathbf{d}_i|\mathbf{x}_i, \mathbf{z}_i, \mathbf{l}_i) &= f(y_i|\mathbf{d}_i, \mathbf{x}_i, \mathbf{l}_i) \times \Pr(\mathbf{d}_i|\mathbf{z}_i, \mathbf{l}_i)\\ &= f(\mathbf{x}_i'\beta + \mathbf{d}_i'\gamma + \mathbf{l}_i'\lambda)\\ &\quad \times \mathbf{g}(\mathbf{z}_i'\alpha_1 + \delta_1 l_{i1}, \ldots, \mathbf{z}_i'\alpha_J + \delta_J l_{iJ}).\end{aligned} \tag{4.21}$$

This model does not have a closed-form log-likelihood, but it can be estimated by numerical integration and simulation-based methods (Gourieroux and Monfort 1997). Specifically, as l_{ij} are unknown, it is assumed that the l_{ij} are i.i.d. draws from (standard normal) distribution and one can numerically integrate over them.

$$\begin{aligned}
\Pr(y_i, \mathbf{d}_i|\mathbf{x}_i, \mathbf{z}_i) &= \int \big[f(\mathbf{x}_i'\beta + \mathbf{d}_i'\gamma + \mathbf{l}_i'\lambda) \\
&\quad \times \mathbf{g}(\mathbf{z}_i'\alpha_1 + \delta_1 l_{i1}, \ldots, \mathbf{z}_i'\alpha_J + \delta_J l_{iJ})\big]\,\mathbf{h}(\mathbf{l}_i)d\mathbf{l}_i \\
&\approx \frac{1}{S}\sum_{s=1}^{S}\big[f(\mathbf{x}_i'\beta + \mathbf{d}_i'\gamma + \tilde{\mathbf{l}}_{is}'\lambda) \\
&\quad \times \mathbf{g}(\mathbf{z}_i'\alpha_1 + \delta_1 \tilde{l}_{i1s}, \ldots, \mathbf{z}_i'\alpha_J + \delta_J \tilde{l}_{iJs})\big], \qquad (4.22)
\end{aligned}$$

where $\tilde{\mathbf{l}}_{is}$ is the sth draw (from a total of S draws) of a pseudo-random number from the density $\mathbf{h}$. Maximizing simulated log-likelihood is equivalent to maximizing the log-likelihood for S sufficiently large.

$$\begin{aligned}
\ln l(y_i, \mathbf{d}_i|\mathbf{x}_i, \mathbf{z}_i) &\approx \sum_{i=1}^{N}\ln\Bigg(\frac{1}{S}\sum_{s=1}^{S}\big[f(\mathbf{x}_i'\beta + \mathbf{d}_i'\gamma + \tilde{\mathbf{l}}_{is}'\lambda) \\
&\quad \times \mathbf{g}(\mathbf{z}_i'\alpha_1 + \delta_1 \tilde{l}_{i1s}, \ldots, \mathbf{z}_i'\alpha_J + \delta_J \tilde{l}_{iJs})\big]\Bigg). \qquad (4.23)
\end{aligned}$$

For identification the scale of each choice equation should be normalized, and the covariances between choice equation errors be fixed. A natural set of normalization restrictions given by $\delta_{jk} = 0\ \forall j \neq k$, i.e., each choice is affected by a unique latent factor, and $\delta_{jj} = 1\ \forall j$, which normalizes the scale of each choice equation. This leads to an element in the covariance matrix being restricted to zero; see Deb and Trivedi (2006a) for details.

Under the unrealistic assumption of correct specification of the model, this approach will generate consistent, asymptotically normal, and efficient estimates. But the restrictions on preferences implied by the MMNL of choice are quite strong and not necessarily appropriate for all data sets. Estimation requires computer intensive simulation based methods that are discussed in Section 4.6.

4.4.4 Endogeneity in Two-Part Models

In considering endogeneity and self-selection in two-part models, we gain clarity by distinguishing carefully between several variants current in the literature. The baseline TPM model is that stated in Section 4.2; the first part is a model of dichotomous outcome whether the count is zero or positive, and the second part is a truncated count model, often the Poisson or NB, for positive counts. In this benchmark model the two parts are independent and all regressors are assumed to be strictly exogenous.

We now consider some extensions of the baseline. The first variant that we consider, referred to as TPM-S, arises when the independence assumption for the two parts is dropped. Instead assume that there is a bivariate distribution of random variables $(\nu_1\ \nu_2)$, representing correlated unobserved factors that affect both the probability of the dichotomous outcome and the conditional count outcome. The two-parts are connected via unobserved heterogeneity. The resulting model is the count data analog of the classic Gronau-Heckman selection model applied to female labor force participation. It is also a special case of the model given in the previous section and can be formally derived by specializing Equations 4.18 to 4.20 to the case of one dichotomous variable and one truncated count distribution. Notice that in this variant the dichotomous endogenous variable will not appear as a regressor in the outcome equation. In practical application of the TPM-S model one is required to choose an appropriate distribution of unobserved heterogeneity. Greene (2007b) gives specific examples and relevant algebraic details. Following Terza (1998) he also provides the count data analog of Heckman two-step estimator.

A second variant of the two-part model is an extension of the TPM-S model described above as it also allows for dependence between the two parts of TPM and further allows for the presence of endogenous regressors in both parts. Hence we call this the TPM-ES model. If dependence between endogenous regressors and the outcome variable is introduced thorough latent factors as in Subsection 4.4.3, then such a model can be regarded a hybrid based on TPM-ES model and the latent factor model. Identification of such a model will require restrictions on the joint covariance matrix of errors, while simulation-based estimation appears to be a promising alternative.

The third and last variant of the TPM is a special case. It is obtained under the assumption that conditional on the inclusion of common endogenous regressor(s) in the two parts, plus the exogenous variables, the two parts are independent. We call this specification the TPM-E model. This assumption is not easy to justify, especially if endogeneity is introduced via dependent latent factors. However, if this assumption is accepted, estimation using moment-based IV estimation of each equation is feasible. Estimation of a class of binary outcome models with endogenous regressors is well established in the literature and has been incorporated in several software packages such as Stata. Both two-step sequential and ML estimators have been developed for the case of a continuous endogenous regressor; see Newey (1987). The estimator also assumes multivariate normality and homoscedasticity, and hence cannot be used for the case of an endogenous discrete regressor. Within the GMM framework the second part of the model will be based on the truncated moment condition

$$\mathrm{E}[y_i \exp(-\mathbf{x}_i'\beta) - 1|\mathbf{z}_i, y_i > 0] = \mathbf{0}. \tag{4.24}$$

The restriction $y_i > 0$ is rarely exploited either in choosing the instruments or in estimation. Hence most of the discussion given in Subsection 4.4.1 remains relevant.

4.4.5 Bayesian Approaches to Endogeneity and Self-Selection

Modern Bayesian inference is attractive whenever the models are parametric and important features of models involve latent variables that can be simulated. There are two recent Bayesian analyses of endogeneity in count models that illustrate key features of such analyses; see Munkin and Trivedi (2003) and Deb, Munkin, and Trivedi (2006a). We sketch the structure of the model developed in the latter.

Deb, Munkin, and Trivedi (2006a) develop a Bayesian treatment of a more general potential outcome model to handle endogeneity of treatment in a count-data framework. For greater generality the entire outcome response function is allowed to differ between the treated and the nontreated groups. This extends the more usual selection model in which the treatment effect only enters through the intercept, as in Munkin and Trivedi (2003). This more general formulation uses the potential outcome model in which causal inference about the impact of treatment is based on a comparison of observed outcomes with constructed counterfactual outcomes. The specific variant of the potential outcome model used is often referred to as the "Roy model," which has been applied in many previous empirical studies of distribution of earnings, occupational choice, and so forth. The study extends the framework of the "Roy model" to nonnegative and integer-valued outcome variables and applies Bayesian estimation to obtain the full posterior distribution of a variety of treatment effects.

Define latent variable Z to measure the difference between the utility generated by two choices that reflect the benefits and the costs associated with them. Assume that Z is linear in the set of explanatory variables $\mathbf{W}$

$$Z = \mathbf{W}\alpha + u, \tag{4.25}$$

such that $d = 1$ if and only if $Z \geq \mathbf{0}$, and $d = 0$ if and only if $Z < \mathbf{0}$.

Assume that individuals choose between two regimes in which two different levels of utility are generated. As before latent variable Z, defined by Equation 4.25 where $u \sim \mathcal{N}(0, 1)$, measures the difference between the utility. In Munkin and Trivedi (2003) $d = 1$ means having private insurance (the treated state) and $d = 0$ means not having it (the untreated state). Two potential utilization variables Y_1, Y_2 are distributed as Poisson with means $\exp(\mu_1)$, $\exp(\mu_2)$, respectively. Variables μ_1, μ_2 are linear in the set of explanatory variables $\mathbf{X}$ and u such as

$$\mu_1 = \mathbf{X}\beta_1 + u\pi_1 + \varepsilon_1, \tag{4.26}$$
$$\mu_2 = \mathbf{X}\beta_2 + u\pi_2 + \varepsilon_2, \tag{4.27}$$

where $\text{Cov}(u, \varepsilon_1|\mathbf{X}) = 0$, $\text{Cov}(u, \varepsilon_2|\mathbf{X}) = 0$, and $\varepsilon = (\varepsilon_1, \varepsilon_2) \sim \mathcal{N}(\mathbf{0}, \Sigma)$, $\Sigma = \text{diag}(\sigma_1, \sigma_2)$. The observability condition for Y is $Y = Y_1$ if $d = 1$ and $Y = Y_2$ if $d = 0$. The counted variable Y, representing utilization of medical services, is Poisson distributed with two different conditional means depending on the insurance status. Thus, there are two regimes generating count variables Y_1,

Y_2, but only one value is observed. Observe the restriction $\sigma_{12} = 0|\mathbf{X}, u$. This is imposed since the covariance parameter is unidentified in this model.

The standard Tanner–Wong data augmentation approach can be adapted to include latent variables μ_{1i}, μ_{2i}, Z_i in the parameter set making it a part of the posterior. Then the Bayesian MCMC approach can be used to obtain the posterior distribution of all parameters. A test to check the null hypothesis of no endogeneity is also feasible. Denote by M_1 the specification of the model that leaves parameters π_1 and π_2 unconstrained, and by M_0 the model that puts $\pi_1 = \pi_2 = 0$ constraint. Then a test of no endogeneity can be implemented using the Bayes factor $B_{0,1} = m(\mathbf{y}|M_0)/m(\mathbf{y}|M_1)$, where $m(\mathbf{y}|M)$ is the marginal likelihood of the model specification M.

In the case when the proportions of zero observations are so large that even extensions of the Poisson model that allow for overdispersion, such as negative binomial and the Poisson-lognormal models, do not provide an adequate fit, the ordered probit (OP) modeling approach might be an option. Munkin and Trivedi (2008) extend the OP model to allow for endogeneity of a set of categorical dummy covariates (e.g., types of health insurance plans), defined by a multinomial probit model (MNP). Let $\mathbf{d}_i = (d_{1i}, d_{2i}, \ldots, d_{J-1i})$ be binary random variables for individual i ($i = 1, \ldots, N$) choosing category j ($j = 1, \ldots, J$) (category J is the baseline) such that $d_{ji} = 1$ if alternative j is chosen and $d_{ji} = 0$ otherwise. The MNP model is defined using the multinomial latent variable structure which represents gains in utility received from the choices, relative to the utility received from choosing alternative J. Let the $(J-1) \times 1$ random vector $\mathbf{Z}_i$ be defined as

$$\mathbf{Z}_i = \mathbf{W}_i\alpha + \varepsilon_i,$$

where $\mathbf{W}_i$ is a matrix of exogenous regressors, such that

$$d_{ji} = \prod_{l=1}^{J} I_{[0,+\infty)}\left(Z_{ji} - Z_{li}\right), \quad j = 1, \ldots, J,$$

where $Z_{Ji} = 0$ and $I_{[0,+\infty)}$ is the indicator function for the set $[0, +\infty)$. The distribution of the error term ε_i is $(J-1)$-variate normal $\mathcal{N}(\mathbf{0}, \Sigma)$. For identification it is customary to restrict the leading diagonal element of Σ to unity.

To model the ordered dependent variable it is assumed that there is another latent variable Y_i^* that depends on the outcomes of $\mathbf{d}_i$ such that

$$Y_i^* = \mathbf{X}_i\beta + \mathbf{d}_i\rho + u_i,$$

where $\mathbf{X}_i$ is a vector of exogenous regressors, and ρ is a $(J-1) \times 1$ parameter vector. Define Y_i as

$$Y_i = \sum_{m=1}^{M} m I_{[\tau_{m-1}, \tau_m)}\left(Y_i^*\right),$$

where $\tau_0, \tau_1, \ldots, \tau_M$ are threshold parameters and $m = 1, \ldots, M$. For identification, it is standard to set $\tau_0 = -\infty$ and $\tau_M = \infty$ and additionally restrict

$\tau_1 = 0$. The choice of insurance is potentially endogenous to utilization and this endogeneity is modeled through correlation between u_i and ε_i, assuming that they are jointly normally distributed with variance of u_i restricted for identification since Y_i^* is latent; see Deb, Munkin, and Trivedi (2006b).

Munkin and Trivedi (2009) extend the Ordered Probit model with Endogenous Selection to allow for a covariate such as income to enter the insurance equation nonparametrically. The insurance equation is specified as

$$Z_i = f(s_i) + \mathbf{W}_i\alpha + \varepsilon_i, \tag{4.28}$$

where $\mathbf{W}_i$ is a vector of regressors, α is a conformable vector of parameters, and the distribution of the error term ε_i is $\mathcal{N}(0, 1)$. Function $f(.)$ is unknown and s_i is income of individual i. The data are sorted by values of s so that s_1 is the lowest level of income and s_N is the largest. The main assumption made on function $f(s_i)$ is that it is smooth such that it is differentiable and its slope changes slowly with s_i such that, for a given constant C, $|f(s_i) - f(s_{i-1})| \leq C|s_i - s_{i-1}|$ — a condition which covers a wide range of functions.

Economic theory predicts that risk-averse individuals prefer to purchase insurance against catastrophic or simply costly evens because they value eliminating risk more than money at sufficiently high wealth levels. This is modeled by assuming that a risk-averse individual's utility is a monotonically increasing function of wealth with diminishing marginal returns. This is certainly true for general medical insurance when liabilities could easily exceed any reasonable levels. However, in the context of dental insurance the potential losses have reasonable bounds. Munkin and Trivedi (2009) find strong evidence of diminishing marginal returns of income on dental insurance status and even a nonmonotonic pattern.

4.5 Panel Data

We begin with a model for scalar dependent variable y_{it} with regressors $\mathbf{x}_{it}$, where i denotes the individual and t denotes time. We will restrict our coverage to the case of t small, usually referred to as "short panel," which is also of most interest in microeconometrics. Assuming multiplicative individual scale effects applied to exponential function

$$\mathrm{E}[y_{it}|\alpha_i, \mathbf{x}_{it}] = \alpha_i \exp(\mathbf{x}_{it}'\beta), \tag{4.29}$$

As $\mathbf{x}_{it}$ includes an intercept, α_i may be interpreted as a deviation from 1 because $\mathrm{E}(\alpha_i|x) = 1$.

In the standard case in econometrics the time interval is fixed and the data are equi-spaced through time. However, the panel framework can also cover the case where the data are simply repeated events and not necessarily equi-spaced through time. An example of such data is the number of epileptic seizures during a two-week period preceding each of four consecutive clinical visits; see Diggle et al. (2002).

4.5.1 Pooled or Population-Averaged (PA) Models

Pooling occurs when the observations $y_{it}|\alpha_i, \mathbf{x}_{it}$ are treated as independent, after assuming $\alpha_i = \alpha$. Consequently cross-section observations can be "stacked" and cross-section estimation methods can then be applied.

The assumption that data are poolable is strong. For parametric models it is assumed that the marginal density for a single (i, t) pair,

$$f(y_{it}|\mathbf{x}_{it}) = f(\alpha + \mathbf{x}_{it}'\beta, \gamma), \tag{4.30}$$

is correctly specified, regardless of the (unspecified) form of the joint density

$$f(y_{it}, \ldots, y_{iT}|\mathbf{x}_{i1}, \ldots, \mathbf{x}_{iT}, \beta, \gamma).$$

The pooled model, also called the population-averaged (PA) model, is easily estimated. A panel-robust or cluster-robust (with clustering on i) estimator of the covariance matrix can then be applied to correct standard errors for any dependence over time for given individual. This approach is the analog of pooled OLS for linear models.

The pooled model for the exponential conditional mean specifies $\mathrm{E}[y_{it}|\mathbf{x}_{it}] = \exp(\alpha + \mathbf{x}_{it}'\beta)$. Potential efficiency gains can be realized by taking into account dependence over time. In the statistics literature such an estimator is constructed for the class of generalized linear models (GLM) that includes the Poisson regression. Essentially this requires that estimation be based on weighted first-order moment conditions to account for correlation over t, given i, while consistency is ensured provided the conditional mean is correctly specified as $\mathrm{E}[y_{it}|\mathbf{x}_{it}] = \exp(\alpha + \mathbf{x}_{it}'\beta) \equiv g(\mathbf{x}_{it}, \beta)$. The efficient GMM estimator, known in the statistics literature as the population-averaged model, or generalized estimating equations (GEE) estimator (see Diggle et al. [2002]), is based on the conditional moment restrictions, stacked over all T observations,

$$\mathrm{E}[\mathbf{y}_i - \mathbf{g}_i(\beta)|\mathbf{X}_i] = \mathbf{0}, \tag{4.31}$$

where $\mathbf{g}_i(\beta) = [g(\mathbf{x}_{i1}, \beta), \ldots, g(\mathbf{x}_{iT}, \beta)]'$ and $\mathbf{X}_i = [\mathbf{x}_{i1}, \ldots, \mathbf{x}_{iT}]'$. The optimally weighted unconditional moment condition is

$$\mathrm{E}\left[\frac{\partial \mathbf{g}_i'(\beta)}{\partial \beta}\{\mathrm{V}[\mathbf{y}_i|\mathbf{X}_i]\}^{-1}(\mathbf{y}_i - \mathbf{g}_i(\beta))\right] = \mathbf{0}. \tag{4.32}$$

Given Σ_i a working variance matrix for $\mathrm{V}[\mathbf{y}_i|\mathbf{X}_i]$, the moment condition becomes

$$\sum_{i=1}^{N} \frac{\partial \mathbf{g}_i'(\beta)}{\partial \beta}\Sigma_i^{-1}(\mathbf{y}_i - \mathbf{g}_i(\beta)) = \mathbf{0}. \tag{4.33}$$

The asymptotic variance matrix, which can be derived using standard GEE/GMM theory (see CT, 2005, Chapter 23.2), is robust to misspecification of Σ_i. For the case of strictly exogenous regressors the GEE methodology is not strictly speaking "recent," although it is more readily implementable nowadays because of software developments.

While the foregoing analysis applies to the case of additive errors, there are multiplicative versions of moment conditions (as detailed in Subsection 4.4.1) that will lead to different estimators. Finally, in the case of endogenous regressors, the choice of the optimal GMM estimator is more complicated as it depends upon the choice of optimal instruments; if $\mathbf{z}_i$ defines a vector of valid instruments, then so does any function $h(\mathbf{z}_i)$.

Given its strong restrictions, the GEE approach connects straightforwardly with the GMM/IV approach used for handling endogenous regressors. To cover the case of endogenous regressors we simply rewrite the previous moment condition as $\mathrm{E}[\mathbf{y}_i - \mathbf{g}_i(\beta)|\mathbf{Z}_i] = \mathbf{0}$, where $\mathbf{Z}_i = [\mathbf{z}_{i1}, \ldots, \mathbf{z}_{iT}]'$ are appropriate instruments.

Because of the greater potential for having omitted factors in panel models of observational data, fixed and random effect panel count models have relatively greater credibility than the above PA model. The strong restrictions of the pooled panel model are relaxed in different ways by random and fixed effects models. The recent developments have impacted the random effects panel models more than the fixed effect models, in part because computational advances have made them more accessible.

4.5.2 Random-Effects Models

A random-effects (RE) model treats the individual-specific effect α_i as an unobserved random variable with specified mixing distribution $g(\alpha_i|\gamma)$, similar to that considered for cross-section models of Section 4.2. Then α_i is eliminated by integrating over this distribution. Specifically the unconditional density for the ith observation is

$$\begin{aligned} f(y_{it}, \ldots, y_{iT_i}|\mathbf{x}_{i1}, \ldots, \mathbf{x}_{iT_i}, \beta, \gamma, \eta) \\ = \int \left[\prod_{t=1}^{T_i} f(y_{it}|\mathbf{x}_{it}, \alpha_i, \beta, \gamma) \right] g(\alpha_i|\eta) d\alpha_i. \end{aligned} \tag{4.34}$$

For some combinations of $\{f(\cdot), g(\cdot)\}$ this integral usually has analytical solution. However, if randomness is restricted to the intercept only, then numerical integration is also feasible as only univariate integration is required. The RE approach, when extended to both intercept and slope parameters, becomes computationally more demanding.

As in the cross-section case, the negative binomial panel model can be derived under two assumptions: first, y_{ij} has Poisson distribution conditional on μ_i, and second, μ_i are i.i.d. gamma distributed with mean μ and variance $\alpha\mu^2$. Then, unconditionally $y_{ij} \sim NB(\mu_i, \mu_i + \alpha\mu_i^2)$. Although this model is easy to estimate using standard software packages, it has the obvious limitation that it requires a strong distributional assumption for the random intercept and it is only useful if the regressors in the mean function $\mu_i = \exp(\mathbf{x}_i'\beta)$ do not vary over time. The second assumption is frequently violated.

Morton (1987) relaxed both assumptions of the preceding paragraph and proposed a GEE-type estimator for the following exponential mean with multiplicative heterogeneity model: $\mathrm{E}[y_{it}|\mathbf{x}_{it}, \nu_i] = \exp(\mathbf{x}_{it}'\beta)\nu_i$; $\mathrm{Var}[y_{it}|\nu_i] = \phi\mathrm{E}[y_{it}|\mathbf{x}_{it}, \nu_i]$; $\mathrm{E}[\nu_i] = 1$ and $\mathrm{Var}[\nu_i] = \alpha$. These assumptions imply $\mathrm{E}[y_{it}|\mathbf{x}_{it}] = \exp(\mathbf{x}_{it}'\beta)$ and $\mathrm{Var}[y_{it}] = \phi\mu_{it} + \alpha\mu_{it}^2$. A GEE-type estimator based on Equation 4.33 is straight-forward to construct; see Diggle et al. (2002).

Another example is Breslow and Clayton (1993) who consider the specification

$$\ln\{\mathrm{E}[y_{it}|\mathbf{x}_{it}, z_{it}]\} = \mathbf{x}_{it}'\beta + \gamma_{1t} + \gamma_{2t}z_{it},$$

where the intercept and slope coefficients $(\gamma_{1t}, \gamma_{2t})$ are assumed to be bivariate normal distributed. Whereas regular numerical integration estimation for this can be unstable, adaptive quadrature methods have been found to be more robust; see Rabe-Hesketh, Skrondal, and Pickles (2002).

A number of authors have suggested a further extension of the RE models mentioned above; see Chib, Greenberg, and Winkelmann (1998). The assumptions of this model are: 1. $y_{it}|\mathbf{x}_{it}, \mathbf{b}_i \sim \mathcal{P}(\mu_{it})$; $\mu_{it} = \mathrm{E}[y_{it}|\mathbf{x}_{it}'\beta + \mathbf{w}_{it}'\mathbf{b}_i]$; and $\mathbf{b}_i \sim \mathcal{N}[\mathbf{b}^*, \Sigma_\mathbf{b}]$ where $(\mathbf{x}_{it}')$ and $(\mathbf{w}_{it}')$ are vectors of regressors with no common elements and only the latter have random coefficients. This model has an interesting feature that the contribution of random effect is not constant for a given i. However, it is fully parametric and maximum likelihood is computationally demanding. Chib, Greenberg, and Winkelmann (1998) use Markov chain Monte Carlo to obtain the posterior distribution of the parameters.

A potential limitation of the foregoing RE panel models is that they may not generate sufficient flexibility in the specification of the conditional mean function. Such flexibility can be obtained using a finite mixture or latent class specification of random effects and the mixing can be with respect to the intercept only, or all the parameters of the model. Specifically, consider the model

$$f(y_{it}|\beta, \pi) = \sum_{j=1}^{m} \pi_j(z_{it}|\gamma) f_j(y_{it}|\mathbf{x}_{it}, \beta_j), \quad 0 < \pi_j(\cdot) < 1, \sum_{j=1}^{m} \pi_j(\cdot) = 1 \tag{4.35}$$

where for generality the mixing probabilities are parameterized as functions of observable variables z_{it} and parameters γ, and the j-component conditional densities may be any convenient parametric distributions, e.g., the Poisson or negative binomial, each with its own conditional mean function and (if relevant) a variance parameter. In this case individual effects are approximated using a distribution with finite number of discrete mass points that can be interpreted as the number of "types." Such a specification offers considerable flexibility, albeit at the cost of potential over-parametrization. Such a model is a straightforward extension of the finite mixture cross-section model. Bago d'Uva (2005) uses the finite mixture of the pooled negative binomial in her study of primary care using the British Household Panel Survey; Bago d'Uva (2006) exploits the panel structure of the Rand Health Insurance Experiment data to estimate a latent class hurdle panel model of doctor visits.

The RE model has different conditional mean from that for pooled and population-averaged models, unless the random individual effects are additive or multiplicative. So, unlike the linear case, pooled estimation in nonlinear models leads to inconsistent parameter estimates if instead the assumed random-effects model is appropriate, and vice-versa.

4.5.3 Fixed-Effects Models

Given the conditional mean specification

$$E[y_{it}|\alpha_i, \mathbf{x}_{it}] = \alpha_i \exp(\mathbf{x}_{it}'\beta) = \alpha_i \mu_{it}, \tag{4.36}$$

a fixed-effects (FE) model treats α_i as an unobserved random variable that may be correlated with the regressors $\mathbf{x}_{it}$. It is known that maximum likelihood or moment-based estimation of both the population-averaged Poisson model and the RE Poisson model will not identify the β if the FE specification is correct. Econometricians often favor the fixed effects specification over the RE model. If the FE model is appropriate then a fixed-effects estimator should be used, but it may not be available if the problem of incidental parameters cannot be solved. Therefore, we examine this issue in the following section.

4.5.3.1 Maximum Likelihood Estimation

Whether, given short panels, joint estimation of the fixed effects $\alpha = (\alpha_1, \ldots, \alpha_N)$ and β is feasible is the first important issue. Under the assumption of strict exogeneity of $\mathbf{x}_{it}$, the basic result that there is no incidental parameter problem for the Poisson panel regression is now established and well understood (CT 1998; Lancaster 2000; Windmeijer 2008). Consequently, corresponding to the fixed effects, one can introduce N dummy variables in the Poisson conditional mean function and estimate (α, β) by maximum likelihood. This will increase the dimensionality of the estimation problem. Alternatively, the conditional likelihood principle may be used to eliminate α and to condense the log-likelihood in terms of β only. However, maximizing the condensed likelihood will yield estimates identical to those from the full likelihood. Table 4.2 displays the first order condition for FE Poisson MLE of β, which can be compared with the pooled Poisson first-order condition to see how the fixed effects change the estimator. The difference is that μ_{it} in the pooled model is replaced by $\mu_{it}\bar{y}_i/\bar{\mu}_i$ in the FE Poisson MLE. The multiplicative factor $\bar{y}_i/\bar{\mu}_i$ is simply the ML estimator of α_i; this means the first-order condition is based on the likelihood concentrated with respect to α_i.

The result about the incidental parameter problem for the Poisson FE model does not extend to the fixed effects NB2 model (whose variance function is quadratic in the conditional mean) if the fixed effects parameters enter multiplicatively through the conditional mean specification. This fact is confusing

TABLE 4.2

Selected Moment Conditions for Panel Count Models

Model	Moment or Model Specification	Estimating Equations or Moment Condition
Pooled Poisson	$\mathrm{E}[y_{it}\vert\mathbf{x}_{it}] = \exp(\mathbf{x}_{it}'\beta)$,	$\sum_{i=1}^{N}\sum_{t=1}^{T}\mathbf{x}_{it}(y_{it}-\mu_{it}) = \mathbf{0}$ where $\mu_{it} = \exp(\mathbf{x}_{it}'\beta)$
Pop. averaged		$\rho_{ts} = \mathrm{Cor}[(y_{it}-\exp(\mathbf{x}_{it}'\beta))$ $(y_{is}-\exp(\mathbf{x}_{is}'\beta))]$.
Poisson RE	$\mathrm{E}[y_{it}\vert\alpha_i,\mathbf{x}_{it}] = \alpha_i\exp(\mathbf{x}_{it}'\beta)$,	$\sum_{i=1}^{N}\sum_{t=1}^{T}\mathbf{x}_{it}\left(y_{it}-\mu_{it}\frac{\bar{y}_i+\eta/T}{\bar{\mu}_i+\eta/T}\right) = \mathbf{0}$ $\bar{\mu}_i = T^{-1}\sum_t\exp(\mathbf{x}_{it}'\beta);\ \eta = \mathrm{Var}(\alpha_i)$
Poisson FE	$\mathrm{E}[y_{it}\vert\alpha_i,\mathbf{x}_{it}] = \alpha_i\exp(\mathbf{x}_{it}'\beta)$	$\sum_{i=1}^{N}\sum_{t=1}^{T}\mathbf{x}_{it}\left(y_{it}-\mu_{it}\frac{\bar{y}_i}{\bar{\mu}_i}\right) = \mathbf{0}$,
GMM (Windmeijer)	$y_{it} = \exp(\mathbf{x}_{it}'\beta+\alpha_i)u_{it}$,	$\sum_{i=1}^{N}\sum_{t=1}^{T}\left[y_{it}\frac{\mu_{it-1}}{\mu_{it}} - y_{it-1}\vert\mathbf{x}_i^{t-1}\right] = \mathbf{0}$
Strict exog	$\mathrm{E}[\mathbf{x}_{it}u_{it+j}] = 0,\ j\geq 0$	
Predetermined reg.	$\mathrm{E}[\mathbf{x}_{it}u_{it-s}] \neq 0,\ s\geq 1$	
GMM (Wooldridge)	$\mathrm{E}\left[\frac{y_{it}}{\mu_{it}} - \frac{y_{it-1}}{\mu_{it-1}}\vert\mathbf{x}_i^{t-1})\right] = 0$	$\sum_{i=1}^{N}\sum_{t=1}^{T}\left[\frac{y_{it}}{\mu_{it}} - \frac{y_{it-1}}{\mu_{it-1}}\vert\mathbf{x}_i^{t-1})\right] = \mathbf{0}$
GMM (Chamberlain)	$\mathrm{E}\left[y_{it}\frac{\mu_{it-1}}{\mu_{it}} - y_{it-1}\vert\mathbf{x}_i^{t-1})\right] = 0$	$\sum_{i=1}^{N}\sum_{t=1}^{T}\left[y_{it}\frac{\mu_{it-1}}{\mu_{it}} - y_{it-1}\vert\mathbf{x}_i^{t-1})\right] = \mathbf{0}$
GMM/endog	$\mathrm{E}\left[\frac{y_{it}}{\mu_{it}} - \frac{y_{it-1}}{\mu_{it-1}}\vert\mathbf{x}_i^{t-2})\right] = 0$	$\sum_{i=1}^{N}\sum_{t=1}^{T}\left[y_{it}\frac{\mu_{it-1}}{\mu_{it}} - y_{it-1}\vert\mathbf{x}_i^{t-2}\right] = \mathbf{0}$
Dynamic feedabck	$y_{it} = \theta y_{it-1} + \exp(\mathbf{x}_{it}'\beta+\alpha_i)$ $+u_{it}$,	$\mathrm{E}\Big[(y_{it}-\theta y_{it-1})\frac{\mu_{it-1}}{\mu_{it}} -$ $-(y_{it-1}-\theta y_{it-2})\vert y_{it-2},\mathbf{x}_i^{t-1}\Big] = \mathbf{0}$

for many practitioners who observe the availability of the fixed effects NB option in several commercial computer packages. Greene (2007b) provides a good exposition of this issue. He points out that the option in the packages is that of Hausman, Hall, and Griliches (1984) who specified a variant of the "fixed effects negative binomial" (FENB) distribution in which the variance function is linear in the conditional mean; that is, $\mathrm{Var}[y_{it}|\mathbf{x}_{it}] = (1+\alpha_i)\mathrm{E}[y_{it}|\mathbf{x}_{it}]$, so the variance is a scale factor multiplied by the conditional mean, and the fixed effects parameters enter the model through the scaling factor. This is the NB model with linear variance (or NB1), not that with a quadratic variance (or NB2 formulation). As fixed effects come through the variance function, not the conditional mean, this is clearly a different formulation from the Poisson fixed effects model. Given that the two formulations are not nested, it is not clear how one should compare FE Poisson and this particular variant of the FENB. Greene (2007b) discusses related issues in the context of an empirical example.

4.5.3.2 Moment Function Estimation

Modern literature considers and sometimes favors the use of moment-based estimators that may be potentially more robust than the MLE. The starting point here is a moment condition model. Following Chamberlain (1992), and mimicing the differencing transformations used to eliminate nuisance parameters in linear models, there has been an attempt to obtain moment condition models based on quasi-differencing transformations that eliminate fixed effects; see Wooldridge (1999, 2002). This step is then followed by application of one of the several available variants of the GMM estimation, such as two-step GMM or continuously updated GMM. Windmeier (2008) provides a good survey of the approach for the Poisson panel model.

Windmeier (2008) considers the following alternative formulations:

$$y_{it} = \exp(\mathbf{x}_{it}'\beta + \alpha_i)u_{it}, \tag{4.37}$$

$$y_{it} = \exp(\mathbf{x}_{it}'\beta + \alpha_i) + u_{it}, \tag{4.38}$$

where, in the first case $\mathrm{E}(u_{it}) = 1$, the $\mathbf{x}_{it}$ are predetermined with respect to u_{it}, and u_{it} are serially uncorrelated and independent of α_i. The table lists the implied restriction. A quasi-differencing transformation eliminates the fixed effects and generates moment conditions whose form depend on whether we start with Equation 4.37 or 4.38. Several variants are shown in Table 4.2 and they can be used in GMM estimation. Of course, these moment conditions only provide a starting point and important issues remain about the performance of alternative variants or the best variants to use. Windmeier (2008) discusses the issues and provides a Monte Carlo evaluation.

It is conceivable that a fixed effects–type formulation may adequately account for overdispersion of counts. But there are other complications that generate overdispersion in other ways, e.g., excess zeros and fat tails. At present little is known about the performance of moment-based estimators when the d.g.p. deviates significantly from the Poisson-type behavior. Moment-based models do not exploit the integer-valued aspect of the dependent variable. Whether this results in significant efficiency loss — and if so, when — is a topic that deserves future investigation.

4.5.4 Conditionally Correlated Random Effects

The standard random effect panel model assumes that α_i and $\mathbf{x}_{it}$ are uncorrelated. Instead we can relax this and assume that they are conditionally correlated. This idea, originally developed in the context of a linear panel model by Mundlak (1978) and Chamberlain (1982), can be interpreted as intermediate between fixed and random effects. That is, if the correlation between α_i and the regressors can be controlled by adding some suitable "sufficient" statistic for the regressors, then the remaining unobserved heterogeneity can be treated as random and uncorrelated with the regressors. While in principle we may introduce a subset of regressors, in practice it is more parsimonious to introduce time-averaged values of time-varying regressors. This is the conditionally correlated random (CCR) effects model. This formulation allows for correlation by assuming a relationship of the form

$$\alpha_i = \bar{\mathbf{x}}_i'\lambda + \varepsilon_i, \tag{4.39}$$

where $\bar{\mathbf{x}}$ denotes the time-average of the time-varying exogenous variables and ε_i may be interpreted as unobserved heterogeneity uncorrelated with the regressors. Substituting this into the above formulation essentially introduces no additional problems except that the averages change when new data are added. To use the standard RE framework, however, we need to make an assumption about the distribution of ε_t and this will usually lead to an integral that would need evaluating. Estimation and inference in the pooled Poisson or NLS model can proceed as before. This formulation can also be used when dynamics are present in the model.

Because the CCR formulation is intermediate between the FE and RE models, it may serve as a useful substitute for not being able to deal with FE in some specifications. For example, a panel version of the hurdle model with FE is rarely used as the fixed effects cannot be easily eliminated. In such a case the CCR specification is feasible.

4.5.5 Dynamic Panels

As in the case of linear models, inclusion of lagged values is appropriate in some empirical models. An example is the use of past research and development expenditure when modeling the number of patents, see Hausman, Hall, and Griliches (1984). When lagged exogenous variables are used, no new modeling issues arise from their presence. However, to model lagged dependence more flexibly and more parsimoniously, the use of lagged dependent variables y_{t-j} $(j \geq 1)$ as regressors is attractive, but it introduces additional complications that have been studied in the literature on autoregressive models of counts (see CT [1998], Chapters 7.4 and 7.5). Introducing autoregressive dependence through the exponential mean specification leads to a specification of the type

$$\mathrm{E}[y_{it}|\mathbf{x}_{it}, y_{it-1}, \alpha_i] = \exp(\gamma y_{it-1} + \mathbf{x}_{it}'\beta + \alpha_i), \tag{4.40}$$

where α_i is the individual-specific effect. If the α_i are uncorrelated with the regressors, and further if parametric assumptions are to be avoided, then this model can be estimated using either the nonlinear least squares or pooled Poisson MLE. In either case it is desirable to use the robust variance formula.

The estimation of a dynamic panel model requires additional assumptions about the relationship between the initial observations ("initial conditions") y_0 and the α_i. For example, using the CCR model we could write $\alpha_i = y_0'\delta + \bar{\mathbf{x}}_i'\lambda + \varepsilon_i$ where y_0 is an initial condition. Then maximum likelihood estimation could proceed by treating the initial condition as given. The alternative of taking the initial condition as random, specifying a distribution for it, and then integrating out the condition is an approach that has been suggested for other dynamic panel models, and it is computationally more demanding; see Stewart (2007). Under the assumption that the initial conditions are nonrandom, the standard random effects conditional maximum likelihood approach identifies the parameters of interest. For a class of nonlinear dynamic panel models, including the Poisson model, Wooldridge (2005) analyzes this model which conditions the joint distribution on the initial conditions.

The inclusion of lagged y_{it} inside the exponential mean function introduces potentially sharp discontinuities that may result in a poor fit to the data. It is not the case that this will always happen, but it might when the range of counts is very wide. Crepon and Duguet (1997) proposed using a better starting point in a dynamic fixed effects panel model; they specified the model as

$$y_{it} = h(y_{it-1}, \theta) \exp(\mathbf{x}'_{it}\beta + \alpha_i) + u_{it} \tag{4.41}$$

where the function $h(y_{it-1}, \theta)$ parametrizes the dependence on lagged values of y_{it}. Crepon and Duguet (1997) suggested switching functions to allow lagged zero values to have a different effect from positive values. Blundell, Griffith, and Windmeijer (2002) proposed a linear feedback model with multiplicative fixed effect α_i,

$$y_{it} = \theta y_{it-1} + \exp(\mathbf{x}'_{it}\beta + \alpha_i) + u_{it}, \tag{4.42}$$

but where the lagged value enters linearly. This formulation avoids awkward discontinuities and is related to the integer valued autoregressive (INAR) models. A quasi-differencing transformation can be applied to generate a suitable estimating equation. Table 4.2 shows the estimating equation obtained using a Chamberlain-type quasi-differencing transformation. Consistent GMM estimation here depends upon the assumption that regressors are predetermined. Combining this with the CCR assumption about α_i is straight forward.

Currently the published literature does not provide detailed information on the performance of the available estimators for dynamic panels. Their development is in early stages and, not surprisingly, we are unaware of commercial software to handle such models.

4.6 Multivariate Models

Multivariate count regression models, especially its bivariate variant, are of empirical interest in many contexts. In the simplest case one may be interested in the dependence structure between counts $y_1, \ldots, y_m$, conditional on vectors of exogenous variables $\mathbf{x}_1, \ldots, \mathbf{x}_m$, $m \geq 2$. For example, y_1 denotes the number of prescribed and y_2 the number of nonprescribed medications taken by individuals over a fixed period.

4.6.1 Moment-Based Models

The simplest and attractive semiparametric approach here follows Delgado (1992); it simply extends the seemingly unrelated regressions (SUR) for linear models to the case of multivariate exponential regression. For example, in the bivariate case we specify $E[\mathbf{y}_1|\mathbf{x}_1] = \exp(\mathbf{x}_1'\beta_1)$ and $E[\mathbf{y}_2|\mathbf{x}_2] = \exp(\mathbf{x}_2'\beta_2)$, assume additive errors and then apply nonlinear least squares, but estimate variances using the heteroscedasticity-robust variance estimator supported by many software packages. This is simply nonlinear SUR and is easily extended to several equations. It is an attractive approach when all conditional means have exponential mean specifications and the joint distribution is not desired. It also permits a very flexible covariance structure and its asymptotic theory is well established. Tests of cross-equation restrictions are easy to implement.

An extension of the model would include a specification for variances and covariance. For example, we could specify $V[\mathbf{y}_j|\mathbf{x}_j] = \alpha_j \exp(\mathbf{x}_j'\beta_j)$, $j = 1, 2$, and $\mathrm{Cov}[\mathbf{y}_1, \mathbf{y}_2|\mathbf{x}_1, \mathbf{x}_1] = \rho \times \exp(\mathbf{x}_1'\beta_1)^{1/2} \exp(\mathbf{x}_1'\beta_2)^{1/2}$. This specification is similar to univariate Poisson quasi-likelihood except improved efficiency is possible using a generalized estimating equations estimator.

4.6.2 Likelihood-Based Models

At issue is the joint distribution of $(y_1, y_2|\mathbf{x}_1, \mathbf{x}_2)$. A different data situation is one in which y_1 and y_2 are paired observations that are jointly distributed, whose marginal distributions $f_1(y_1|\mathbf{x}_1)$ and $f_2(y_2|\mathbf{x}_2)$ are parametrically specified, but our interest is in some function of y_1 and y_2. They could be data on twins, spouses, or paired organs (kidneys, lungs, eyes), and the interest lies in studying and modeling the difference. When the bivariate distribution of (y_1, y_2) is known, standard methods can be used to derive the distribution of any continuous function of the variables, say $H(y_1, y_2)$.

A problem arises, however, when an analytical expression for the joint distribution is either not available at all or is available in an explicit form only under some restrictive assumptions. This situation arises in case of multivariate Poisson and negative binomial distributions that are only appropriate for positive dependence between counts, thus lacking generality. Unrestricted multivariate distributions of discrete outcomes often do not have closed form expressions, see Marshall and Olkin (1990), CT (1998), and Munkin and Trivedi (1999). The first issue to consider is how to generate flexible specifications of multivariate count models. The second issue concerns estimation and inference.

4.6.2.1 Latent Factor Models

One fruitful way to generate flexible dependence structures between counts is to begin by specifying latent factor models. Munkin and Trivedi (1999) generate a more flexible dependence structure using a correlated unobserved heterogeneity model. Suppose y_1 and y_2 are, respectively, $\mathcal{P}(\mu_1|\nu_1)$ and $\mathcal{P}(\mu_2|\nu_2)$

$$\mathrm{E}[y_j|\mathbf{x}_j,\nu_j] = \mu_j = \exp(\beta_{0j} + \lambda_j\nu_j + \mathbf{x}_j'\beta_j), \quad j = 1, 2 \tag{4.43}$$

where ν_1 and ν_2 represent correlated latent factors or unobserved heterogeneity and (λ_1, λ_2) are factor loadings. Dependence is induced if ν_1 and ν_2 are correlated. Assume (ν_1, ν_2) to be bivariate normal distributed with correlation ρ, $0 \leq \rho \leq 1$. Integrating out (ν_1, ν_2), we obtain the joint distribution

$$f(y_1, y_2|\mathbf{x}_1, \mathbf{x}_2, \nu_1, \nu_2) = \int f_1(\mathbf{y}_1|\mathbf{x}_1, \nu_1) f_2(\mathbf{y}_2|\mathbf{x}_2, \nu_2) g(\nu_1, \nu_2) d\nu_1 d\nu_2, \tag{4.44}$$

where the right-hand side can be replaced by simulation-based numerical approximation

$$\frac{1}{S}\sum_{s=1}^{S} f_1\big(\mathbf{y}_1|\mathbf{x}_1, \nu_1^{(s)}\big) f_2\big(\mathbf{y}_2|\mathbf{x}_2, \nu_2^{(s)}\big), \tag{4.45}$$

The method of simulation-based maximum likelihood (SMLE) estimates the unknown parameters using the likelihood based on such an approximation. As shown in Munkin and Trivedi (1999), while SMLE of $(\beta_{01}, \beta_1, \beta_{02}, \beta_2, \lambda_1, \lambda_2)$ is feasible it is not computationally straightforward. Recently two alternatives to SMLE have emerged. The first uses Bayesian Monte Carlo Markov Chain (MCMC) approach to estimation; see Chib and Winkelmann (2001). MCMC estimation is illustrated in Subsection 4.6.3. The second uses copulas to generate a joint distribution whose parameters can be estimated without simulation.

4.6.2.2 Copulas

Copula-based joint estimation is based on Sklar's theorem which provides a method of generating joint distributions by combining marginal distributions using a copula. Given a continuous m-variate distribution function $F(y_1, \ldots, y_m)$ with univariate marginal distributions $F_1(y_1), \ldots, F_m(y_m)$ and inverse (quantile) functions $F_1^{-1}, \ldots, F_m^{-1}$, then $y_1 = F_1^{-1}(u_1) \sim F_1, \ldots, y_m = F_m^{-1}(u_m) \sim F_m$, where $u_1, \ldots, u_m$ are uniformly distributed variates. By Sklar's theorem, an m-copula is an m-dimensional distribution function with all m univariate margins being $U(0, 1)$, i.e.,

$$F(y_1, \ldots, y_m) = F(F_1^{-1}(u_1), \ldots, F_m^{-1}(u_m)) = C(u_1, \ldots, u_m; \theta), \tag{4.46}$$

is the unique copula associated with the distribution function. Here $C(\cdot)$ is a given functional form of a joint distribution function and θ is a dependence parameter. Zero dependence implies that the joint distribution is the product of marginals. A leading example is a Gaussian copula based on any relevant marginal such as the Poisson.

Sklar's theorem implies that copulas provide a "recipe" to derive joint distributions when only marginal distributions are given. The approach is attractive because copulas (1) provide a fairly general approach to joint modeling of count data; (2) neatly separate the inference about marginal distribution from inference on dependence; (3) represent a method for deriving joint distributions given the fixed marginals such as Poisson and negative binomial; (4) in a bivariate case copulas can be used to define nonparametric measures of dependence that can capture asymmetric (tail) dependence as well as correlation or linear association; (4) are easier to estimate than multivariate latent factor models with unobserved heterogeneity. However, copulas and latent factor models are closely related; see Trivedi and Zimmer (2007) and Zimmer and Trivedi (2006).

The steps involved in copula modeling is specification of marginal distributions and a copula. There are many possible choices of copula functional forms, see Nelsen (2006). The resulting model can be estimated by a variety of methods such as joint maximum likelihood of all parameters, or two-step estimation in which marginal models are estimated first and θ is estimated at the second step. For details see Trivedi and Zimmer (2007).

An example of copula estimation is Cameron et al. (2004) who use the copula framework to analyze the empirical distribution of two counted measures, y_1 denoting self-reported doctor visits, and y_2 denoting independent report of doctor visits. They derive the distribution of $y_1 - y_2$, by first obtaining the joint distribution $f[y_1, y_2]$. Zimmer and Trivedi (2006) use a trivariate copula framework to develop a joint distribution of two counted outcomes and one binary treatment variable.

There is growing interest in Bayesian analysis of copulas. A recent example is Pitt, Chan, and Kohn (2006), who use a Gaussian copula to model the joint distribution of six count measures of health care. Using a multivariate density of the Gaussian copula Pitt, Chan, and Kohn develop a MCMC algorithm for estimating the posterior distribution for discrete marginals, which is then applied to the case where marginal densities are zero-inflated geometric distributions.

4.7 Simulation-Based Estimation

Simulation-based estimation methods, both classical and Bayesian, deal with distributions that do not have closed form solutions. Such distributions are usually generated when general assumptions are made on unobservable variables that need to be integrated out. The classical estimation methods include both parametric and semiparametric approaches. Hinde (1982) and Gouriéroux and Monfort (1991) discuss a parametric Simulated Maximum Likelihood (SML) approach to estimation of mixed-Poisson regression models. Application to some random effects panel count models has been implemented by Crepon and Duguet (1997). Delgado (1992) treats a multivariate count model as a multivariate nonlinear model and suggests a semiparametric generalized least squares estimator. Gurmu and Elder (2007) develop a flexible semiparametric specification using generalized Laguerre polynomials, and propose a semiparametric estimation method without distributional specification of the unobservable heterogeneity. Another approach (Cameron and Johansson 1997) is based on series expansion methods putting forward a squared polynomial series expansion.

Bayesian estimation of both univariate and multivariate Poisson models is a straightforward Gibbs sampler in the case when regressors do not enter the mean parameters of the Poisson distribution. However, since an objective of economists is to calculate various marginal and treatment effects, such covariates must be introduced. This leads to a necessity to use Metropolis-Hastings steps in the MCMC algorithms. In the era when high speed computers were not available Bayesian estimation of various models relied on deriving a closed form posterior distributions whenever possible. When such closed forms do not exist as in the case of the Poisson model, the posterior can be numerically approximated (El-Sayyad 1973). However, since an inexpensive computer power became available a path of utilizing MCMC methods has been taken. Chib, Greenberg, and Winkelmann (1998) propose algorithms based on MCMC methods to deal with panel count data models with random effects. Chib and Winkelmann (2001) develop an MCMC algorithm of a multivariate correlated count data model. Munkin and Trivedi (2003) extend a count data model to account for a binary endogenous treatment variable. Deb, Munkin, and Trivedi (2006a) introduce a Roy-type count model with the proposed algorithm being more efficient (with respect to computational time and convergence) than the existing MCMC algorithms dealing with Poisson-lognormal densities.

4.7.1 The Poisson-Lognormal Model

Whereas the Poisson-gamma mixture model, i.e., the NB distribution, has proved very popular in application, different distributional assumptions on unobserved heterogeneity might be more consistent with real data. One such example is the Poisson lognormal model.

The Poisson lognormal model is a continuous mixture in which the marginal count distribution is still assumed to be Poisson and the distribution of the multiplicative unobserved heterogeneity term ν is lognormal. Let us reparameterize ν such that $\nu = \exp(\varepsilon)$, where $\varepsilon \sim \mathcal{N}(\mathbf{0}, \sigma^2)$, and let the mean of the marginal Poisson distribution be a function of a vector of exogenous variables $\mathbf{X}$.

The count variables y is distributed as Poisson with mean $\exp(\mu)$, where μ is linear in $\mathbf{X}$ and ε

$$\mu = \mathbf{X}\beta + \varepsilon, \tag{4.47}$$

where $\text{Cov}(\varepsilon|\mathbf{X}) = 0$. Then conditionally on unobserved heterogeneity term ε, the marginal count distribution is defined as

$$f(y|\mathbf{X}, \beta, \varepsilon) = \frac{\exp\left[-\exp\left(\mathbf{X}\beta + \varepsilon\right)\right] \exp\left[y\left(\mathbf{X}\beta + \varepsilon\right)\right]}{y!}. \tag{4.48}$$

The unconditional density $f(y|\mathbf{X}, \beta, \sigma)$ does not have a closed form since the integral

$$\int_{-\infty}^{\infty} f(y|\mathbf{X}, \beta, \varepsilon) f\,(\varepsilon|\sigma)\, d\varepsilon \tag{4.49}$$

cannot be solved. Since in many applications the lognormal distribution is a more appealing assumption on unobserved heterogeneity than gamma, a reliable estimation method of such a model is needed. Estimation by Gaussian quadrature is very feasible; Winkelmann (2004) provides a good illustration.

4.7.2 SML Estimation

Assume that we have N independent observations. An SML estimator of $\theta = (\beta, \sigma)$ is defined as

$$\widehat{\theta}_{SN} = \arg\max_{\theta} \sum_{i=1}^{N} \log\left\{\frac{1}{S}\sum_{s=1}^{S} f(y_i|\mathbf{X}_i, \beta, \varepsilon_i^s)\right\}, \tag{4.50}$$

where ε_i^s $(s = 1, \ldots, S)$ are drawn from density $f\,(\varepsilon|\sigma)$. In our case this density depends on unknown parameter σ. Instead of introducing an importance sampling function we reparameterize the model such that $\mu = \mathbf{X}\beta + \sigma u$, where $u \sim \mathcal{N}\,(\mathbf{0}, 1)$. Then

$$f(y|\mathbf{X}, \beta, \sigma) = \int_{-\infty}^{\infty} \frac{\exp([-\exp(\mathbf{X}\beta + \sigma u)] \exp[y(\mathbf{X}\beta + \sigma u)]}{y!} \times \frac{1}{\sqrt{2\pi}} \exp\left(-\frac{u^2}{2}\right) du. \tag{4.51}$$

In this example the standard normal density of u is a natural candidate for the importance sampling function. Then the SML estimates maximize

$$\sum_{i=1}^{N} \log\left\{\frac{1}{S}\sum_{s=1}^{S} \frac{\exp([-\exp\left(\mathbf{X}_i\beta + \sigma u_i^s\right)] \exp\left[y_i\left(\mathbf{X}_i\beta + \sigma u_i^s\right)\right]}{y!}\right\}, \quad j = 1, 2 \tag{4.52}$$

where u_i^s are drawn from $N\,[0, 1]$.

Since log is an increasing function, the sum over i and log do not commute. Then if S is fixed and N tends to infinity $\widehat{\theta}_{SN}$ is not consistent. If both S and N tend to infinity then the SML estimator is consistent.

4.7.3 MCMC Estimation

Next we discuss the choice of the priors and outline the MCMC algorithm. For each observation i derive the joint density of the observable data and latent variables. We adopt the Tanner–Wong data augmentation approach and include latent variables μ_i $(i = 1, \ldots, N)$ in the parameter set making it a part of the posterior. Conditional on μ_i the full conditional density of β is a tractable normal distribution.

Denote $\Delta_i = (\mathbf{X}_i, \beta, \sigma)$. Then the joint density of the observable data and latent variables for observation i is

$$f(y_i, \mu_i|\Delta_i) = \frac{\exp[y_i\mu_i - \exp(\mu_i)]}{y_i!}\frac{1}{\sqrt{2\pi\sigma^2}}\exp[-0.5\sigma^{-2}(\mu_i - \mathbf{X}_i\beta)^2]. \quad (4.53)$$

The posterior density kernel is the product of $f(y_i, \mu_i|\Delta_i)$ for all N observations and the prior densities of the parameters.

We choose a normal prior for parameter β, center it at zero and choose relatively large variance

$$\beta \sim \mathcal{N}(\mathbf{0}_k, 10\mathbf{I}_k). \quad (4.54)$$

The priors for the variance parameter is

$$\sigma^{-2} \sim \mathcal{G}\left(\frac{n}{2}, \left(\frac{c}{2}\right)^{-1}\right) \quad \text{where } n = 5 \text{ and } c = 10.$$

First, we block the parameters as $\mu_i, \beta, \sigma^{-2}$. The steps of the MCMC algorithm are the following:

1. The full conditional density for μ_i is proportional to

$$p(\mu_i|\Delta_i) = \frac{\exp[y_i\mu_i - \exp(\mu_i)]}{y_i!}\exp[-0.5\sigma^{-2}(\mu_i - \mathbf{X}_i\beta)^2]. \quad (4.55)$$

Sample μ_i using the Metropolis–Hasting algorithm with normal distribution centered at the modal value of the full conditional density for the proposal density. Let

$$\widehat{\mu}_i = \arg\max \log p(\mu_i|\Delta_i) \quad (4.56)$$

and $\mathbf{V}_{\widehat{\mu}_i} = -(\mathbf{H}_{\widehat{\mu}_i})^{-1}$ be the negative inverse of the Hessian of log $p(\mu_i|\Delta_i)$ evaluated at the mode $\widehat{\mu}_i$. Choose the proposal distribution $q(\mu_i) = \phi(\mu_i|\widehat{\mu}_i, \mathbf{V}_{\widehat{\mu}_i})$. When a proposal value μ_i^* is drawn, the chain moves to the proposal value with probability

$$\alpha(\mu_i, \mu_i^*) = \min\left\{\frac{p(\mu_i^*|\Delta_i)q(\mu_i)}{p(\mu_i|\Delta_i)q(\mu_i^*)}, 1\right\}. \quad (4.57)$$

If the proposal value is rejected, the next state of the chain is at the current value μ_i.

2. Specify prior distributions $\beta \sim \mathcal{N}\left[\underline{\beta}, \underline{H}_{\beta}^{-1}\right]$. The the conditional distribution of β is $\beta \sim \mathcal{N}[\overline{\beta}, \overline{\mathbf{H}}_{\beta}^{-1}]$ where

$$\overline{\mathbf{H}}_{\beta} = \underline{\mathbf{H}}_{\beta} + \sum_{i=1}^{N} \mathbf{X}_i' \sigma^{-2} \mathbf{X}_i \tag{4.58}$$

$$\overline{\beta} = \overline{\mathbf{H}}_{\beta}^{-1} \left[\underline{\mathbf{H}}_{\beta} \underline{\beta} + \sum_{i=1}^{N} \mathbf{X}_i' \sigma^{-2} \mu_i \right]. \tag{4.59}$$

3. Finally, specify the prior $\sigma^{-2} \sim \mathcal{G}(n/2, (c/2)^{-1})$. Then the full conditional of σ^{-2} is

$$\mathcal{G}\left(\frac{n+N}{2}, \left[\frac{c}{2} + \sum_{i=1}^{N} \frac{(\mu_i - \mathbf{X}_i \beta)^2}{2} \right]^{-1} \right). \tag{4.60}$$

This concludes the MCMC algorithm.

4.7.4 A Numerical Example

To examine properties of our SML estimator and MCMC algorithm and their performance, we generate several artificial data sets. In this section we report our experience based on one specific data generating process (d.g.p.). We generate 1000 observations using the following structure with assigned parameter values: $X_i = (1, x_i)$ and $x_i \sim \mathcal{N}(0, 1)$; $\beta = (2, 1)$, $\sigma = 1$. Such parameter values generate a count variable with mean of 19. The priors for parameters are selected to be uninformative but still proper, i.e., $\beta \sim \mathcal{N}(\mathbf{0}, 10 I_2)$ and $\sigma^{-1} \sim \mathcal{G}(\frac{n}{2}, (\frac{c}{2})^{-1})$ with $n = 5$ and $c = 10$.

Table 4.3 gives SML estimates and the posterior means and standard deviations for the parameters based on 10,000 replications preceded by 1000 replications of the burn-in phase. It also gives the true values of the parameters in the d.g.p. As can be seen from the table the true values of the parameters fall close to the centers of the estimated confidence intervals. However, if the true values of β is selected such that the mean of the count variable is increased to 50, the estimates of the SML estimator display a considerable bias when the number of simulations is limited to $S = 500$.

TABLE 4.3

MCMC Estimation for Generated Data

Parameter	True Value of d.g.p.	MCMC	SML
β_0 (*Constant*)	2	1.984	1.970
		0.038	0.036
β_1 (x)	1	0.990	0.915
		0.039	0.027
σ	1	1.128	1.019
		0.064	0.026

4.7.5 Simulation-Based Estimation of Latent Factor Model

We now consider some issues in the estimation of the latent factor model of Subsection 4.6.1. The literature indicates that S should increase faster than $\sqrt{N}$, but this does not give explicit guidance in choosing S. In practice some tests of convergence should be applied to ensure that S was set sufficiently high. Using a small number of draws (often 50–100) works well for models such as the mixed multinomial logit, multinomial probit, etc. However, more draws are required for models with endogenous regressors. Thus computation can be quite burdensome if the standard methods are used. For the model described in Subsection 4.4.3, Deb and Trivedi (2006b) find the standard simulation methods to be quite slow. They adapt a simulation acceleration technique that uses quasi-random draws based on Halton sequences (Bhat 2001; Train 2002). This method, instead of using S pseudo-random points, makes draws based on a nonrandom selection of points within the domain of integration. Under suitable regularity conditions, the integration error using pseudo-random sequences is in the order of N^{-1} as compared to pseudo-random sequences where the convergence rate is $N^{-1/2}$ (Bhat 2001). For variance estimation, they use the robust Huber–White formula.

4.8 Software Matters

In the past decade the scope of applying count data models has been greatly enhanced by availability of good software and fast computers. Leading microeconometric software packages such as Limdep, SAS, Stata, and TSP provide a good coverage of the basic count model estimation for single equation and Poisson-type panel data models. See Greene (2007a) for details of Limdep, and Stata documentation for coverage of Stata's official commands; also see Kitazawa (2000) and Romeu (2004). The present authors are especially familiar with Stata official estimation commands. The Poisson, ZIP, NB, and ZINB are covered in the Stata reference manuals. Stata commands support calculation of marginal effects for most models. Researchers should also be aware that there are other add-on Stata commands that can be downloaded from Statistical Software Components Internet site at Boston College Department of Economics. These include commands for estimating hurdle and finite mixture models due to Deb (2007), goodness-of-fit and model evaluation commands due to Long and Freese (2006), quantile count regression commands due to Miranda (2006), and commands due to Deb and Trivedi (2006b) for simulation-based estimation of multinomial latent factor model discussed in Subsection 4.4.3. Stata 11, released in late 2009, facilitates implementing GMM estimation of cross-section and panel data models based on the exponential mean specification.

4.8.1 Issues with Bayesian Estimation

The main computational difficulty with the simulated maximum likelihood approach is the fact that when the number of simulations is small the parameters estimates are biased. This is true for even simple one equation models. When the model becomes multivariate and multidimensional a much larger number of simulations is required for consistent estimation. Sometimes it can be very time consuming with the computational time increasing exponentially with the number of parameters. In Bayesian Markov chain Monte Carlo the computational time increases proportionally to the dimension of the model. Besides, the approach does not suffer from the bias problem of the SML. However, there are computational problems with the Markov chain Monte Carlo methods as well. Such problems arise when the produced Markov chains display a high level of serial correlation leading to the posterior distribution being saturated in a closed neighborhood with the Markov chain not visiting the entire support of the posterior distribution. When the serial correlation is high but reasonably smaller, the solution is to use a relatively larger number of replications for a precise estimation of the posterior. However, when the serial correlations are close to one such a problem must have a model specific solution.

Bayesian model specification requires a choice of priors which can result in a completely different posterior. When improper priors are selected this can lead to improper posterior. In general, Bayesian modeling does not restrict itself to only customized models and new programs must be written for various model specifications. Many programs for the well-developed existing models are written in MATLAB. Koop, Porier, and Tobias (2007) give an excellent overview of different methods and models and provide a rich library of programs. This book can serve as a good MATLAB reference for researchers dealing with Bayesian modeling and estimation.

References

Bago d'Uva, T. 2005. Latent class models for use of primary care: evidence from a British panel. *Health Economics* 14: 873–892.

Bago d'Uva, T. 2006. Latent class models for use of health care. *Health Economics* 15: 329–343.

Besag, J. 1974. Spatial interaction and the statistical analysis of lattice systems. *Journal of Royal Statistical Society 36B*: 192–225.

Bhat, C. R. 2001. Quasi-random maximum simulated likelihood estimation of the mixed multinomial logit model. *Transportation Research: Part B* 35: 677–693.

Blundell, R., R. Griffith, and F. Windmeijer. 2002. Individual effects and dynamics in count data models. *Journal of Econometrics* 102: 113–131.

Bohning, D., and R. Kuhnert. 2006. Equivalence of truncated count mixture distributions and mixtures of truncated count distributions, *Biometrics* 62(4): 1207–1215.

Breslow, N. E., and D. G. Clayton. 1993. Approximate inference in generalized linear mixed models. *Journal of American Statistical Association* 88: 9–25.

Cameron, A. C., and P. Johansson. 1997. Count data regressions using series expansions with applications. *Journal of Applied Econometrics* 12(3): 203–223.

Cameron, A. C., T. Li, P. K. Trivedi, and D. M. Zimmer. 2004. Modeling the differences in counted outcomes using bivariate copula models: with application to mismeasured counts. *Econometrics Journal* 7(2): 566–584.

Cameron, A. C., and P. K. Trivedi. 1998. *Regression Analysis of Count Data*. New York: Cambridge University Press.

Cameron, A. C., and P. K. Trivedi. 2005. *Microeconometrics: Methods and Applications*. Cambridge, U.K.: Cambridge University Press.

Cameron, A. C., and P. K. Trivedi. 2009. *Microeconometrics Using Stata*. College Station, TX: Stata Press.

Chamberlain, G. 1982. Multivariate regression models for panel data. *Journal of Econometrics* 18: 5–46.

Chamberlain, G. 1992. Comment: sequential moment restrictions in panel data. *Journal of Business and Economic Statistics* 10: 20–26.

Chang, F. R., and P. K. Trivedi. 2003. Economics of self-medication: theory and evidence. *Health Economics* 12: 721–739.

Chib, S., E. Greenberg, and R. Winkelmann. 1998. Posterior simulation and Bayes factor in panel count data models. *Journal of Econometrics* 86: 33–54.

Chib, S., and R. Winkelmann. 2001. Markov chain Monte Carlo analysis of correlated count data. *Journal of Business and Economic Statistics* 19: 428–435.

Crepon, B., and E. Duguet. 1997. Research and development, competition and innovation: pseudo-maximum likelihood and simulated maximum likelihood method applied to count data models with heterogeneity. *Journal of Econometrics* 79: 355–378.

Davidson, R., and J. G. MacKinnon. 2004. *Econometric Theory and Methods*, Oxford, U.K.: Oxford University Press.

Davis, R. A., W. T. M. Dunsmuir, and S. B. Streett. 2003. Observation-driven models for Poisson counts. *Biometrika* 90: 777–790.

Deb, P. 2007. FMM: Stata module to estimate finite mixture models. Statistical Software Components S456895. Boston College Department of Economics.

Deb, P., M. K. Munkin, and P. K. Trivedi. 2006a. Private insurance, selection, and the health care use: a Bayesian analysis of a Roy-type Model. *Journal of Business and Economic Statistics* 24: 403–415.

Deb, P., M. K. Munkin, and P. K. Trivedi. 2006b. Bayesian analysis of the two-part model with endogeneity: application to health care expenditure. *Journal of Applied Econometrics* 21(6): 1081–1099.

Deb, P., and P. K. Trivedi. 1997. Demand for medical care by the elderly: a finite mixture approach. *Journal of Applied Econometrics* 12: 313–326.

Deb, P., and P. K. Trivedi. 2002. The structure of demand for medical care: latent class versus two-part models. *Journal of Health Economics* 21: 601–625.

Deb, P., and P. K. Trivedi. 2006a. Specification and simulated likelihood estimation of a non-normal treatment-outcome model with selection: application to health care utilization. *Econometrics Journal* 9: 307–331.

Deb, P., and P. K. Trivedi. 2006b. Maximum simulated likelihood estimation of a negative-binomial regression model with multinomial endogenous treatment. *Stata Journal* 6: 1–10.

Delgado, M. A. 1992. Semiparametric generalized least squares in the multivariate nonlinear regression model. *Econometric Theory* 8: 203–222.

Demidenko, E. 2007. Poisson regression for clustered data. *International Statistical Review* 75(1): 96–113.

Diggle, P., P. Heagerty, K. Y. Liang, and S. Zeger. 2002. *Analysis of Longitudinal Data*. Oxford, U.K.: Oxford University Press.

El-Sayyad, G. M. 1973. Bayesian and classical analysis of poisson regression. *Journal of the Royal Statistical Society*, Series B (Methodological) 35(3): 445–451.

Fruhwirth-Schnatter, S. 2006. *Finite Mixture and Markov Switching Models*. New York: Springer-Verlag.

Gouriéroux, C., and A. Monfort. 1991. Simulation based inference in models with heterogeneity. *Annales d'Economie et de Statistique* 20/21: 69–107.

Gourieroux, C., and A. Monfort. 1997. *Simulation Based Econometric Methods*. Oxford, U.K.: Oxford University Press.

Greene, W. H. 2007a. *LIMDEP 9.0 Reference Guide*. Plainview, NY: Econometric Software, Inc.

Greene, W. H. 2007b. Functional form and heterogeneity in models for count data. *Foundations and Trends in Econometrics* 1(2): 113–218.

Griffith, D. A., and R. Haining. 2006. Beyond mule kicks: the Poisson distribution in geographical analysis. *Geographical Analysis* 38: 123–139.

Guo, J. Q., and P. K. Trivedi. 2002. Flexible parametric distributions for long-tailed patent count distributions. *Oxford Bulletin of Economics and Statistics* 64: 63–82.

Gurmu, S., and J. Elder. 2007. A simple bivariate count data regression model. *Economics Bulletin* 3 (11): 1–10.

Gurmu, S., and P. K. Trivedi. 1996. Excess zeros in count models for recreational trips. *Journal of Business and Economic Statistics* 14: 469–477.

Hardin, J. W., H. Schmiediche, and R. A. Carroll. 2003. Instrumental variables, bootstrapping, and generalized linear models. *Stata Journal* 3: 351–360.

Hausman, J. A., B. H. Hall, and Z. Griliches. 1984. Econometric models for count data with an application to the patents–R and D relationship. *Econometrica* 52: 909–938.

Hinde, J. 1982. Compound Poisson regression models. In R. Gilchrist ed., 109-121, GLIM 82: *Proceedings of the International Conference on Generalized Linear Models*. New York: Springer-Verlag.

Jung, R. C., M. Kukuk, and R. Liesenfeld. 2006. Time series of count data: modeling, estimation and diagnostics. *Computational Statistics & Data Analysis* 51: 2350–2364.

Kaiser, M., and N. Cressie. 1997. Modeling Poisson variables with positive spatial dependence. *Statistics and Probability Letters* 35: 423–32.

Karlis, D., and E. Xekalaki. 1998. Minimum Hellinger distance estimation for Poisson mixtures. *Computational Statistics and Data Analysis* 29: 81–103.

Kitazawa, Y. 2000. TSP procedures for count panel data estimation. Fukuoka, Japan: Kyushu Sangyo University.

Koenker, R. 2005. *Quantile Regression*. New York: Cambridge University Press.

Koop, G., D. J. Poirier, and J. L. Tobias. 2007. *Bayesian Econometric Methods*. Volume 7 of Econometric Exercises Series. New York: Cambridge University Press.

Lancaster, T. 2000. The incidental parameters problem since 1948. *Journal of Econometrics* 95: 391–414.

Long, J. S., and J. Freese. 2006. *Regression Models for Categorical Dependent Variables Using Stata*, 2nd ed. College Station, TX: Stata Press.

Lourenco, O. D., and P. L. Ferreira. 2005. Utilization of public health centres in Portugal: effect of time costs and other determinants. Finite mixture models applied to truncated samples. *Health Economics* 14: 939–953.

Lu, Z., Y. V. Hui, and A. H. Lee. 2003. Minimum Hellinger distance estimation for finite mixtures of Poisson regression models and its applications. *Biometrics* 59(4): 1016–1026.

MacDonald, I. L., and W. Zucchini. 1997. *Hidden Markov and Other Models for Discrete-Valued Time Series*. London: Chapman & Hall.

Machado, J., and J. Santos Silva. 2005. Quantiles for counts. *Journal of American Statistical Association* 100: 1226–1237.

Marshall, A. W., and I. Olkin. 1990. Multivariate distributions generated from mixtures of convolution and product families. In H.W. Block, A.R. Sampson, and T.H. Savits, eds, *Topics in Statistical Dependence*, IMS Lecture Notes-Monograph Series, Volume 16: 371–393.

Miranda, A. 2006. QCOUNT: Stata program to fit quantile regression models for count data. Statistical Software Components S456714. Boston College Department of Economics.

Miranda, A. 2008. Planned fertility and family background: a quantile regression for counts analysis. *Journal of Population Economics* 21: 67–81.

Morton, R. 1987. A generalized linear model with nested strata of extra-Poisson variation. *Biometrika* 74: 247–257.

Mullahy, J. 1997. Instrumental variable estimation of Poisson regression models: application to models of cigarette smoking behavior. *Review of Economics and Statistics* 79: 586–593.

Mundlak, Y. 1978. On the pooling of time series and cross section data. *Econometrica* 56: 69–86.

Munkin, M., and P. K. Trivedi. 1999. Simulated maximum likelihood estimation of multivariate mixed-Poisson regression models, with application. *Econometric Journal* 1: 1–21.

Munkin, M. K., and P. K. Trivedi. 2003. Bayesian analysis of self-selection model with multiple outcomes using simulation-based estimation: an application to the demand for healthcare. *Journal of Econometrics* 114: 197–220.

Munkin, M. K., and P. K. Trivedi, 2008. Bayesian analysis of the ordered Probit model with endogenous selection, *Journal of Econometrics* 143: 334–348.

Munkin, M. K., and P. K. Trivedi, 2009. A Bayesian analysis of the OPES Model with a non-parametric component: application to dental insurance and dental care. Forthcoming in *Advances in Econometrics, Volume 23: Bayesian Econometrics*, edited by Siddhartha Chib, Gary Koop, and Bill Griffiths. Elsevier Press.

Nelsen, R. B. 2006. *An Introduction to Copulas*. 2nd ed. New York: Springer.

Newey, W. 1987. Efficient estimation of limited dependent variable models with endogenous explanatory variables. *Journal of Econometrics* 36: 231–250.

Pitt, M., D. Chan, and R. Kohn. 2006. Efficient Bayesian inference for Gaussian copula regression. *Biometrika* 93: 537–554.

Rabe-Hesketh, S., A. Skrondal, and A. Pickles. 2002. Reliable estimation of generalized linear mixed models using adaptive quadrature. *Stata Journal* 2: 1–21.

Romeu, A. 2004. *ExpEnd:* Gauss code for panel count data models. *Journal of Applied Econometrics* 19: 429–434.

Skrondal, A., and S. Rabe-Hesketh. 2004. *Generalized Latent Variable Modeling: Multilevel, Longitudinal and Structural Equation Models.* London: Chapman & Hall.

Stewart, M. 2007. The inter-related dynamics of unemployment and low-wage employment. *Journal of Applied Econometrics* 22(3): 511–531.

Terza, J. 1998. Estimating count data models with endogenous switching: sample selection and endogenous switching effects. *Journal of Econometrics* 84: 129–139.

Train, K. 2002. *Discrete Choice Methods with Simulation.* New York: Cambridge University Press.

Trivedi, P. K., and D. M. Zimmer. 2007. Copula modeling: an introduction for practitioners. *Foundations and Trends in Econometrics* 1(1): 1–110.

Vuong, Q. 1989. Likelihood ratio tests for model selection and non-nested hypotheses. *Econometrica* 57: 307–333.

Wang, K., K. K. W. Yau, and A. H. Lee. 2002. A hierarchical Poisson mixture regression model to analyze maternity length of hospital stay. *Statistics in Medicine* 21: 3639–3654.

Windmeijer, F. 2008. GMM for panel count data models. *Advanced Studies in Theoretical and Applied Econometrics* 46: 603–624.

Windmeijer, F., and J. M. C. Santos Silva. 1997. Endogeneity in count data models. *Journal of Applied Econometrics* 12: 281–294.

Winkelmann, R. 2004. Health care reform and the number of doctor visits – an econometric analysis. *Journal of Applied Econometrics* 19: 455–472.

Winkelmann, R. 2005. *Econometric Analysis of Count Data.* 5th ed. Berlin: Springer-Verlag.

Winkelmann, R. 2006. Reforming health care: Evidence from quantile regressions for counts. *Journal of Health Economics* 25: 131–145.

Wooldridge, J. M. 1997. Multiplicative panel data models without the strict exogeneity assumption. *Econometric Theory* 13: 667–678.

Wooldridge, J. M. 1999. Distribution-free estimation of some nonlinear panel data models. *Journal of Econometrics* 90: 77–97.

Wooldridge, J. M. 2002. *Econometric Analysis of Cross Section and Panel Data,* 2001. Cambridge, MA: MIT Press.

Wooldridge, J. M. 2005. Simple solutions to the initial conditions problem in dynamic, nonlinear panel data models with unobserved heterogeneity. *Journal of Applied Econometrics* 20: 39–54.

Xiang, L., K. K. W. Yau, Y. Van Hui, and A. H. Lee. 2008. Minimum Hellinger distance estimation for k-component Poisson mixture with random effects. *Biometrics* 64(2): 508–518.

Zimmer, D. M., and P. K. Trivedi. 2006. Using trivariate copulas to model sample selection and treatment effects: application to family health care demand. *Journal of Business and Economic Statistics* 24(1): 63–76.

5

An Introduction to Textual Econometrics

Stephen Fagan and Ramazan Gençay

CONTENTS

5.1 Introduction

> If we are not to get lost in the overwhelming, bewildering mass of statistical data that are now becoming available, we need the guidance and help of a powerful theoretical framework. (Frisch 1933)

This quote from Ragnar Frisch's Editor's Note in the first issue of *Econometrica* in 1933 has never seemed more timely. The phrase "data rich, information poor" is often used to characterize the current state of our digitized world. Over recent decades, data storage and availability has been growing at an exponential rate, and currently, data sets on the order of terabytes are not uncommon. While a portion of this new data is in the form of numerical or categorical data in well-structured databases, the vast majority is in the form of unstructured textual data. These news stories, government reports, blog entries, e-mails, Web pages, and the like are the medium of information flow throughout the world. It is this unstructured data that most decision-makers turn to for information.

In the context of numerical and categorical data, Frisch's desire for powerful tools for data analysis has, to a great extant, been satiated. The field of econometrics has expanded at a rate that has well matched the increasing availability of numerical data. However, in the context of the vast amount of textual data that has become available, econometrics has barely scratched the surface of its potential. Of course, the problems involved in the analysis of textual data are much greater than those of other forms of data. The complexity and nuances of language, as well as its very-high-dimensional character, have made it an illusive target for quantitative analysis. Nevertheless, decision-making is at the core of economic behavior, and given the importance of textual data in the decision-making process, there is a clear need for more powerful econometric tools and techniques that will permit the important features of this data to be included in our economic and financial models.

While textual econometrics is still early in its development, some important progress has been made that has empirically demonstrated the importance textual data. Research has shown that even at a coarse level of sophistication, automated textual processing can extract useful knowledge from large collections of textual documents. Most of this new work has taken one of two approaches: either the textual data have been greatly simplified into a small number of dimensions so that traditional econometric techniques can be brought to bear, or new techniques have been adopted that are better able to deal with the complexity and high-dimensional character of the data. Both approaches have proven useful and will be discussed in this chapter.

This early work is important for two reasons: First, it has shown that textual data is an economically significant source of information and is not beyond quantitative analysis. Second, while the techniques employed in these studies can capture only a fraction of the linguistic sophistication contained in the documents, they do offer good baselines against which we can compare future work. The development of linguistically sophisticated techniques is a multidisciplinary endeavor pursued by researchers in linguistics, natural language processing, text mining, and other areas. New techniques are constantly being developed, and their application in econometrics will need to be tested to determine whether an increase in complexity is made worthwhile by an increased ability to quantify the information embedded in textual data.

The aim of this chapter is to introduce the reader to this new field of textual econometrics. It proceeds as follows: Section 5.2 provides a review of some recent applications of textual analysis in the economics and finance literature, Section 5.3 describes some special properties of textual data, Section 5.4 identifies various sources of textual data, Sections 5.5 through 5.7 describe the important tasks of preprocessing, creating feature vectors, and reducing dimensionality. Section 5.8 considers the application of automated document classification. Section 5.9 directs readers to popular software that can process textual data, and Section 5.10 concludes. Throughout the chapter, references are given for further reading and public resources that will be useful for the interested researcher.

5.2 Textual Analysis in Economics and Finance

Since the introduction of event studies into economics (Fama et al. 1969), researchers have been investigating the economic impact of media releases on markets (Chan 2003; Mitchell and Mulherin 1994; Niederhoffer 1971). While manual classification of stories as positive or negative was used in many such studies, very little of the actual information content could be extracted for use in statistical analysis. To a large degree, it was often the fact that there was new information, not the information itself, that was being analyzed.

The mere fact that communication is occurring, either to or between market participants, can be useful in understanding markets. For example, Coval and Shumway (2001) look at the volume of noise in a Chicago Board of Trade futures pit and found that following a rise in the sound levels, prices become more volatile, depth declines, and information asymmetry increases. In the written domain, Wysocki (1999) examines the relationship between posting volume on Internet stock message boards and stock market activity, finding that market characteristics determine posting levels and at the same time, posting levels predict future volume and returns.

Using more linguistic content, Antweiler and Frank (2004) investigate whether there is any predictive information in Internet stock message boards. Having manually classified a training set of 1000 messages as either BUY, SELL, or HOLD, they use the Naive Bayes classification algorithm to classify a larger set of 1.5 million messages from Yahoo!Finance and Raging Bull. The classifier accepts the document as a "bag of words" without retaining any linguistic structure, and uses simple word frequencies to decide on an appropriate indicator. By aggregating these indicators within each period, they construct a bullishness sentiment indicator that has predictive power for market volatility, but not for returns.

While Antweiler and Frank (2004) use all of the words (but not their linguistic relationships) of their documents, they ultimately reduce the dimensionality of textual data significantly by identifying the tone or sentiment of documents. Similarly, Tetlock, Saar-Tsechansky, and Macskassy (2008) create a measure of news sentiment by identifying the fraction of negative words in over 350,000 firm-specific news stories from the *Wall Street Journal* and the Dow Jones News Service.[1] Using the traditional regression methodology, they find that the information captured by this simple variable has predictive power for earnings and equity returns.

These two papers represent the two approaches to textual data: either it has been greatly simplified into a small number of dimensions so that traditional

[1] Their source for the news stories was the Factiva database (`www.factiva.com`).

econometric techniques can be brought to bear, or new techniques have been adopted that are better able to deal with the complexity and high-dimensional character of the data.

Bhattacharya et al. (2008) manually read and classified 171,488 news stories written during the Internet IPO frenzy to test whether the media "hyped" Internet stocks by giving them more positive news relative to a group of non-Internet IPOs during the same period. They find that there was indeed significant media hype, but after controlling for other return-related factors, find that the aggregated news affect from a given day would last only two days and explained only a small portion of the difference between Internet and non-Internet IPO returns.

Tetlock (2007) looks for interactions between the content of a popular *Wall Street Journal* column and the stock market over a 16-year period. After reducing the dimensionality of the textual data through dictionary classifications and principal component analysis, he uses a vector autoregressive methodology to identify the relationship between media pessimism and market returns and volumes. Findings indicate that the WSJ column does impact returns and volume, and that his textual variable depends on prior market activity.

Like Antweiler and Frank (2004), Das and Chen (2007) study the impact of stock message boards on market characteristics. Rather than using a single classification algorithm to label messages as BUY, SELL, or HOLD, they employ five different classification algorithms which get to vote for the final label. These classifiers use different data sets, extracted using various grammatical parsing and statistical techniques, to decide on a label. Within the forecasting literature, combining forecasts has been found to improve forecasting performance (Armstrong 2001; Bates and Granger 1969), and compared to the Bayes text classifier, the combined algorithms have better classification accuracy.

Financial economics is not the only field doing textual econometrics, and in fact only a minority of the work to date has been published in economic or finance journals. The majority of the work has been done in the area of Knowledge Discovery in Databases (KDD). KDD is a multidisciplinary field (drawing primarily on developments of computing sciences) whose goal is to extract nontrivial, implicit, previously unknown, and potentially useful information from databases. A critical step in the KDD process often uses data mining[2] techniques to extract hidden patterns from data. In fact KDD and data mining are already used in many economic applications including marketing, fraud detection, and credit scoring, just to name a few.

Economic and financial data are popular for KDD research because of their abundance and the complex relationships within large economic datasets. A large portion of this research focuses on stock market forecasting since the difficulty of this task is well established. Consequently, techniques that work in this area are likely to represent true innovations. Examples of this work

---

[2] While the phrase "data mining" has negative associations of "data snooping" within the field of econometrics, the term also refers to a respected, well-developed field of computing science.

include that by Mittermayer (2004), Fung, Yu, and Lu (2005), Kroha, Baeza-Yates, and Krellner (2006), Rachlin et al. (2007), and Schumaker and Chen (2009). Often, this research is more data-driven than work in economics with, for example, features chosen for their predictive ability (e.g., associations with trending prices) rather than their linguistic properties (e.g., negative affectivity). This may be an important difference since, for example, markets can fall on "good news" if the news was not as "good" as expected.

Beyond market forecasting, there has been a small amount of research on the impact of textual data on macroeconomic conditions (Gao and Beling 2003) and in the area of labor laws (Ticom and de Lima 2007). However, there are many other areas of economics and finance where the effects of textual information may prove illuminating. Such areas include corporate finance, bankruptcy and default, public policy, consumer behavior, among others.

5.3 Properties of Textual Data

Many researchers and philosophers have argued that human language and human intelligence are intimately linked (Dennett 1994). In fact, one of the long-standing proposals for testing genuine artificial intelligence is whether a computer could engage a human in conversation with sufficient ability that the human cannot tell, from language use alone, whether they are conversing with a human or not (Turing 1950). While this level of automated linguistic ability is still a distant goal, many less ambitious tasks have been automated through the exploitation of structural and statistical regularities of language (Jurafsky and Martin 2000). Clearly, the research outlined in the previous section was not based on a deep level of automated understanding of language, and yet many of the results are interesting and useful. On the one hand, this indicates that there is still much to be gained from further advances in language sciences and textual econometrics. However, it is necessary to be aware of the challenges that textual data presents.

The primary function of language is to encode information that other language users can extract. This is achieved through the sequential ordering of linguistic primitives, either simple sounds (phonemes) in spoken language or written symbols. In the written domain, these symbols are combined to form words, and these words combine to form phrases and sentences, which combine to form larger linguistic entities. The word is the smallest meaningful unit, and much of the current textual econometric research deals with language at the level of words, ignoring the important structural relationships between words.

While words are the smallest syntactic unit, we are ultimately interested in the intended meaning of the word. However, there is a many-to-many relationship between words and meanings. The fact that a word can have more than one meaning (or sense, as they are sometimes called) is referred to as *polysemy*. On average, English words have 1.4 meanings, with some words having more than 70 (Fellbaum 1998). In addition to multiple meanings, words can also have multiple linguistic functions, or part-of-speech (PoS). Thus, using words as independent variables rather than their intended sense is akin to introducing measurement error into a model.

On the other side of the word-meaning relationship, a sense/meaning can often be expressed with more than one word. Two words sharing a common meaning are called *synonyms*, and in the English language, an average of 1.75 words express a single meaning (Fellbaum 1998). This phenomenon is intensified by the fact that most of us are taught not to be repetitive in our writing style, and at the same time when writing about a given topic, a small set of senses will be used multiple times. This means that synonyms can exhibit strong correlations, and lead to problems akin to multicollinearity.

Moving up to the level of sentences creates another layer of language-to-meaning difficulties. Ambiguity is one such difficulty. For example, consider the sentence "Jill saw Jack with binoculars." Does Jill or Jack have the binoculars? The prepositional phrase "with binoculars" is not unambiguously attached to either Jill of Jack. The use of metaphors and sarcasm, which are not meant to be taken literally, is also difficult to account for. Another, though not so pressing, difficulty is the fact that languages evolve over time, and grammar rules are not always strictly obeyed. Overcoming these difficulties is actively being pursued by language researchers, and there have been many developments including part-of-speech identification, grammatical parsing, word-sense disambiguation, automated translation, and others.

Such difficulties have led researchers to work with textual data at the word level. This approach, sometimes called the "bag-of-words" approach, treats documents as unordered sets of words and ignores the sequential and grammatical structure. The assumption that word occurrences are independent features of a document is clearly false and results in the loss of the vast majority of the information contained in the document. However, this assumption has been defended on pragmatic grounds. As Fung, Yu, and Lu (2005) state,

> Research shows that this assumption will not harm the system performance. Indeed, maintaining the dependency of features is not only extremely difficult, but also may easily degrade the system performance. (page 6)

Antweiler and Frank (2004) use the same defense:

> As an empirical matter it has been found that a surprisingly small amount is gained at substantial cost by attempting to exploit grammatical structure in the algorithms. (page 1264)

Simple word choice can be a useful predictor of a document's tone and the author's sentiment. That is, words alone can capture some of the emotive content of the text. Words also permit researchers to capture documents on a given subject. However, much is lost, and as advances are made in the language sciences, we should encourage the use of new, performance-enhancing techniques. Some of these advances are discussed in later sections of this chapter.

Given the importance of words in current textual econometrics, it is important to consider some of their stylized facts. To begin, the distribution of words in a natural language can be approximated by a Zipfian (a.k.a. Yale) distribution (Zipf 1932). This characterization of language is usually referred to as Ziph's law or the rank-size law and it states that the frequency (f) of a word is inversely proportional to its statistical rank (r). Thus, within a large corpus (i.e., a large collection of texts) there is a constant k such that

$$f \cdot r = k \tag{5.1}$$

Graphically, Ziph's law predicts that a scatter plot of log(f) against log(r) will form a straight line with a slope of -1. A consequence of this property is that in a natural language, there are a few very frequent words, a relatively small group of medium frequency words, and a very large number of rarely occurring words. For example, in the Brown Corpus, consisting of over one million words, half of the word volume consists of repeated uses of only 135 words.

Zipf's law has many implications for the statistical properties of textual data. For example, observational data on word occurrence will be very sparse, with only a few words having many examples. This can impact classification and prediction problems since even in large collections of documents, there can be words that occur in only a single document. Thus, classification algorithms must be robust to overfitting since, otherwise, training documents will be classified according to their unique words.

While Zipf's law is fairly accurate over most corpora, Mandelbrot (1954) noted that the fit is poor for both very-high- and very-low-frequency words, and proposed the following alternative characterization of the frequency-rank relationship:

$$f = P(r + \rho)^{-B} \tag{5.2}$$

where P, ρ, and B are parameters describing the use of words in the text. Both Mandelbrot's and Zipf's characterization of the distribution of words is consistent with Ziph's argument that language properties develop as an efficient compromise that minimizes the efforts of listeners (who prefer large vocabularies to reduce ambiguity) and speakers (who prefer smaller vocabularies to reduce effort). In a similar argument, Zipf proposes that the number of meanings (m) of a word is related to its frequency by

$$m \propto \sqrt{f} \tag{5.3}$$

or, given Zipf's law,

$$m \propto \frac{1}{\sqrt{r}}. \tag{5.4}$$

An important partitioning of a lexicon (i.e., a dictionary or collection of words) is between "content" words and "function" words. Function words are (usually little) words that have important grammatical roles, including determiners (e.g., the, a, that, my, ...), prepositions (e.g., of, at, in, ...), and others. There are relatively few of these words (around 300 in English), but they occur very frequently. Content words, on the other hand, constitute the vast majority of words in a language. They are the nouns (e.g., Jim, house, question, ...), adjectives (e.g., happy, old, slow, ...), and full verbs (e.g., run, grow, save, ...), and others that present the informational content of texts. As expected these types of words have very different distributional properties.

While function words appear to occur fairly uniformly throughout documents, content words appear to cluster. Zipf (1932) noted this phenomenon by examining the distance (D) between occurrences of a given word, and then calculating the frequency (F) of these distances. He found that the number of observations of a given distance between word occurrences was inversely related to the magnitude of the distance. That is,

$$F \propto \frac{1}{D^p} \tag{5.5}$$

for values of p around 1.2. This implies that most content words occur near other occurrences of the same word. Thus, content words have persistence over time.

Several models have been proposed to model word distributions. Zipf's observation about the persistence of content words makes the Poisson distribution a poor choice because of its assumption of independence between word occurrences. Better models include mixture models of multiple Poissons and the K mixture model proposed by Katz (1996). For an expanded discussion of word distribution models, and other statistical properties of natural languages, interested readers should read the text by Manning and Schutze (1999).

5.4 Textual Data Sources

The first step in any textual analysis project is the collection of the relevant data. Textual econometrics will rarely, if ever, exclusively use textual data, but will combine textual data with other more traditional data types. The specific source for textual data will vary with the project, but there are many resources that researchers in the area should be aware of.

News sources are of particular importance for textual econometrics since markets, as information aggregators, will move when new and relevant information is released. A widely used, though fairly old, set of news stories is called *Reuters Corpus Volume 1*, or *RCV1*. It has been documented by Lewis et al. (2004) and can be obtained from the National Institute of Standards and Technology.[3] *RCV1* contains about 810, 000 Reuters English language news stories from 1996-08-20 to 1997-08-19. A multilingual version, *RCV2*, is also available. Another source for collected textual data sets, or corpora as they are often called, is the Linguistic Data Consortium.[4] These sources contain relatively old material because of copyright issues, and researchers will likely want to purchase or collect more recent data. A commercial source of textual data that has been used in econometric research is the Dow Jones Factive[5] group, which collects news from over 25, 000 sources including the *Wall Street Journal*, the *Financial Times*, as well as the Dow Jones, Reuters, and Associated Press news services. There are other news-source databases to which many academic institutions subscribe such as LexisNexis,[6] Business Source Complete,[7] the Canadian Business & Current Affairs Database,[8] among others. However, these databases are aimed at specific topic searches, and not large scale downloads. Many database providers may seize access to the data when it detects such unusual activity, so it is advisable to get the vendor's permission before downloading large quantities of data.

Collecting your own data is also a viable option through the use of public Internet resources. Many Web sites, blogs, and message boards have archives that may be downloaded manually or using a special type of software known as Web crawlers (also called Web robots, Web spiders, ...). Many open source Web crawlers[9] are available with an array of document collection properties, but the essence of each is that it browses the Internet in a structured way while creating copies of the visited Web pages and storing them for future processing. The starting URL(s) is specified and the crawler travels through other Web pages through the hyperlinks contained in previously visited pages according to a specified set of rules. There are many variants of Web crawlers that researchers may find useful including focused and topical crawlers that attempt to only visit Web pages on a given topic.

There are many other sources of textual data that are publicly available including academic literature, firm press releases and shareholder reports, analyst reports, political speeches, and government/institutional reports.

---

[3] `trec.nist.gov/data/reuters/reuters.html`

[4] `www.ldc.upenn.edu`

[5] `www.factiva.com`

[6] `www.lexisnexis.com`

[7] `www.ebscohost.com/titleLists/bt-complete.htm`

[8] `www.proquest.com/en-US/catalogs/databases/detail/cbca.shtml`

[9] One such Web crawler is the DataparkSearch Engine, available at `www.dataparksearch.org`, which will search and extract documents from within a Web site or group of Web sites.

5.5 Document Preprocessing

The technology that permits the quantification of textual data has been largely developed within the field of natural language processing (NLP) (Charniak 1996; Jurafsky and Martin 2000; Manning and Schutze 1999). Many tasks that seem trivial to a language using human are in fact frustratingly difficult to program a computer to do. The level of NLP pre-processing can vary widely depending on the particular research domain and goals. Various levels of preprocessing are described below:

- *Document format standardization:* The current standard for document formatting is the XML (Extensible Markup Language)[10] format. An advantage of XML is that it permits the placement of delimiting tags around various parts of a document such as <DOC> ··· </DOC> which indicates where a document begins and ends. Other tags indicate document components, such as titles and section headings, that may be of special importance. In cases where document components are easily identified, plain text files are easiest to work with.
- *Tokenization:* This process is breaking stream of characters into groups called tokens. The order of the original sequence is maintained, and only white-space is removed. For example, the stream "You are reading this sentence." would be transformed into the following sequence of tokens: [You][are][reading][this][sentence[.] Some multi-word expressions, such as "data set," may be treated as single tokens. These word groups include collocations, fixed expressions, and idioms.
- *Misspelling correction:* Commonly misspelt words may be automatically replaced to improve task performance. However, unsupervised replacement of all unfamiliar words (i.e., those not in a specified dictionary) can reduce performance (Malouf and Mullen 2008).
- *Sentence boundary detection:* If the intended linguistic features involve more than token occurrences, then the next step is to identify where sentences begin and end. There are some hand-crafted algorithms for sentences boundary detection that exploit language regularities which can achieve greater than 90% accuracy. Given sufficient training data, classification learning algorithms can achieve more than 98% accuracy (Weiss et al. 2005).
- *Part-of-speech (PoS) tagging:* Each token has a linguistic function (called its part-of-speech) that can be identified (or "tagged"). This will be useful if you are only interested in certain types of tokens, such as adjectives or nouns, or if you wish to do further NLP processing. The number of classes of PoS objects varies considerably depending on the level of detail. As an example, the CLAWS PoS Tagger[11] takes the sentence, "You are reading this sentence." and outputs "`[You_PNP]`

[10] `www.w3.org/XML`

[11] `ucrel.lancs.ac.uk/claws`

[are_VBB] [reading_VVG] [this_DT0] [sentence_NN1] [._PUN]" where PNP indicates a personal pronoun, VBB indicates the "base forms" of the verb "BE," VVG identifies the -ing form of a lexical verb, DT0 indicates a general determiner, NN1 indicates a singular noun, and PUN indicates punctuation. Another well known and publicly available PoS tagger is the Brill tagger.[12]

- *Phrase identification:* Also known as "text-chunking," identifies important word sequences. Primarily these sequences are noun phrases (e.g., "the economic crisis"), verb phrases (e.g., "has worsened"), or prepositional phrases (e.g., "because of"). Phrases can be used as features, and also they can be used in parsing and named entity recognition.
- *Named entities:* Identifying proper noun phrases (named entities) such as people, organizations, and locations, is important in many textual processing applications. This is a subtask of phrase recognition, but as 90% of new lexemes encountered NLP systems are proper nouns (do Prado and Ferneda 2008), there is considerable focus on identifying and classifying them.
- *Parsing:* Parsing is the process of linking words within a sentence by grammatical relationships. For example, the Stanford Parser[13] identifies the following relationships in the sentence, "You are reading this sentence."
    1. nsubj(reading-3, You-1) – "You" is a nominal subject of a clause of which "reading" is the governor.
    2. aux(reading-3, are-2) – "are" is an auxiliary of a clause whose main verb is "reading".
    3. det(sentence-5, this-4) – "this" is the determiner of a noun phrase whose head is "sentence".
    4. dobj(reading-3, sentence-5) – "sentence" is the direct object of a verb phrase whose main verb is "reading".

  The output of a parser is often in the form of a tree representing the dependency between the parts of the sentence. Parsing permits us to automatically associate descriptions and actions to named entities. Without parsing, textual data such as "So XYZ's earnings were not bad this year." and "Not so! XYZ's earnings were bad this year." would likely be treated the same despite their very different meaning.
- *Normalization/lemmatization/stemming:* Performance may be improved by aggregating different types of tokens. For example we may want to treat the words "book" and "books" as instances of the same thing, and so we may transform both into a single standard form. There are many types and degrees of such standardization. Inflectional

[12] www.tech.plym.ac.uk/soc/staff/guidbugm/software/RULE_BASED_TAGGER_V.1.14.tar.Z

[13] http://nlp.stanford.edu/software/lex-parser.shtml

stemming involves aggregating across grammatical variation from factors such as tense and plurality. Lemmatization replaces words with their primitive form. "Stemming to a Root" removes all prefixes and suffixes from a word. This last form of stemming is very strong and since it often removes important information, it should not often be used. While such normalization can be done immediately after tokenization, if further NLP processing such as parsing is intended, it should be postponed until later.
- *Synonyms and coreferences:* Collapsing synonyms and coreferences involves aggregating all synonyms to their common sense (meaning). An important tool in dealing with synonyms is WordNet,[14] which organizes English nouns, verbs, adjectives, and adverbs into synonym sets.

5.6 Quantifying Textual Information

With documents in a standardized format, the most critical question of quantitative textual analysis must be considered: How do we transform textual data into numerical data? The general strategy is to define a feature vector for each document, where each vector element is associated with a linguistic feature (such as a word, type of word, phrase, relationship, etc.) and the numerical entry in this vector element measures the extant to which the feature is present in the document. The choice of the type of features is critical and may depend on the intended task, the properties of the textual data, or the level of linguistic sophistication required. The number of features generated from textual data can easily reach into the tens and even hundreds of thousands. Consequently, combinations of dimensionality reduction and appropriate classification and prediction methods are also needed. The following list indicates some of the possible choices of types of features:

- *Word occurrence:* For this type of feature, the feature vector will have a length that is the size of the dictionary. That is, for every distinct word in the training data (or corpus), there is a corresponding feature element. The feature vector for a given document will have a 1 in the nth element if the document contains the nth word in the dictionary, otherwise it contains a 0.
- *Word frequencies:* Rather than just having a 1 or 0 in each entry, this feature type uses some measure of the frequency of words in the document to weight the entry. There are several potential frequency measures that can be used. The simplest method is just the number of times each word occurs in the document; however, there are

[14] `wordnet.princeton.edu`

more effective measures. There are two primitives that are used in the construction of term weighting measures: term frequency ($tf_{i,j}$: the number of occurrences of word w_i in document d_j) and document frequency (df_i: the number of documents in the collection in which w_i occurs). A common weighting scheme based on these primitives is called the inverse document frequency (*idf*)

$$idf_{i,j} = \begin{cases} (1 + \log(tf_{i,j})) \log \frac{N}{df_i} & \text{if} \quad tf_{i,j} \geq 1 \\ 0 & \text{if} \quad tf_{i,j} = 0 \end{cases} \tag{5.6}$$

where N is the number of documents. The idea of this type of weighting scheme is that words that occur frequently are informative, but words that occur in many documents are less informative. There are several other weighting schemes based on this idea that use different functions for term frequency, document frequency, and normalization. Both word-occurrence and word-frequency type features are often referred to as the bag-of-words representation since all structural properties of the textual data are lost.

- *n-Grams/multi-word features:* An n-gram is a sequence of n words. Thus, the word-occurrence and word-frequency type features are examples of 1-gram, or unigram, features. We could extend these types of features to include pairs of words that occur together to generate 2-gram, or bigram, features. Tan, Wang, and Lee (2002) show that bigrams plus unigrams improve performance in a Web page classification task compared to unigrams alone.
- *Word+PoS features:* Occurrence or frequency features could also be generated for words paired with their parts of speech.
- *Parsed features:* Word n-grams together with their parsed dependencies would provide considerable linguistic sophistication to the feature vectors. Such vectors would very large, with lengths on the order of hundreds of thousands.

5.7 Dimensionality Reduction

With such large feature vectors, dimensionality reduction (also called feature selection) can improve classification performance as well as increase computational speed. An introduction to feature selection in high-dimensional settings is given by Guyon and Elisseeff (2003), and experimental results for various feature selection strategies is given by Yang and Pedersen (1997) and Formen (2003).

The idea behind dimensionality reduction is that not all features are equally informative, so we would like to score each potential feature by some "informativeness" metric, and then select only the best *k* features. Yang and Pedersen (1997) have shown that in some cases removal of up to 98% of the features can improve classification performance. Before reviewing some scoring procedures, it is useful to review some of the informal filters that are often applied.

In bag-of-words feature sets, words can be aggregated into broad categories to capture more general features rather than specific meanings. For example, Tetlock (2007) classifies all words in his documents as belonging to one of 77 categories using the Harvard-IV-4 dictionary. He then further reduces the dimensionality down to one feature using principal component analysis (PCA) to extract the single factor that captures the maximum variation in the dictionary categories. In Tetlock, Saar-Tsechansky, and Macskassy (2008), only one of the 77 categories is used to measure the negative sentiment of a document, thereby again reducing the dimensionality to one.

A common feature filter is to eliminate rare words on the grounds that they will not help with classification. In many cases, up to half of the distinct words in a collection will occur only once. Removing these features will greatly improve the processing speed as well as reduce potential for over-fitting.

Similarly, removing the most common words will not likely harm classification performance since these function words (such as "the", "a", and "of") are purely grammatical rather than informative. A frequency threshold can be used to identify these words, or a "stopword" list may be provided. Also, as mentioned earlier, collapsing synonyms and coreferences, as well as lemmatization and stemming can aggregate words into broader classes thereby reducing dimensionality and potentially improving classification and prediction performance.

There are many formal feature selection metrics that can be used to rank potential features. According to Formen (2003), the Bi-Normal Separation metric outperforms other more common metrics. Other metrics include Information Gain, Mutual Information, Chi-Square, Odds Ratio, and others. Formal descriptions of these metrics can be found in Formen (2003) and Yang and Pedersen (1997).

5.8 Classification Problems

The growth of unsolicited and unwanted e-mails that are sent out in a bulk or automatic fashion, also known as spam, has posed a great threat to Internet communications. Today, estimates put the volume of spam in the range of 88%–92% of all e-mails (MAAWG 2010). In order to maintain the useability of e-mail communications, spam detection systems were developed to classify e-mails as spam or not. This is one of the most successful examples of automated textual classification, but similar technology can be applied to a wide array of classification and prediction problems.

5.8.1 Classification

Textual classification (also known as categorization or supervised learning) is the task of assigning documents $\mathcal{D} = \{d_1, \ldots, d_{|\mathcal{D}|}\}$ to predefined[15] classes $\mathcal{C} = \{c_1, \ldots, c_{|\mathcal{C}|}\}$. Thus, a classifier is a function

$$\mathcal{F} : \mathcal{D} \times \mathcal{C} \rightarrow \{T, F\} \tag{5.7}$$

where $\mathcal{F}(d_i, c_j) = T$ if and only if document d_i belongs to class c_j. Under this general specification, a document may simultaneously belong to multiple classes, as may be appropriate when, for example, an e-mail is not spam, about a particular project, and from a particular sender. Such a problem is usually broken down into $|\mathcal{C}|$ simpler binary classification problems. That is, for each class $c_i \in \mathcal{C}$, we define a binary classifier

$$\mathcal{F}_i : \mathcal{D} \rightarrow \{T, F\} \tag{5.8}$$

where $\mathcal{F}_i(d_j) = T$ if and only if document d_j belongs to class c_i. When exactly one class can be assigned to each document, as in the case of BUY, SELL, and HOLD recommendations, the classifier will be of the form

$$\mathcal{F} : \mathcal{D} \rightarrow \mathcal{C} \tag{5.9}$$

As before, depending on the classification algorithm chosen, this type of problem may be reduced to a set of binary classifiers with rules for dealing with multiple class assignments.

If $\mathcal{F}$ is the correct or authoritative classifier, then we wish to approximate this function with $\widehat{\mathcal{F}}$. Approximating classifiers has been extensively studied in the field of machine learning (Mitchell 1997). The specific classification scheme of $\widehat{\mathcal{F}}$ is determined by a set of training documents for which the correct classifications are known. These training documents can be classified by a domain expert according to their linguistic properties, or they can be classified according to some specific data that is aligned with the text document. Fung, Yu, and Lu (2005) uses this latter approach and aligns news stories in a time series with stock market performance.

5.8.2 Classifier Evaluation

A classifier can be evaluated across many dimensions. Its ability to correctly classify documents is of paramount importance; however, its speed and scalability in both the training and classification phases are also important. Additionally, a good classifier should be relatively easy to use and understand. The focus in this section, however, is limited to the evaluation of a classifier's primary task.

Recall that the classifier $\mathcal{F}$ maps $\mathcal{D} \times \mathcal{C}$ into the $\{T, F\}$ such that $\mathcal{F}(d_i, c_j) = T$ if and only if document d_i belongs to class c_j. In this case, we call d_i a positive

[15] When the classes are not known in advance, then the task is referred to as clustering.

example of class c_j. When $\mathcal{F}(d_i, c_j) = F$ then document d_i is not a member of the class c_j and so we call d_i a negative example of c_j. To capture the correctness of classifications from a trained classifier $\widehat{\mathcal{F}}$, we introduce the following four basic evaluation functions:

$$TP_{\widehat{\mathcal{F}}}(d_i, c_j) = \begin{cases} 1 & \text{if} \quad \widehat{\mathcal{F}}(d_i, c_j) = T \text{ and } \mathcal{F}(d_i, c_j) = T \\ 0 & \text{otherwise} \end{cases} \tag{5.10}$$

$$TN_{\widehat{\mathcal{F}}}(d_i, c_j) = \begin{cases} 1 & \text{if} \quad \widehat{\mathcal{F}}(d_i, c_j) = F \text{ and } \mathcal{F}(d_i, c_j) = F \\ 0 & \text{otherwise} \end{cases} \tag{5.11}$$

$$FP_{\widehat{\mathcal{F}}}(d_i, c_j) = \begin{cases} 1 & \text{if} \quad \widehat{\mathcal{F}}(d_i, c_j) = T \text{ and } \mathcal{F}(d_i, c_j) = F \\ 0 & \text{otherwise} \end{cases} \tag{5.12}$$

$$FN_{\widehat{\mathcal{F}}}(d_i, c_j) = \begin{cases} 1 & \text{if} \quad \widehat{\mathcal{F}}(d_i, c_j) = F \text{ and } \mathcal{F}(d_i, c_j) = T \\ 0 & \text{otherwise} \end{cases} \tag{5.13}$$

Thus, every classification by $\widehat{\mathcal{F}}$ will either be a true positive (*TP*), a true negative (*TN*), a false positive (*FP*), or a false negative (*FN*). The *TP*s and *TN*s indicate correct classifications, and the *FP*s and *FN*s indicate incorrect classifications.

A simple, and sometimes overused, performance measure of classifiers is accuracy (A) measured as the proportion of correct classifications.

$$A = \frac{\sum_{i=1}^{|\mathcal{D}|} \sum_{j=1}^{|\mathcal{C}|} \left(TP_{\widehat{\mathcal{F}}}(d_i, c_j) + TN_{\widehat{\mathcal{F}}}(d_i, c_j)\right)}{\sum_{i=1}^{|\mathcal{D}|} \sum_{j=1}^{|\mathcal{C}|} \left(TP_{\widehat{\mathcal{F}}}(d_i, c_j) + TN_{\widehat{\mathcal{F}}}(d_i, c_j) + FP_{\widehat{\mathcal{F}}}(d_i, c_j) + FN_{\widehat{\mathcal{F}}}(d_i, c_j)\right)} \tag{5.14}$$

The converse of accuracy is the error rate (E) measured as

$$E = 1 - A \tag{5.15}$$

Accuracy, and error, are often useful performance measures, but they do not always capture the intended notion of correctness. For example, when trying to classify rare events, positive and negative examples will be strongly imbalanced with far more negative than positive examples. In this case, a universal rejector (i.e., $\widehat{\mathcal{F}}(d_i, c_j) = F, \forall d_i, c_j$) will have a high accuracy while being of no practical use.

To address such concerns, two other performance measures have become popular: precision and recall. Precision, with respect to a class c_j, is the proportion of documents assigned to class c_j that actually belong to that class.

$$P_{\widehat{\mathcal{F}}}(c_j) = \frac{\sum_{i=1}^{|\mathcal{D}|} TP_{\widehat{\mathcal{F}}}(d_i, c_j)}{\sum_{i=1}^{|\mathcal{D}|} \left(TP_{\widehat{\mathcal{F}}}(d_i, c_j) + FP_{\widehat{\mathcal{F}}}(d_i, c_j)\right)} \tag{5.16}$$

So, given a particular class, precision is the ratio of correct positive classifications to the total number of positive classifications. Precision can be aggregated across classes in two ways. First, microaveraged precision averages the precision of $\widehat{\mathcal{F}}$ for each class, weighted by the number of positive documents.

$$P_{\widehat{\mathcal{F}}}^{\text{Micro}} = \frac{\sum_{j=1}^{|\mathcal{C}|}\sum_{i=1}^{|\mathcal{D}|} TP_{\widehat{\mathcal{F}}}(d_i, c_j)}{\sum_{j=1}^{|\mathcal{C}|}\sum_{i=1}^{|\mathcal{D}|}\left(TP_{\widehat{\mathcal{F}}}(d_i, c_j) + FP_{\widehat{\mathcal{F}}}(d_i, c_j)\right)} \tag{5.17}$$

Alternatively, macroaveraged precision averages the precision for each class with equal weights.

$$P_{\widehat{\mathcal{F}}}^{\text{Macro}} = \frac{\sum_{j=1}^{|\mathcal{C}|} P_{\widehat{\mathcal{F}}}(c_j)}{|\mathcal{C}|} \tag{5.18}$$

The second performance measure is recall. Recall, for a given class c_j, is the proportion of documents that truly belong to c_j that are classified as belonging to that class by $\widehat{\mathcal{F}}$.

$$R_{\widehat{\mathcal{F}}}(c_j) = \frac{\sum_{i=1}^{|\mathcal{D}|} TP_{\widehat{\mathcal{F}}}(d_i, c_j)}{\sum_{i=1}^{|\mathcal{D}|}\left(TP_{\widehat{\mathcal{F}}}(d_i, c_j) + FN_{\widehat{\mathcal{F}}}(d_i, c_j)\right)} \tag{5.19}$$

So, given a particular class, precision is the ratio of the number of correct positive classifications by the total number of truly positive class documents. As with precisions, we can use microaveraging to define a measure of recall across all classes.

$$R_{\widehat{\mathcal{F}}}^{\text{Micro}} = \frac{\sum_{j=1}^{|\mathcal{C}|}\sum_{i=1}^{|\mathcal{D}|} TP_{\widehat{\mathcal{F}}}(d_i, c_j)}{\sum_{j=1}^{|\mathcal{C}|}\sum_{i=1}^{|\mathcal{D}|}\left(TP_{\widehat{\mathcal{F}}}(d_i, c_j) + FN_{\widehat{\mathcal{F}}}(d_i, c_j)\right)} \tag{5.20}$$

Alternatively, we may define a macroaverage measure of recall.

$$R_{\widehat{\mathcal{F}}}^{\text{Macro}} = \frac{\sum_{j=1}^{|\mathcal{C}|} R_{\widehat{\mathcal{F}}}(c_j)}{|\mathcal{C}|} \tag{5.21}$$

Most classifiers can be set up to tradeoff precision for recall, or vice versa. Consequently, it is useful to present a combined measure of performance, the F_1 score, which is the harmonic mean of precision and recall:

$$F_1 = \frac{2 \times P \times R}{P + R} \tag{5.22}$$

where P and R are either micro- or macroaveraged.

5.8.3 Classification Algorithms

There is a multitude of classification algorithms to choose from including decision trees, Bayesian classifiers, Bayesian belief networks, rule-based classifiers, backpropagation, genetic algorithms, k-nearest neighbor classifiers, and others. Detailed treatments of these methods can be found in many textbooks including the one by Han and Kamber (2006). For text classification, however, support vector machines (SVMs) deserve special mention since they consistently rank as or among the best classification methods (Joachims 1998) and can handle very-high-dimensional data.

A complete description of SVMs is given by Burges (1998) and Vapnik (1998), and there are many public resources[16] available to those who wish to learn about and use these classifiers. Thorsten Joachims has created a popular implementation of the SVM algorithm called SVMlight,[17] which is publicly available. The basic SMV is a binary classifier that finds a hyperplane with the maximum margin between positive and negative training documents. There are now SMVs for multi-class classification as well as for regression. SMVs have three unique features:

1. Not all training documents are used to train the SVM. Instead, only documents near the classification boarder are used to train the SMV.
2. Not all features from the training documents are used, so excessive feature reduction is not needed.
3. SMVs can construct irregular boarders between positive and negative training documents.

Another classification technique that is worthy of mention is the use of ensemble methods that combine several different classification methods. Two examples of ensemble methods are bagging (i.e., bootstrap aggregation) and boosting (a series of classifications that weight previously misclassified training examples more heavily). Han and Kamber (2006) describe implementations of these techniques, and Das and Chen (2007) employ an ensemble voting method.

As a final note, regression techniques are also available for textual data represented as a feature vector. Traditional regression methods can be used when the number of features has been greatly reduced as in Tetlock (2007). For larger feature vectors, support vector regressions (SVRs) can be used.

5.9 Software

There are many software packages available to facilitate textual econometric research. In addition to those listed throughout this chapter, some popular statistical packages have textual analysis modules:

- *SAS text miner*[18]*:* This package provides tools for transforming textual data into a usable a format, as well as for classifying documents, finding relationships and associations between documents, and clustering documents into categories.

[16] www.support-vector-machines.org

[17] svmlight.joachims.org

[18] www.sas.com/technologies/analytics/datamining/textminer

- *SPSS text mining for clementine*[19]: This package uses NLP techniques to extract key concepts, sentiments, and relationships from textual data. Feature vectors can be created and used in SPSS for predictive modeling.

There are many other packages available. One package that is very comprehensive, freely available, and well documented is the Natural Language Toolkit[20] for the Python programming language. It contains open source Python modules, linguistic data, and documentation for many of the tasks described in this chapter. In addition to the documentation available, a book has been written by Bird, Klein, and Loper (2009) as a guide to the toolkit. Other programming resources are described by Bilisoly (2008), Konchandy (2006), and Chakrabarti (2003).

5.10 Conclusion

The aim of this chapter has been to introduce econometricians to tools and techniques that allow textual data to be analyzed in a quantitative and statistical manner. This new area of textual econometrics is in its early stages of development and draws heavily from the fields of natural language processing and text mining (Feldman and Sanger 2007; Weiss et al. 2005). Early work in the field has proven that useful information is embedded in textual data that can be extracted using these techniques. As these tools improve and the areas of application expand, textual econometrics is a field bound to expand.

5.11 Acknowledgment

The authors wish to thank Anoop Sarkar of SFU's Natural Language Laboratory for introducing them to this interesting field.

References

Antweiler, W., and M. Frank. 2004. Is all that talk just noise? the information content of internet stock message boards. *Journal of Finance* LIX(3): 1259–1294.

Armstrong, J. 2001. Combining forecasts. *Principles of Forecasting*. Norwell, MA: Kluwer, pp. 417–439.

[19] `www.spss.com/text_mining_for_clementine`

[20] `www.nltk.org.`

Bates, J., and C. Granger. 1969. The combination of forecasts. *Operations Research Quarterly* 20: 451–468.

Bhattacharya, U., N. Galpin, R. Ray, and X. Yu. 2009. The role of the media in the internet IPO bubble. *Journal of Financial and Quantitative Analysis* 44(3): 657–682.

Bilisoly, R. 2008. *Practical Text Mining with Perl*. Hoboken, NJ: Wiley.

Bird, S., E. Klein, and E. Loper. 2009. *Natural Language Processing with Python: Analyzing Text with the Natural Language Toolkit*. Sebastopol, CA: O'Reilly Media.

Burges, C. 1998. A tutorial on support vector machines for pattern recognition. *Data Mining and Knowledge Discovery* 2: 121–168.

Chakrabarti, S. 2003. *Mining the Web: Discovering Knowledge from Hypertext Data*. San Francisco: Morgan Kaufmann Publishers.

Chan, W. 2003. Stock price reaction to news and no-news: Drift and reversal after headlines. *Journal of Financial Economics* 70(2): 223–260.

Charniak, E. 1996. *Statistical Language Learning*. Cambridge, MA: MIT Press.

Coval, J., and T. Shumway. 2001. Is sound just noise? *Journal of Finance* LVI(5): 1887–1910.

Das, S., and M. Chen. 2007. Yahoo! for Amazon: Sentiment extraction from small talk on the web. *Management Science* 53(9): 1375–1388.

Dennett, D. 1994. The role of language in intelligence. In J. Khalfa (ed.), *What is Intelligence? The Darwin College Lectures*. Cambridge, U.K.: Cambridge University Press.

do Prado, H., and E. Ferneda. 2008. *Emerging Technologies of Text Mining: Techniques and Applications*. New York: Information Science Reference.

Fama, E., L. Fisher, M. Jensen, and R. Roll. 1969. The adjustment of stock prices to new information. *International Economic Review* 10.

Feldman, R., and J. Sanger. 2007. *The Text Mining Handbook: Advanced Approaches in Analyzing Unstructured Data*. New York: Cambridge University Press.

Fellbaum, C. 1998. *WordNet: An Electronic Lexical Database*. Cambridge, MA: MIT Press.

Formen, G. 2003. An extensive empirical study of feature selection metrics for text classification. *Journal of Machine Learning Research* 3: 1289–1305.

Frisch, R. 1933. Editor's note. *Econometrica* 1(1): 1–4.

Fung, G., J. Yu, and H. Lu. 2005. The predicting power of textual information on financial markets. *IEEE Intelligent Informatics Bulletin* 5(1): 1–10.

Gao, L., and P. Beling. 2003. Machine quantification of text-based economic reports for use in predictive modeling. *IEEE International Conference on Systems, Man and Cybernetics* 4: 3536–3541.

Guyon, I., and A. Elisseeff. 2003. An introduction to variable and feature selection. *Journal of Machine Learning Research* 3: 1157–1182.

Han, J., and M. Kamber. 2006. *Text Mining: Concepts and Techniques*. 2nd ed. San Francisco: Morgan Kaufmann.

Joachims, T. 1998. Text categorization with support vector machines: Learning with many relevant features. *Proceedings of the Tenth European Conference on Machine Learning*. Heidelberg: Springer-Verlag, pp. 137–142.

Jurafsky, D., and J. Martin. 2000. *Speech and Language Processing: An Introduction to Natural Language Processing, Computational Linguistics, and Speech Recognition*. Upper Saddle River, NJ: Prentice Hall.

Katz, S. 1996. Distribution of content words and phrases in text and language modelling. *Natural Language Engineering* 2: 15–59.

Konchandy, M. 2006. *Text Mining Application Programming*. Boston: Charles River Media.

Kroha, P., R. Baeza-Yates, and B. Krellner. 2006. Text mining of business news for forecasting. *Proceedings of the 17th International Conference on Database and Expert Systems Applications*. Berlin: Springer, pp. 171–175.

Lewis, D., Y. Yang, T. Rose, and F. Li. 2004. RCV1: A new benchmark collection for text categorization research. *Journal of Machine Learning Research* 5: 361–397.

MAAWG. 2010. Report 12 of the Messaging Anti-Abuse Working Group: Third and fourth quarter 2009. *Email Metrics Program: The Network Operators Perspective*. August 20, 2010.

Malouf, R., and T. Mullen. 2008. Taking sides: User classification for informal online political discourse. *Internet Research* 18: 177–190.

Mandelbrot, B. 1954. Structure formelle des textes et communication. *Word*, 10: 1–27.

Manning, C., and H. Schutze. 1999. *Foundations of Statistical Natural Language Processing*. Cambridge, MA: MIT Press.

Mitchell, T. 1997. *Machine Learning*. New York: McGraw-Hill.

Mitchell, M., and J. Mulherin. 1994. The impact of public information on the stock market. *Journal of Finance* 49(3): 923–950.

Mittermayer, M. 2004. Forecasting intraday stock price trends with text mining techniques. *Proceedings of the 37th Hawaii International Conference on System Sciences*, pp. 1–10. IEEE Computer Society Press.

Niederhoffer, V. 1971. The analysis of world events and stock prices. *Journal of Business* 44(2), 193–219.

Rachlin, G., M. Last, D. Alberg, and A. Kandel. 2007. ADMIRAL: A data mining based financial trading system. *Proceedings of the 2007 IEEE Symposium on Computational Intelligence and Data Mining*, pp. 720–725. IEEE Computer Society Press.

Schumaker, R., and H. Chen. 2009. Textual analysis of stock market prediction using breaking financial news: The azfintext system. *Association for Computing Machinery Transactions on Information Systems* 27(2): 1–19.

Tan, C., Y. Wang, and C. Lee. 2002. Using bi-grams to enhance text categorization. *Information Processing and Management* 38(4): 529–546.

Tetlock, P. 2007. Giving content to investor sentiment: The role of media in the stock market. *Journal of Finance* LXII(3): 1139–1168.

Tetlock, P., M. Saar-Tsechansky, and S. Macskassy. 2008. More than words: Quantifying language to measure firms' fundamentals. *Journal of Finance* LXIII(3): 1437–1467.

Ticom, A., and B. de Lima. 2007. Text mining and expert systems applied in labor laws. *Seventh International Conference on Intelligent Systems Design and Applications*, pp. 788–792. IEEE Computer Society Press.

Turing, A. 1950. Computing machinery and intelligence. *Mind* LIX: 433–460.

Vapnik, V. 1998. *Statistical Learning Theory*. New York: John Wiley & Sons.

Weiss, S., N. Indurkhya, T. Zhang, and F. Damerau. 2005. *Text Mining: Predictive Methods for Analyzing Unstructured Information*. New York: Springer.

Wysocki, P. 1999. Cheap talk on the Web: The determinants of posting on stock message boards. *University of Michigan Business School Working Paper* (No. 98025), Ann Arbor.

Yang, Y., and J. Pedersen. 1997. A comparative study of feature selection in text categorization. *Proceedings of the Fourteenth International Conference on Machine Learning* 3: 412–420.

Zipf, G. 1932. *Selective Studies and the Principle of Relative Frequency in Language*. Boston: Cambridge University Press.

6

Large Deviations Theory and Econometric Information Recovery

Marian Grendár and George Judge

CONTENTS

6.1 Estimation and Inference Base

Econometricians rarely have at their disposal enough information to formulate a model in terms of a parametric family of distributions. Consequently, traditional methods of parametric estimation and inference that are based either on a likelihood or on a posterior distribution are prone to committing specification errors that lead to problems of inference. On the other hand, economic data are partial and incomplete and usually there is seldom a large enough data sample to rely on purely nonparametric methods.

Looking for a compromise, econometricians have turned to a traditional method of estimation and inference known as the *method of moments* (MM); cf., e.g., Mittelhammer, Judge, and Miller (2000). This formulation permits a researcher to specify only some moment properties/features of the data-sampling distribution F, with probability density function (pdf) $r(x)$, of a random variable $X \in \mathcal{R}^d$. This is accomplished through estimating functions $u(X; \theta) \in \mathcal{R}^J$ of parameter $\theta \in \Theta \subseteq \mathcal{R}^k$ (see Godambe and Kale 1991). The estimating functions are used to form a set $\Delta(\Theta) = \bigcup_{\theta \in \Theta} \Delta(\theta)$ of parametrized pdf's $\rho(x; \theta)$, defined through unbiased estimating equations (EE)

$$\Delta(\theta) = \left\{ \rho(x; \theta) : \int \rho(x; \theta) u(x; \theta) = 0 \right\}.$$

When $J = k$, there is usually a unique solution θ of the "just determined" set of EEs. Given a sample $X_1^n = X_1, \ldots, X_n$, the solution can be estimated by solving an empirical counterpart of the EEs: $\frac{1}{n} \sum_{i=1}^n u(x_i; \hat{\theta}_{\mathrm{MM}}) = 0$. The resulting estimator $\hat{\theta}_{\mathrm{MM}}$ is known as the *method of moments* estimator. Asymptotic distributional properties of the estimator are well known (Mittelhammer, Judge, and Miller 2000) and provide a basis for inference.

In econometric modeling, it often happens that there are more EEs than unknown parameters; $J > k$. A considerable amount of work has been devoted to extending the method of moments for this overdetermined case. As a result the Generalized Method of Moments (Hansen 1982) evolved (see also Hall 2005).

More recently (cf. Bickel et al. 1993; Mittelhammer, Judge, and Miller 2000; Owen 2001; among others), a new basis has emerged for regularizing the overdetermined EEs. This approach is based on minimization of a discrepancy, or divergence measure of a pdf ρ with respect to the true sampling distribution pdf r:

$$\phi(\rho \,\|\, r) = \int \phi\left(\frac{\rho(x)}{r(x)}\right) r(x) dx,$$

where ϕ is a convex function. If ρ is assumed to belong to model set $\Delta(\Theta)$ then the minimization problem

$$\hat{\rho}(\hat{\theta}) = \arg \inf_{\theta \in \Theta} \inf_{\rho(x;\, \theta) \in \Delta(\theta)} \phi(\rho \,\|\, r)$$

can be equivalently expressed in the convex dual form. Thanks to the convex duality, the optimal $\hat{\theta}$ can be obtained as

$$\hat{\theta} = \arg \inf_{\theta \in \Theta} \sup_{\gamma \in \mathcal{R}, \lambda \in \mathcal{R}^J} \gamma - \mathrm{E}[\phi^*(\gamma + \lambda' u(x; \theta))], \tag{6.1}$$

where ϕ^* is the convex conjugate of ϕ. In order to make Equation 6.1 operational, it is necessary to connect it with the sample data X_1^n. Indeed, in

Equation 6.1 the expectation is taken with respect to the true sampling distribution r. It is natural to replace the expectation by its sample-based estimate, and this leads to the empirical minimum divergence (EMD) estimator:

$$\hat{\theta}_{\text{EMD}} = \arg \inf_{\theta \in \Theta} \sup_{\gamma \in \mathcal{R}, \lambda \in \mathcal{R}^J} \gamma - \frac{1}{n} \sum_{i=1}^{n} [\phi^*(\gamma + \lambda' u(x_i; \theta))]. \tag{6.2}$$

Kitamura (2006) notes that the estimator (Equation 6.2) is the generalized minimum contrast estimator considered by Bickel et al. (1993).

There are two possible ways of using the parametric model $\Delta(\Theta)$, specified by EE.

1. One option is to use the EEs to define a feasible set of possible parametrized sampling distributions $q(x; \theta)$. In order to distinguish this way, the model set will be denoted $\Phi(\Theta)$. *The objective of information recovery is to select a representative sampling distribution from* $\Phi(\Theta)$. *This modeling strategy and associated problem will be referred to as a sampling distribution (SD) problem.*
2. Alternatively EEs can be used to form a set into which a parametrized empirical distribution should, in a researcher's view, belong. The sample X_1^n is used to estimate the sampling distribution r. In this case the model $\Delta(\Theta)$ will be denoted $\Pi(\Theta)$. *The objective of information recovery is the selection of a representative parametrized empirical distribution from* $\Pi(\Theta)$. *This modeling strategy and associated problem will be referred to as an empirical distribution (ED) problem.*

There are two choices of ϕ that are popular: (1) $\phi(x) = -\log(x)$ which leads[1] to the L-divergence (Grendár and Judge 2009a) $L(\rho \,\|\, r) = -\int r \log \rho$ of pdf ρ with respect to r, and (2) $\phi(x) = x \log(x)$ which leads to the I-divergence $I(\rho \,\|\, r) = \int \rho \log \frac{\rho}{r}$, which is also known as the Kullback Leibler divergence or the negative of relative entropy (Cover and Thomas 1991; Csiszár 1998). They both are members of the Cressie–Read (cf. Cressie and Read 1984; Cressie and Read 1988) (CR) parametric family of discrepancy measures

$$\phi(\rho \,\|\, r; \alpha) = \frac{1}{\alpha(\alpha+1)} \int \left(\left(\frac{\rho(x)}{r(x)} \right)^{\alpha} - 1 \right) \rho(x) dx.$$

In the former case, the resulting EMD estimator is known as the empirical likelihood (EL) estimator (Mittelhammer, Judge, and Miller 2000; Owen 2001; Qin and Lawless 1994):

$$\hat{\theta}_{\text{EL}} = \arg \inf_{\theta \in \Theta} \sup_{\lambda \in \mathcal{R}^J} \frac{1}{n} \sum_{i=1}^{n} \log \left(1 - \lambda' u(x_i; \theta)\right). \tag{6.3}$$

[1] From the point of view of optimization of $\phi(\rho \,\|\, r)$ wrt ρ.

In the latter case the empirical maximum maximum entropy (EMME) or the maximum entropy empirical likelihood results (Back and Brown 1990; Imbens, Spady, and Johnson 1998; Kitamura and Stutzer 1997; Mittelhammer, Judge, and Miller 2000; Owen 2001):

$$\hat{\theta}_{\text{EMME}} = \arg\sup_{\theta\in\Theta} \inf_{\lambda\in\mathcal{R}^J} \frac{1}{n}\sum_{i=1}^{n} \exp(-\lambda' u(x_i; \theta)).$$

Asymptotic distributional properties of both estimators are known (cf., e.g., Mittelhammer, Judge, and Miller 2000; Owen 2001) and an inferential basis follows. Other members of the CR class of discrepancies appear in the literature; cf., e.g., Schennach (2007). It is also possible to select an optimal EMD estimator from the CR class, where optimality may be suitably defined by minimum mean squared error criterion or some other loss function (Judge and Mittelhammer 2004; Mittelhammer and Judge 2005). A survey of the known small and large sample properties of EMD estimators can be found in, e.g., Schennach (2007) and Grendár and Judge (2008); see also Owen (2001) and Mittelhammer, Judge, and Miller (2000).

In practice, an econometrician usually does not have enough information to guarantee that the model set (either Φ, or Π) contains the true data-sampling distribution r. Given that most econometric-statistical models are misspecified, it is of basic interest to know which of the methods of information recovery is consistent in the misspecified case. This is the place where the large deviations (LD) theory (cf., e.g., Dembo and Zeitouni 1998; Cover and Thomas 1991) is of great use, as it permits us to find out, in both the ED and SD settings, the methods that are consistent under misspecification. This way LD provides a guidance in information recovery, as it shows which methods are ruled out and in what sense.

6.1.1 Purpose-Objectives of This Chapter

In the context of the above estimation and inference base this chapter provides a nontechnical[2] introduction to LD theory with a focus on the implications of some of the key LD theorems for information recovery in *Econometrics* and *Statistics*. LD theory is a subfield of probability theory where, informally, the typical concern is about the asymptotic behavior, on a logarithmic scale, of the probability of a given event. In more technical words, LD theory studies the exponential decay of the probability of an event. For example, consider the event that an empirical measure belongs to a specified set. The rate of exponential decay of the probability of this event is determined by an extremal value of a certain quantity, called the rate function, over the set; cf. the Sanov theorem (ST), Subsection 6.2.1. This permits one to estimate the probability of this event with precision of the first order in the exponent. Even more importantly, ST leads to the conditional law of large numbers (CLLN), which says,

[2] Theorems are stated without proof, but references to the literature where the proofs can be found are provided. Theorems are intentionally not stated at greatest possible generality.

given that the event has occurred, only empirical measures arbitrarily close to those that minimize the rate function, can occur as the sample size goes to infinity.

Although LD theory also studies other types of events, we concentrate on ST as a means for establishing CLLN. CLLN has profound implications for the relative entropy maximization (REM/MaxEnt) – since, in the i.i.d. case, the rate function is just the Kullback Leibler information divergence. These implications carry over in a parametric context to EEs, where they provide a probabilistic justification (cf. Kitamura and Stutzer 2002) to empirical Max-MaxEnt estimation method (or exponential tilt estimator).

The lack of a comparable probabilistic justification for the empirical likelihood approach has motivated a study of LD theorems for data-sampling distributions: Bayesian Sanov theorem (BST) and its corollary: Bayesian law of large numbers (BLLN). The other objective of this chapter is to expose, in general, the relatively new LD theorems for sampling distributions (also known as Bayesian LD theorems) and their implications. Bayesian LLN provides a probabilistic justification to the maximum nonparametric likelihood (MNPL) method, which carries over in a parametric context, where it justifies the estimation method known as empirical likelihood (EL). The BLLN also shows that from an estimation point of view MNPL as well as EL can be seen as asymptotic instances of Bayesian maximum a posteriori probability (MAP) method, in nonparametric and parametric context, respectively.

6.1.2 Organization of the Chapter

The chapter is divided into two large sections (Sections 6.2 and 6.4), which are connected by an Intermezzo (Section 6.3). Section 6.2 is devoted to explaining key LD theorems for empirical measures, culminating with CLLN and its implications for ED selection problems. In particular, in Subsection 6.2.1, the Sanov theorem is stated, explained, and illustrated with an example. Then we demonstrate that the law of large numbers (LLN) is a direct consequence of ST that implies how the simplest problem of selection of ED should be solved. This is intended to facilitate understanding of how the CLLN provides a probabilistic justification to relative entropy maximization and maximum probability methods, in context of the general ED problem. Next we step by step extend the ED problem into parametric and then to empirical parametric ED problems. CLLN is used to determine regularization methods. Not surprisingly, the methods are MaxMaxEnt, empirical MaxMaxEnt (also known as maximum entropy empirical likelihood or exponential tilt), respectively. Finally, the continuous case of the empirical parametric ED problem is addressed. Intermezzo (Section 6.3) summarizes implications of CLLN for the parametric ED selection problems and prepares a transition to the opposite SD selection problems. Next, Section 6.4 presents several LD theorems for sampling distributions (including the Bayesian Sanov theorem and the Bayesian law of large numbers), and discusses implications of BLLN for the SD problem in its basic as well as in its parametric forms. A summary is followed by some literature notes.

6.2 Large Deviations for Empirical Distributions

In order to discuss the ST CLLN, it is necessary to introduce some basic terminology; cf. Csiszár (1998).

Let $\mathcal{P}(\mathcal{X})$ be a set of all probability mass functions on the finite, m-element set $\mathcal{X} = \{x_1, x_2, \ldots, x_m\}$. The support of $p \in \mathcal{P}(\mathcal{X})$ is a set $S(p) = \{x : p(x) > 0\}$.

Let $x_1, x_2, \ldots, x_n$ be a random sample from a pmf $q \in \mathcal{P}(\mathcal{X})$. Let ν^n denote the empirical measure induced by a random sample of length n. Formally, the empirical measure $\nu^n = [n_1, n_2, \ldots, n_m]/n$, where n_i is the number of occurrences of ith element of $\mathcal{X}$ in the random sample. When there is a need to stress the size n of the random sample that induces the empirical measure, we will speak about the n-empirical measure. Finally, note that there are $\Gamma(\nu^n) = n!(\prod_{i=1}^{m} n_i!)^{-1}$ different random samples of length n that induce the same empirical measure ν^n.

As previously mentioned, we are interested in the event that the random sample drawn from a fixed pmf q induces the empirical measure ν^n from a set $\Pi \subseteq \mathcal{P}(\mathcal{X})$. The probability of this event is therefore

$$\pi(\nu^n \in \Pi; q) = \sum_{\nu^n \in \Pi} \pi(\nu^n; q),$$

where

$$\pi(\nu^n; q) = \Gamma(\nu^n) \prod_{i=1}^{m} q_i^{n_i}.$$

The Large Deviations (LD) rate function of probability of this event is the information divergence. The information divergence (I-divergence, ±-relative entropy, Kullback Leibler distance, etc.) $I(p \| q)$ of $p \in \mathcal{P}(\mathcal{X})$ with respect to $q \in \mathcal{P}(\mathcal{X})$ is

$$I(p \| q) = \sum_{i=1}^{m} p_i \log \frac{p_i}{q_i},$$

where $0 \log 0 = 0$ and $\log b/0 = +\infty$, by convention. The information projection $\hat{p}$ of q on Π is

$$\hat{p} = \arg \inf_{p \in \Pi} I(p \| q).$$

The I-divergence at an I-projection of q on Π is denoted $I(\Pi \| q)$. Finally, recall that $I(p \| q) \geq 0$, where $I(p \| q) = 0$ iff $p = q$.

Topological qualifiers, e.g. openness, will be used with respect to the topology induced on $\mathcal{P}(\mathcal{X})$ by the standard topology on $\mathcal{R}^m$.

6.2.1 Sanov Theorem

The Sanov theorem, which is the basic LD result on the asymptotic behavior of the probability that an n-empirical measure from a specified set Π occurs, may be stated as:

Sanov Theorem *Let Π be an open set and let $S(q) = \mathcal{X}$. Then*

$$\lim_{n\to\infty} \frac{1}{n} \log \pi(\nu^n \in \Pi; q) = -I(\Pi \| q).$$

Phrased informally, ST tells us that the rate of the exponential convergence of the probability $\pi(\nu^n \in \Pi; q)$ toward zero, is determined by the information divergence at (any of) the I-projection(s) of q on Π. The other probability mass functions (pmf's) do not influence the rate of convergence. Obviously, the greater the value $I(\Pi \| q)$, the faster the convergence of probability $\pi(\nu^n \in \Pi; q)$ to zero, as $n \to \infty$. The Sanov theorem thus permits us to speak about and measure how rare a set Π is with respect to a sampling distribution.

We use Example 6.1 to illustrate ST.

Example 6.1

Let $\mathcal{X} = [1, 2, 3, 4]$. Let q be the uniform pmf. Let $\Pi = \{p : \sum_{i=1}^{4} p_i x_i = 3.0\}$. The Table 6.1 illustrates the convergence of the normalized log-probability $\frac{1}{n} \log \pi(\nu^n \in \Pi; q)$ to $-I(\Pi \| q)$, as $n \to \infty$.

The probability that a 1000-empirical measure with a mean of 3.0 was drawn from q is $\pi(\nu^{1000} \in \Pi; q) = 4.1433e{-47}$. An approximate estimate of the probability can be obtained by means of ST, as $\pi(\nu^{1000} \in \Pi; q) \approx \exp(-1000\, I(\Pi \| q)) = 3.4121e{-45}$. Note that the approximation is comparable to the exact probability.

6.2.2 Law of Large Numbers

Since the probability $\pi(\nu^n \in A; q)$ goes to zero for any $A \subset \mathcal{P}(\mathcal{X})$ such that $I(A \| q) > 0$, the use of ST leads to the following law of large numbers (LLN):

Law of Large Numbers *Let $B(\hat{p}, \epsilon)$ be the closed ϵ-ball defined by the total variation metric and centered at the I-projection $\hat{p} \equiv q$ of q on $\mathcal{P}(\mathcal{X})$. Then,*

$$\lim_{n\to\infty} \pi(\nu^n \in B(\hat{p}, \epsilon); q) = 1.$$

TABLE 6.1

Convergence of the Normalized Log-Probability to the Negative of Minimum Value of the I-Divergence

n	$1/n \log \pi(\nu^n \in \Pi; q)$
10	−0.3152913
100	−0.1350160
1000	−0.1068000
$-I(\Pi \| q)$	−0.1023890

Thus, the information divergence that, through its infimal value, gave the rate of exponential decay also provides, through the "point" of its infimum, the pmf around which n-empirical measures concentrate. This LLN may be interpreted as saying that, asymptotically, the only possible empirical measures are those that are arbitrary close to the I-projection $\hat{p}$ of q on $\mathcal{P}(\mathcal{X})$, i.e., in this case, to the data-sampling distribution q.

Next, we consider the example showing how LLN implies that the simplest problem of selection of ED has to be solved using the REM/MaxEnt method. The ED problem concerns selection of empirical measure from Π, when the information-quadruple $(\mathcal{X}, q, n, \Pi)$ and nothing else is available. The next Example illustrates the ED problem, which is simplest in the sense that Π is identical with the entire $\mathcal{P}(\mathcal{X})$.

Example 6.2

Let $\mathcal{X} = \{1, 2, 3, 4\}$ and let $q = [0.1, 0.4, 0.2, 0.3]$. Let a random sample of size $n = 10^9$ be drawn from q. The sample is not available to us. We are told only that the sample mean is somewhere in the interval $[1, 4]$; i.e., $\Pi = \{p : \sum p_i x_i \in [1, 4]\}$. Given the information-quadruple $(\mathcal{X}, q, n, \Pi)$ we are asked to select an n-empirical measure from Π.

Since any 10^9-empirical measure fits the given interval of mean values, there are $N = \binom{n+m-1}{m-1}$ empirical measures from which we are asked to make a choice. The problem that Example 6.2 asks us to solve is in the form of an under-determined, ill-posed inverse problem.

The information that the unknown n-empirical measure was drawn from q is crucial for comprehending that the ill-posed problem has a simple solution implied by LLN. Though there are N n-empirical measures in Π, LLN implies that only those n-empirical measures that are close to the I-projection $\hat{p} \equiv q$ of the sampling distribution q on $\mathcal{P}(\mathcal{X})$, i.e., close to q, are possible. Hence, LLN regularizes the ill-posed inverse problem. Since the relative entropy maximization method (REM/MaxEnt) selects just the I-projection, REM must be used to solve this problem.

Let us conclude by noting that a consistency requirement would imply that the same method should also be used for "small" n. Thus, if the sample size were $n = 10$ instead of 10^9, as in Example 6.2, consistency would imply that one should select the I-projection of q on the set $\mathcal{P}_{10}(\mathcal{X})$ of all possible 10-empirical measures. Less stringently viewed, LLN implies that any method that asymptotically becomes identical to REM can be used for solving this instance of the ED problem.

6.2.3 CLLN, Maximum Entropy, and Maximum Probability

To demonstrate that LLN is a special case of the CLLNs, it is instructive to express the claim of LLN in the following form:

$$\lim_{n\to\infty} \pi(\nu^n \in B(\hat{p}, \epsilon) \mid \nu^n \in \mathcal{P}(\mathcal{X});\ q) = 1.$$

Compare this to the result of CLLN:

Conditional Law of Large Numbers *Let Π be a convex, closed set that does not contain q. Let $B(\hat{p}, \epsilon)$ be a closed ϵ-ball defined by the total variation metric and centered at the I-projection $\hat{p}$ of q on Π. Then,*

$$\lim_{n\to\infty} \pi(\nu^n \in B(\hat{p}, \epsilon) \mid \nu^n \in \Pi; q) = 1.$$

Thus, one can see that CLLN generalizes LLN. Interpretation of CLLN is similar to that of LLN except that the conditioning set Π is no longer the entire $\mathcal{P}(\mathcal{X})$ but a convex, closed subset that does not contain q; hence the model Π is misspecified. In other words, given the conditioning set Π, CLLN demonstrates that empirical measures asymptotically conditionally concentrate on the I-projection of q on Π, provided that the set satisfies certain technical requirements.

CLLN follows directly from the Sanov theorem. The conditional probability in question is

$$\pi(\nu^n \in B(p, \epsilon) \mid \nu^n \in \Pi;\ q) = \frac{\pi(\nu^n \in B(p, \epsilon); q)}{\pi(\nu^n \in \Pi; q)}.$$

Hence, by ST, if $B(p, \epsilon)$ is such that $\hat{p} \notin B(p, \epsilon)$ then

$$\frac{1}{n} \log \pi(\nu^n \in B(p, \epsilon) \mid \nu^n \in \Pi;\ q) \to -\underbrace{(I(B(p, \epsilon) \,\|\, q) - I(\Pi \,\|\, q))}_{>0}$$

and consequently $\pi(\nu^n \in B(p, \epsilon) \mid \nu^n \in \Pi;\ q) \to 0$. Since, by assumption, there is a unique I-projection of q on Π, the conditional probability concentrates on it.

In Subsection 6.2.2, LLN was invoked to solve Example 6.2, a simple instance of the ED problem. The following extension of Example 6.2 provides a more general instance of the problem in the sense that Π is now a subset of $\mathcal{P}(\mathcal{X})$.

Example 6.3
Let $\mathcal{X} = \{1, 2, 3, 4\}$ and let $q = [0.1, 0.4, 0.2, 0.3]$. Let a random sample of size $n = 10^9$ be drawn from q. The sample is not available to us. What we are told, only, is that the sample mean is 3.0. Note that it is different from the expected value of X under the pmf q, 2.7; i.e., the model is misspecified. Hence, the feasible set to which n-types belong is $\Pi = \{p : \sum_{i=1}^{4} p_i x_i = 3.0\}$. Given the information-quadruple $(\mathcal{X}, q, n, \Pi)$, how should one go about selecting an n-empirical measure from Π?

The same discussion as that following Example 6.2 applies to this example except that now it is CLLN instead of LLN that regularizes the ill-posed ED problem.

Example 6.3 (cont'd)
CLLN dictates that one selects an empirical measure close to the I-projection $\hat{p} = [0.057,\ 0.310,\ 0.209,\ 0.424]$ of q on Π, which is now different from q. The I-projection can be obtained in the standard way of solving the constrained relative

entropy maximization task (cf., e.g., Golan, Judge, and Miller 1996): $\hat{p} = \arg\max_{p\in\Pi} - \sum p_i \log \frac{p_i}{q_i}$, *where* $\Pi = \{p : \sum_{i=1}^{4} p_i x_i = 3.0, \sum_{i=1}^{4} p_i = 1\}$.

The following is an example of the ED problem with an economic relevance.

Example 6.4

(Cox, Daniell, and Nicole 1998) studied the UK National Lottery, where every week, 6 numbers from 49 are drawn, at random. The Lottery makes available the following information about a draw: the winning ticket s, the total number n of sold tickets, number of winners n_r *in each category (i.e., matched r-tuple),* $r = 3, 4, 5, 6$. *The info is available for W weeks. The authors assume that the distribution* $q(t)$ *of tickets is uniform.*

Given the above information, the objective is to select a representative empirical distribution of tickets. This is an instance of the problem of ED selection, where the feasible set of empirical pmf's is formed by the available information as follows: for each draw (week) w, winning ticket s and category $r = 3, 4, 5$,

$$n_r(w) = \sum_t \delta_r(t, s)n(t),$$

where $n(t)$ *is the unknown number of people who bought the ticket t and* $\delta_r(t, s) = 1$, *if t and s have common just r numbers, 0 otherwise. There are 3W constraints.*

The authors used information from $W = 113$ *weeks and regularized the* Π*-problem by relative entropy maximization method. In the Table 1.1 (of Cox, Daniell, and Nicole 1998) one can find, for instance, that ticket with numbers* 26 34 44 46 47 49 *has estimated* $n(t) = 0.41$ *while the ticket* 7 17 23 32 40 42 *has estimated* $n(t) = 45.62$; *on an appropriate scale.*

This work triggered a new economic interest in lotteries; cf., for instance, Farrell et al. (2000).

In conclusion, LD theory for empirical distributions, through the CLLNs provides a probabilistic justification for using REM to solve the ED selection problem. Alternatively, any method of solving the ED problem that asymptotically does not behave like REM, violates CLLN. For instance, using the maximum Tsallis entropy method (maxTent) to solve the Π-problem would go against CLLN. However, using the Maximum Probability method (MaxProb) (Boltzmann 1877; Vincze 1972; Grendár and Grendár 2001; Niven 2007) satisfies CLLN since it asymptotically turns into REM (cf. Grendár and Grendár 2001, 2004). The MaxProb method suggests that one may solve the ED problem by selecting the μ-projection of q on Π, i.e., the n-empirical measure

$$\nu^n_{\text{MaxProb}} = \arg\sup_{\nu^n\in\Pi} \pi(\nu^n; q).$$

It is worth noting that the convergence of the most probable empirical measure(s), obtained with MaxProb, to the distribution that maximizes relative entropy, obtained with REM/MaxEnt, provides another, deeper reading of CLLN, namely that the empirical measures conditionally concentrate on the asymptotically most probable empirical measure.

Finally, a word of caution: The information-divergence minimization method (REM/MaxEnt) is also used to regularize problems like spectrum estimation or recovering of X-ray attenuation functions or optical images (cf. Jones and Byrne 1990), that cannot be cast into the form of the Π-problem. In such cases, the LD justification used for REM cannot be invoked, and one has to rely on other arguments; cf. Jones and Byrne (1990) and Csiszár (1996).

6.2.4 Parametric ED Problem and Maximum Maximum Entropy Method

The feasible set Π of the ED problem can be, in general, defined by means of J moment-consistency constraints $\Pi = \{p : \sum_{i=1}^{m} p_i u_j(x_i) = a_j, j = 1, 2, \ldots, J\}$, where $u_j(\cdot)$ is a real-valued function of X called the u-moment and $a \in \mathcal{R}^J$ is given. In this case, the I-projection of q on Π is easy to find, as it belongs to the exponential family of distributions

$$\mathcal{E}(X, \lambda, u) = k(\lambda)q(X)\exp\left(-\sum_{j=1}^{J}\lambda_j u_j(X)\right),$$

where $k(\lambda) = (\sum_{i=1}^{m} q(x_i)\exp(-\sum_{j=1}^{J}\lambda_j u_j(x_i)))^{-1}$ is the normalizing constant. Example 6.3 presents a simple Π with a single u-moment of the form $u(X) = X$.

The u-moment function $u(X)$ can be viewed as a special case of a general, parametric $u(X, \theta)$-moment function, where θ is a parameter, $\theta \in \Theta \subseteq \mathcal{R}^k$. The parametric u-moments define the parametric feasible set $\Pi(\Theta) = \bigcup_{\theta\in\Theta}\Pi(\theta)$, where

$$\Pi(\theta) = \left\{p(\cdot;\theta) : \sum_{i=1}^{m} p(x_i;\theta)u_j(x_i,\theta) = 0,\ j = 1, 2, \ldots, J\right\}. \tag{6.4}$$

Example 6.5 illustrates the extension of the ED problem to the parametric ED problem.

Example 6.5

Let $\mathcal{X} = \{1, 2, 3, 4\}$ and let $q = [0.1, 0.4, 0.2, 0.3]$. Let a random sample of size $n = 10^9$ be drawn from q. The sample is unavailable to us. What we are told, only, is that the sample mean is in the interval $[3.0, 4.0]$. Here, $u(X, \theta) = X - \theta$, where $\theta \in [3.0, 4.0]$; so that $\Pi(\theta) = \{p(\cdot;\theta) : \sum p(x_i;\theta)(x_i - \theta) = 0\}$ and $\theta \in \Theta = [3.0, 4.0]$. Note that the interval does not contain the expected value of X with respect to q, $E_q X = 2.7$; i.e., the model $\Pi(\Theta)$ is misspecified. The objective is to select a parametrized n-empirical measure $\nu^n(\theta)$ from $\Pi(\Theta)$, given the available information.

Let us link the parametric Π-problem to the estimating equations (EE) approach to estimation. The general, parametric $u(X, \theta)$-moment function is, in this context, commonly known as an estimating function. Unbiased estimating functions are the most commonly considered estimating functions in

Econometrics. Thus, the parametric ED problem becomes a problem of estimating the unknown value of θ. In other words, given a scheme for selecting $\hat{p}(\cdot;\hat{\theta})$, we are more interested in $\hat{\theta}$ than in the corresponding $\hat{p}(\cdot;\hat{\theta})$. It is clear that the selected $\hat{\theta}$ will depend on q.

LD theory provides a clue as to how one should solve the parametric ED problem. If $\Pi(\Theta)$ is a convex, closed set that does not contain q (i.e., the model is misspecified), CLLN can be invoked to claim that such a parametric Π-problem should be solved by selecting the I-projection $\hat{p}(\cdot;\hat{\theta})$ of the sampling distribution q on $\Pi(\Theta)$, i.e.,

$$\hat{p}(\cdot;\theta) = \arg \inf_{p(\cdot;\theta)\in\Pi(\theta)} I(p(\cdot;\theta) \,\|\, q)$$

with $\theta = \hat{\theta}_{\mathrm{MME}}$, where

$$\hat{\theta}_{\mathrm{MME}} = \arg \inf_{\theta\in\Theta} I(\hat{p}(\cdot;\theta) \,\|\, q).$$

Because of the double maximization of the entropy, we will call the method associated with this prescription Maximum Maximum Entropy method (MaxMaxEnt). If $\Pi(\Theta)$ is defined by the (Equation 6.4) the estimator $\hat{\theta}_{\mathrm{MME}}$ can be expressed as

$$\hat{\theta}_{\mathrm{MME}} = \arg \sup_{\theta\in\Theta} \inf_{\lambda\in\mathcal{R}^J} \log \sum_{i=1}^{m} q(x_i;\theta) \exp\left(-\lambda' u(x_i;\theta)\right).$$

Example 6.5 (cont'd)

Note that MaxMaxEnt when applied to Example 6.5 selects the same pmf as did MaxEnt in Example 6.3. Indeed, since the information divergence is a convex function in the first argument, the minimum of the information divergence over $\theta \in [3.0, 4.0]$ *is attained for* $\theta = 3.0$. *Phrased in EE terms, the MaxMaxEnt estimator of the unknown true value of* θ, *based on the available information is* $\hat{\theta}_{MME} = 3.0$.

6.2.5 Empirical ED Problem

The setting of the Π-problem is idealized. In practice, the data-sampling distribution q is rarely known. Let us continue assuming that the other components of the information-quadruple that constitute the ED problem, i.e., Π, n (the size of the sample that is unavailable to us) and $\mathcal{X}$, are known to us. To make the setup and problem more realistic, imagine that we draw a random sample $X_1^N = X_1, X_2, \ldots, X_N$ of size N from the true data-sampling distribution q. Let the sample induce the N-empirical measure ν^N. When q in the information-quadruple is replaced by ν^N, we speak about the empirical ED problem. CLLN implies that the empirical ED problem should be solved by any method whose choice becomes asymptotically (i.e., as $n \to \infty$) identical with the I-projection of ν^N on Π.

6.2.6 Empirical Parametric ED Problem and Empirical MaxMaxEnt

The discussion of Subsection 6.2.5 extends directly to the empirical parametric ED problem, which CLLN implies should be solved by selecting

$$\hat{p}(\cdot;\theta) = \arg \inf_{p(\cdot;\theta)\in\Pi(\theta)} I(p(\cdot;\theta)\parallel\nu^N)$$

with $\theta = \hat{\theta}_{\mathrm{EMME}}$, where

$$\hat{\theta}_{\mathrm{EMME}} = \arg\inf_{\theta\in\Theta} I(\hat{p}(\cdot;\theta)\parallel\nu^N).$$

The estimator $\hat{\theta}_{\mathrm{EMME}}$ is known in *Econometrics* under various names such as maximum entropy empirical likelihood and exponential tilt. We call it the empirical MaxMaxEnt estimator (EMME). Note that thanks to the convex duality, the estimator $\hat{\theta}_{\mathrm{EMME}}$ can equivalently be obtained as

$$\hat{\theta}_{\mathrm{EMME}} = \arg\sup_{\theta\in\Theta}\inf_{\lambda\in\mathcal{R}^J}\log\sum_{i=1}^{m}\nu^N(x_i;\theta)\exp\left(-\lambda' u(x_i;\theta)\right). \tag{6.5}$$

Example 6.6 illustrates the extension of the parametric ED problem (cf. Example 6.5) to the empirical parametric ED problem.

Example 6.6

Let $\mathcal{X} = \{1,2,3,4\}$. *Let a random sample of size* $N = 100$ *from data-sampling distribution* q *induces N-type* $\nu^N = [7\ 42\ 24\ 27]/100$. *Let in addition a random sample of size* $n = 10^9$ *be drawn from* q, *but it remains unavailable to us. We are told only that the sample mean is in the interval* $[3.0, 4.0]$. *Thus* $\Pi(\theta) = \{p(\cdot;\theta) : \sum_{i=1}^{4} p(x_i;\theta)(x_i-\theta) = 0\}$ *and* $\theta\in\Theta = [3.0, 4.0]$. *The objective is to select an n-empirical measure from* $\Pi(\Theta)$, *given the available information.*

CLLN dictates that we solve the problem by EMME. Since n is very large, we can without much harm ignore rational nature of n-types (i.e., $\nu^n(\cdot;\theta)\in\mathcal{Q}^m$*) and seek the solution among pmf's* $p(\cdot;\theta)\in\mathcal{R}^m$. *CLLN suggests the selection of* $\hat{p}(\hat{\theta}_{\mathrm{EMME}})$. *Since the average* $\sum_{i=1}^{4}\nu_i^N x_i = 2.71$, *is outside of the interval* $[3.0, 4.0]$, *convexity of the information divergence implies that* $\hat{\theta}_{\mathrm{EMME}} = 3.0$, *i.e., the lower bound of the interval.*

Kitamura and Stutzer (2002) were the first to recognize that LD theory, through CLLN, can provide justification for the use of the EMME estimator. The CLLNs demonstrate that selection of I-projection is a consistent method, which in the case of a parametric, possibly misspecified model $\Pi(\Theta)$, establishes consistency under misspecification of the EMME estimator.

Let us note that ST and CLLN have been extended also to the case of continuous random variables; cf. Csiszár (1984); this extension is outside the scope of this chapter. However, we note that the theorems, as well as Gibbs conditioning principle (cf. Dembo and Zeitouni 1998) and Notes on literature), when applied to the parametric setting, single out

$$\hat{\theta}_{\text{EMME}} = \arg\sup_{\theta\in\Theta} \inf_{\lambda\in\mathcal{R}^J} \frac{1}{N}\sum_{l=1}^{N} \exp\left(-\lambda' u(x_l;\theta)\right) \tag{6.6}$$

as an estimator that is consistent under misspecification. The estimator is the continuous-case form of Empirical MaxMaxEnt estimator. Note that the above definition (Equation 6.6) of the EMME reduces to Equation 6.5, when X is a discrete random variable. In conclusion it is worth stressing that in ED-setting the EMD estimators from the CR class (cf. Section 6.1) other than EMME are not consistent, if the model is not correctly specified.

A setup considered by Qin and Lawless (1994) (see also Grendár and Judge 2009b) serves for a simple illustration of the empirical parametric ED problem for a continuous random variable.

Example 6.7

Let there be a random sample from a (unknown to us) distribution $f_X(x)$ on $\mathcal{X} = \mathcal{R}$. We assume that the data were sampled from a distribution that belongs to the following class of distributions (Qin and Lawless 1994): $\Pi(\theta) = \{p(x;\theta) : \int_{\mathcal{R}} p(x;\theta)(x-\theta)\,dx = 0, \int_{\mathcal{R}} p(x;\theta)(x^2 - (2\theta^2+1))\,dx = 0, p(x;\theta) \in \mathcal{P}(\mathcal{R})\}$, and $\theta \in \Theta = \mathcal{R}$. However, the true sampling distribution need not belong to the model $\Pi(\Theta)$. The objective is to select a $p(\theta)$ from $\Pi(\Theta)$. The large deviations theorems mentioned above single out $\hat{p}(\hat{\theta}_{\text{EMME}})$, which can be obtained by means of the nested optimization (Equation 6.6).

For further discussions and application of EMME to asset pricing estimation, see Kitamura and Stutzer (2002).

6.3 Intermezzo

Since we are about to leave the area of LD for empirical measures for the, in a sense, opposite area of LD for data-sampling distributions, let us pause and recapitulate the important points of the above discussions.

The Sanov theorem, which is the basic result of LD for empirical measures, states that the rate of exponential convergence of probability $\pi(\nu^n \in \Pi; q)$ is determined by the infimal value of information divergence (Kullback-Leibler divergence) $I(p \,\|\, q)$ over $p \in \Pi$. Though seemingly a very technical result, ST has fundamental consequences, as it directly leads to the law of large numbers and, more importantly, to its extension, the CLLNs (also known as the conditional limit theorem). Phrased in the form implied by Sanov theorem, LLN says that the empirical measure asymptotically concentrates on the I-projection $\hat{p} \equiv q$ of the data-sampling q on $\Pi \equiv \mathcal{P}(\mathcal{X})$. When applying LLN, the feasible set of empirical measures Π is the entire $\mathcal{P}(\mathcal{X})$. It is of interest to know the point of concentration of empirical measures when Π is a subset of $\mathcal{P}(\mathcal{X})$. Provided that Π is a convex, closed subset of $\mathcal{P}(\mathcal{X})$, this guarantees that the I-projection is unique. Consequently, CLLN shows that the empirical measure asymptotically conditionally concentrates around the I-projection $\hat{p}$ of the data-sampling distribution of q on Π. *Thus, the CLLNs regularizes the ill-posed problem of ED selection. In other words, it provides a firm probabilistic justification for the application of the relative entropy maximization method in solving the ED problem.* We have gradually considered more complex forms of the problem, recalled the associated conditional laws of large numbers, and showed how CLLN also provides a probabilistic justification for the empirical MaxMaxEnt method (EMME). It is also worth recalling that any method that fails to behave like EMME asymptotically would violate CLLN if it were used to obtain a solution to the empirical parametric ED problem.

6.4 Large Deviations for Sampling Distributions

Now, we turn to a corpus of "opposite" LD theorems that involves LD theorems for data-sampling distributions, which assume a Bayesian setting. First, the Bayesian Sanov theorem (BST) will be presented. We will then demonstrate how this leads to the Bayesian law of large numbers (BLLN). These LD theorems for sampling distributions will be linked to the problem of selecting a sampling distribution (SD problem, for short). We then demonstrate that if the sample size n is sufficiently large the problem should be solved with the maximum nonparametric likelihood (MNPL) method. As with the problem of empirical distribution (ED) selection, requiring consistency implies that the SD problem should be solved with a method that asymptotically behaves like MNPL. The Bayesian LLN implies that, for finite n, there are at least two such methods, MNPL itself and maximum a posteriori probability. Next, it will be demonstrated that the Bayesian LLN leads to solving the parametric SD problem with the empirical likelihood method when n is sufficiently large.

6.4.1 Bayesian Sanov Theorem

In a Bayesian context assume that we put a strictly positive prior probability mass function $\pi(q)$ on a countable[3] set $\Phi \subset \mathcal{P}(\mathcal{X})$ of probability mass functions (sampling distributions) q. Let r be the "true" data-sampling distribution, and let X_1^n denote a random sample of size n drawn from r. Provided that $r \in \Phi$, the posterior distribution

$$\pi(q \in Q \mid X_1^n = x_1^n; r) = \frac{\sum_Q \pi(q) \prod_{i=1}^n q(x_i)}{\sum_\Phi \pi(q) \prod_{i=1}^n q(x_i)}$$

[3] We restrict presentation to this case, in order to not obscure it by technicalities; cf. Grendár and Judge (2009a) for Bayesian LD theorems in a more general case and more complete discussions.

is expected to concentrate in a neighborhood of the true data-sampling distribution r as n grows to infinity. Bayesian nonparametric consistency considerations focus on exploration of conditions under which it indeed happens; for entries into the literature we recommend Ghosh and Ramamoorthi (2003); Ghosal, Ghosh, and Ramamoorthi (1999); Walker (2004); and Walker, Lijoi, and Prünster (2004), among others. Ghosal, Ghosh, and Ramamoorthi (1999) define consistency of a sequence of posteriors with respect to a metric or discrepancy measure d as follows: The sequence $\{\pi(\cdot \mid X_1^n; r), n \geq 1\}$ is said to be d-consistent at r, if there exists a $\Omega_0 \subset \mathcal{R}^\infty$ with $r(\Omega_0) = 1$ such that for $\omega \in \Omega_0$, for every neighborhood U of r, $\pi(U \mid X^n; r) \to 1$ as n goes to infinity. If a posterior is d-consistent for any $r \in \Phi$, then it is said to be d-consistent. Weak consistency and Hellinger consistency are usually studied in the literature.

Large deviations techniques can be used to study Bayesian nonparametric consistency. The Bayesian Sanov theorem identifies the rate function of the exponential decay. This in turn identifies the sampling distributions on which the posterior concentrates, as those distributions that minimize the rate function. In the i.i.d. case the rate function can be expressed in terms of the L-divergence. The L-divergence (Grendár and Judge 2009a) $L(q \parallel p)$ of $q \in \mathcal{P}(\mathcal{X})$ with respect to $p \in \mathcal{P}(\mathcal{X})$ is defined as

$$L(q \parallel p) = -\sum_{i=1}^{m} p_i \log q_i.$$

The L-projection $\hat{q}$ of p on $A \subseteq \mathcal{P}(\mathcal{X})$ is

$$\hat{q} = \arg \inf_{q \in A} L(q \parallel p).$$

The value of L-divergence at an L-projection of p on A is denoted by $L(A \parallel p)$. Finally, let us stress that in the discussion that follows, r need not be from Φ; i.e., we are interested in Bayesian nonparametric consistency under misspecification.

In this context the Bayesian Sanov theorem (BST) provides the rate of the exponential decay of the posterior probability.

Bayesian Sanov Theorem *Let $Q \subset \Phi$. As $n \to \infty$,*

$$\frac{1}{n} \log \pi(q \in Q \mid x_1^n; r) \to -\{L(Q \parallel r) - L(\Phi \parallel r)\}, \quad a.s.\ r^\infty.$$

In effect BST demonstrates that the posterior probability $\pi(q \in Q \mid x_1^n; r)$ decays exponentially fast (almost surely), with the decay rate specified by the difference in the two extremal L-divergences.

6.4.2 BLLNs, Maximum Nonparametric Likelihood, and Bayesian Maximum Probability

The Bayesian law of large numbers (BLLN) is a direct consequence of BST.

Bayesian Law of Large Numbers *Let $\Phi \subseteq \mathcal{P}(\mathcal{X})$ be a convex, closed set. Let $B(\hat{q}, \epsilon)$ be a closed ϵ-ball defined by the total variation metric and centered at the L-projection $\hat{q}$ of r on Φ. Then, for $\epsilon > 0$,*

$$\lim_{n\to\infty} \pi(q \in B(\hat{q}, \epsilon) \mid q \in \Phi, x_1^n; r) = 1, \quad a.s.\ r^\infty.$$

Thus, there is asymptotically *a posteriori* (a.s. r^∞) zero probability of a data-sampling distribution other than those arbitrarily close to the L-projection $\hat{q}$ of r on Φ.

BLLN is Bayesian counterpart of the CLLNs. When $\Phi = \mathcal{P}(\mathcal{X})$ the BLLN reduces to a special case, which is a counterpart of the law of large numbers. In this special case the L-projection $\hat{q}$ of the true data-sampling r on $\mathcal{P}(\mathcal{X})$ is just the data-sampling distribution r. Hence the BLLN can be in this case interpreted as indicating that, asymptotically, *a posteriori* the only possible data-sampling distributions are those that are arbitrary close to the "true" data-sampling distribution r.

The following example illustrates how BLLN, in the case where $\Phi \equiv \mathcal{P}(\mathcal{X})$, implies that the simplest problem of selecting of sampling distribution, has to be solved with the maximum nonparametric likelihood method. The SD problem is framed by the information-quadruple $(\mathcal{X}, \nu^n, \Phi, \pi(q))$. The objective is to select a sampling distribution from Φ.

Example 6.8

Let $\mathcal{X} = \{1, 2, 3, 4\}$, and let $r = [0.1, 0.4, 0.2, 0.3]$ be unknown to us. Let a random sample of size $n = 10^9$ be drawn from r, and let ν^n be the empirical measure that the sample induced. We assume that the mean of the true data-sampling distribution r is somewhere in the interval $[1, 4]$. Thus, r can be any pmf from $\mathcal{P}(\mathcal{X})$. Given the information $\mathcal{X}$, ν^n, $\Phi \equiv \mathcal{P}(\mathcal{X})$ and our prior $\pi(\cdot)$, the objective is to select a data-sampling distribution from Φ.

The problem presented in Example 6.8 is clearly an underdetermined, ill-posed inverse problem. Fortunately, BLLN regularizes it in the same way LLN did for the simplest empirical distribution selection problem, cf. Example 6.2 (Subsection 6.2.2). BLLN says that, given the sample, asymptotically *a posteriori* the only possible data-sampling distribution is the L-projection $\hat{q} \equiv r$ of r on $\Phi \equiv \mathcal{P}(\mathcal{X})$. Clearly, the true data-sampling distribution r is not known to us. Yet, for sufficiently large n, the sample-induced empirical measure ν^n is close to r. Hence, recalling BLLN, it is the L-projection of ν^n on Φ what we should select. Observe that this L-projection is just the probability distribution that maximizes $\sum_{i=1}^m \nu_i^n \log q_i$, the nonparametric likelihood.

We suggest the consistency requirement relative to potential methods for solving the SD problem. Namely, any method used to solve the problem should be such that it asymptotically conforms to the method implied by the Bayesian law of large numbers. We know that one such method is the maximum nonparametric likelihood. Another method that satisfies the consistency requirement and is more sound than MNPL, in the case of finite n, is the method of maximum a posteriori probability (MAP), which selects

$$\hat{q}_{\mathrm{MAP}} = \arg\sup_{q \in \Phi} \pi(q \mid \nu^n; r).$$

MAP, unlike MNPL, takes into account the prior distribution $\pi(q)$. It can be shown (cf. Grendár and Judge 2009a) that under the conditions for BLLN, MAP and MNPL asymptotically coincide and satisfy BLLN.

Although MNPL and MAP can legitimately be viewed as two different methods (and hence one should choose between them when n is finite), we prefer to view MNPL as an asymptotic instance of MAP (also known as Bayesian MaxProb), much like the view in (Grendár and Grendár 2001) that REM/MaxEnt is an asymptotic instance of the maximum probability method.

As CLLN regularizes ED problems, so does the Bayesian LLN for SD problems such as the one in Example 6.9.

Example 6.9

Let $\mathcal{X} = \{1, 2, 3, 4\}$, and let $r = [0.1, 0.4, 0.2, 0.3]$ be unknown to us. Let a random sample of size $n = 10^9$ be drawn from r, and let $\nu^n = [0.7, 0.42, 0.24, 0.27]$ be the empirical measure that the sample induced. We assume that the mean of the true data-sampling distribution r is 3.0; i.e., $\Phi = \{q : \sum_{i=1}^{4} q_i x_i = 3.0\}$. Note that the assumed value is different from the expected value of X under r, 2.7. Given the information $\mathcal{X}$, ν^n, Φ and our prior $\pi(\cdot)$, the objective is to select a data-sampling distribution from Φ.

The BLLN prescribes the selection of a data-sampling distribution close to the L-projection $\hat{p}$ of the true data-sampling distribution r on Φ. Note that the L-projection of r on Φ, defined by linear moment consistency constraints $\Phi = \{q : \sum q(x_i) u_j(x_i) = a_j, j = 1, 2, \ldots, J\}$, where u_j is a real-valued function and $a_j \in \mathcal{R}$, belongs to the Λ-family of distributions (cf. Grendár and Judge 2009a),

$$\Lambda(r, u, \lambda, a) = \left\{ q : q(x) = r(x) \left[1 - \sum_{j=1}^{J} \lambda_j (u_j(x) - a_j) \right]^{-1}, x \in \mathcal{X} \right\}.$$

Since r is unknown to us, it is reasonable to replace r with the empirical measure ν^n induced by the sample X_1^n. Consequently, the BLLN instructs us to select the L-projection of ν^n on Φ, i.e., the data-sampling distribution that maximizes nonparametric likelihood. When n is finite, it is the maximum a posteriori probability data-sampling distribution(s) that should be selected. Thus, given certain technical conditions, BLLN provides a strong probabilistic justification for using the maximum a posteriori probability method and its asymptotic instance, the maximum nonparametric likelihood method, to solve the problem of selecting an SD.

Example 6.9 (cont'd)
Since n is sufficiently large, MNPL and MAP will produce a similar result. The L-projection $\hat{q}$ of ν^n on Φ belongs to the Λ family of distributions. The correct values $\hat{\lambda}$ of the parameters λ can be found by means of the convex dual problem (cf., e.g., Owen 2001):

$$\hat{\lambda} = \arg \inf_{\lambda \in \mathcal{R}^J} - \sum_i \nu_i^n \log \left(1 - \sum_j \lambda_j (u_j(x_i) - a_j)\right).$$

For the setting of Example 6.9, the L-projection $\hat{q}$ of ν^n on Φ can be found to be [0.043, 0.316, 0.240, 0.401].

6.4.3 Parametric SD Problem and Empirical Likelihood

Note that the SD problem is naturally in an empirical form. As such, there is only one step from the SD problem to the parametric SD problem, and this step means replacing Φ with a parametric set $\Phi(\Theta)$, where $\theta \in \Theta \subseteq \mathcal{R}^k$. The most common such set $\Phi(\theta)$, considered in *Econometrics*, is that defined by unbiased EEs, i.e., $\Phi(\Theta) = \bigcup_{\theta \in \Theta} \Phi(\theta)$, where

$$\Phi(\theta) = \left\{ q(x;\theta) : \sum_{i=1}^{m} q(x_i;\theta) u_j(x_i;\theta) = 0,\ j = 1, 2, \ldots, J \right\}.$$

The objective in solving the parametric SD problem is to select a representative sampling distribution(s) when only the information $(\mathcal{X}, \nu^n, \Phi(\Theta), \pi(q))$ is given. Provided that $\Phi(\Theta)$ is a convex, closed set and that n is sufficiently large, BLLN implies that the parametric Φ-problem should be solved with the maximum nonparametric likelihood method, i.e., by selecting

$$\hat{q}(\cdot;\theta) = \arg \inf_{q(\cdot;\theta) \in \Phi(\theta)} L(q(\cdot;\theta) \parallel \nu^n),$$

with $\theta = \hat{\theta}$, where

$$\hat{\theta}_{\text{EL}} = \arg \inf_{\theta \in \Theta} L(\hat{q}(\cdot;\theta) \parallel \nu^n).$$

The resulting estimator $\hat{\theta}_{\text{EL}}$ is known in the literature as the empirical likelihood (EL) estimator.

If n is finite/small, BLLN implies that the problem should be regularized with MAP method/estimator. It is worth highlighting that in the semiparametric EE setting, the prior $\pi(q)$ is put over $\Pi(\Theta)$, and the prior in turn induces a prior $\pi(\theta)$ over the parameter space Θ; cf. Florens and Rolin (1994).

BST and BLLN are also available for the case of continuous random variables; cf. (Grendár and Judge 2009a). In the case of EEs for continuous random variables, BLLN provides a consistency-under-misspecification argument for the continuous-form of EL estimator (see Equation (6.3)). BLLN also supports the Bayesian MAP estimator

$$\hat{q}_{\text{MAP}}(x; \hat{\theta}_{\text{MAP}}) = \arg \sup_{q(x;\theta)\in\Phi(\theta)} \sup_{\theta\in\Theta} \pi(q(x;\theta) \mid x_1^n).$$

Since EL and the MAP estimators are consistent under misspecification, this provides a basis for the EL as well for the Bayesian MAP estimation methods. In conclusion it is worth stressing that in SD setting the other EMD estimators from the CR class (cf. Section 6.1) are not consistent, if the model is not correctly specified. The same holds, in general, for the posterior mean.

Example 6.10

As an illustration of application of EL in finance, consider a problem of estimation of the parameters of interest in rate diffusion models. In Laffers (2009), parameters of Cox, Ingersoll, and Ross (1985) model, for an Euro overnight index average data, were estimated by empirical likelihood method, with the following set of estimating functions, for time t (Zhou 2001):

$$\begin{gathered}
r_{t+1} - \mathrm{E}(r_{t+1} \mid r_t), \\
r_t[r_{t+1} - \mathrm{E}(r_{t+1} \mid r_t)], \\
\mathrm{V}(r_{t+1} \mid r_t) - [r_{t+1} - \mathrm{E}(r_{t+1} \mid r_t)]^2, \\
r_t\{\mathrm{V}(r_{t+1} \mid r_t) - [r_{t+1} - \mathrm{E}(r_{t+1} \mid r_t)]^2\}.
\end{gathered}$$

There, r_t denotes the interest rate at time t, V denotes the variance. In Laffers (2009) also a Monte Carlo study of small sample properties of EL estimator was conducted; cf. also Zhou (2001).

6.5 Summary

The Empirical Minimum Divergence (EMD) approach to estimation and inference, described in Section 6.1, is an attractive alternative to the generalized method of Moments. EMD comprises two components: a parametric model, which is usually specified by means of EEs, and a divergence (discrepancy) measure of a pdf with respect to the true sampling distribution. The divergence is minimized among parametrized pdf's from the model set, and this way a pdf is selected. The selected parametrized pdf depends on the true, yet unknown in practice, sampling distribution. Since the assumed discrepancy measures are convex and the model set is a convex set, the optimization problem has its convex dual equivalent formulation; cf. Equation 6.1. The convex dual problem (Equation 6.1) can be tied to the data by replacing the expectation by its empirical analogue; cf. (Equation 6.2). This way the data are taken into account and the EMD estimator results.

A researcher can choose between two possible ways of using the parametric model, defined by EEs. One option is to use the EEs to define a feasible set $\Phi(\Theta)$ of possible parametrized sampling distributions. Then the objective of EMD procedure is to select a parametrized sampling distribution (SD) from the model set $\Phi(\Theta)$, given the data. This modeling strategy and the objective deserve a name, and we call it the parametric SD problem. The other option is to let the EEs define a feasible set $\Pi(\Theta)$ of possible parametrized empirical distributions and use the observed, data-based empirical pmf in place of a sampling distribution. If this option is followed, then, given the data, the objective of the EMD procedure is to select a parametrized empirical distribution from the model set $\Pi(\Theta)$, given the data; we call it the parametric empirical ED problem. The empirical attribute stems for the fact that the data are used to estimate the sampling distribution.

In addition to the possibility of choosing between the two strategies, a researcher who follows the EMD approach to estimation and inference can select a particular divergence measure. Usually, divergence measures from Cressie–Read (CR) family are used in the literature. Prominent members of the CR-based class of EMD estimators are: maximum empirical likelihood estimator (MELE), empirical maximum maximum entropy estimator (EMME), and Euclidean empirical likelihood (EEL) estimator. Properties of EMD estimators have been studied in numerous works. Of course, one is not limited to the "named" members of CR family. Indeed, in the literature an option of letting the data select "the best" member of the family, with respect to a particular loss function, has been explored.

Consistency is perhaps the least debated property of estimation methods. EMD estimators are consistent, provided that the model is well-specified; i.e., the feasible set (being it Φ or Π) contains the true data-sampling distribution r. However, models are rarely well-specified. It is thus of interest to know which of the EMD methods of information recovery is consistent under misspecification. And here the large deviations (LD) theory enters the scene. LD theory helps to both define consistency under misspecification and to identify methods with this property. Large deviations are rather a technical subfield of the probability theory. Our objective has been to provide a nontechnical introduction to the basic theorems of LD, and step-by-step show the meaning of the theorems for consistency-under-misspecification requirement.

Since there are two modeling strategies, there are also two sets of LD theorems. LD theorems for empirical measures are at the base of classic (orthodox) LD theory. The theorems suggest that the relative entropy maximization method (REM, aka MaxEnt) possesses consistency-under-misspecification in the nonparametric form of the ED problem. The consistency extends also to the empirical parametric ED problem, where it is the empirical maximum maximum entropy method that has the desired property. LD theorems for sampling distributions are rather recent. They provide a consistency-under-misspecification argument in favor of the Bayesian maximum a posteriori probability, maximum nonparametric likelihood, and empirical likelihood methods in nonparametric and semiparametric form of the SD problem, respectively.

6.6 Notes on Literature

1. The LD theorems for empirical measures discussed here can be found in any standard book on LD theory. We recommend Dembo and Zeitouni (1998), Ellis 2005, Csiszár (1998), and Csiszár and Shields (2004) for readers interested in LD theory and closely related method of types, which is more elucidating. An accessible presentation of ST and CLLN can be found in Cover and Thomas (1991). Proofs of the theorems cited here can be found in any of these sources. A physics-oriented introduction to LD can be found in Aman and Atmanspacher (1999) and Ellis (1999).
2. Sanov theorem (ST) was considered for the first time in Sanov (1957), extended by Bahadur and Zabell (1979). Groeneboom, Oosterhoff, and Ruymgaart (1979) and Csiszár (1984) proved ST for continuous random variables; cf. Csiszár (2006) for a lucid proof of continuous ST. Csiszár, Cover, and Choi (1987) proved ST for Markov chains. Grendár and Niven (2006) established ST for the Pólya urn sampling. The first form of CLLNs known to us is that of Bártfai (1972). For developments of CLLN see Vincze (1972), Vasicek (1980), van Campenhout and Cover (1981), Csiszár (1984,1985,1986), Brown and Smith (1986), Harremoës (2007), among others.
3. Gibbs conditioning principle (GCP) (cf. Csiszár 1984; Lanford 1973), and (see also Csiszár 1998; Dembo and Zeitouni 1998), which was not discussed in this chapter, is a stronger LD result than CLLN. GCP reads:

   Gibbs conditioning principle: *Let $\mathcal{X}$ be a finite set. Let Π be a closed, convex set. Let $n \to \infty$. Then, for a fixed t,*

$$\lim_{n\to\infty} \pi(X_1 = x_1, \ldots, X_t = x_t \mid \nu^n \in \Pi; q) = \prod_{l=1}^{t} \hat{p}_{x_l}.$$

   Informally, GCP says that, if the sampling distribution q is confined to produce sequences which lead to types in a set Π, then elements of any such sequence of fixed length t will behave asymptotically conditionally as if they were drawn identically and independently from the I-projection $\hat{p}$ of q on Π — provided that the last is unique. There is no direct counterpart of GCP in the Bayesian Φ-problem setting. In order to keep symmetry of the exposition, we decided to not discuss GCP in detail.
4. Jaynes' views of maximum entropy method can be found in Jaynes (1989). In particular, the entropy concentration theorem (cf. Jaynes 1989)

is worth mentioning. It says, using our notation, that, as $n \to \infty$, $2n\Delta H(\nu^n) \sim \chi^2_{m-J-1}$ and $H(p) = -\sum p_i \log p_i$ is the Shannon entropy.

For a mathematical treatment of the maximum entropy method see Csiszár (1996, 1998). Various uses of MaxEnt are discussed in Solana-Ortega and Solana (2005). For a generalization of MaxEnt which is of direct relevance to *Econometrics*, see Golan, Judge, and Miller (1996), and also Golan (2008).

Maximization of the Tsallis entropy (MaxTent) leads to the same solution as maximization of Rényi entropy. Bercher proposed a few arguments in support of MaxTent; cf. Bercher (2008) for a survey.

For developments of the maximum probability method cf. Boltzmann (1877), Vincze (1972), Vincze (1997), Grendár and Grendár (2001), Grendár and Grendár (2004), Grendár and Niven (2006), Niven (2007). For the asymptotic connection between MaxProb and MaxEnt see Grendár and Grendár (2001, 2004).

5. While the LD theorems for empirical measures have already found their way into textbooks, discussions of LD for data-sampling distributions are rather recent. To the best of our knowledge, the first Bayesian posterior convergence via LD was established by Ben-Tal, Brown, and Smith (1987). In fact, their Theorem 1 covers a more general case where it is assumed that there is a set of empirical measures rather than a single such a measure ν^n. The authors extended and discussed their results in Ben-Tal, Brown, and Smith (1988). For some reasons, these works remained overlooked. More recently, ST for data-sampling distributions was established in an interesting work by Ganesh and O'Connell (1999). The authors established BST for finite $\mathcal{X}$ and well-specified model. In Grendár and Judge (2009a), Bayesian ST and the Bayesian LLN were developed for $\mathcal{X} = \mathcal{R}$ and a possibly misspecified model.
6. Relevance of LD for empirical measures for empirical estimator choice was recognized by Kitamura and Stutzer (1997), where LD justification of empirical MaxMaxEnt was discussed.
7. Finding empirical likelihood or empirical MaxMaxEnt estimators is a demanding numeric problem; cf., e.g., Mittelhammer and Judge (2001). In Brown and Chen (1998) an approximation to EL via the Euclidean likelihood was suggested, which makes the computations easier. Chen, Variyath, and Abraham (2008) proposed the Adjusted EL which mitigates a part of the numerical problem of EL. Recently, it was recognized that empirical likelihood and related methods are susceptible to the empty set problem that requires a revision of the available empirical evidence on EL-like methods; cf. Grendár and Judge (2009b).
8. Properties of estimators from EMD class were studied in numerous works; cf. Back and Brown (1990), Baggerly (1998), Baggerly (1999), Bickel et al. (1993), Chen et al. (2008), Corcoran (2000), DiCiccio, Hall, and Romano (1991), DiCiccio, Hall, and Romano (1990), Grendár and Judge (2009a), Imbens (1993), Imbens, Spady, and Johnson (1998), Jing and Wood (1996), Judge and Mittelhammer (2004), Judge and

Mittelhammer (2007), Kitamura and Stutzer (1997), Kitamura and Stutzer (2002), Lazar (2003), Mittelhammer and Judge (2001), Mittelhammer and Judge (2005), Mittelhammer, Judge, and Schoenberg (2005), Newey and Smith (2004), Owen (1991), Qin and Lawless (1994), Schennach (2005), Schennach (2004), Schennach (2007), Grendár and Judge (2009a), Grendár and Judge (2009b), among others.

6.7 Acknowledgments

Valuable feedback from Doug Miller, Assad Zaman, and an anonymous reviewer is gratefully acknowledged.

References

Amann, A., and H. Atmanspacher. 1999. Introductory remarks on large deviations statistics. *J. Sci. Explor.* 13(4):639–664.

Back, K., and D. Brown. 1990. Estimating distributions from moment restrictions. Working paper, Graduate School of Business, Indiana University.

Baggerly, K. A. 1998. Empirical likelihood as a goodness-of-fit measure. *Biometrika.* 85(3):535–547.

Baggerly, K. A. 1999. Studentized empirical likelihood and maximum entropy. Technical report, Rice University, Dept. of Statistics, Houston, TX.

Bahadur, R., and S. Zabell. 1979. Large deviations of the sample mean in general vector spaces. *Ann. Probab.* 7:587–621.

Bártfai, P. 1972. On a conditional limit theorem. *Coll. Math. Soc. J. Bolyai.* 9:85–91.

Ben-Tal, A., D. E., Brown, and R. L. Smith. 1987. Posterior convergence under incomplete information. Technical report 87–23. University of Michigan, Ann Arbor.

Ben-Tal, A., D. E., Brown, and R. L. Smith. 1988. Relative entropy and the convergence of the posterior and empirical distributions under incomplete and conflicting information. Technical report 88–12. University of Michigan Ann Arbor.

Bercher, J.-F. 2008. Some possible rationales for Rényi-Tsallis entropy maximization. In *International Workshop on Applied Probability, IWAP 2008.*

Bickel, P. J., C. A. J., Klassen, Y., Ritov, and J. Wellner. 1993. *Efficient and Adaptive Estimation for Semiparametric Models.* Baltimore: Johns Hopkins University Press.

Boltzmann, L. 1877. Über die Beziehung zwischen dem zweiten Hauptsatze der mechanischen Wärmetheorie und der Wahrscheilichkeitsrechnung respektive den Sätzen über das Wärmegleichgewicht. *Wiener Berichte* 2(76):373–435.

Brown, B. M., and S. X. Chen. 1998. Combined and least squares empirical likelihood. *Ann. Inst. Statist. Math.* 90:443–450.

Brown, D. E., and R. L. Smith. 1986. A weak law of large numbers for rare events. Technical report 86–4. University of Michigan, Ann Arbor.

Chen, J., A. M., Variyath, and B. Abraham. 2008. Adjusted empirical likelihood and its properties. *J. Comput. Graph. Stat.* 17(2):426–443.

Corcoran, S. A. 2000. Empirical exponential family likelihood using several moment conditions. *Stat. Sinica.* 10:545–557.

Cover, T., and J. Thomas. 1991. *Elements of Information Theory*. New York: Wiley.
Cox, J. C., J. E., Ingersoll, and S. A. Ross. 1985. A theory of the term structure of interest rates. *Econometrica* 53:385–408.
Cox, S. J., G. J., Daniell, and D. A. Nicole. 1998. Using maximum entropy to double ones expected winnings in the UK National Lottery. *JRSS Ser. D*. 47(4):629–641.
Cressie, N., and T. Read. 1984. Multinomial goodness of fit tests. *JRSS Ser. B*. 46:440–464.
Cressie, N., and T. Read. 1988. *Goodness-of-Fit Statistics for Discrete Multivariate Data*. New York: Springer-Verlag.
Csiszár, I. 1984. Sanov property, generalized I-projection and a conditional limit theorem. *Ann. Probab.* 12:768–793.
Csiszár, I. 1985. An extended maximum entropy principle and a Bayesian justification theorem. In *Bayesian Statistics 2*, 83–98. Amsterdam: North-Holland.
Csiszár I. 1996. MaxEnt, mathematics and information theory. In *Maximum Entropy and Bayesian Methods*. K. M. Hanson and R. N. Silver (eds.), pp. 35–50. Dordrecht: Kluwer Academic Publishers.
Csiszár I. 1998. The method of types. *IEEE IT*. 44(6):2505–2523.
Csiszár, I. 2006. A simple proof of Sanov's theorem. *Bull. Braz. Math. Soc.* 37(4):453–459.
Csiszár, I., T., Cover, and B. S. Choi. 1987. Conditional limit theorems under Markov conditioning, *IEEE IT*. 33:788–801.
Csiszár, I., and P. Shields. 2004. Information theory and statistics: a tutorial. *Found. Trends Comm. Inform. Theory*. 1(4):1–111.
Dembo, A., and O. Zeitouni. 1998. *Large Deviations Techniques and Applications*. New York: Springer-Verlag.
DiCiccio, T. J., P. J. Hall, and J. Romano. 1990. Nonparametric confidence limits by resampling methods and least favorable families. *I.S.I. Review*. 58:59–76.
DiCiccio, T. J., P. J. Hall, and J. Romano. 1991. Empirical likelihood is Bartlett-correctable. *Ann. Stat.* 19:1053–1061.
Ellis, R. S. 1999. The theory of large deviations: from Boltzmann's 1877 calculation to equilibrium macrostates in 2D turbulence. *Physica D*. 106–136.
Ellis, R. S. 2005. *Entropy, Large Deviations and Statistical Mechanics*. 2nd ed. New York: Springer-Verlag.
Farrell, L., R., Hartley, G., Lanot, and I. Walker. 2000. The demand for Lotto: the role of conscious selection. *J. Bus. Econ. Stat.* 18(2):228–241.
Florens, J.-P., and J.-M. Rolin. 1994. Bayes, bootstrap, moments. Discussion paper 94.13. Institute de Statistique, Université catholique de Louvain.
Ganesh, A., and N. O'Connell. 1999. An inverse of Sanov's Theorem. *Stat. Prob. Lett.* 42:201–206.
Ghosal, A., J. K., Ghosh, and R. V. Ramamoorthi. 1999. Consistency issues in bayesian nonparametrics. In *Asymptotics, Nonparametrics and Time Series: A Tribute to Madan Lal Puri*, pp. 639–667. New York: Marcel Dekker.
Ghosh, J. K., and R. V. Ramamoorthi. 2003. *Bayesian Nonparametrics*. New York: Springer-Verlag.
Godambe, V. P., and B. K. Kale. 1991. Estimating functions: an overview. In *Estimating Functions*. V. P. Godambe (ed.), pp. 3–20. Oxford, U.K.: Oxford University Press.
Golan, A. 2008. Information and entropy econometrics: a review and synthesis. *Foundations and Trends in Econometrics*, 2(12):1–145.
Golan, A., G., Judge, and D. Miller. 1996. *Maximum Entropy Econometrics. Robust Estimation with Limited Data*. New York: Wiley.
Grendár M. Jr., and M. Grendár. 2001. What is the question that MaxEnt answers? A probabilistic interpretation. In *Bayesian Inference and Maximum Entropy Methods in*

Science and Engineering. A. Mohammad-Djafari (ed.), pp. 83-94. Melville, NY: AIP. Online at arxiv:math-ph/0009020.

Grendár. M., Jr., and M. Grendár. 2004. Asymptotic identity of μ-projections and I-projections. *Acta Univ. Belii. Math.* 11:3–6.

Grendár, M., and G. Judge. 2008. Large deviations theory and empirical estimator choice. *Econometric Rev.* 27(4–6):513–525.

Grendár, M., and G. Judge. 2009a. Asymptotic equivalence of empirical likelihood and Bayesian MAP. *Ann. Stat.* 37(5A):2445–2457.

Grendár, M., and G. Judge. 2009b. Empty set problem of maximum empirical likelihood methods. *Electron. J. Stat.* 3:1542–1555.

Grendár, M., and R. K. Niven. 2006. The Pólya urn: limit theorems, Pólya divergence, maximum entropy and maximum probability. On-line at: arXiv:cond-mat/0612697.

Groeneboom, P., J., Oosterhoff, and F. H. Ruymgaart. 1979. Large deviation theorems for empirical probability measures. *Ann. Probab.* 7:553–586.

Hall, A. R. 2005. *Generalized Method of Moments*. Advanced Texts in Econometrics. Oxford, U.K.: Oxford University Press.

Hansen, L. P. 1982. Large sample properties of generalized method of moments estimators. *Econometrica* 50:1029–1054.

Harremoës, P. 2007. Information topologies with applications. In *Entropy, Search and Complexity*, I. Csiszár et al. (eds.), pp.113–150. New York: Springer.

Imbens, G. W. 1993. A new approach to generalized method of moments estimation. Harvard Institute of Economic Research working paper 1633.

Imbens, G. W., R. H., Spady, and P. Johnson. 1998. Information theoretic approaches to inference in moment condition models. *Econometrica* 66(2):333–357.

Jaynes, E. T. 1989. *Papers on Probability, Statistics and Statistical Physics*. 2nd ed. R. D. Rosenkrantz (ed.). New York: Springer.

Jing, B.-Y., and T. A. Wood. 1996. Exponential empirical likelihood is not Bartlett correctable. *Ann. Stat.* 24:365–369.

Jones, L. K., and C. L. Byrne. 1990. General entropy criteria for inverse problems, with applications to data compression, pattern classification and cluster analysis. *IEEE IT* 36(1):23–30.

Judge G. G., and R. C. Mittelhammer. 2004. A semiparametric basis for combining estimation problems under quadratic loss. *JASA* 99:479–487.

Judge, G. G., and R. C. Mittelhammer. 2007. Estimation and inference in the case of competing sets of estimating equations. *J. Econometrics* 138:513–531.

Kitamura, Y. 2006. Empirical likelihood methods in econometrics: theory and practice. In *Advances in Economics and Econometrics: Theory and Applications, Ninth world congress*. Cambridge, U.K.: CUP.

Kitamura, Y., and M. Stutzer. 1997. An information-theoretic alternative to generalized method of moments estimation. *Econometrica* 65:861–874.

Kitamura, Y., and M. Stutzer. 2002. Connections between entropic and linear projections in asset pricing estimation. *J. Econometrics* 107:159–174.

Lafférs, L. 2009. Empirical likelihood estimation of interest rate diffusion model. Master's thesis, Comenius University.

Lanford, O. E. 1973. Entropy and equilibrium states in classical statistical mechanics. In *Statistical Mechanics and Mathematical Problems*, A. Lenard (ed.), LNP 20, pp. 1–113. New York: Springer.

Lazar, N. 2003. Bayesian empirical likelihood. *Biometrika* 90:319–326.

Mittelhammer, R. C., and G. G. Judge. 2001. Robust empirical likelihood estimation of models with non-orthogonal noise components. *J. Agricult. Appl. Econ.* 35: 95–101.

Mittelhammer, R. C., and G. G. Judge. 2005. Combining estimators to improve structural model estimation and inference under quadratic loss. *J. Econometrics* 128(1):1–29.

Mittelhammer, R. C., Judge, G. G., and D. J. Miller. 2000. *Econometric Foundations*. Cambridge, U.K.: CUP.

Mittelhammer, R. C., Judge, G. G., and R. Schoenberg. 2005. Empirical evidence concerning the finite sample performance of EL-type structural equations estimation and inference methods. In *Identification and Inference for Econometric Models. Essays in Honor of Thomas Rothenberg*. D. Andrews, and J. Stock (eds.). Cambridge, U.K.: Cambridge University Press.

Newey, W., and R. J. Smith. 2004. Higher-order properties of GMM and generalized empirical likelihood estimators. *Econometrica* 72:219–255.

Niven, R. K. 2007. Origins of the combinatorial basis of entropy. In *Bayesian Inference and Maximum Entropy Methods in Science and Engineering*. K. H. Knuth et al. (eds.). pp. 133–142. Melville, NY: AIP.

Owen, A. B. 1991. Empirical likelihood for linear models. *Ann. Stat.* 19:1725–1747.

Owen, A. B. 2001. *Empirical Likelihood.* New York: Chapman-Hall/CRC.

Qin, J., and J. Lawless. 1994. Empirical likelihood and general estimating equations. *Ann. Stat.* 22:300–325.

Sanov, I. N. 1957. On the probability of large deviations of random variables. *Mat. Sbornik*. 42:11–44. (in Russian).

Schennach, S. M. 2004. Exponentially tilted empirical likelihood. Working paper, Department of Economics, University of Chicago.

Schennach, S. M. 2005. Bayesian exponentially tilted empirical likelihood. *Biometrika* 92(1):31–46.

Schennach, S. M. 2007. Point estimation with exponentially tilted empirical likelihood. *Ann. Stat.* 35(2):634–672.

Shannon, C. E. 1948. A mathematical theory of communication. *Bell Sys. Tech. J.* 27:379–423 and 27:623–656.

Solana-Ortega, A., and V. Solana. 2005. Entropic inference for assigning probabilities: some difficulties in axiomatics and applications, In *Bayesian Inference and Maximum Entropy Methods in Science and Engineering*. A. Mohammad-Djafari (ed.). pp. 449–458. Melville, NY: AIP.

van Campenhout J. M., and T. M. Cover. 1981. Maximum entropy and conditional probability. *IEEE IT* 27:483–489.

Vasicek O. A. 1980. A conditional law of large numbers. *Ann. Probab.* 8:142–147.

Vincze, I. 1972. On the maximum probability principle in statistical physics. *Coll. Math. Soc. J. Bolyai*. 9:869–893.

Vincze, I. 1997. Indistinguishability of particles or independence of the random variables? *J. Math. Sci.* 84:1190–1196.

Walker. S. 2004. New approaches to bayesian consistency. *Ann. Stat.* 32:2028–2043.

Walker, S., A., Lijoi, and I. Prünster. 2004. Contibutions to the understanding of bayesian consistency. Working paper no. 13/2004, International Centre for Economic Research, Turin.

Zhou, H. 2001. Finite sample properties of EMM, GMM, QMLE, and MLE for a square-root interest rate diffusion model. *J. Comput. Finance* 5:89–122.

7

Nonparametric Kernel Methods for Qualitative and Quantitative Data

Jeffrey S. Racine

CONTENTS

7.1 Introduction

Nonparametric kernel methods have become an integral part of the applied econometrician's toolkit. Their appeal, for applied researchers at least, lies in their ability to reveal structure in data that might be missed by classical parametric methods. Basic kernel methods are now found in virtually all popular statistical and econometric software programs. Such programs contain routines for the estimation of an unknown density function defined over a real-valued continuous random variable, or for the estimation of an unknown bivariate regression model defined over a real-valued continuous regressor. For example, the R platform for statistical computing and graphics (R Development Core Team 2008) includes the function `density` that computes a univariate kernel density estimate supporting a variety of kernel functions and bandwidth methods, while the `locpoly` function in the R "KernSmooth" package (Wand and Ripley 2008) can be used to estimate a bivariate regression function and its derivatives using a local polynomial kernel estimator with a fast binned bandwidth selector.

Those familiar with traditional nonparametric kernel smoothing methods such as that embodied in `density` or `locpoly` will appreciate that these methods presume that the underlying data are real-valued and continuous in nature, which is frequently not the case as one often encounters categorical along with continuous data types in applied settings. A popular traditional method for handling the presence of both continuous and categorical data is called the "frequency" approach. For this approach the data are first broken up into subsets ("cells") corresponding to the values assumed by the categorical variables, and then one applies, say, `density` or `locpoly` to the continuous data remaining in each cell. Unfortunately, nonparametric frequency approaches are widely acknowledged to be unsatisfactory because they often lead to substantial efficiency losses arising from the use of sample splitting, particularly when the number of cells is large.

Recent developments in kernel smoothing offer applied econometricians a range of kernel-based methods for categorical data only (i.e., unordered and ordered factors), or for a mix of continuous and categorical data. These methods have the potential to recover the efficiency losses associated with nonparametric frequency approaches since they do not rely on sample splitting. Instead, they smooth the categorical variables in an appropriate manner; see Li and Racine (2007) and the references therein for an in-depth treatment of these methods, and see also the references listed in the bibliography.

In this chapter we shall consider a range of kernel methods appropriate for the mix of categorical and continuous data one often encounters in applied settings. Though implementations of hybrid methods that admit the mix of categorical and continuous data types are quite limited, there exists an R package titled "np" (Hayfield and Racine 2008) that implements a variety of hybrid kernel methods, and we shall use this package to illustrate a few of the methods that are discussed in the following sections. Since many readers will no doubt be familiar with the classical approaches embodied in the functions `density` or `locpoly` or their peers, we shall begin with some recent developments in the kernel smoothing of categorical data only.

7.2 Kernel Smoothing of Categorical Data

The kernel smoothing of categorical data would appear to date from the seminal work of Aitchison and Aitken (1976) who proposed a novel method for kernel estimation of a probability function defined over multivariate binary data types. The wonderful monograph by Simonoff (1996) also contains chapters on the kernel smoothing of categorical data types such as sparse contingency tables and so forth. Econometricians are more likely than not interested in estimation of conditional objects, so we shall introduce the kernel smoothing of categorical objects via the estimation of a probability function and then immediately proceed to the estimation of a conditional mean. The estimation of a conditional mean with categorical covariates offers a unique springboard for presenting recent developments that link kernel smoothing to Bayesian methods. This exciting development offers a deeper understanding of kernel methods while also delivering novel methods for bandwidth selection and provides bounds ensuring that kernel smoothing will dominate frequency methods on mean square error (MSE) grounds.

7.2.1 Kernel Smoothing of Univariate Categorical Probabilities

Suppose we were interested in estimating a univariate *probability* function where the data are categorical in nature. The nonparametric nonsmooth approach would construct a frequency estimate, while the nonparametric smooth approach would construct a kernel estimate. For those unfamiliar with the term "frequency" estimate, we mean simply the estimator of a probability computed via the sample frequency of occurrence. For example, if a random variable is the result of a Bernoulli trial (i.e., zero or one with fixed probability from trial to trial) then the frequency estimate of the probability of a zero (one) is simply the number of zeros (ones) divided by the number of trials.

First, consider the estimation of a probability function defined for $X_i \in \mathcal{S} = \{0, 1, \ldots, c-1\}$. The nonsmooth "frequency" (nonkernel) estimator of $p(x)$ is given by

$$\tilde{p}(x) = \frac{1}{n}\sum_{i=1}^{n} \mathbf{1}(X_i, x),$$

where $\mathbf{1}(A)$ is an indicator function taking on the value 1 if A is true, zero otherwise. It is straightforward to show that

$$E\,\tilde{p}(x) = p(x),$$
$$\operatorname{Var}\tilde{p}(x) = \frac{p(x)(1-p(x))}{n},$$

hence,

$$\text{MSE}(\tilde{p}(x)) = n^{-1}p(x)(1 - p(x)) = O(n^{-1}),$$

which implies that

$$\tilde{p}(x) - p(x) = O_p\left(n^{-1/2}\right).$$

Now, consider the kernel estimator of $p(x)$,

$$\hat{p}(x) = \frac{1}{n}\sum_{i=1}^{n} l(X_i, x, \lambda), \tag{7.1}$$

where $l(\cdot)$ is a kernel function defined by, say,

$$l(X_i, x, \lambda) = \begin{cases} 1 - \lambda & \text{if } X_i = x \\ \lambda/(c-1) & \text{otherwise,} \end{cases} \tag{7.2}$$

and where $\lambda \in [0, (c-1)/c]$ is a "smoothing parameter" or "bandwidth." The requirement that λ lie in $[0, (c-1)/c]$ ensures that $\tilde{p}(x)$ is a proper probability estimate lying in $[0, 1]$. It is easy to show that

$$\begin{aligned} E\,\hat{p}(x) &= p(x) + \lambda\left\{\frac{1 - cp(x)}{c-1}\right\}, \\ \text{Var}\,\hat{p}(x) &= \frac{p(x)(1-p(x))}{n}\left(1 - \lambda\frac{c}{(c-1)}\right)^2. \end{aligned} \tag{7.3}$$

This estimator was proposed by Aitchison and Aitken (1976) for discriminant analysis with multivariate binary data; see also Simonoff (1996).

The above expressions indicate that the kernel smoothed estimator may possess some finite-sample bias; however, its finite-sample variance is less than its frequency counterpart. This suggests that the kernel estimator can dominate the frequency estimator on MSE grounds, which turns out to be the case; see Ouyang, Li, and Racine (2006) for extensive simulations. Results similar to those outlined in Subsection 7.3.1 for categorical Bayesian methods could be extended to this setting, though we do not attempt this here for the sake of brevity.

Note that when $\lambda = 0$ this estimator collapses to the frequency estimator $\tilde{p}(x)$, while when λ hits its upper bound, $(c-1)/c$, this estimator is the rectangular (i.e., discrete uniform) estimator which yields equal probabilities across all outcomes.

Using a bandwidth that balances bias and variance such as that proposed by Ouyang, Li, and Racine (2006), it can be shown that

$$\hat{p}(x) - p(x) = O_p\left(n^{-1/2}\right).$$

It can also be shown that

$$\sqrt{n}(\hat{p}(x) - p(x)) \to N\{0, p(x)(1 - p(x))\} \text{ in distribution.} \tag{7.4}$$

See Ouyang, Li, and Racine (2006) for details. For the sake of brevity we shall gloss over bandwidth selection methods, and direct the interested reader to Ouyang, Li, and Racine (2006) and Li and Racine (2007) for a detailed description of data-driven bandwidth selection methods for this object.

We have considered the univariate estimator by way of introduction. A multivariate version follows trivially by replacing the univariate kernel function with a multivariate product kernel function. We would let X now denote an r-dimensional discrete random vector taking values on $\mathcal{S}$, the support of X. We use x^s and X_i^s to denote the sth component of x and X_i $(i = 1, \ldots, n)$, respectively. The product kernel function is then given by

$$L_\lambda(X_i, x) = \prod_{s=1}^{r} l(X_i^s, x^s, \lambda_s) = \prod_{s=1}^{r} \{\lambda_s/(c_s - 1)\}^{I_{x_i^s \neq x^s}} (1 - \lambda_s)^{I_{x_i^s = x^s}}, \tag{7.5}$$

where $I_{x_i^s \neq x^s} = I(X_i^s \neq x^s)$, and $I_{x_i^s = x^s} = I(X_i^s = x^s)$. The kernel estimator is identical to that in Equation 7.1 except that we replace $l(X_i, x, \lambda)$ with $L_\lambda(X_i, x)$. All results (rate of convergence, asymptotic normality, etc.) remain unchanged.

7.2.1.1 A Simulated Example

In the following R code chunk we simulate $n = 250$ draws from five trials of a Bernoulli process having probability of success 1/2 from trial to trial, hence $x \in \{0, \ldots, 5\}$ and $c = 6$.

```
R> library("np")

Nonparametric Kernel Methods for Mixed Datatypes
   (version 0.30-7)

R> library(xtable)
R> set.seed(12345)
R> n <- 250
R> x <- sort(rbinom(n,5,.5))
R> ## Compute the non-smoothed (frequency) probability
   estimates
R> ptilde <- table(x)/n
R> ## Compute the smoothed probability estimates
R> phat <- unique(fitted(npudens(~factor(x))))
```

It can be seen that the nonsmooth frequency and the smooth kernel estimates are quite close for this example as expected, while the kernel estimators shrink slightly toward the uniform probability estimate

TABLE 7.1

Nonparametric Frequency ($\tilde{p}(x)$, Nonsmooth) and Nonparametric Smoothed ($\hat{p}(x)$) Probability Estimates.

x	$\tilde{p}(x)$	$\hat{p}(x)$
0	0.024	0.029
1	0.132	0.133
2	0.272	0.268
3	0.360	0.353
4	0.168	0.168
5	0.044	0.049

$p = 1/c = 1/6 = 0.1667$. We shall discuss the relationship between the kernel estimator and Bayesian methods in Subsection 7.3.1.

7.2.2 Kernel Smoothing of Bivariate Categorical Conditional Means

Now suppose by way of example that we observe $\{Y_i, X_i\}$ pairs generated by $y = g(x) + \epsilon$, where $g(x)$ is defined by

$$Y_i = X_i + \epsilon_i \tag{7.6}$$

where $X_i \in \mathcal{S} = \{0, 1, \ldots, c-1\}$ and $\epsilon_i \sim N(0, 1)$ represent i.i.d. draws.

The nonsmooth "frequency" (nonkernel) estimator of $g(x)$ (which is also the least squares estimator) is given by

$$\tilde{g}(x) = \frac{\sum_{i=1}^{n} Y_i \mathbf{1}(X_i, x)}{\sum_{i=1}^{n} \mathbf{1}(X_i, x)},$$

which simply returns the sample mean of those Y_i for which $X_i = x \in \mathcal{S} = \{0, 1, \ldots, c-1\}$. It can be shown that

$$\tilde{g}(x) - g(x) = O_p\left(n^{-1/2}\right).$$

Now, consider the kernel estimator of $g(x)$,

$$\hat{g}(x) = \frac{\sum_{i=1}^{n} Y_i l(X_i, x, \lambda)}{\sum_{i=1}^{n} l(X_i, x, \lambda)}, \tag{7.7}$$

where $l(\cdot)$ is, say, the kernel function defined in Equation 7.2.

Note that when $\lambda = 0$ this estimator collapses to the frequency estimator $\tilde{g}(x)$, while when λ hits its upper bound, $(c-1)/c$, this estimator yields equal fitted values across all $x \in \mathcal{S} = \{0, 1, \ldots, c-1\}$, namely, the overall (unconditional) mean of Y_i.

Using a bandwidth that balances bias and variance, it can be shown that

$$\hat{g}(x) - g(x) = O_p\left(n^{-1/2}\right),$$

TABLE 7.2

Nonparametric Frequency ($\tilde{g}(x)$, Nonsmooth) and Nonparametric Smoothed ($\hat{g}(x)$) Regression Estimates

x	$\tilde{g}(x)$	$\hat{g}(x)$
0	−0.587	−0.484
1	0.860	0.871
2	2.092	2.094
3	3.055	3.054
4	4.072	4.066
5	5.574	5.524

and that

$$\sqrt{n}\,(\hat{g}(x) - g(x))\,/\sqrt{\hat{\Omega}(x)} \to N(0, 1) \text{ in distribution,}$$

where $\hat{\Omega}(x) = \hat{\sigma}^2(x)/\hat{p}(x)$, and where $\hat{\sigma}^2(x) = n^{-1}\sum_i [Y_i - \hat{g}(X_i)]^2 l(X_i, x, \lambda)/\hat{p}(x)$ is a consistent estimator of $\sigma^2(x) = E(u_i^2 \mid X_i = x)$. See Ouyang, Li, and Racine (2008) for details.

7.2.2.1 A Simulated Example

In the following R code chunk we simulate $n = 250$ draws for x from five trials of a Bernoulli process having probability of success 1/2 from trial to trial, hence $x \in \{0, \ldots, 5\}$ and $c = 6$, then simulate $y = x + \epsilon$ where $\epsilon \sim N(0, 1)$.

```
R> set.seed(12345)
R> n <- 250
R> x <- sort(rbinom(n,5,.5))
R> y <- x + rnorm(n)
R> ## Regression on dummy variables (same as unconditio-
   nal group means)
R> gtilde <- unique(predict(model.par <- lm(y~factor(x))
   ))
R> ## Nonparametric regression on a factor (shrink
   towards overall mean)
R> ghat <- unique(predict(model.np <- npreg(y~factor(x))
   ))
```

We have considered the univariate estimator by way of introduction. A multivariate version follows trivially by replacing the univariate kernel function with a multivariate product kernel function defined in Equation 7.5. The kernel estimator is identical to that in Equation 7.7 except that we replace $l(X_i, x, \lambda)$ with $L_\lambda(X_i, x)$. All results (rate of convergence, asymptotic normality, etc.) remain unchanged.

7.3 Categorical Kernel Methods and Bayes Estimators

Kiefer and Racine (2009) have recently investigated the relationship between nonparametric categorical kernel methods and hierarchical Bayes models of the type considered by Lindley and Smith (1972). By exploiting certain similarities among the approaches, they gain a deeper understanding of the nature of kernel-based methods and leverage some theoretical apparatus developed for hierarchical Bayes models which is immediately relevant for kernel-based techniques. We outline their approach below as it provides additional insight and also delivers a new approach toward bandwidth selection for categorical kernel methods.

7.3.1 Kiefer and Racine's (2009) Analysis

In order to facilitate a direct comparison with Kiefer and Racine's (2009) notation, we now let the sample realizations $\{X_i, Y_i\}$ be written instead as $\{X_{ji}, Y_{ji}\}$, $j = 1, \ldots, n_i, i = 1, \ldots, c$. We let y_i be the frequency estimator of μ_i defined as

$$y_i = \frac{1}{n_i} \sum_{k=1}^{c} \sum_{j=1}^{n_k} Y_{jk} \mathbf{1}(X_{jk} = i), \tag{7.8}$$

i.e., the sample mean of Y when $X = i$ (a "cell" mean). Let $y_{\bar{i}}$ be defined as

$$y_{\bar{i}} = \frac{1}{(n - n_i)} \sum_{k=1}^{c} \sum_{j=1}^{n_k} Y_{jk} \mathbf{1}(X_{jk} \neq i),$$

i.e., the sample mean of Y over all values of X other than $X = i$ ($\bar{i}$ is taken to be the complement of i), while the frequency estimator of $E(Y)$ (the "overall" mean) is

$$y_{\cdot} = \frac{1}{n} \sum_{k=1}^{c} \sum_{j=1}^{n_k} Y_{jk} = \frac{n_i y_i + (n - n_i) y_{\bar{i}}}{n}.$$

Adopting Kiefer and Racine's (2009) notation, the kernel estimator of μ_i could be written as

$$y_{i,\lambda} = \hat{g}(i) = \frac{n^{-1} \sum_{k=1}^{c} \sum_{j=1}^{n_k} Y_{jk} L(X_{jk}, i, \lambda)}{p_{i,\lambda}}.$$

In order to facilitate a comparison of the Bayesian approach of Lindley and Smith (1972) and the kernel approach, we wish to express $y_{i,\lambda}$ as a weighted average of y_i and $y_.$. The kernel estimator $y_{i,\lambda}$ can be rewritten as follows,

$$\begin{aligned}
y_{i,\lambda} &= \frac{n^{-1}\sum_{k=1}^{c}\sum_{j=1}^{n_k} Y_{jk} L(X_{jk}, i, \lambda)}{p_{i,\lambda}} \\
&= \frac{n^{-1}(n_i y_i (1-\lambda) + (n-n_i) y_{\bar{i}} \lambda/(c-1))}{n^{-1}(n_i(1-\lambda) + (n-n_i)\lambda/(c-1))} \\
&= \frac{n_i y_i (1-\lambda) + (n y_. - n_i y_i)\lambda/(c-1)}{n_i(1-\lambda) + (n-n_i)\lambda/(c-1)} \\
&= \left[\frac{n_i/n(1-\lambda c/(c-1))}{n_i/n(1-\lambda c/(c-1)) + \lambda/(c-1)}\right] y_i \\
&\quad + \left[\frac{\lambda/(c-1)}{n_i/n(1-\lambda c/(c-1)) + \lambda/(c-1)}\right] y_. \\
&= (1-\Phi_i) y_i + \Phi_i y_.,
\end{aligned}$$

where the third equality follows from Equation 7.8 by noting that

$$n y_. - n_i y_i = (n - n_i) y_{\bar{i}},$$

where

$$1 - \Phi_i = \left[\frac{n_i/n(1-\lambda c/(c-1))}{n_i/n(1-\lambda c/(c-1)) + \lambda/(c-1)}\right]$$

and

$$\Phi_i = \left[\frac{\lambda/(c-1)}{n_i/n(1-\lambda c/(c-1)) + \lambda/(c-1)}\right],$$

and where $\lambda \in [0, (c-1)/c]$ implies that $\Phi_i \in [0, 1]$.

When $\lambda = 0$ (i.e., $\Phi_i = 0 \forall i$), $y_{i,\lambda} = y_i$ (the frequency estimator), while when $\lambda = (c-1)/c$ (i.e., $(1 - \lambda c/(c-1)) = 0$ or $\Phi_i = 1 \forall i$), $y_{i,\lambda} = y_.$, $i = 1, \ldots, c$ (the global mean). Note that this is exactly the same result using the notation in Equation 7.7.

Kiefer and Racine (2009) consider hierarchical models of the form

$$y_{ji} = \mu_i + \epsilon_{ji}, \quad j = 1, \ldots, n_i, \quad i = 1, \ldots, c,$$

where n_i is the number of observations drawn from group i, and where there exist c groups.

For the ith group,

$$\begin{pmatrix} y_{1i} \\ \vdots \\ y_{n_i i} \end{pmatrix} = \iota_{n_i} \mu_i + \epsilon_i, \quad i = 1, \ldots, c,$$

where ι_{n_i} is a vector of ones of length n_i, $\epsilon_i = (\epsilon_{1i}, \ldots, \epsilon_{n_i i})'$, and, for the sample, $\mathbf{y} = A\mu + \epsilon$ where $\mathbf{y}$ is the n-vector of observations, A is the $(n \times c)$ design matrix, and $\mu = (\mu_1, \ldots, \mu_c)'$, the vector of group means. The goal is to understand the connection between hierarchical Bayes models and kernel estimators of multivariate means.

Kiefer and Racine (2009) consider a three-stage hierarchical Bayes model. The first stage is given by

$$\mathbf{y} \sim (A_1\theta_1, C_1).$$

As a function of θ_1 and C_1 for given y, this first stage specification can be regarded as the likelihood function for the normally distributed case, otherwise as a quasi likelihood based on two moments (Heyde 1997). We return to A_1 below.

The second stage,

$$\theta_1 \sim (A_2\theta_2, C_2),$$

can be regarded as a prior distribution for θ_1 given $A_2\theta_2$ and C_2 in the normal case (where it is conjugate) or as an approximation to the prior if not normal, or from a frequency viewpoint as a second stage in the data generating process (DGP). The first stage "parameters" are themselves generated by a random process in this view. This interpretation focuses attention on the hyperparameters θ_2 (and C_2) rather than θ_1, which strictly speaking is not a parameter in the frequency sense.

The third stage,

$$\theta_2 \sim (A_3\theta_3, C_3),$$

can again be regarded as a prior on the second stage parameter θ_2, or as an additional stage in the DGP.

Interest lies in estimating the $c \times 1$ vector of means θ_1. Following Lindley and Smith (1972) we are thinking of normal distributions at each stage. For our purposes we can also regard the stages as approximate distributions characterized by two moments noting the calculations are exact only for the normal. The point of the stages is that the dimension of the conditioning parameter is reduced at each step. We are using the Bayesian hierarchical setup to obtain insight into the kernel estimator. Lindley and Smith (1972) suggest specifications proportional to identity matrices and inverted gamma densities for the factors of proportion (and related generalizations). They suggest using modal estimators in the expressions for the posterior means of interest.

For the problem at hand, we try to stick with the notation of Lindley and Smith as closely as possible. The first stage is

$$A_1 = \{a_{ji}\} \text{ with } a_{ji} \in \{0, 1\}, \sum_{i=1}^{c} a_{ki} = 1, \sum_{k=1}^{n} a_{ki} = n_i,$$

$$\theta_1 = \mu = \begin{bmatrix} \mu_1 \\ \vdots \\ \mu_c \end{bmatrix},$$

$$C_1 = \sigma^2 I_n.$$

A_1 is the $n \times c$ design matrix with $A_1' A_1$ the $c \times c$ diagonal matrix with n_i, the number of observations in the ith group, as the ith diagonal element, μ is a $c \times 1$ vector of (population) group means, σ^2 is the within-group variance (i.e., $\text{Var}(y_{ij})$), and I_n is the $n \times n$ identity matrix. Next, the second stage will become

$$A_2 = \iota_c,$$

$$\theta_2 = \mu_.,$$

$$C_2 = \tau^2 I_c,$$

where $\mu_.$ is the (population) "overall mean," and $\tau^2 = \text{Var}(\mu_i)$. Note that $A_2\theta_2 = \iota\mu_.$ is simply a $c \times 1$ vector with elements being the overall mean $\mu_.$ to which the Bayes (and kernel) estimators can shrink. Finally, we let the scalar

$$C_3^{-1} \to 0$$

so that the prior on $\mu_.$ is improper. Note that the impropriety is confined to one dimension. The frequency analysis corresponds to an improper prior on the c-vector θ_1, so that we expect inadmissibility of the frequency estimator through a Stein effect if $c > 2$. By adding a third stage, we reduce the improper prior to one dimension. The results are seen below.

The three stage Bayes estimate is (Lindley and Smith 1972, p. 7, Eq. 16)

$$\theta_1^* = D_0 d_0$$

where

$$D_0^{-1} = \left(A_1' C_1^{-1} A_1 + C_2^{-1} - C_2^{-1} A_2 \left(A_2' C_2^{-1} A_2\right)^{-1} A_2' C_2^{-1}\right)$$

$$d_0 = \left(A_1' C_1^{-1} y\right).$$

θ_1^* is the posterior mean and is an optimal estimator under quadratic loss. Writing

$$\Lambda = A_1' C_1^{-1} A_1 = \frac{1}{\sigma^2}\begin{bmatrix} n_1 & 0 & 0 & \dots \\ 0 & n_2 & 0 & \dots \\ \vdots & & \ddots & \\ \vdots & 0 & 0 & n_c \end{bmatrix},$$

we see that

$$\begin{aligned} D_0^{-1} &= (\Lambda + \tau^{-2} I_c - \tau^{-2}\iota(\iota'^{-2}\iota)^{-1}\iota'^{-2}) \\ &= (\Lambda + \tau^{-2} I_c - \tau^{-2}\iota\iota'/c), \\ d_0 &= A_1' C_1^{-1}\mathbf{y} \\ &= \begin{pmatrix} \frac{y_1 n_1}{\sigma^2} \\ \vdots \\ \frac{y_c n_c}{\sigma^2} \end{pmatrix}. \end{aligned}$$

Recall that y_i is the mean for group i. Thus the vector of posterior means satisfies

$$(\Lambda + \tau^{-2} I_c - \tau^{-2}\iota(\iota'\iota)^{-1}\iota')\,\theta_1^* = d_0$$

or, element-wise

$$(\sigma^{-2} n_j + \tau^{-2})\theta_{1j}^* - \tau^{-2}\theta_{1.}^* = \sigma^{-2} n_j y_j,$$

where $\theta_{1.}^* = \sum_{j=1}^{c} \theta_{1j}^*/c$. Thus

$$\theta_{1j}^* = (\sigma^{-2} n_j y_j + \tau^{-2}\theta_{1.}^*)/(\sigma^{-2} n_j + \tau^{-2})$$

and the Bayes estimator for the jth mean is a weighted average of the group mean and the overall posterior mean. This, in general, cannot be expressed as a weighted average of the group mean and the overall mean.

We consider the "balanced case" (n_i equal for all i) in what follows. Let $n_i = n^*$ for all i. The kernel estimator of the ith component of μ can be written as

$$
\begin{aligned}
y_{i,\lambda} &= \left[\frac{n^*(1-\lambda c/(c-1))}{n^*(1-\lambda c/(c-1))+n\lambda/(c-1)}\right] y_i \\
&\quad + \left[\frac{n^*\lambda/(c-1)}{n^*(1-\lambda c/(c-1))+n^*\lambda/(c-1)}\right] y_. \\
&= \left[\frac{n^*}{n^*+n^*/((c-1)/\lambda-c)}\right] y_i + \left[\frac{n^*/((c-1)/\lambda-c)}{n^*+n^*/((c-1)/\lambda-c)}\right] y_.,
\end{aligned} \tag{7.9}
$$

where λ is a smoothing parameter to be set by the researcher.

Further, the Bayes estimator of the ith component of μ is given by (in the balanced case)

$$
\begin{aligned}
\mu_i^* &= \left[\frac{n^*}{n^*+\kappa^{-1}}\right] y_i + \left[\frac{\kappa^{-1}}{n^*+\kappa^{-1}}\right] y_. \\
&= v y_i + (1-v) y
\end{aligned} \tag{7.10}
$$

where $v = n^*/(n^*+\kappa^{-1})$ is the common value of the v_i term from above. The correspondence between the two methods is given by

$$n^*/((c-1)/\lambda - c) = \kappa^{-1},$$

hence

$$\kappa = \frac{1}{n^*}((c-1)/\lambda - c).$$

Alternatively, λ can be expressed as

$$\lambda = (c-1)/(c+n^*\kappa). \tag{7.11}$$

This gives some intuition for the choice of the smoothing parameter λ if one chooses not to adopt the Bayesian approach explicitly. λ should be larger as the groups are thought to be more homogeneous (smaller κ or τ^2) and smaller as the groups are thought to be less similar.

Next, we turn to another frequency property, that of MSE. It is known that the MSE of the Bayes/kernel estimator (identical in the balanced case) improves over that of the frequency estimator y_i if and only if (Lindley and Smith 1972, p. 3, Eq. 2)

$$\hat{\tau}^2 \leq 2\tau^2 + \sigma^2,$$

where

$$\hat{\tau}^2 = \sum_i \frac{(y_i - y_.)^2}{c-1}. \tag{7.12}$$

This allows us to obtain an upper bound for λ that will ensure (in probability) that $\text{MSE}(y_{i,\lambda}) \leq \text{MSE}(y_i)$. Substituting, we have

$$\hat{\tau}^2 \leq 2\frac{\sigma^2}{n}((c-1)/\lambda - c) + \sigma^2,$$

which is equivalent to

$$\frac{n(\hat{\tau}^2 - \sigma^2)}{2\sigma^2} + c \leq \frac{c-1}{\lambda},$$

which implies that

$$\lambda \leq \frac{2\sigma^2(c-1)}{n(\hat{\tau}^2 - \sigma^2) + 2c\sigma^2}. \tag{7.13}$$

The only unknown in this formula is σ^2, which can be estimated directly from the data via

$$\hat{\sigma}^2 = \frac{\sum_{i=1}^{c}\sum_{j=1}^{n^*}(y_{ij} - y_i)^2}{n-c}. \tag{7.14}$$

It is widely known that the smoothing parameter must obey $\lambda \to 0$ as $n \to \infty$ for consistent estimation while, as noted earlier, λ is restricted to lie in $[0, (c-1)/c]$ (see Equation 7.2). Note that Equation 7.13 tells us that an *oversmoothed* kernel estimator can be consistent but can be beaten by the frequency estimator on MSE grounds (i.e., when λ is overly large).

The results obtained above yield a number of implications for applied kernel estimation with categorical data. The first is that they provide bounds for bandwidth selection that are previously unknown in the literature. The second is that they deliver a simple plug-in method of bandwidth selection with an empirical Bayes flavor (Efron and Morris 1973) that possesses appealing finite-sample properties and, in addition, is computationally trivial. Recall that $[0, (c-1)/c]$ is the range of λ when using the kernel function defined in Equation 7.2. We now incorporate the result summarized in Equation 7.13 to obtain tighter bounds on λ.

Note that when $\hat{\tau}^2 = \sigma^2$, Equation 7.13 equals $(c-1)/c$, the upper bound possible for λ, hence the bound is nonbinding in this case. It is also nonbinding when $\hat{\tau}^2 \leq \sigma^2$. However, when $\hat{\tau}^2 > \sigma^2$, then in order to outperform the frequency estimator on MSE grounds, the kernel estimator must obey $\lambda < (c-1)/c$ with the upper bound now given by Equation 7.13. On MSE grounds, the range of λ is no longer $[0, (c-1)/c]$, rather it is

$$\left[0, \min\left\{\frac{c-1}{c}, \frac{2\sigma^2(c-1)}{n(\hat{\tau}^2 - \sigma^2) + 2c\sigma^2}\right\}\right]. \tag{7.15}$$

In other words, Equation 7.13 tells us that when the idiosyncratic variation (i.e., $\sigma^2 = \text{Var}(y_{ij})$) is greater than the intergroup variation (i.e., $\hat{\tau}^2 = \text{Var}(y_i)$), there exists a λ in the feasible range (i.e., $[0, (c-1)/c]$) that will outperform the frequency estimator on MSE grounds (e.g., that given by Equation 7.11). On the other hand, when the idiosyncratic variation is less than the intergroup variation, imposing this (reduced) bound on λ (rather than $(c-1)/c$) avoids situations where the frequency estimator may outperform the smoothed estimator. Note that Equation 7.11 always satisfies the bound.

Equation 7.11 suggests a computationally trivial formula for a plug-in bandwidth selector for the kernel estimator of a multivariate mean that might serve as an alternative to that proposed in Ouyang, Li, and Racine (2008).

7.3.1.1 A Simulated Example

Next we simulate $y = \epsilon$, where $\epsilon \sim N(0, 1)$, and use leave-one-out cross-validation to select the unknown bandwidth.

```
R> set.seed(12345)
R> n <- 250
R> x <- sort(rbinom(n,5,.5))
R> y <- rnorm(n)
R> ## Regression on dummy variables (same as unconditio-
   nal group means)
R> gtilde <- unique(predict(model.par <- lm(y~factor(x))
   ))
R> ## Nonparametric regression on a factor (shrink towa-
   rds overall mean)
R> ghat <- unique(predict(model.np <- npreg(y~factor(x))
   ))
```

Note that, for this example, the unconditional mean of y is $y_. = 0.05$. It can be seen from the above example that the kernel estimator correctly shrinks the nonparametric frequency estimator towards the overall mean in accordance with the findings of Kiefer and Racine (2009).

We now discuss recent developments in the kernel estimation of objects involving the mix of categorical and continuous data types often found in applied settings.

TABLE 7.3

Nonparametric Frequency ($\tilde{g}(x)$, Nonsmooth) and Nonparametric Smoothed ($\hat{g}(x)$) Regression Estimates.

x	$\tilde{g}(x)$	$\hat{g}(x)$
0	−0.587	0.050
1	−0.140	0.050
2	0.092	0.050
3	0.055	0.050
4	0.072	0.050
5	0.574	0.050

7.4 Kernel Methods with Mixed Data Types

So far we have presumed that the categorical variable is of the "unordered" ("nominal" data type). We shall now distinguish between categorical (discrete) data types and real-valued (continuous) data types. Also, for categorical data types we could have unordered or ordered ("ordinal" data type) variables. For an ordered discrete variable $\tilde{x}^d$, we could use Wang and van Ryzin (1981) kernel given by

$$\tilde{l}(\tilde{X}_i^d, \tilde{x}^d, \lambda) = \begin{cases} 1-\lambda, & \text{if } \tilde{X}_i^d = \tilde{x}^d, \\ \dfrac{(1-\lambda)}{2}\lambda^{|\tilde{X}_i^d - \tilde{x}^d|}, & \text{if } \tilde{X}_i^d \neq \tilde{x}^d. \end{cases}$$

We shall now refer to the unordered kernel defined in Equation 7.2 as $\bar{l}(\cdot)$ so as to keep each kernel type separate notationally speaking. We shall denote the traditional kernels for continuous data types such as the Epanechnikov of Gaussian kernels by $W(\cdot)$.

A generalized product kernel for one continuous, one unordered, and one ordered variable would be defined as follows,

$$K(\cdot) = W(\cdot) \times \bar{l}(\cdot) \times \tilde{l}(\cdot). \tag{7.16}$$

Using such product kernels, we can modify any existing kernel-based method to handle the presence of categorical variables, thereby extending the reach of kernel methods. We define $K_\gamma(X_i, x)$ to be this product, where $\gamma = (h, \lambda)$ is the vector of bandwidths for the continuous and categorical variables.

7.4.1 Kernel Estimation of a Joint Density Defined over Categorical and Continuous Data

Estimating a joint probability/density function defined over mixed data follows naturally using these generalized product kernels. For example, for one unordered discrete variable $\bar{x}^d$ and one continuous variable x^c, our kernel estimator of the PDF would be

$$\hat{f}(\bar{x}^d, x^c) = \frac{1}{nh_{x^c}} \sum_{i=1}^{n} \bar{l}(\bar{X}_i^d, \bar{x}^d) W\left(\frac{X_i^c - x^c}{h_{x^c}}\right).$$

This extends naturally to handle a mix of ordered, unordered, and continuous data (i.e., both quantitative and qualitative data). This estimator is particularly well suited to "sparse data" settings. Li and Racine (2003) demonstrate that

$$\sqrt{nh^p}\left(\hat{f}(z) - f(z) - \hat{h}^2\mathcal{B}_1(z) - \hat{\lambda}\mathcal{B}_2(z)\right) \to N(0, V(z)) \text{ in distribution}, \tag{7.17}$$

where $\mathcal{B}_1(z) = (1/2)tr\{\nabla^2 f(z)\}[\int W(v)v^2 dv]$, $\mathcal{B}_2(z) = \sum_{x' \in \mathcal{D}, d_{x,x'}=1}[f(x', y) - f(x, y)]$, and $V(z) = f(z)[\int W^2(v)dv]$.

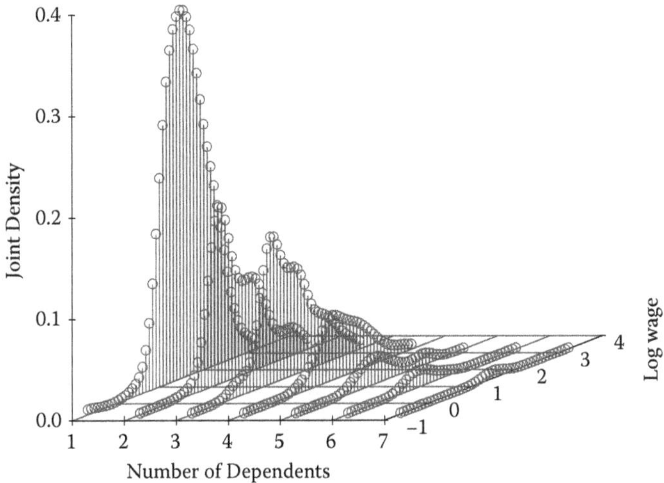

FIGURE 7.1
Nonparametric kernel estimate of a joint density defined over one continuous and one discrete variable.

7.4.1.1 An Application

We consider Wooldridge's (2002) "wage1" dataset having $n = 526$ observations, and model the joint density of two variables, one continuous ("lwage") and one discrete ("numdep"). "lwage" is the logarithm of average hourly earnings for an individual. "numdep" the number of dependents $(0, 1, \dots)$. We use likelihood cross-validation to obtain the bandwidths, and the resulting estimate is presented in Figure 7.1.

Note that this is indeed a case of "sparse" data for some cells (see Table 7.4), and the traditional approach would require estimation of a nonparametric univariate density function based upon only two observations for the last cell $(c = 6)$.

TABLE 7.4
Summary of the Number of Dependents in the Wooldridge (2002) "wage1" Dataset ("numdep")

	numdep
0	252
1	105
2	99
3	45
4	16
5	7
6	2

7.4.2 Kernel Estimation of a Conditional PDF

Let $f(\cdot)$ and $\mu(\cdot)$ denote the joint and marginal densities of (X, Y) and X, respectively, where we allow Y and X to consist of continuous, unordered, and ordered variables. For what follows we shall refer to Y as a dependent variable (i.e., Y is explained), and to X as covariates (i.e., X is the explanatory variable). We use $\hat{f}$ and $\hat{\mu}$ to denote kernel estimators thereof, and we estimate the conditional density $g(y \mid x) = f(x, y)/\mu(x)$ by

$$\hat{g}(y \mid x) = \frac{\hat{f}(x, y)}{\hat{\mu}(x)}. \tag{7.18}$$

The kernel estimators of the joint and marginal densities $f(x, y)$ and $\mu(x)$ are described in the previous sections; see Hall, Racine, and Li (2004) for details on the theoretical underpinnings of a data-driven method of bandwidth selection for this method.

7.4.2.1 The Presence of Irrelevant Covariates

Hall, Racine, and Li (2004) proposed the estimator defined in Equation 7.18, but choosing appropriate smoothing parameters in this setting can be tricky, not least because plug-in rules take a particularly complex form in the case of mixed data. One difficulty is that there exists no general formula for the optimal smoothing parameters. A much bigger issue is that it can be difficult to determine which components of X are relevant to the problem of conditional inference. For example, if the jth component of X is independent of Y then that component is irrelevant to estimating the density of Y given X, and ideally should be dropped before conducting inference. Hall, Racine, and Li (2004) show that a version of least-squares cross-validation overcomes these difficulties. It automatically determines which components are relevant and which are not, through assigning large smoothing parameters to the latter and consequently shrinking them toward the uniform distribution on the respective marginals. This effectively removes irrelevant components from contention, by suppressing their contribution to estimator variance; they already have very small bias, a consequence of their independence of Y. Cross-validation also gives us important information about which components are relevant; the relevant components are precisely those that cross-validation has chosen to smooth in a traditional way, by assigning them smoothing parameters of conventional size. Cross-validation produces asymptotically optimal smoothing for relevant components, while eliminating irrelevant components by oversmoothing.

Hall, Racine, and Li (2004) demonstrate that, for irrelevant conditioning variables in X, their bandwidths in fact ought to behave exactly the opposite, namely, $h \to \infty$ as $n \to \infty$ for optimal smoothing. The same has been demonstrated for regression as well; see Hall, Li, and Racine (2007) for further details. Note that this result is closely related to the Bayesian results described in detail in Section 7.3.

7.4.3 Kernel Estimation of a Conditional CDF

Li and Racine (2008) propose a nonparametric conditional CDF kernel estimator that admits a mix of discrete and categorical data along with an associated nonparametric conditional quantile estimator. Bandwidth selection for kernel quantile regression remains an open topic of research, and they employ a modification of the conditional PDF-based bandwidth selector proposed by Hall, Racine, and Li (2004).

We use $F(y \mid x)$ to denote the conditional CDF of Y given $X = x$, while $f(x)$ is the marginal density of X. We can estimate $F(y \mid x)$ by

$$\hat{F}(y \mid x) = \frac{n^{-1} \sum_{i=1}^{n} G\left(\frac{y - Y_i}{h_0}\right) K_\gamma(X_i, x)}{\hat{f}(x)}, \tag{7.19}$$

where $G(\cdot)$ is a kernel CDF chosen by the researcher, say, the standard normal CDF, h_0 is the smoothing parameter associated with Y, and $K_\gamma(X_i, x)$ is a product kernel such as that defined in Equation 7.16 where each univariate continuous kernel has been divided by its respective bandwidth for notational simplicity.

Li and Racine (2008) demonstrate that

$$(nh_1 \ldots h_q)^{1/2} \left[\tilde{F}(y \mid x) - F(y \mid x) - \sum_{s=1}^{q} h_s^2 B_{1s}(y \mid x) - \sum_{s=1}^{r} \lambda_s B_{2s}(y \mid x) \right]$$

$$\rightarrow N(0, V(y \mid x)) \text{ in distribution,} \tag{7.20}$$

where $V(y \mid x) = \kappa^q F(y \mid x)[1 - F(y \mid x)]/\mu(x)$, $B_{1s}(y \mid x) = (1/2)\kappa_2[2F_s(y \mid x) \times \mu_s(x) + \mu(x) F_{ss}(y \mid x)]/\mu(x)$, $B_{2s}(y \mid x) = \mu(x)^{-1} \sum_{z^d \in D} I_s(z^d, x^d)[F(y \mid x^c, z^d) \times \mu(x^c, z^d) - F(y \mid x)\mu(x)]/\mu(x)$, $\kappa = \int W(v)^2 dv$, $\kappa_2 = \int W(v) v^2 dv$, and D is the support of X^d.

7.4.4 Kernel Estimation of a Conditional Quantile

Estimating regression functions is a popular activity for practitioners. Sometimes, however, the regression function is not representative of the impact of the covariates on the dependent variable. For example, when the dependent variable is left (or right) censored, the relationship given by the regression function is distorted. In such cases, conditional quantiles above (or below) the censoring point are robust to the presence of censoring. Furthermore, the conditional quantile function provides a more comprehensive picture of the conditional distribution of a dependent variable than the conditional mean function.

Once we can estimate conditional CDFs, estimating conditional quantiles follows naturally. That is, having estimated the conditional CDF we simply invert it at the desired quantile as described below. A conditional αth quantile of a conditional distribution function $F(\cdot\,|\,x)$ is defined by ($\alpha \in (0, 1)$)

$$q_\alpha(x) = \inf\{y:\ F(y\,|\,x) \geq \alpha\} = F^{-1}(\alpha\,|\,x).$$

Or equivalently, $F(q_\alpha(x)\,|\,x) = \alpha$. We can directly estimate the conditional quantile function $q_\alpha(x)$ by inverting the estimated conditional CDF function, i.e.,

$$\hat{q}_\alpha(x) = \inf\{y:\ \hat{F}(y\,|\,x) \geq \alpha\} \equiv \hat{F}^{-1}(\alpha\,|\,x).$$

Li and Racine (2008) demonstrate that

$$(nh_1 \ldots h_q)^{1/2}[\hat{q}_\alpha(x) - q_\alpha(x) - B_{n,\alpha}(x)] \rightarrow N(0, V_\alpha(x)) \text{ in distribution}, \quad (7.21)$$

where $V_\alpha(x) = \alpha(1-\alpha)\kappa^q/[f^2(q_\alpha(x)\,|\,x)\mu(x)] \equiv V(q_\alpha(x)\,|\,x)/f^2(q_\alpha(x)\,|\,x)$ (since $\alpha = F(q_\alpha(x)\,|\,x)$).

7.4.5 Binary Choice and Count Data Models

Another application of kernel estimates of PDFs with mixed data involves the estimation of conditional mode models. By way of example, consider some discrete outcome, say $Y \in \mathcal{S} = \{0, 1, \ldots, c-1\}$, which might denote by way of example the number of successful patent applications by firms. We define the conditional mode of $y\,|\,x$ by

$$m(x) = \max_y g(y\,|\,x). \quad (7.22)$$

In order to estimate a conditional mode $m(x)$, we need to model the conditional density. Let us call $\hat{m}(x)$ the estimated conditional mode, which is given by

$$\hat{m}(x) = \max_y \hat{g}(y\,|\,x), \quad (7.23)$$

where $\hat{g}(y\,|\,x)$ is the kernel estimator of $g(y\,|\,x)$ defined in Equation 7.18.

7.4.6 Kernel Estimation of Regression Functions

The local constant (Nadaraya 1965; Watson 1964) and local polynomial (Fan 1992) estimators are perhaps the most well-known of all kernel methods. Racine and Li (2004) and Li and Racine (2004) propose local constant and local polynomial estimators of regression functions defined over categorical and continuous data types. To extend these popular estimators so that they can handle both categorical and continuous regressors requires little more than replacing the traditional kernel function with the generalized kernel given in Equation 7.16. That is, the local constant estimator defined in Equation 7.7 would then be

$$\hat{g}(x) = \frac{\sum_{i=1}^n Y_i K_\gamma(X_i, x)}{\sum_{i=1}^n K_\gamma(X_i, x)}. \quad (7.24)$$

Racine and Li (2004) demonstrate that

$$\sqrt{n\hat{h}^p}\left(\hat{g}(x) - g(x) - \hat{B}(\hat{h}, \hat{\lambda})\right) / \sqrt{\hat{\Omega}(x)} \rightarrow N(0, 1) \text{ in distribution.} \tag{7.25}$$

See Racine and Li (2004) for further details.

7.5 Summary

We survey recent developments in the kernel estimation of objects defined over categorical and continuous data types. We focus on theoretical underpinnings, and focus first on kernel methods for categorical data only. We pay close attention to recent theoretical work that draws links between kernel methods and Bayesian methods and also highlight the behavior of kernel methods in the presence of irrelevant covariates. Each of these developments leads to kernel estimators that diverge from more traditional kernel methods in a number of ways, and sets the stage for mixed data kernel methods which we briefly discuss. We hope that readers are encouraged to pursue these methods, and draw the readers attention to an R package titled "np" (Hayfield and Racine 2008) that implements a range of the approaches discussed above. A number of relevant examples can also be found in Hayfield and Racine (2008), and we direct the interested reader to the applications contained therein.

References

Aitchison, J., and C. G. G. Aitken. 1976. Multivariate binary discrimination by the kernel method. *Biometrika* 63(3): 413–420.

Efron, B., and C. Morris. 1973. Stein's estimation rule and its competitors–an empirical Bayes approach. *Journal of the American Statistical Association* 68(341): 117–130.

Fan, J. 1992. Design-adaptive nonparametric regression. *Journal of the American Statistical Association* 87: 998–1004.

Hall, P., Q. Li, and J. S. Racine. 2007. Nonparametric estimation of regression functions in the presence of irrelevant regressors. *The Review of Economics and Statistics* 89: 784–789.

Hall, P., J. S. Racine, and Q. Li. 2004. Cross-validation and the estimation of conditional probability densities. *Journal of the American Statistical Association* 99(468): 1015–1026.

Hayfield, T., and J. S. Racine. 2008. Nonparametric econometrics: the np package. *Journal of Statistical Software* 27(5). http://www.jstatsoft.org/v27/i05/

Heyde, C. 1997. *Quasi-Likelihood and Its Application*. New York: Springer-Verlag.

Kiefer, N. M., and J. S. Racine. 2009. The smooth colonel meets the reverend. *Journal of Nonparametric Statistics* 21: 521–533.

Li, Q., and J. S. Racine. 2003. Nonparametric estimation of distributions with categorical and continuous data. *Journal of Multivariate Analysis* 86: 266–292.

Li, Q., and J. S. Racine. 2004. Cross-validated local linear nonparametric regression. *Statistica Sinica* 14(2): 485–512.

Li, Q., and J. S. Racine. 2007. *Nonparametric Econometrics: Theory and Practice*. Princeton, NJ: Princeton University Press.

Li, Q., and J. S. Racine. 2008. Nonparametric estimation of conditional CDF and quantile functions with mixed categorical and continuous data. *Journal of Business and Economic Statistics*. 26(4): 423–434.

Lindley, D. V., and A. F. M. Smith. 1972. Bayes estimates for the linear model. *Journal of the Royal Statistical Society* 34: 1–41.

Nadaraya, E. A. 1965. On nonparametric estimates of density functions and regression curves. *Theory of Applied Probability* 10: 186–190.

Ouyang, D., Q. Li, and J. S. Racine. 2006. Cross-validation and the estimation of probability distributions with categorical data. *Journal of Nonparametric Statistics* 18(1): 69–100.

Ouyang, D., Q. Li, and J. S. Racine. 2008. Nonparametric estimation of regression functions with discrete regressors. *Econometric Theory*. 25(1): 1–42.

R Development Core Team. 2008. *R: A Language and Environment for Statistical Computing*, R Foundation for Statistical Computing, Vienna, Austria. ISBN 3-900051-07-0. http://www.R-project.org

Racine, J. S. and Q. Li. 2004. Nonparametric estimation of regression functions with both categorical and continuous data. *Journal of Econometrics* 119(1): 99–130.

Simonoff, J. S. 1996. *Smoothing Methods in Statistics*. New York: Springer Series in Statistics.

Wand, M., and B. Ripley, 2008. *KernSmooth: Functions for Kernel Smoothing*. R package version 2.22-22. http://CRAN.R-project.org/package=KernSmooth

Wang, M. C., and J. van Ryzin, 1981. A class of smooth estimators for discrete distributions. *Biometrika* 68: 301–309.

Watson, G. S. 1964. Smooth regression analysis. *Sankhya* 26:(15): 359–372.

Wooldridge, J. M. 2002. *Econometric Analysis of Cross Section and Panel Data*. Cambridge, MA: MIT Press.

8

The Unconventional Dynamics of Economic and Financial Aggregates

Karim M. Abadir and Gabriel Talmain

CONTENTS

8.1 Introduction

Time series models have provided econometricians with a rich toolbox from which to choose. Linear ARIMA models have been very influential and have enhanced our understanding of many empirical features of economics and finance. As with any scientific endeavor, data have emerged that show the need for refinements and improvements over existing models.

Nonlinear models have gained popularity in recent times, but which one do we choose from? Once we move away from linear models, there is a huge variety on offer. Surely, economic theory should provide the guiding light, insofar as economics and finance are the subject in question. Abadir and Talmain (2002) provided one possible answer. This chapter is mainly a summary of the econometric aspects of the line of research started by that paper.

The main result of that literature is that macroeconomic and aggregate financial series follow a nonlinear long-memory process that requires new econometric tools. It also shows that integrated series (which are a special case of the new process) are not the norm in our subject, and proposes a new approach to econometric modeling.

8.2 The Economic Origins of the Nonlinear Long-Memory

Abadir and Talmain (AT) started with a micro-founded macro model. It was a standard real business cycle (RBC) model, except that it allowed for heterogeneity: the "representative firm" assumption was dropped. They worked out the intertemporal general equilibrium solution for the economy, and the result was an explicit dynamic equation for GDP and all the variables that move along with it.

It was well known, long before AT, that heterogeneity and aggregation led to long-memory; e.g., see Robinson (1978) and Granger (1980) for a start of the literature on linear aggregation of ARIMA models, and Granger and Joyeux (1980) and Hosking (1981) for the introduction of long-memory models.[1] But in economics, there is an inherent nonlinearity which makes linear aggregation results incomplete. Let us illustrate the nonlinearity in the simplest possible aggregation context; see AT for the more general CES-type aggregation.

Decompose GDP, denoted by Y, into the outputs $Y(1), Y(2), \ldots$ of firms (alternatively, sectors) in the economy as

$$Y := Y(1) + Y(2) + \cdots = \mathrm{e}^{y(1)} + \mathrm{e}^{y(2)} + \cdots,$$

where we write the expression in terms of $y(i) := \log Y(i)$ $(i = 1, 2, \ldots)$ to consider percentage changes in $Y(i)$ (and to make sure that models to be chosen for $y(i)$ keep $Y(i) > 0$, but this can be achieved by other methods too). With probability 1,

$$\mathrm{e}^{y(1)} + \mathrm{e}^{y(2)} + \cdots \neq \mathrm{e}^{y(1)+y(2)+\cdots},$$

where the right-hand side is what linear aggregation entails. The right-hand side is the aggregation considered in the literature, typically with $y(i) \sim \mathrm{ARIMA}(p_i, d_i, q_i)$, but it is not what is needed in macroeconomics. AT (especially p. 765) show that important features are missed by linearization when aggregating dynamic series.

One implication of the nonlinear aggregation is that the auto-correlation function (ACF) ρ_τ of the logarithm of GDP and other variables moving with it take the common form

$$\rho_\tau := \frac{\mathrm{cov}(y_t, y_{t-\tau})}{\sqrt{\mathrm{var}(y_t)\mathrm{var}(y_{t-\tau})}} = \frac{1 - a\,[1 - \cos(\omega\tau)]}{1 + b\tau^c}, \tag{8.1}$$

[1] A time series is said to have long memory if its autocorrelations dampen very slowly, more so than the exponential decay rate of stationary autoregressive models but faster than the permanent memory of unit roots. Unlike the latter, long-memory series revert to their (possibly trending) means.

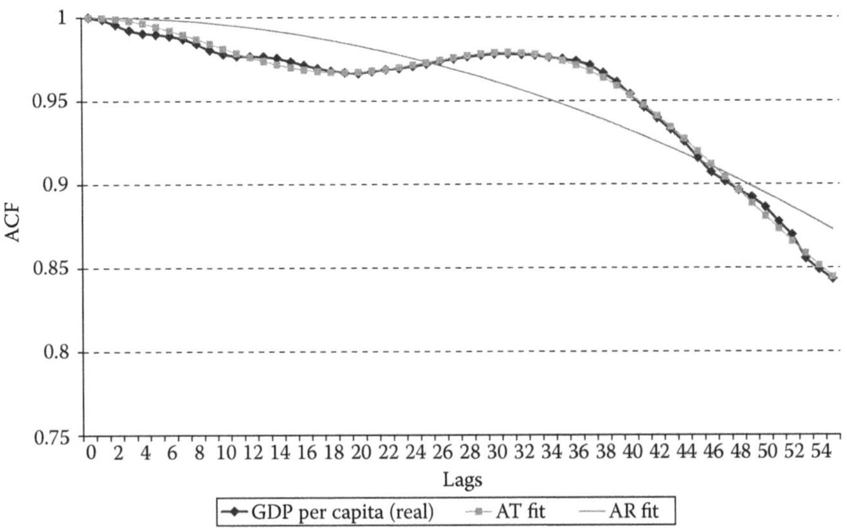

FIGURE 8.1
ACF of the log of U.S. real GDP per capita over 1929–2004.

where the subscript of y denotes the time period and $a,\ b,\ c,\ \omega$ depend on the parameters of the underlying economy but differ across variables.[2] Abadir, Caggiano, and Talmain (2006) tried this on all the available macroeconomic and aggregate financial data, about twice as many as (and including the ones) in Nelson and Plosser (1982). The result was an overwhelming rejection of AR-type models and the shape they imply for ACFs, as opposed to the one implied by Equation 8.1. For example, for the ACF of the log of U.S. real GDP per capita over 1929–2004, Figure 8.1 presents the fit of the best AR(p) model (it turns out that $p = 2$ with one root of almost 1) by the undecorated solid line, compared to the fit of Equation 8.1 by nonlinear LS. Linear models, like ARIMA, are simply incapable of allowing for sharp turning points that we can see in the decay of memory. The empirical ACFs found that there is typically an initial period where persistence is high, almost like a unit-root with a virtually flat ACF, then a sudden loss of memory. We can illustrate this also in the time domain in Figure 8.2, where we see that the log of real GDP per capita is evolving around a linear time trend, well within small variance bands that don't expand over time (unlike unit-root processes whose variance expands linearly to infinity as time passes).

ACFs of this shape have important implications for macroeconomic policymakers, as Abadir, Caggiano, and Talmain (2006) show. For example, if an economy is starting to slow down, such ACFs predict that it will produce a long sequence of small signs of a slowdown followed by an abrupt decline. When only the small signs have appeared, no-one fitting a linear (e.g., AR)

[2] The restrictions $b,\ c,\ \omega > 0$ apply, but the restriction on a cannot be expressed explicitly.

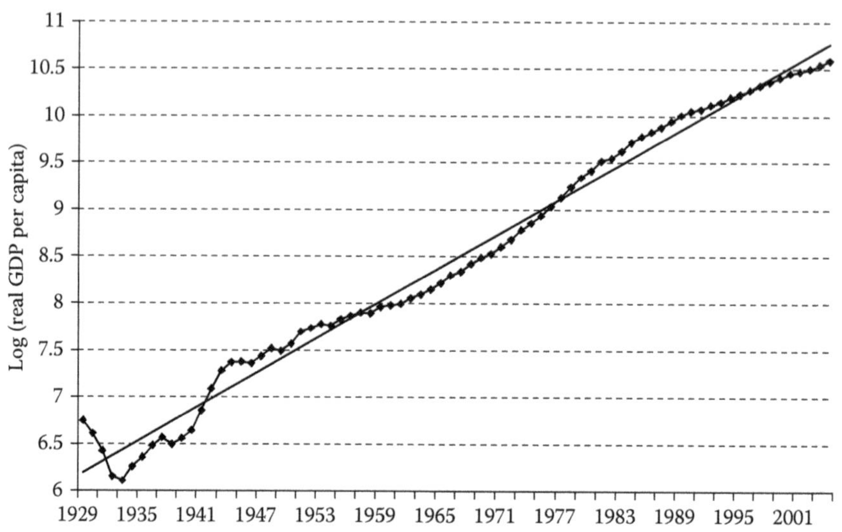

FIGURE 8.2
Time series of the log of U.S. real GDP per capita over 1929–2004.

model would be able to guess the substantial turning point that is about to occur. Another implication is that any stimulus that is applied to the economy should be timed to start well before the abrupt decline of the economy has taken place, and will take a long time to have an impact (and will eventually wear off unlike in unit root models). Consequently, a gradualist macroeconomic policy will not yield the desired results because it will be a case of too little and too late. In other words, a gradualist approach can be compatible with linear models but will be disastrous in the context of the ACFs that arise from macroeconomic data and that are compatible with the nonlinear dynamics generated by the general-equilibrium model of AT.

The ACF shape has important implications for econometric methods also. The long-memory cycles it generates require the consideration of singularities at frequencies other than 0 in spectral analysis. In fact, if a is close to 1 in the ACF (Equation 8.1), Fourier inversion produces a spectrum $f(\lambda)$ that is approximately proportional to $|\lambda - \omega|^{c-1}$; that is, at frequency ω, there is a singularity when $c \in (0, 1)$. For I(d) series having $d \in (0, \frac{1}{2})$, the spectrum has a singularity *at the origin* that is proportional to $|\lambda|^{-2d}$, giving the correspondence $c = 1 - 2d$ in the special case of $\omega = 0$. This correspondence holds also *in the tails* of the ACFs of the two processes when $\omega = 0$.

I(d) models are a special case of the new process. We therefore need to go beyond I(d) models and consider the estimation of spectral densities near singularities that are not necessarily located at the origin, as a counterpart (when $a \approx 1$) to the ACF-domain estimation mentioned earlier. Giraitis, Hidalgo, and Robinson (2001) and Hidalgo (2005) give a frequency-domain method of estimating ω and d, when $d \in (0, \frac{1}{2})$.

For $a \approx 1$, we introduce the following definition.

Definition 8.1 *A process is said to be of cyclical long-memory, respectively with parameters* $\omega \in [0, \pi]$ *and* $d \in (0, \frac{1}{2})$, *if it has a spectrum* $f(\lambda)$ *that is proportional to* $|\lambda - \omega|^{-2d}$ *as* $\lambda \to \omega$ *and is bounded elsewhere. Such a process is denoted by* $\text{CM}(\omega, d)$, *with the special case* $\text{CM}(0, d) = \text{I}(d)$.

It is no wonder that a statistical model with cycles arises from a real business cycle model. Note that integrated processes cannot generate cycles that have long memory because their spectrum is bounded at $\omega \neq 0$. They can only generate short transient cycles that are not sufficiently long for macroeconomics.

When a is not close to 1 in the ACF (Equation 8.1), the result of the Fourier inversion is approximately a linear combination of one $\text{I}(d)$ and one $\text{CM}(\omega, d)$ when $\omega \neq 0$. Here, too, the approximation arises from the inversion focusing more on the tail of the ACF and neglecting to some extent the initial concave part of the ACF in Equation 8.1.

But if the individual series are not of the *integrated* type, can we talk of *co-integrated* series? It is an approximation that many not be adequate enough. What about the modification of co-integration modeling for variables that have this new type of dynamics? Abadir and Talmain (2008) propose a solution. We summarize it in the next section, and present an additional definition to complement Definition 8.1.

8.3 Modeling Co-Movements for Series with Nonlinear Long-Memory

This section contains three parts. We start with the specification, estimation, and inference in a model where the residual's dynamics are allowed to have the ACF in Equation 8.1. We then explore some empirical implications of such a model. Finally, we introduce a special case of the model that implies an extension of co-integration to allow for co-movements of CM processes.

8.3.1 Econometric Model

Suppose we have a sample of $t = 1, \ldots, T$ observations. To simplify the exposition, consider the model

$$z = X\beta + u, \tag{8.2}$$

where z is $T \times 1$ and β is $k \times 1$. The matrix X can contain lagged dependent variables, so that we cover autoregressive distributed-lag models (e.g., used in co-integration analysis) as one of the special cases. The vector u contains the residual dynamics of the adjustment of z toward its fundamental value $X\beta$. By definition, $\boldsymbol{u}$ is centered around zero and is mean-reverting, otherwise z will not revert to its fundamental value. We write $\boldsymbol{u} \sim \mathrm{D}(\mathbf{0}, \boldsymbol{\Sigma})$, where $\boldsymbol{\Sigma}$ is the $T \times T$ autocovariance matrix of the u's. The autocorrelation matrix of $\boldsymbol{u}$ is denoted by $\boldsymbol{R}$, and Abadir and Talmain (2008) use Equation 8.1 to parameterize the typical ijth element $\rho_{|i-j|}$ of $\boldsymbol{R}$. There are two implications to u_t being mean-reverting (which is a testable assumption). First, $\boldsymbol{\Sigma}$ is proportional to $\boldsymbol{R}$. Second, the ML estimator of β and the ACF parameters in $\boldsymbol{u}$ is consistent. The asymptotic distribution will depend on the properties of the variables, but if the estimated residuals are found to satisfy $c > \frac{1}{2}$ (implying square-summability of ρ_τ), then standard t, F, LR tests are justified asymptotically.[3] This condition on c is sufficient but not necessary, and we have found it to hold in practice when dealing with macro and financial series.

The quasi maximum likelihood (QML) procedure of Abadir and Talmain (2008) estimates jointly the parameters β and the ACF parameters in ρ_τ of $\boldsymbol{u}$. They remove the sample mean of each variable in Equation 8.2 to avoid multicollinearity in practice, with the constant term in $\boldsymbol{X}$ redefined accordingly. They also assume that $\boldsymbol{X}$ is weakly exogenous (see Engle, Hendry, and Richard 1983) for the parameters of Equation 8.2.

For any given $\boldsymbol{R}$, define

$$\widehat{\beta}_R := \left(X'R^{-1}X\right)^{-1} X'R^{-1}z \tag{8.3}$$

as a function of $\boldsymbol{R}$. Denoting the determinant of a matrix $\boldsymbol{M}$ by $|\boldsymbol{M}|$, Abadir and Talmain (2008) show that the QML estimator (QMLE) of $\boldsymbol{R}$ is obtained by maximizing the concentrated log-likelihood

$$-\log\left|\left(z - X\widehat{\beta}_R\right)' R^{-1} \left(z - X\widehat{\beta}_R\right) R\right| \tag{8.4}$$

with respect to the parameters of the ACF: the optimization of the joint likelihood (for Σ and β) now depends on only four parameters that are given in Equation 8.1 and that determine the whole autocorrelation matrix $\boldsymbol{R}$. Once the optimal value $\widehat{\boldsymbol{R}}$ of $\boldsymbol{R}$ is obtained, the QMLE of β is $\widehat{\beta} \equiv \widehat{\beta}_{\widehat{R}}$.

8.3.2 Empirical Implications

One is often interested in detecting the presence of co-movements between series. This may be for the purpose of empirically validating theoretical work, producing predictions, or determining optimal policies. In practice, one is often frustrated by the results produced by co-integration analysis. The theory of purchasing power parity (PPP) is typically tested using co-integration.

[3] This is a case where the results of Tsay and Chung (2000) on the divergent behavior of t-statistics do not apply, since the condition on c corresponds to the case of the series having $d < \frac{1}{4}$.

Generally, the findings are that PPP does not hold in the short run and deviations from PPP are cycling around the theoretical value at very low frequency, implying that the estimated reversion to PPP is, if at all, unrealistically slow.

Even when the series have less memory, dynamic modeling of co-movements can spring surprises. According to the uncovered interest parity (UIP) theory, no contemporaneous variable should be able to predict the future excess returns in investing in a foreign asset. However, researchers have consistently found a strong negative relation between future excess returns and the forward premium on a currency. With the usual interpretation of the forward rate as a predictor of the future spot exchange rate, this would imply the irrational result that a currency is expected to depreciate in periods when assets denominated in this currency actually do produce systematic excess returns!

These "anomalies" or "paradoxes" are what one would find if the true nature of the relation between the variables is of the type in Equation 8.2, but the possibility of unconventional dynamics for $\boldsymbol{u}$ has been neglected. Co-integration would try to force a noncyclical zero-frequency pattern on this residual term which, in reality, is slowly cycling. By allowing for the possibility of long-memory cycles, the methodology described above brings to light the true nature of the residuals and, thus, of the true relation between the co-moving variables. The "long-run" relation between economic variables often involves long cycles of adjustment.

8.3.3 Special Case: co-CM

The model in Equation 8.2 avoids the question of the individual ω, d in each of the series contained in $z, \mathbf{X}$. It just states that the dynamics of adjustment to the fundamental value (through changes in $\boldsymbol{u}$) is of the general AT type. A way in which this can arise is through the following special CM case of the AT process, where we use a bivariate context to simplify the illustration and to show how it generalizes the notion of co-integration.

Definition 8.2 *Two processes are said to be linearly co-CM if they are both* CM(ω, d) *and there exists a linear combination that is* CM(ω, s) *with* $s < d$.

This follows by the same spectral methods used in Granger (1981, Section 4). The definition can be extended to allow for nonlinear co-CM, for example, if $z_t = g(x_t) + u_t$ with g a nonlinear function. For the effect on the ACF (hence on ω, d) of parametric nonlinear transformations, see Abadir and Talmain (2005).

In Equation 8.2, it was not assumed that $s < d$. In fact, in the UIP application in Abadir and Talmain (2008), we had the ACF equivalent of $s = d$ because it was a trivial co-CM case where the right-hand side variable had a zero coefficient and $z_t = u_t$.

8.4 Further Developments

Work is currently being carried out on a number of developments of these models and the tools required to estimate them and test hypotheses about their parameters. The topic is less than a decade old, at the time of writing this chapter, but we hope to have demonstrated its potential importance.

A simple time-domain parameterization of the CM(ω, d) process has been developed in preliminary work by Abadir, Distaso, and Giraitis. The frequency-domain estimation of this process is also being considered, generalizing the FELW estimator of Abadir, Distaso, and Giraitis (2007) to the case where ω is not necessarily zero.

8.5 Acknowledgments

We gratefully acknowledge support from ESRC grant RES-062-23-0790.

References

Abadir, K. M., G. Caggiano, and G. Talmain. 2006. Nelson-Plosser revisited: the ACF approach. Working Paper Series 18-08 (revised version, 2008), Rimini Centre for Economic Analysis, Rimini, Italy.

Abadir, K. M., W. Distaso, and L. Giraitis. 2007. Nonstationarity-extended local Whittle estimation. *Journal of Econometrics* 141: 1353–1384.

Abadir, K. M., and G. Talmain. 2002. Aggregation, persistence and volatility in a macro model. *Review of Economic Studies* 69: 749–779.

Abadir, K. M., and G. Talmain. 2005. Autocovariance functions of series and of their transforms. *Journal of Econometrics* 124: 227–252.

Abadir, K. M., and G. Talmain. 2008. Macro and financial markets: the memory of an elephant? Working Paper Series 17-08, Rimini Centre for Economic Analysis.

Engle, R. F., D. F. Hendry, and J. -F. Richard. 1983. Exogeneity. *Econometrica* 51: 277–304.

Giraitis, L., J. Hidalgo, and P. M. Robinson. 2001. Gaussian estimation of parametric spectral density with unknown pole. *Annals of Statistics* 29: 987–1023.

Granger, C. W. J. 1980. Long memory relationships and the aggregation of dynamic models. *Journal of Econometrics* 14: 227–238.

Granger, C. W. J. 1981. Some properties of time series data and their use in econometric model specification. *Journal of Econometrics* 16: 121–130.

Granger, C. W. J., and R. Joyeux. 1980. An introduction to long-range time series models and fractional differencing. *Journal of Time Series Analysis* 1: 15–29.

Hidalgo, J. 2005. Semiparametric estimation for stationary processes whose spectra have an unknown pole. *Annals of Statistics* 33: 1843–1889.

Hosking, J. R. M. 1981. Fractional differencing. *Biometrika* 68: 165–176.

Nelson, C. R., and C. I. Plosser. 1982. Trends and random walks in macroeconomic time series: some evidence and implications. *Journal of Monetary Economics* 10: 139–162.

Robinson, P. M. 1978. Statistical inference for a random coefficient autoregressive model. *Scandinavian Journal of Statistics* 5: 163–168.

Tsay, W. -J., and C.-F. Chung. 2000. The spurious regression of fractionally integrated processes. *Journal of Econometrics* 96: 155–182.

9

Structural Macroeconometric Modeling in a Policy Environment

Martin Fukač and Adrian Pagan

CONTENTS

9.1 Introduction

Since the basic ideas of structural macroeconometric modeling were laid out by the Cowles Commission, there has been substantial effort invested in turning their vision into a practical and relevant tool. Research and development has proceeded across a broad front but basically can be characterized as responses to four issues.

1. The design of models to be used in a policy environment.
2. Estimation of the parameters in these models.
3. Match of these models to the data, i.e., how to evaluate their ability to adequately represent the outcomes from an actual economy.
4. Prediction and policy analysis with the models.

Econometric texts and articles typically deal with the last three topics while the first tends to be neglected. Consequently this chapter will focus on how the design of models in policy use has evolved over the past 60 years. In concentrating on the models that have been adopted in institutions concerned with policy making, we have not dealt with models either developed in the private sector or by individuals, e.g., the model set out initially by Fair (1974) which has gone through several generations of change. Moreover, although our primary focus is on model design, it is impossible to ignore questions of estimation and data matching, as often these are driven by the design of the models, so that we will need to spend some time on the second and third of the issues.

Model design has evolved in a number of ways. At a primal level it is due to the fact that the academic *miniature* model upon which they are based, and which aims to capture the essential forces at work in the economy, has changed over time. We can distinguish five of these miniature models:

1. Ramsey model – Ramsey (1928)
2. IS-LM, Aggregate Demand-Supply (AD-AS) models – Hicks (1937).
3. Solow-Swan Model – Solow (1956), Swan (1956)
4. Stochastic Ramsey Model (Real Business Cycle Model/Dynamic Stochastic General Equilibrium -DSGE- models) – King Plosser, and Rebelo (1988)
5. New Keynesian model – Clarida, Gali, and Gertler (1999)

Essentially, these models were meant to provide a high-level interpretation of macroeconomic outcomes. Mostly they were too simple for detailed policy work and so needed to be *adapted* for use. Although providing some broad intellectual foundations they need to be augmented for practical application. The adaptions have led to four generations of models distinguished later which loosely relate to the miniature models given above.

Coexisting with these *interpretative models* have been *summative* models that aim to fit a given set of data very closely and which employ various statistical approaches to do this, e.g., vector autoregressions (VARs). Mostly these models are used for forecasting. Sometimes the summative and interpretative models have been identical, but increasingly there has been a divorce between them, resulting in a multiplicity of models in any policy institution today. To some extent this reflects developments in computer hardware and software since the cost of maintaining a variety of models has shrunk quite dramatically in the past few decades. The greater range of models also means that how we are to judge or evaluate a given model will differ depending upon what it seeks to achieve. Consequently, this often accounts for why proponents of a particular representative of each of the classes are reluctant to evaluate their models with criteria that might be appropriate for another of the classes.

The four generations of models we will distinguish in the succeeding sections are often represented as being vastly different. Sometimes the differences that are stressed are superficial, reflecting characteristics such as size and underlying motivation. It would be unfortunate if this attitude prevailed as it obscures the fact that each generation has drawn features from previous generations as well as adding new ones. Evolution rather than revolution is a better description of the process describing the move from one generation to another. To see this it will help to structure the discussion according to how each generation has dealt with five fundamental questions:

1. How should the dynamics evident in the macroeconomy be incorporated into models? Specifically, are these to be external (imposed) or internal (model consistent)?
2. How does one incorporate expectations and what horizon do they refer to?
3. Do stocks and flows need to be integrated? If so, is this best done by having an equilibrium viewpoint in which all economic variables gravitate to a steady-state point or growth path?
4. Are we to use theoretical ideas in a loose or tight way?
5. How are nominal rather than real quantities to be determined?

The sections that follow outline the essential characteristics of each of the four generations of models distinguished in this chapter by focusing on the questions just raised. This enables one to see more clearly what is common and what is different between them.

9.2 First Generation (1G) Models

These are the models of the 1950s and 1960s. If one had to associate a single name with them it would be Klein. If one had to associate a single institution it would be the University of Pennsylvania. A very large number of modelers in many countries went to the latter and were supervised by the former.

The miniature model that underlies representatives of this generation was effectively that associated with the IS/LM framework. Accordingly, the modeling perspective was largely about the determination of demand. Adaption of the miniature model to policy use involved disaggregation of the components of the national income identity. Such a disaggregation inevitably led to these models becoming large.

Dynamics in the models were of two types. One alternative was to allow for a dynamic relation between y_t and x_t by making y_t a function of $\{x_{t-j}\}_{j=0}^{p}$. If p was large, as might be the case for the effect of output (x_t) upon investment (y_t), then some restrictions were imposed upon the shape of the lagged effects of a change in x_t upon y_t. A popular version of this was termed "Almon lags"– Almon (1965). But mostly dynamics were imposed using a different strategy — that associated with the partial adjustment model (PAM). With real variables in logs (some nominal variables such as interest rates, however, were left in levels form) this had the structure[1]

$$\Delta z_t = \gamma(z_t^* - z_{t-1}), \tag{9.1}$$

where z_t^* was some target for z_t which was made observable by relating it to a function of x_t. The specification of the function linking z_t^* and x_t was generally loosely derived from theoretical ideas. As an example, targeted consumption c_t^* was related to income (y_t) and other things expected to influence consumption, such as interest rates (r_t). Thus,

$$c_t^* = ay_t + br_t. \tag{9.2}$$

In these models there was often an awareness of the importance of expectations in macroeconomics, reflecting their long history in macroeconomic discussion. To model these expectations, one assumed they could be measured as a combination of the past history of a small set of variables (generally) present in the model, with the weights attached to those variables being estimated directly using the observations on the variables expectations were being formed about.

Because the supply side in these models was mostly ignored, there was not a great deal of attention paid to stocks and flows. Wallis (1995), in an excellent review of these and the 2G models discussed later, notes that there was an implicit assumption underlying them that variables evolved deterministically over longer periods of time, although there was not any discussion about whether such paths were consistent and their relative magnitudes did not seem to play a major role in model construction and design.

To build a link between the real and nominal sides of the economy modelers generally viewed prices as a mark up over (mostly) wages, and the markup was often influenced by business conditions. A dynamic account of wages was provided by the Phillips curve. Later versions just assumed that the Phillips curve applied to inflation itself and so had the form

$$\pi_t = \alpha_1 \pi_{t-1} + \delta u_t + \varepsilon_t, \tag{9.3}$$

where π_t was price inflation and u_t was the unemployment rate. There was a lot of debate about whether there was a trade-off between inflation and unemployment, i.e., was $\delta \neq 0, \alpha_1 < 1$? Sometimes one saw this relation augmented as

$$\pi_t = \alpha_1 \pi_{t-1} + \delta(u_t - \overline{u}) + \gamma_2(p_{t-1} - ulc_{t-1}) + \varepsilon_t, \tag{9.4}$$

[1] In many of the early models variables were expressed in terms of their levels and it was only later that log quantities were used more extensively.

where p_t was the log of the price level and ulc_t was unit labor cost. Without some modification like this there was no guarantee that the level of prices and wages would remain related.

Estimation of these models was mostly done with single equation methods and so evaluation largely involved applying a range of specification tests to the individual equations. These equations could be represented as

$$y_t = \phi_1 y_{t-1} + \phi_2 z_t + \phi_3 z_{t-1} + \varepsilon_t, \tag{9.5}$$

where z_t might be endogenous variables and ε_t was an "error term." Tests therefore considered the residuals $\hat{\varepsilon}_t$ as a way of gaining information about specification problems with this equation. Although useful, this evaluation process did not tell one much about the fit of the complete model, which was a key item of interest if the model is to be used for forecasting. For that it needs to be recognized that z_t is not given but also needs to be solved for. System and single equation performance might therefore be very different.

Once a complete system was found one could find a numerical value for what one would expect z_t to be from the model (given some exogenous variables) either analytically or by simulation methods (when the system was nonlinear). The software developed to do so was an important innovation of this generation of models. Chris Higgins, one of Klein's students, and later Secretary of the Australian Treasury, felt that any assurance on system performance required that modelers should "simulate early and simulate often." For that, computer power and good software were needed. It was also clear that, in multistep forecasts, you had to allow for the fact that both y_{t-1} and z_{t-1} needed to be generated by the model. Hence dynamic simulation methods arose, although it is unclear if these provided any useful extra information about model specification over that available from the static simulations, since the residuals from dynamic simulations are just transformations of the $\hat{\varepsilon}_t$.[2] Perhaps the major information gained from a dynamic simulation of the effects following from a change in an exogenous variable was what happened as the policy horizon grew. If the change was transitory, i.e., lasted for only a single period, then one would expect the effects to die out. In contrast, if it was permanent, one would expect stabilization of the system at some new level. It was easy to check that this held if one only has a single equation, e.g., in the PAM scheme $0 < \gamma < 1$ was needed. Thus each of the individual equations could be checked for stability. But this did not guarantee system stability because, *inter alia*, z_t might depend upon y_{t-1}, thereby making the stability condition much more complex. An advantage of a dynamic simulation was that it could provide the requisite information regarding the presence or absence of stability relatively cheaply and easily.

---

[2] Wallis (1995) has a good discussion of these issues.

9.3 Second Generation (2G) Models

These began to emerge in the early 1970s and stayed around for 10–20 years. Partly stimulated by inflation, and partly by the oil price shocks of the early 1970s, the miniature model that became their centerpiece was the AD/AS model — which recognized the need for a supply side in the model. When adapted for use this involved introducing a production function to place a constraint on aggregate supply, particularly over longer horizons. A leading light in the development of these models for policy use was John Helliwell with his RDX2 model of the Canadian economy (Helliwell et al. 1971), but others emerged such as the Fed-MIT-Penn (FMP) model (Brayton and Mauskopf 1985) which was also called MPS, see Gramlich (2004).

These models retained much of the structure of the previous generation in that demand was captured by disaggregated equations stemming from the national income identity. Now these were supplemented with equations which introduced much stronger supply side features. There was also some movement toward deriving the relationships as the consequence of optimization problems solved by agents — in particular the consumption decision and the choice of factors of production were often described in this way. Thus for consumption an intertemporal dimension was introduced through the use of life-cycle ideas. These implied that consumption depended on financial wealth (y_t) and current labor income (w_t), i.e., $c_t^* = a w_t + b y_t$. Dynamics were again introduced through a distributed lag on the static relationships determining the desired levels z_t^*. The advance on previous work was the use of an error correction mechanism (ECM),

$$\Delta z_t = \delta \Delta z_t^* + \alpha(z_{t-1} - z_{t-1}^*). \tag{9.6}$$

As Wallis (1995) observes the ECM originated in Phillips' control work of the 1950s and was applied by Sargan (1964) when modeling inflation, but its widespread use began with Davidson et al. (1978).

Now, with the introduction of a production function, and a household's decisions coming loosely from a life cycle perspective, the presence of household wealth and the capital stock meant that there were dynamics present in the model which stemmed from depreciation and savings. Consequently, dynamic stability of the complete system became a pressing issue. Gramlich (1974) comments on his work with the MPS model that "... the aspect of the model that still recalls frustration was that whenever we ran dynamic full-model simulations, the simulations would blow up."Once again one needed to keep an eye on system performance when modifying the individual equations. It might be a necessary condition that the individual equations of the system were satisfactory, but it was not a sufficient one.

Like the previous generation of models there was considerable diversity within this class and it grew larger over time. Often this diversity was the result of a slow absorption into practical models of new features that were becoming important in academic research. For example, since many of these models had an array of financial assets — certainly a long and a short rate–rational (or model consistent) expectations were increasingly introduced into the financial markets represented in them. By the end of the era of 2G models, this development was widely accepted. But, when determining real quantities, expectations were still mainly formulated in an ad hoc way. One reason for this was the size of the models. The UK models were almost certainly the most advanced in making expectations model-consistent. By 1985 this work had produced a number of models, such as the London Business School and National Institute models, which had implemented solutions; see the review in Wallis and Whitley (1991). A significant factor in this movement was the influence of the Macro-Economic Modelling Bureau at the University of Warwick (see Wallis 1995).

Dynamics in prices were again operationalized through the Phillips curve, but with some modifications. Now either a wage or price Phillips curve had the form

$$\pi_t = \alpha_1 \pi_{t-1} + \delta(u_t - \overline{u}) + \varepsilon_t, \tag{9.7}$$

where $\overline{u}$ was the nonaccelerating inflation rate of unemployment (NAIRU), and, often, $\alpha_1 = 1$. The NAIRU was a prescribed value and it became the object of attention. Naturally questions arose of whether one could get convergence back to it once a policy changed. In models with rational expectations dynamic stability questions such as these assume great importance. If expectations are to be model consistent, then one needed the model to converge to some quantity. Of course one might circumvent this process by simply making the model converge to some prespecified terminal conditions, but that did not seem entirely satisfactory. By the mid 1980s, however, it appeared that many of the models had been designed (at least in the UK) to exhibit dynamic stability, and would converge to a steady state (or an equilibrium deterministic path).

9.4 Third Generation (3G) Models

9.4.1 Structure and Features

Third generation (3G) models reversed what had been the common approach to model design by first constructing a steady-state model (more often a steady-state deterministic growth path, or balanced growth path) and then later asking if extra dynamics needed to be grafted on to it in order to broadly represent the data. Since one of the problems with 2G models was getting stocks to change in such a way as to eventually exhibit constant ratios to flows, it was much more likely that there would be stock-flow consistency if decisions about expenditure items came from well-defined optimization choices for households and firms, and if rules were implemented to describe the policy decisions of monetary and fiscal authorities. In relation to the latter external debt was taken to be a fixed proportion of GDP and fiscal policy was varied to attain this. Monetary authorities needed to respond vigorously enough to expected inflation — ultimately more than one-to-one to movements in inflation.

There are many versions of 3G models, with an early one being an Australian model by Murphy (1988) and a multi-country model (MSG) by McKibbin and Sachs (McKibbin 1988; McKibbin and Sachs 1989). Murphy's model was more fully described in Powell and Murphy (1995). 3G models became dominant in the 1990s, being used at the Reserve Bank of New Zealand (FPS, Black et al. 1997), the Federal Reserve (FRB-US, Brayton and Tinsley 1996) and, more recently, the Bank of Japan Model (JEM, Fujiwara et al. 2004). Probably the most influential of these was QPM (quarterly projection model) built at the Bank of Canada in the early to mid-1990s, and described in a series of papers (e.g., Black et al., 1994; Coletti et al., 1996). Its steady-state model (QPS) was basically an adaption of the Ramsey model for policy use. To this point in time the latter miniature model had played a major role in theoretical economics but a rather more limited one in applied macroeconomics. An important variation on Ramsey was the use of an overlapping generations perspective that modified the discount rate by the probability of dying, as advocated in Blanchard (1985) and Yaari (1965).

As a simple example of the change in emphasis between 2G and 3G models, take the determination of equilibrium consumption. It was still the case that consumption ultimately depends on financial wealth and labor income, but now the coefficients attached to these were explicitly recognized to be functions of a deeper set of parameters — the steady-state real rate of return, utility function parameters and the discount factor. Because these parameters also affect other decisions made by agents, one cannot easily vary any given relationship, such as between consumption and wealth, without being forced to account for the impact on other variables of such a decision.

Thus a steady-state model was at the core of 3G models. How was it to be used? In a strict steady-state (SSS) dynamics have ceased and values of the variables consistent with these equations will be constant (more generally one could allow for a constant steady-state growth path, but we will leave this qualification for later sections). But the model generating the steady state has embedded in it *intrinsic dynamics* that describe the transition from one steady-state position to another. These dynamics come from the fact that the capital stock depreciates and assets accumulate. Consequently, solving the model produces a *transitional* steady-state solution for the model variables, i.e., these variables will vary over time due to the fact that movements from one point to another are not instantaneous. In addition to this feature, in 3G models some variables were taken to be exogenous, i.e., treated as determined outside the model economy. Since it is unlikely that these will be at their steady-state values over any period of time, the endogenous variable solutions using either the pure or transitional steady-state model will need to reflect the time variation of those exogenous variables. One might refer to the latter values as the *short-run steady-state (SRSS) solutions*.

In adapting the steady-state model for use, it was necessary to recognize that the intrinsic dynamics were rarely sufficient to track the movements of variables in actual economies. Thus it became necessary to augment the intrinsic dynamics. Generally this involved a second stage optimization. The model with the augmented dynamics constituted QPM. The intrinsic dynamics in QPS might therefore be called the *first-stage dynamics*, while the extra dynamics introduced into QPM could be labeled the *second-stage dynamics*. To implement this second stage, one might have simply specified an ECM relating z_t to the SRSS values z_t^* and, in some 3G models, this was how it was done, e.g., Murphy (1988). But in QPM the extra dynamics were introduced in a quasi-theoretical way by choosing z_t to minimize the objective function

$$\frac{1}{2}\sum_{j=0}^{\infty}\beta^j E_t\{(z_{t+j}-z_{t+j}^*)^2+\phi(\Delta z_{t+j}-E(\Delta z_{t+j}))^2\}, \tag{9.8}$$

where $E_t(.)$ is the expected value conditional upon the information available at t. Setting $E(\Delta z_{t+j})=0$ would produce an optimal rule for determining z_t (the Euler equation) of

$$(1+\phi+\beta\phi)z_t+\beta\phi E_t z_{t+1}-z_t^*=0 \tag{9.9}$$

and an ultimate solution for z_t of the form

$$z_t=\lambda z_{t-1}+\frac{\lambda}{(1-\phi)}E_t\sum_{j=0}^{\infty}(\beta\lambda)^j z_{t+j}^*, \tag{9.10}$$

where λ depends on β and ϕ. Thus z_t can be constructed by weighting together past and future expected values of z_t and z_t^*. Because expectations in 3G models were effectively of the perfect foresight variety, model-consistent expectations would mean that $E_t z_{t+j}^*=z_{t+j}^*$. But, in the practice, the expectations were taken to be modeled as a function of the steady-state model solution, a finite number of lagged values of z_t, and the solution for z_{t+j}^* from QPM itself. The weights attached to these components were prescribed by the modelers.

Nickell (1985) and Rotemberg (1982) noted that the quadratic optimization scheme described above would result in an ECM connecting z_t and z_t^*, when z_t^* was a scalar and followed an autoregressive process. Hence, effectively QPM was imposing a set of ECM equations that determined the outcomes for z_t by reference to the short-run steady-state values z_t^*.

As in 2G models nominal quantities were handled by making prices a markup on marginal costs and then structuring the relation to handle dynamics and expectations. As marginal costs were primarily wages, a Cobb–Douglas production function and perfect competition meant that the wage share in GDP (or real unit labor costs) was a constant in equilibrium. With these ideas, and expectations handled as described above, one might think of the 3G Phillips curve as effectively having the form

$$\pi_t = \alpha_1 E_t \pi_{t-1} + (1 - \alpha_1) E_t \pi_{t+1} + \delta \Delta mc_t + \omega (p_{t-1} - mc_{t-1}), \qquad (9.11)$$

where mc_t was the log of nominal marginal cost and $mc_{t-1} - p_{t-1}$ was lagged real unit labor costs. Thus inflation was determined from past inflation, future expectations of inflation, current growth in nominal costs and the extent to which real unit labor costs were not constant.

9.4.2 Estimation and Evaluation

There was little formal estimation of the parameters of these models. Ratios such as consumption to income were often the main source of information used in setting values. When it was necessary to specify parameters determining dynamic responses there seems to have been significant interaction between modelers, policy advisers, and policy makers over whether the outcomes from the model with particular parameter values accorded with their views. Sometimes this involved studying the speed of adjustment after a shock while at other times estimates of quantities such as the sacrifice ratio would help in deciding on the balance between future and backward looking expectations (α_1 in the Phillips curve). Consequently, data did play some role in quantifying parameters, for example in QPM, but it was only used informally via the experience that had accumulated of the Canadian economy. Conceptually, one might think of this process as involving the use of a criterion function to match data (generally filtered) with simulated output from the models. The criterion function could then also be used to discriminate between different sets of parameter values. The exception to this strategy was when standard estimation methods were applied to the ECMs used in quantifying the second stage dynamics.

Evaluation of these models was rarely done. Indeed there was even an hostility toward data (see Colletti et al. 1996, p. 14), where they say about modeling in the Bank of Canada:

> There had been a systematic tendency to overfitting equations and too little attention paid to capturing the underlying economics. It was concluded that the model should focus on capturing the fundamental economics necessary to describe how the macro economy functions and, in particular, how policy works, and that it should be calibrated to reflect staff judgement on appropriate properties rather than estimated by econometric techniques.

Leaving this debate aside, given the way the models were used it would have been very difficult to perform a satisfactory evaluation of them. The reason was the method of producing a series on the short-run steady-state path z_t^*. The description given above was in fact too simplified. An implication of that account was that the steady-state solutions for the logs of the endogenous variables would be constructed from the exogenous variables by using a set of weights that are functions of the model parameters and that the latter would be assumed to be invariant over time. Such a scenario would generally imply constancy in a number of ratios. For example, the investment to capital stock ratio would be a constant since, in steady state, it equals a parameter — the depreciation rate of capital. But, after examining the data, it was evident that these ratios were rarely constant, and often wandered far away from any fixed point. So, although one did need to assume some fixed values for the steady-state ratios (equivalently the model parameters), it also became necessary to make some allowance for the substantial time variation seen in ratios over any given data period. Failure to do so would constitute a gross mismatch of the data and model predictions. Consequently, a two-part strategy evolved to deal with this problem. It firstly involved smoothing the observed ratios with some filter to produce an adjusted ratio that changed slowly. Secondly, this adjusted ratio was forced to converge to whatever long-run ratio was prespecified in the steady-state model. Essentially this strategy meant that the steady-state model parameters were allowed to vary smoothly over time with the restriction that they converged to a set of final steady-state choices.[3]

Cast in terms of our discussion above, the time variation in z_t^* comes not only from exogenous variables, transition paths, etc., but can also occur due to time-varying model parameters. Without this latter source of variation a comparison on how well z_t^* (the model SRSS) tracks z_t (the data) would seem a useful diagnostic for how well the two paths match, but, if one can vary the parameters of the model in a complex way so as to get a better fit to the data, such a comparative exercise becomes meaningless. Thus one cannot satisfactorily evaluate the success of the static model constructed in the first stage of a 3G modeling exercise.[4]

Turning to the second stage of 3G model construction, if a series on z_t^* was available one might think about checking the dynamic representation chosen in that stage. But here resort was often made to polynomial adjustment schemes that introduced much higher order lags than the first order of the stylized ECM connecting z_t and z_t^* described above. In doing that one could almost be certain of getting a good fit to any historical series on z_t. For 3G models therefore the only satisfactory evaluation method probably resided in whether their clients were happy with the information provided.

In the description above attention was centered upon the "gap" between z_t and z_t^* and it therefore became natural to convert all the variables in the model to "gap" format, particularly when the model was used in forecasting mode. This enabled one to improve forecasting performance by augmenting the equations for z_t with variables that were zero in the steady state. Hence,

---

[3] Some parameters were held constant since not all ratios exhibited a substantial degree of time variation.

[4] In more technical terms, if there is such great variation in the time history of the ratio that it needs to be described as an $I(1)$ process, then the methods used were essentially eliminating a unit root in the observed ratio through a filtering operation such as the Hodrick–Prescott filter. Of course if the model predicts that the ratio is $I(0)$, and the data that it is $I(1)$, it might be thought that some modification of the steady-state model is needed. Simply ignoring the mismatch by eliminating the $I(1)$ behavior via filtering seems unsatisfactory.

in the case where z_t was the log of the price level, one could add on an output gap to the equation that came from the second stage optimization. Over time this emphasis on "gaps" gave rise to the miniature models known as New Keynesian, and today these small models are often used for policy analysis and some forecasting, e.g., Berg, Karam, and Laxton (2006). In some ways the philosophy underlying 3G models had much in common with that stream of computable general equilibrium (CGE) modeling stemming from Johansen (1960). In that literature models were log-linearized around some "steady-state" values and the computation of these steady states (often termed the benchmark data set) involved substantial manipulation of data on input-output tables, etc. Of course the CGE models were not in "real time" and so transition paths were essentially irrelevant. It was simply assumed that enough time had elapsed for a new steady state to be attained once a policy change was made.

Another feature of 3G models was that shocks became the focus of attention. In the academic literature shocks had become a dominant feature of models and, with the advent of policy rules, one could no longer think about changing variables such as government expenditure or the money supply, since these were now endogenous variables. Only exogenous shocks to them might be varied. However, although the language was stochastic, often the solution methods were essentially deterministic, and so there was no "clean" incorporation of shocks into the models.

An issue that arose when these models were applied to a small-open economy with the rest of the world being treated as exogenous was what modification needed to be made to ensure that agents did not borrow indefinitely at the fixed external rate of interest; see Schmidt-Grohe and Uribe (2003) for a discussion of strategies for dealing with this issue. In practice, two of these adjustments tended to be used to design models that ruled out such behavior. In the first, the infinitely lived consumer of the Ramsey model was replaced by agents with finite lives. This formulation could be shown to be equivalent to a model with a representative consumer whose discount rate depended on the probability of death as in Blanchard (1985) and Yaari (1965). A second approach was to have the risk premium attached to foreign debt rising with the level of foreign borrowing, so that eventually agents would not wish to borrow from foreign sources to finance consumption. The ratio of foreign debt to GDP therefore became a crucial element in the latter models and decision rules had to be constructed to ensure that this prescribed ratio was achieved in steady state.

9.5 Fourth Generation (4G) Models

A fourth generation of models has arisen in the early 2000s. Representatives are TOTEM (Bank of Canada, Murchinson and Rennison 2006); MAS (the Modelling and Simulation model of the Bank of Chile, Medina and

Soto 2006); GEM (the Global Economic Model of the IMF, Laxton and Pesenti 2003); BEQM (Bank of England Quarterly Model, Harrison et al. 2005); NEMO (Norwegian Economic Model at the Bank of Norway (Brubakk et al. 2006); The New Area Wide Model at the European Central Bank, Kai, Coenen, and Warne 2008); the RAMSES model at the Riksbank (Adolfson et al. 2007); AINO at the Bank of Finland (Kilponnen and Ripatti 2005); SIGMA (Erceg, Guerrieri, and Gust 2006) at the U.S. Federal Reserve; and KITT (Kiwi inflation targeting technology) at the Reserve Bank of New Zealand (Beneš et al. 2009).

9.5.1 Extensions of 3G Model Features

In some ways these new models represent a completion of the program for adapting the Ramsey model for macroeconometric use. As with 3G models they are designed to have an underlying steady-state representation. But other features of their design are different to what was standard with 3G models. Four of these are of particular importance.

Firstly, shocks are now becoming explicitly part of the model rather than being appended at the end of the modeling process. A shock is what remains unpredictable relative to an information set specified within the model, and so it is necessary to be explicit about what this information is. In addition, how persistent the shocks are becomes important to describing the complete dynamics of the model, and this makes it necessary to decide on the degree of persistence. Given that shocks are unobservable (they are essentially defined by the model itself) this inevitably points to the need to quantify the parameters of the model from data.

Secondly, there is now no second-stage process to introduce dynamics. Instead, the adjustment cost terms used to rationalize slow adjustment in 3G models now appear directly in the primary objective functions that lead to the agent's decision rules, i.e., the short- and long-run responses are found simultaneously rather than sequentially. Of course the logic of the two-stage process used in 3G models was a recognition that adjustment costs (and the parameters associated with them) do not affect the steady-state solutions, and it was only the transition paths between steady states that depended on those parameters. In fact, recognition of this feature was the motivation for adapting 3G models to an existing forecasting environment by treating the construction of dynamics in two steps.

Thirdly, the structural equations of the model are now kept in Euler equation form rather than using a partially solved out version as was characteristic of 3G models. Thus the optimal intertemporal rule describing consumption decisions appears in most 4G models as

$$C_t = \beta E_t(C_{t+1} R_{t+1}), \tag{9.12}$$

which contrasts with the 3G model approach that combines this relation with the wealth accumulation identity to express consumption as a function of financial wealth and labor income. One reason for doing so is that it is easier to modify the model design through its Euler equations. An example is the extra dynamics introduced into consumption decisions by the use of habit persistence. This can take a number of forms, but often results in the addition of C_{t-1} to the equation to give

$$C_t = \beta E_t\left(C_{t-1}^{h} C_{t+1}^{1-h} R_{t+1}\right). \tag{9.13}$$

Finally, because shocks were an integral part of some of these models, solution methods needed to be shaped to account for them. Indeed, with this focus on shocks one had to be careful when referring to "forward" and "backward" expectations; all expectations are now formed using information available at time t, and so technically all depend on past observations (unless there are exogenous variables in the system). Thus the important feature becomes the relative weights to be attached to the available information at time t when forming expectations at different periods. A second consequence of the shift to a "shocks" perspective is that the distinction between "parameters" and "shocks" becomes blurry. Thus a depreciation rate might now be regarded as a random variable that evolves stochastically over time with an expected value equal to whatever specified value for it appears in the steady-state model. Thus this provides a formal way of allowing the model parameters to change, something that was only done in an *ad hoc* way in 3G models.

9.5.2 New Features of 4G Models

The modifications above are essentially adjustments to the basic strategies employed in the design of 3G models and are intended to produce a more precise and satisfactory statement of the design criteria. But there are also additions. Four can be mentioned.

1. Although the models are ultimately about aggregates the theoretical structure is now often based on studying the actions of heterogeneous units and providing an account of how these are to be aggregated. This heterogeneity is used in many contexts. Thus analysis often begins with different types of labor services, many intermediate goods being produced and used to make a final good, many types of imported goods, firms being differentiated in their price setting policies, etc. The question is then how one performs an aggregation of the micro decisions. The solution is an extensive use of methods popular in CGE modeling. These involve the presence of an "aggregator." This intermediary uses CES functions as a way of combining together the many separate items into a composite commodity. Thus aggregate output in a two sector model, Y_t, would be the following combination of the sectoral outputs Y_{it}

$$Y_t = \left[Y_{1t}^{-\rho} + Y_{2t}^{-\rho}\right]^{-1/\rho}, \quad \rho = \frac{1-\lambda}{\lambda}. \tag{9.14}$$

A continuum of micro-units over (0, 1) is generally used in place of a finite number as above, and, in such a case, Y_t would be represented as

$$Y_t = \left[\int_0^1 Y_{it}^{\frac{\lambda-1}{\lambda}} di\right]^{\frac{\lambda}{\lambda-1}}. \tag{9.15}$$

Profit maximization by the sectoral producers means that the amount of the sectoral output produced would depend on Y_t and the relative price $\frac{P_{it}}{P_t}$, with the functional form being

$$Y_{it} = Y_t \left(\frac{P_{it}}{P_t}\right)^{-\lambda}. \tag{9.16}$$

As well, the aggregate price level relates to the sectoral ones as

$$P_t^{1-\lambda} = \int_0^1 (P_{it})^{1-\lambda} di. \tag{9.17}$$

Models are then built for P_{it} and Y_{it} and aggregated with these functions. The method is well known from Dixit and Stiglitz (1977). Because of the use of CES functions any underlying heterogeneity has an impact only through the presence of parameters that describe the nature of the heterogeneity, i.e., the distribution of the micro decisions (say on P_{it}). Basing the model design on a microeconomic structure can potentially expand the range of information available for parameter estimation through the use of studies of microeconomic decision making.

2. In the case of firms following different pricing strategies the aggregation scheme just described forms the basis of the Calvo pricing model. In this some firms can optimally reset their prices each period and others need to follow a simple rule of thumb. Consequently, the heterogeneity in decisions about P_{it} can be summarized by a single parameter — the fraction of firms (ξ) who are able to optimally adjust their price at each point in time. The aggregate Phillips curve can then be shown to have the form

$$\begin{aligned} \pi_t - \overline{\pi} = {} & \frac{1}{1+\beta}(\pi_{t-1} - \overline{\pi}) + \frac{(1-\xi)(1-\beta\xi)}{\xi(1+\beta)}(\text{rmc}_t - \overline{\text{rmc}}) + \\ & \frac{\beta}{1+\beta} E_t(\pi_{t+1} - \overline{\pi}) + \varepsilon_t, \end{aligned} \tag{9.18}$$

where rmc_t is real marginal cost (or unit labor costs with a Cobb–Douglas production function) and $\overline{\pi}$ is a target rate of inflation. An appealing argument for building the curve up from a micro-unit level was that it allowed for monopolistic and monopsonistic behavior at that level rather than the competitive markets of the 3G models. Thus the rather awkward assumption used in QPM that there was a mark-up of prices over marginal costs in the short run, but that it went to zero in steady state (owing to the competitive markets assumption), can be dispelled.

It should be observed though that, although widespread, it is not always the case that the Calvo pricing structure is used in 4G models. Sometimes the approach used by Rotemberg (1982) is adopted. But the nature of the resulting Phillips curve is similar.

3. The steady state used in 3G models saw real variables such as output, capital, etc. as either a constant or following a deterministic growth path. This reflected the fact that labor augmenting technical change was taken as growing at a constant rate, basically following Solow (1956) and Swan (1956). Although initially in 4G models technology was treated as stationary, many models now allow the technical change to have a stochastic permanent component as well as a deterministic one. Thus the "steady-state" solution evolves stochastically over time. With some variables now having permanent components questions arise over how one should treat this fact when operationalizing the model, and we return to that later in the section.
4. Now that the models are treated as stochastically focused, when log-linearized they can be represented as structural equations of the form[5]

$$B_0 z_t^M = B_1 z_{t-1}^M + C E_t z_{t+1}^M + F\varepsilon_t, \tag{9.19}$$

where z_t^M and ε_t are the model variables and shocks, respectively. The solution to Equation 9.19 when ε_t has no serial correlation is[6]

$$z_t^M = A z_{t-1}^M + G\varepsilon_t. \tag{9.20}$$

Because it is possible that some of the model variables are not observed, it is useful to connect those variables that are observable, z_t^D, to the model variables via an observation equation

$$z_t^D = H z_t^M + \eta_t, \tag{9.21}$$

where η_t is what needs to be added on to the model solution to replicate the data. Here η_t will be termed the "tracking shocks."Altug (1989) pioneered this approach assuming that the η_t were i.i.d. and uncorrelated with model shocks. Ireland (2004) has a generalization of this where η_t can be serially correlated. Sometimes the η_t are referred to as "errors in variables," but many of the variables modeled, such as interest rates and exchange rates, are very accurately measured, and any mismatch is due to difficulties with the model rather than measurement issues. Equations 9.20 and 9.21 constitute a State Space Form (SSF) and is pivotal to estimation methods for those models in which not all model variables are observable, i.e., when the dimension of z_t^D is less than z_t^M.

---

[5] Of course the system may have higher order lags. Any exogenous variables are placed in z_t and assumed to evolve as a VAR.

[6] If the shocks ε_t follow a VAR(1) process then the solution to the system is a VAR(2), as shown in Kapetanios, Pagan, and Scott (2007). Note that, while A is a function solely of B_0, B_1, C, G will depend on these parameters plus any parameters describing the persistence in the shocks ε_t; see Binder and Pesaran (1995). This demarcation can be a very useful result.

9.5.3 Quantifying the Parameters of 4G Models

There is no one method of estimating the parameters that appear in 4G models. In some cases the approach used is the same as in 3G models. Broadly, this involved first estimating any parameters that appear in the steady state with observable ratios of variables, i.e., a method of moments estimator was implicitly being utilized. Secondly, parameters associated with the transitional paths were generally quantified by utilizing opinions about desirable model performance. Increasingly the latter strategy has been replaced by variants of maximum likelihood estimation.

9.5.3.1 Identification of the Parameters

The equations to be estimated are in Equation 9.19. A first complication in estimating this system comes from the presence of $E_t z_{t+1}$.[7] Now it is clear from Equation 9.20 that $E_t z_{t+1} = A z_t$ and so Equation 9.19 becomes

$$\Phi z_t = B_1 z_{t-1} + G\varepsilon_t, \tag{9.22}$$

where $\Phi = (B_0 - CA)$, which is a standard set of simultaneous equations. One could therefore ask whether Φ, B_1, and G are identifiable. But, since it is the parameters θ that appear in the 4G model which are ultimately of interest, i.e., those in B_0, B_1, and C, looking at identification of Φ, B_1, and G would just be a stepping stone toward examining whether θ is identified. In both instances one has to distinguish between whether there are different values of the parameters in a *given model* which would reproduce the second moments of the z_t (assuming it is stationary) and whether there is just *one model* that is consistent with those second moments. These are different questions. As Preston (1978) emphasized, the first is a question of *structural identification*, and so the conditions are effectively those of the Cowles Commission, as generalized by Rothenberg (1971). In contrast, the second depends upon what transformations are allowed in forming new models. If one can reallocate the dynamics across the equations of a given model to form a new model then they are like those in Hannan (1971). Even if the existing dynamics are to be retained, i.e., B_0, B_1, and C are fixed it may still be possible to recombine the shocks ε_t to $\zeta_t = U\varepsilon_t$, where U is nonsingular, so that G in Equation 9.22 becomes GU^{-1}. This results in a different set of impulse responses to the new shocks ζ_t, even though the second moments for z_t are identical to the model with the ε_t shocks. Such a recombination strategy is employed in the VAR sign restrictions literature to give new shocks which obey certain restrictions — in particular U is chosen there so that the shocks remain mutually uncorrelated. Now the new shocks essentially mean a new model has been found but it is one that is observationally equivalent to the old one (since the second moments of z_t are the same). This distinction between these two identification ideas is still not well understood. Many demonstrations of identification problems,

---

[7] We will ignore the distinction between observable and unobservable variables and drop the "M" for the moment.

such as Canova and Sala (2009), are concerned with structural identification, but recent work by Komunjer and Ng (2009) has been more about model identification. Whether there is a unique model might be of interest but, for estimation purposes, it is structural identification that is paramount.

In most situations B_0, B_1, and C are relatively sparse and so standard simultaneous equation identification conditions can be applied to identify Φ, B_1, and G, since there will be enough instruments to apply to each of the structural equations. Of course it may be that the instruments are weak and, in a finite sample, there is effectively a lack of identification. Indeed, many of the examples of identification difficulties that pertain to the structural equations of 4G models, such as the New Keynesian Phillips curve, do seem to be concerned with the presence of weak instruments – Mavroeidis (2004), and Nason and Smith (2005).

Often it is useful to ask whether Φ, B_1, and G can be identified before proceeding to query the identification status of θ. Since these parameters determine the impulse responses to shocks that might be sufficient for much policy analysis. However, it may be that, even when Φ and B_1 are identified, the mapping between these and the 4G model parameters, θ, might not be one to one, i.e., θ is not identified. If policy experiments involved changing the steady-state solutions then we will mostly want to identify θ rather than the impulse responses. Some of the experiments done to look at identification failures are examples of not being able to uniquely recover θ from $\{\Phi, B_1, G\}$. Generally, therefore one can think that there are two aspects to identification. One involves the ability to identify $\{\Phi, B_1, G\}$ from the data and the other is whether the model allows one to recover θ from these matrices. This distinction has been promoted in Iskrev (2007, 2009).

It should be noted that structural identification of 4G models is largely based on exclusion restrictions as proposed by the Cowles Commission. In most instances these models are therefore strongly overidentified. Even if they are not, there is a separate set of exclusion restrictions that need to be taken into account, namely, those that come from the standard assumption in these models that the shocks ε_t are contemporaneously uncorrelated. Those restrictions produce extra instruments that can be used for estimation that were not present in the analysis provided by the Cowles researchers, since they took the errors in their structural equations to be correlated.

9.5.3.2 Maximum Likelihood and Bayesian Estimation of Parameters

In studying identification issues A may be taken as known but, in estimation, a decision has to be made whether it should be found from a regression on Equation 9.20 ($\hat{A}$) or forced to be consistent with the 4G model, in which case A depends on values of the structural parameters θ. In the former case one can utilize limited information methods of estimation, allowing each structural equation to be estimated separately. For the latter a complete systems estimator is needed. Which one is to be used depends on the degree of robustness for the parameter estimates that one wants. Using Equation 9.20 to form an estimate of A (and hence measuring expectations) will be much more robust to system mis-specification, i.e., $\hat{A}$ will be a consistent estimator of A provided the system generating the data can be represented by a VAR of the selected order, and it does not depend upon knowing the structural specification that generated the data. However, a more efficient estimator of A is available by utilizing the mapping between A and θ. As has been known for a long time, such efficiency can come at the expense of bias and inconsistency of estimators, unless the complete system is an adequate representation of the data. As Johansen (2005) has pointed out, this is a price of MLE, and it should not be assumed that the 4G model has the property of being a correct specification.

Making A depend upon θ has led to full information (FI) maximum likelihood (FIML) becoming a standard way of estimating smaller 4G models (those that are generally referred to as DSGE models). This contrasts with the earlier generations of models where limited information (LI) estimation methods prevailed, i.e., the equations (or subsets of them) were estimated separately, and the influence of the complete model was minimal. It is interesting to note that the wheel has almost come full circle as the recommendation by the Cowles Commission was to use FIML, but they were frustrated by the fact that computers were not powerful enough at that time for such an estimator to be effectively employed.

In practice the FIML estimator has increasingly been replaced by a Bayesian full information (BFI) estimator. In this estimates of θ comparable to FIML can be found by maximizing a criterion function $L(\theta) + \ln p(\theta)$, where $p(\theta)$ is the prior on θ and $L(\theta)$ is the log-likelihood. The resulting estimate of θ is often referred to as the mode of the posterior. It is clear that the FIML and the Bayesian FI mode (BFI) will converge as the sample size grows and the prior information becomes dominated. Hence any difficulties arising with FIML involving misspecification of the system cannot be avoided by using a Bayesian estimator. This seems to be widely misunderstood as one often sees comments that Bayesian methods do not require correct specification of the model.

An advantage of the Bayesian method is that there is often information about the range of possible values for θ, either from constraints such as the need to have a steady-state or from past knowledge that has accumulated among researchers. Imposing this information upon the MLE is rarely easy. It can be done by penalty functions, but often these make estimation quite difficult. Adding on $\ln p(\theta)$ to the log-likelihood generally means that the function being maximized is quite smooth in θ, and so estimation becomes much easier. We think that this advantage has been borne out in practice; the number of parameters being estimated in 4G models, like that of Smets and Wouters (2003) and the new area wide model, is quite large, and one suspects that ML estimation would be quite difficult. There is, however, a cost to Bayesian methods. Although sometimes it is portrayed as a way of "filling in the potholes" of the likelihood surface—for example, in Fernández-Villaverde (2009)—often it is more like a "highway redesign." Unlike penalty functions the use of a prior can severely change the shape of the function being optimized. In particular, if $L(\theta)$ is flat in θ then the choice of prior will become very important in determining the estimated parameter values, and so one needs to have methods for detecting that.

To illustrate that Bayesian methods can easily hide the fact that the data has little to say about the estimation of certain parameters, take the exchange rate equation in the model of Lubik and Schorfheide (2007)

$$\Delta e_t - \pi_t = -(1-\alpha)\Delta q_t - \pi_t^*, \tag{9.23}$$

where e_t is the log of the exchange rate, q_t is the observable (exogenous) terms of trade and π_t^* is the (unobservable) foreign inflation rate. The latter are assumed to be generated as AR(1) processes with parameters ρ_q and ρ_{π^*}, respectively, and uncorrelated shocks (as befits the exogeneity assumption in force for Δq_t). Under these assumptions Equation 9.23 is actually a regression equation, with $\Delta e_t - \pi_t$ as dependent variable, Δq_t as the regressor and with first order serially correlated errors. Hence the FIML and LIML estimators should be close. However, there will be a difference between a Bayesian estimator based on limited and full information when there are informative priors about the other parameters of the system.

Table 9.1 gives the LIML estimates of the parameters of Equation 9.23 using UK data from Lubik and Schorfheide. Also reported are the BFI estimator, which estimates the complete system (this involves imposing a zero correlation between all shocks of the system), and a LI Bayesian (BLI) estimator that imposes only a zero correlation between Δq_t and π_t^*. Two BLI estimators are given depending on whether the prior is assumed to be Beta(a_1, a_2) or $N(a_1, a_2)$. The BFI estimator is performed with Beta priors (for parameters appearing in the remainder of the system priors are those in LubiK and Schorfheide). For the estimation of α the Beta prior has $a_1 = .2, a_2 = .05$, while for $\rho_{\pi^*}, a_1 = .8, a_2 = .5$. The normal priors set $\{a_1 = 0, a_2 = .05\}$ and $\{a_1 = .8, a_2 = .5\}$, respectively.

Now it is clear how important the prior is in changing the results. The β prior used for ρ_{π^*} is fairly close to what is traditionally used in estimating 4G models. With just the BFI results you would never discover that the value most consistent with the data is negative. As the BLI estimates show this is not a question of using a more efficient estimator. To get the Bayesian estimator to reveal the lack of information about α in the data it is necessary to choose the prior so as to encompass a wide range of values for the parameter, but often one sees a very restricted parameter range specified so as to get "sensible values,"mostly ruling out certain signs for the estimates. However, a "wrong

TABLE 9.1

FIML and Bayesian Estimates of the Parameters of Equation 9.23

	α		ρ_{π^*}	
	Mean Est.	**95% Range**	**Mean Est.**	**95% Range**
FIML/LIML	−0.11	−0.56–0.34	0.07	−0.13–0.32
BFI-Beta	0.19	0.12–0.27	0.39	0.39–0.67
BLI-Beta	0.19	0.06–0.31	0.44	0.29–0.59
BLI-normal	0.01	−0.07–0.08	0.08	−0.15–0.31

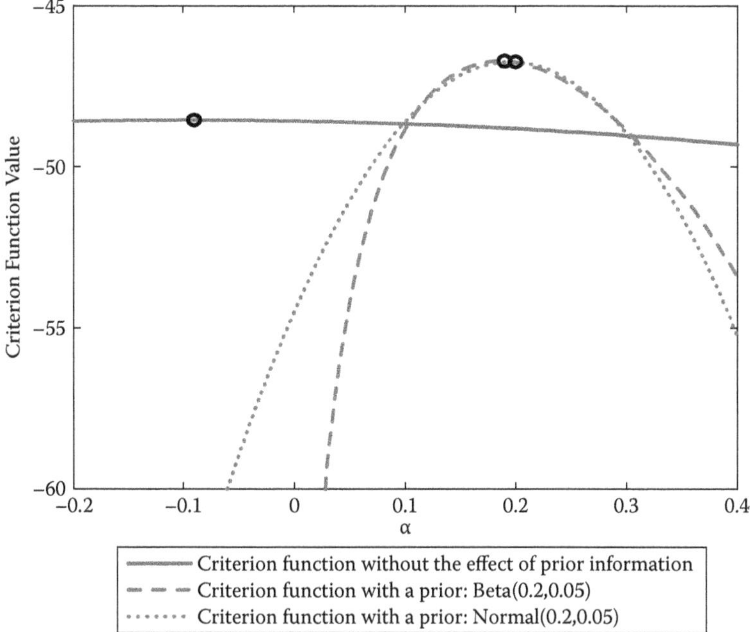

FIGURE 9.1
Log-likelihood and Bayesian mode criterion for α.

sign" can be very informative. In times past it was often taken to suggest that there are specification problems with the equation. In the case of α, a negative value is certainly unattractive, since it is meant to be an import share, but the proper way to interpret the MLE estimate is really that one can not estimate the parameter with any precision, rather than it is negative. What is disturbing about this example is that one does not get any such feeling from the Bayesian estimates, unless one allows for the possibility that the coefficient can easily be negative, as with the last prior. A different way of seeing how the prior has reshaped the surface is in Figure 9.1, which shows how the log-likelihood and the criterion generating the Bayesian modal estimate change with two priors. Notice how the prior can lead to the conclusion that this is a parameter whose value can be determined very precisely.

Although there is nothing surprising in these outcomes, the point is that the Bayesian estimates suggest the opposite, i.e., there seems to be a good deal of information in the sample, as shown by the fact that the mean of the prior for ρ_{π^*} is not contained in the 90% confidence interval for either of the Bayesian estimators. Thus a commonly suggested criterion that there are issues if the posterior and prior distributions coincide would not flag any warnings here. It leads one to ask why one would not just compare the Bayesian modal estimate and its implied ranges for the parameter value to those coming from the MLE as a check on undue influence from the prior? Oddly enough, this information is rarely supplied by those estimating 4G models with Bayesian methods.

It has been observed above that not all the model variables may be observed. This has the effect of potentially making the solved solution *in the observed variables* a VARMA rather than a VAR process.[8] This has to be allowed for when forming the likelihood. It is here that expressing the model and observation information in a state space form is very useful, since the likelihood can be computed recursively (at least when the shocks are normal) using the information provided by the Kalman filter. Most computer programs estimating 4G models use this method, e.g., the DYNARE program. Assuming that the process is a VAR in the observables can lead to quite large biases in the estimates of impulse responses unless there are enough observations to estimate a high order VAR (as that can approximate a VARMA process). For example, Kapetanios, Pagan, and Scott (2007) found that, for a model which was a smaller version of the 4G model BEQM, one needed a VAR(50) to recover the true impulse responses. Otherwise the biases were large when the sample size was that commonly available, around 200 observations, and the VAR order was chosen with standard statistical order selection methods such as BIC and AIC. Of course a VAR(50) is not something that is estimable in sample sizes like 200.

But there are limits to this strategy. One cannot have too many unobserved variables. Strong assumptions may need to be made about variances in order to achieve identification of these parameters if there is a big discrepancy, something that does not seem to be appreciated by many of those applying the methods. For example, it is not enough to follow Smets and Wouters (2003, p. 1140) who say "Identification is achieved by assuming that four of the ten shocks follow a white noise process. This allows us to distinguish those shocks from the persistent 'technology and preference' shocks and the inflation objective shock."

To see the problem that arises with having an excess of unobservables consider the simplest case where there is one observed variable y_t but two unobserved components y_{1t} and y_{2t}. One of these components (y_{1t}) follows an AR(1) with parameter ρ_1 and innovation variance σ_1^2, and the other is white noise ($\rho_2 = 0$) with variance σ_2^2. Then we would have

$$(1 - \rho_1 L)y_t = (1 - \rho_1 L)y_{1t} + (1 - \rho_1 L)y_{2t}, \tag{9.24}$$

and it is clear that, as $\frac{\sigma_2^2}{\sigma_1^2}$ becomes large, it becomes impossible to identify ρ_1. In this case the likelihood is flat in ρ_1, and any prior placed on ρ_1 will effectively determine the value of ρ_1 that results. To avoid this situation a prior would need to be placed on the relative variance and not just the values of ρ_1 and ρ_2, as Smets and Wouters argue. To illustrate this we simulated some data from the setup above and then estimated ρ_1 with a Beta prior centered at

[8] There is a large literature on this and related issues now, e.g., Fernandez-Villaverde et al. (2007). Simple conditions under which this occurs are set out in Fukač and Pagan (2007). Thus in the basic Real Business Cycle model in King, Plosser, and Rebelo (1988), variables such as consumption can be eliminated and the model will remain a VAR, but the capital stock cannot be.

TABLE 9.2

An Example of a Too-Many-Unobservables Model Estimation (Estimates of ρ_1 and 90% Confidence Interval)

	True σ_2^2/σ_1^2		
Prior	**1**	**2**	**5**
$\rho_1 = 0.85$	0.67	0.71	0.80
	[0.49–0.84]	[0.53–0.89]	[0.64–0.94]
$\rho_1 = 0.50$	0.46	0.48	0.49
	[0.32–0.60]	[0.31–0.62]	[0.35–0.65]
$\rho_1 = 0.30$	0.28	0.28	0.29
	[0.12–0.41]	[0.13–0.46]	[0.12–0.44]

Note: We use a Beta prior on ρ_1, with a standard error 0.1. The true value is $\rho_1 = 0.3$. For σ_1 and σ_2 we use an inverse gamma with a mean 1 and standard error 4 as a prior.

different values. The true value of ρ_1 is .3 and Table 9.2 shows the posterior mode for different values of $\frac{\sigma_2^2}{\sigma_1^2}$. It is clear that recovering the true value of ρ_1 is extremely difficult if the type of prior used in many 4G models is adopted.

9.5.4 Handling Permanent Components

Increasingly it has been recognized that there are likely to be permanent components in the data and these must be introduced into the model design in some way. Most commonly this is done by making the log of the level of technology, A_t, an integrated process. Then to keep ratios such as the real capital-output and consumption-output constant in equilibrium, it follows that the permanent components of capital, output, and consumption, must be identical. Obviously the fact that ratios are to be taken as constant in equilibrium implies co-integration between the logs of the variables making up the ratio, and the co-integration vectors have the specific form of $(1 \quad -1)$. To see the implication of such co-integration assume that production is done via a Cobb–Douglas production function of the form $Y_t = K_t^{\alpha}(A_t H_t P_t)^{1-\alpha}$, where H_t is hours worked and P_t is the potential work force. In most models P_t is taken to grow exogenously and it is H_t that fluctuates with the latter being regarded as a stationary process with some average (steady-state) value of H^*. Potential output is then naturally defined as the permanent component of Y_t, Y_t^P. Under the restriction mentioned above that $Y_t^P = A_t^P = K_t^P$,

$$Y_t^P = (A_t^P)^{\alpha}(A_t^P H^* P_t)^{1-\alpha} = A_t^P (H^* P_t)^{1-\alpha}. \tag{9.25}$$

Taking logs and defining an output gap as the transitory component

$$\ln Y_t - \ln Y_t^P = \ln A_t - \ln A_t^P + (1-\alpha)\ln(H_t/H^*), \tag{9.26}$$

shows that the output gap depends upon the transitory component of technology, as well as the deviations of hours from its steady-state value H^*. In the special case when $\ln A_t$ is a pure random walk, the transitory component of $\ln A_t$ is zero. This special case is used quite extensively.

Models exist in the literature where there is more than one permanent component. The presence of more than one generally arises from noticing that the ratios of certain variables cannot be reasonably treated as a constant in the long run. In some instances this lack of stability is due to changes in relative prices. In these cases it is often the nominal rather than the real ratios that appear to be relatively constant, suggesting that the case be handled within a 4G model by employing a second unobservable permanent component that drives the relative prices. An example of a 4G model which incorporates such an adjustment is the KITT model (Beneš et al. 2009).

How does one handle permanent components in the solution and estimation of 4G models? Two strategies are available. One involves formulating the optimization problems used to get the Euler equations of the 4G models in such a way that any $I(1)$ variable appears as a ratio to its permanent component. In this variant, the utility function would be expressed in terms of $\frac{C_t}{C_t^P}$. An example of where this was done is Del Negro and Schorfheide (2008). The second strategy has been to reexpress the Euler equations derived from functions of the levels of the variables in terms of such ratios, e.g., the consumption Euler equation in Equation 9.13 would become

$$\frac{C_t}{C_t^P} = E_t\left[\left(\frac{C_{t-1}}{C_{t-1}^P}\right)^h\left(\frac{C_{t+1}}{C_{t+1}^P}\right)^{1-h} R_{t+1}\frac{(C_{t-1}^P)^h(C_{t+1}^P)^{(1-h)}}{C_t^P}\right]. \tag{9.27}$$

After log linearization this is

$$\zeta_t = h\zeta_{t-1} + E_t\{(1-h)\zeta_{t+1} + R_{t+1} + (1-h)\Delta c_{t+1}^P\} - h\Delta c_t^P, \tag{9.28}$$

where $\zeta_t = \ln C_t - \ln C_t^P$. An assumption now needs to be made concerning how $\Delta a_t = \Delta c_t^P$ is to be generated. In the special case where $\Delta a_t = \Delta \ln A_t = \varepsilon_t^a$ and ε_t^a is white noise, $E_t\Delta c_{t+1}^P = E_t\Delta a_{t+1}^P = E_t\varepsilon_{t+1}^a = 0$.

Which of these two strategies is best is a question that has not been examined much. Certainly they lead to different specifications for the Euler equations of any model. The presence of permanent components in technology makes it highly unlikely that any 4G model can be represented as a VAR and so estimation using the on-model approach is best done within the framework of an SSF. This simply involves specifying $\Delta\zeta_t = \Delta c_t - \Delta c_t^p$ as an observation equation, with Δc_t^P being latent and Δc_t being observed.

Notice that what the above strategy does is to replace any $I(1)$ series with their transitory components or "gaps." Essentially it is performing a multivariate Beveridge–Nelson decomposition of the $I(1)$ variables into their permanent and transitory components. However, often one sees a second strategy, which involves an "off-model" approach wherein permanent components are removed from variables by a filter that is not model consistent. By far the most popular would be the Hodrick–Prescott (HP) filter. Econometrically using off-model filters is a bad idea. To see this consider the consequences of working with HP filtered data. To assess these we note that the HP filter is a two-sided filter which, when applied to a variable y_t, produces a transitory component of $\Sigma_{j=-T}^{j=T}\omega_j \Delta y_{t-j}$.[9] Now, if this component is used as a regressor, the fact that it involves Δy_{t+j} at time t means that one would get inconsistent estimators of the parameters attached to the gaps. Moreover, the correlation of the regressor with Δy_{t+j} is likely to contaminate estimators of other parameters. Even if one used a one-sided version of this filter it is well known – (see Harvey and Jaeger [1993] and Kaiser and Maravell [2002]) that the filter is designed to extract a permanent component from a series that is $I(2)$, not one that is $I(1)$, and hence it is not model consistent unless $\ln A_t$ is $I(2)$; see Fukač and Pagan (2010) for more details. Few 4G modelers are prepared to make that assumption.

9.5.5 Evaluation Issues

Evaluation really has two dimensions to it. One concentrates on the operating characteristics of the model and whether these are "sensible." The other is more about the ability of the model to match the data along a variety of dimensions. The two themes are not really independent but it is useful to make the distinction. Thus it might be that while a model could produce reasonable impulse responses, it may not produce a close match to the data, and conversely.

9.5.5.1 Operating Characteristics

Standard questions that are often asked about the operating characteristics of the model are whether the impulse responses to selected shocks are reasonable and what the relative importance of various shocks are to the explanation of (say) output growth. Although the latter is often answered by recourse to variance decompositions, perhaps a better question to ask is how important the assumptions made about the dynamics of shocks are to the solutions, as it seems crucial to know how much of the operating characteristics and fit to data comes from the economics and how much from exogenous assumptions. This concern stems back at least to Cogley and Nason (1993) who argued that standard RBC models produced weak dynamics if shocks were not highly serially correlated. It would seem important that one investigate this question by examining the impact of setting the serial correlation in the shocks to zero.

The appropriate strategy for assessing operating characteristics depends on whether the model parameters have been formally or informally quantified. If done informally researchers such as Amano et al. (2002) and Canova (1994) have asked the question of whether there is a set of such parameters that would be capable of generating some of the outcomes seen in the data, e.g.,

[9] Simulating data from y_t when it is a pure random walk, and then regressing the measured HP transitory component ($\lambda = 1600$) on to $\Delta y_{t\pm j}$, $j = 0, \ldots, 10$, gives an R^2 of .98 and $\omega_0 = .47$, ω_j $(j = 1.,, 10) = \{.42, .37, .32, .27, .23, .18, .15, .15, .11, .09, .06\}$. It is also the case that $\omega_j = w_{|j|-1}$, $j = -1, \ldots, -10$. When future values of Δy_t were dropped from the regression the R^2 dropped to .5, emphasising the importance of future values of Δy_t in the determination of the HP transitory component.

ratios ϕ such as (say) the consumption-income ratio. This ratio is a function of the model parameters θ. The existing value used for θ in the model, θ^*, is then taken as one element in a set and a search is conducted over the set to see what sort of variation would occur in the resulting value of ϕ. If it is hard to reproduce the observed value of ϕ in the data, $\hat{\phi}$, then the model might be regarded as suspect. In this approach possible values of model parameters are selected to trace out a range of values of ϕ. An efficient way of doing this search is a pseudo-Bayesian one in which trial values of θ are selected from a multivariate density constructed to be consistent with the potential range of values of θ. This enables the corresponding density for ϕ to be determined. If the observed value $\hat{\phi}$ lies too far in the tails of the resulting density of ϕ, one would regard the model as inadequately explaining whatever feature is summarized by ϕ. A second approach treats the parameter values entered into the model, θ^*, as constant and asks whether the estimate $\hat{\phi}$ is close to the value $\phi^* = \phi(\theta^*)$ implied by the model. This is simply an encompassing test of the hypothesis that $\phi = \phi^*$.

9.5.5.2 Matching Data

Since 4G models are structural models there are many tests that could be carried out regarding their implied co-integrating and co-trending vectors, adequacy of the individual equations, etc. Moreover many of the old techniques used in 1G and 2G models, such as an examination of the tracking performance of the model, might be applied. But there are some issues which are specific to 4G models that need to be addressed in designing such tests.

In the first and second generation of models a primary way of assessing their quality was via historical simulation of them under a given path for any exogenous variables. It would seem important that we see such model tracking exercises for 4G models, as the plots of the paths are often very revealing about model performance, far more than might be gleaned from any examination of just a few serial correlation coefficients and bivariate correlations, which has been the standard way of looking at 4G model output to date. It is not that one should avoid computing moments for comparison, but it seems to have been overdone in comparison to tests that focus more on the uses of these models such as forecasting (which is effectively what the tracking exercise is about).

Now a problem arises in doing such exercises for 4G models. If the model's shocks are taken to be an integral part of it then there is no way to assess the model's tracking ability, since the shocks always adjust to produce a perfect match to the data. Put another way, there is no such thing as a residual in 4G models. The only exception to that is when we explicitly allow for tracking shocks, as described earlier, and this technology has sometimes been used to examine the fit. The main difficulty in doing so is the assumption used in setting up the SSF that the tracking shocks and model shocks are uncorrelated (since one cannot generally estimate such a parameter from the likelihood). Some relaxation of this assumption is needed, i.e., an auxiliary criterion needs to be supplied that can be used to set a value for the correlation. Watson (1993) suggested that one find the correlation that minimized the gap between the spectra of the model and the data, as that produces the tracking outcome most favorable to the 4G model. Oddly enough Watson's approach does not seem to have been used much, although it is obviously a very appealing way of getting some feel for how well a 4G model is performing.

Rather than focus on tracking one might ask whether the dynamics are adequately captured by the model. One way to examine this is to compare the VAR implied by the model with that in the data. Canova, Finn, and Pagan (1994) proposed this. In small models this seems to be a reasonable idea but, in large models, it is unlikely to be very useful, as there are just too many coefficients to fit in the VAR. Consequently, the test is likely to lack power. Focusing on a subset of the VAR coefficients might be instructive. Thus Fukač and Pagan (2010) suggest a comparison of $E_t z_{t+1}$ generated from the model with that from a VAR. As there are only a few expectations in most 4G models this is likely to result in a more powerful test and has the added advantage of possessing some economic meaning. They found that the inflation expectations generated by the Lubik and Schorfheide (2007) model failed to match those from a VAR fitted to UK data.

A different way of performing "parameter reduction" that has become popular is due to Del Negro et al. (2007) — the so-called DSGE-VAR approach. To explain this in a simple way consider the AR(1) equation

$$z_t = \rho z_{t-1} + e_t, \tag{9.29}$$

where e_t has variance of unity. Now suppose that a 4G model implies that $\rho = \rho_0$, and that the variance of the shock is correctly maintained to be unity. Then we might think about estimating ρ using a prior $N(\rho_0, \frac{1}{\lambda T})$, where T is the sample size. As λ increases we will end up with the prior concentrating upon ρ_0 while, as it tends to zero, the prior becomes very diffuse. In terms of the criterion used to get a Bayesian modal estimate this would mean that the likelihood will be a function of ρ but the other component of the criterion — the log of the prior — would depend on λ. Hence we could choose different λ and see which produces the highest value of the criterion (or even the highest value of the density of z_t when ρ is replaced by its various model estimates as λ varies). For a scalar case this is not very interesting as we would presumably choose the λ that reproduces the OLS estimate of ρ (at least in large samples) but in a multivariate case this is not so. Basically the method works by reducing the VAR parameters down to a scalar measure, just as in computing expectations. As λ varies one is effectively conducting a sensitivity analysis.

9.6 Conclusion

The chapter has looked at the development of macroeconometric models over the past sixty years. In particular the models that have been used for analysing policy options. We argue that there have been four generations of these. Each generation has evolved new features that have been partly drawn from the developing academic literature and partly from the perceived weaknesses in the previous generation. Overall, the evolution has been governed by a desire to answer a set of basic questions and sometimes by what can be achieved using new computational methods. We have spent a considerable amount of time on the final generation of models, exploring some of the problems that have arisen in how these models are implemented and quantified. It is unlikely that there will be just four generations of models. Those who work with them know that they constantly need to be thinking about the next generation in order to respond to developments in the macroeconomy, to new ideas about the interaction of agents within the economy, and to new data sources and methods of analyzing them.

References

Adolfson, M., S. Laseen, J. Linde, and M. Villani. 2007. RAMSES: A New General Equilibrium Model for Monetary Policy Analysis. *Riksbank Economic Review* No. 2, 2007.

Almon, S. 1965. The Distributed Lag between Capital Appropriations and Expenditures. *Econometrica* 32:178–196.

Altug, S. 1989. Time to Build and Aggregate Fluctuations: Some New Evidence. *International Economic Review* 30:889–920.

Amano, R., K. McPhail, H. Pioro, and A. Rennison. 2002. Evaluating the Quarterly Projection Model: A Preliminary Investigation. *Bank of Canada Working Paper 2002-20.*

Beneš, J., A. Binning, M. Fukač, K. Lees and T. Matheson. 2009. *K.I.T.T.: Kiwi Inflation Targeting Technology*. Reserve Bank of New Zealand, Wellington.

Berg, A., P. Karam, and D. Laxton. 2006. A Practical Model Based Approach to Monetary Analysis – Overview. *IMF Working Paper 06/80.*

Binder, M., and M. H. Pesaran. 1995. Multivariate Rational Expectations Models and Macroeconomic Modelling: A Review and Some New Results. In *Handbook of Applied Econometrics: Macroeconomics.* M. H. Pesaran and M Wickens (eds)., Oxford, U.K.: Basil Blackwell.

Black, R., D. Laxton, D. Rose, and R. Tetlow. 1994. The Bank of Canada's New Quarterly Projection Model Part 1: The Steady-State Model: SSQPM. *Technical Report No 72,* Bank of Canada, Ottawa.

Black, R., V. Cassino, A. Drew, E. Hansen, B. Hunt, D. Rose, and A. Scott. 1997. The Forecasting and Policy System: The Core Model. *Reserve Bank of New Zealand Research Paper 43*, Wellington.

Blanchard, O. J. 1985. Debt, Deficits and Finite Horizons. *Journal of Political Economy* 93:223–247.

Brayton, F., and E. Mauskopf. 1985. The Federal Reserve Board MPS Quarterly Econometric Model of the U.S. Economy. *Economic Modelling* 2:170–292.

Brayton, F. and P. Tinsley. 1996. A Guide to FRB-US: A Macroeconomic Model of the United States. *Finance and Economics Discussion Paper Federal Reserve Bank Board of Governors 96/42*, Washington, D.C.

Brubakk, L., T. A. Husebo, J. Maih, K. Olsen, and M. Ostnor. 2006. Finding NEMO: Documentation of the Norwegian Economy Model. Norges Bank Staff Memo No. 2006/6, Oslo.

Canova, F. 1994. Statistical Inference in Calibrated Models. *Journal of Applied Econometrics* 9:S123–S144.

Canova, F., M. Finn, and A. Pagan. 1994. Evaluating a Real Business Cycle Model. In *Non-Stationary Time Series Analysis and Co-Integration.* C. Hargreaves (ed.). 225–255. Oxford, U.K.: Oxford University Press.

Canova, F., and L. Sala. 2009. Back to Square One: Identification Issues in DSGE Models. *Journal of Monetary Economics* 56:431–449.

Clarida, R., J. Gali, and M. Gertler. 1999. The Science of Monetary Policy. *Journal of Economic Literature* XXXVII:1661–1707.

Cogley, T., and J. M. Nason. 1993. Impulse Dynamics and Propagation Mechanisms in a Real Business Cycle Model. *Economics Letters* 43:77–81.

Coletti, D., B. Hunt, D. Rose, and R. Tetlow. (1996). The Bank of Canada's New Quarterly Projection Model, Part 3: The Dynamic Model: QPM *Technical Report 75*, Bank of Canada, Ottawa.

Davidson, J. E. H., D. F. Hendry, F. Srba, and S. Yeo. 1978. Econometric Modelling of the Aggregate Time-Series Relationship between Consumers' Expenditure and Income in the United States. *Economic Journal* 88:661–692.

Del Negro, M., and F. Schorfheide. 2008. Inflation Dynamics in a Small Open Economy Model under Inflation Targeting: Some Evidence from Chile. In *Monetary Policy under Uncertainty and Learning* K. Schmidt-Hebbel and C. E. Walsh (eds). 11th Annual Conference of the Central Bank of Chile, Santiago.

Del Negro, M., F. Schorfheide, F. Smets, and R. Wouters. 2007. On the Fit and Forecasting Performance of New Keynesian Models. *Journal of Business and Economic Statistics* 25:123–143.

Dixit, A. K., and J. E. Stiglitz. 1977. Monopolistic Competition and Optimum Product Diversity. *American Economic Review* 67:297–308.

Erceg, C. J., L. Guerrieri, and C. Gust. 2006. SIGMA: A New Open Economy Model for Policy Analysis. *International Journal of Central Banking* 2:1–50.

Fair, R. C. 1974. *A Model of Macroeconomic Activity, Volume 1: The Theoretical Model.* Cambridge, MA: Ballinger Publishing Company.

Fernández-Villaverde, J. 2009. The Econometrics of DSGE Models. *NBER Discussion Paper 14677.*

Fernández-Villaverde, J., J. Rubio-Ramirez, T. Sargent, and M. W. Watson. 2007. The ABC (and D) of Structural VARs. *American Economic Review* 97:1021–1026.

Fujiwara, I., N. Hara, Y. Hirose, and Y. Teranishi. 2004. The Japanese Economic Model: JEM. *Bank of Japan Working Paper No. O4-E-3*, Tokyo.

Fukač, M., and A. R. Pagan. 2007. Commentary on an Estimated DSGE Model for the United Kingdom. *Federal Reserve Bank of St Louis Review*, 89: 233–240.

Fukač, M., and A. R. Pagan. 2010. Limited Information Estimation and Evaluation of DSGE Models. *Journal of Applied Econometrics* 25:55–70.

Gramlich, E. M. 2004. Remarks. Paper presented to the Conference on Models and Monetary Policy, Federal Reserve Bank Board of Governors. http://www.federalreserve.gov/boarddocs/speeches/2004/20040326/default.htm

Hannan, E. J. 1971. The Identification Problem for Multiple Equation Systems with Moving Average Errors. *Econometrica* 39:751–765.

Harrison, R., K. Nikolov, M. Quinn, G. Ramsay, A. Scott, and R. Thomas. 2005. The Bank of England Quarterly Model. London: Bank of England. http://www.bankofengland.co.uk/publications/beqm/.

Harvey, A. C., and A. Jaeger. 1993. De-Trending, Stylized Facts and the Business Cycle. *Journal of Applied Econometrics* 8:231–247.

Helliwell, J. F., G. R. Sparks, F. W. Gorbet, H. T. Shapiro, I. A. Stewart, and D. R. Stephenson. 1971. The Structure of RDX2 Bank of Canada Staff Study No. 7 Bank of Canada, Ottawa.

Hicks, J. R. 1937. Mr. Keynes and the "Classics", A Suggested Interpretation. *Econometrica* 5:147–159.

Ireland, P. 2004. A Method for Taking Models to the Data. *Journal of Economic Dynamics and Control* 28:1205–1226.

Iskrev, N. 2007. Evaluating the Information Matrix in Linearized DSGE Models. *Economics Letters* 99:607–610.

Iskrev, N. 2009. Local Identification in DSGE Models. *Banco de Portugal Working Paper* No. 7 Banco de Portugal, Lisbon.

Johansen, L. 1960. *A Multi-Sectoral Study of Economic Growth*. 2nd ed. Amsterdam: North-Holland.

Johansen, S. 2005. What Is the Price of Maximum Likelihood. Paper presented to the Model Evaluation Conference, Oslo, May 2005.

Kai, Ch., G. Coenen, and A. Warne. 2008. The New Area-Wide Model of the Euro Area: A Micro-Founded Open-Economy Model for Forecasting and Policy Analysis. *ECB Working Paper* No. 944.

Kaiser, R., and A. Maravall. 2002. A Complete Model-Based Interpretation of the Hodrick-Prescott Filter: Spuriousness Reconsidered. *Working Paper 0208*, Banco de España, Madrid.

Kapetanios,G., A. R. Pagan, and A. Scott. 2007. Making a Match: Combining Theory and Evidence in Policy-Oriented Macroeconomic Modeling, *Journal of Econometrics* 136:505–594.

Kilponen, J., and A. Ripatti. 2005. Labour and Product Market Competition in a Small Open Economy – Simulation Results Using a DGE Model of the Finnish Economy. *Working Paper* No. 5, Bank of Finland, Helsinki.

King, R. G., C. I. Plosser, and S. T. Rebelo. 1988. Production, Growth, and Business Cycles: I. The Basic Neoclassical Model., *Journal of Monetary Economics* 21:195–232.

Komunjer, I., and S. Ng. 2009. Dynamic Identification of DSGE Models. Columbia University. www.columbia.edu/sn2294/papers/kn_spectral.pdf (accessed 16 March 2010).

Laxton, D. and P. Pesenti. 2003. Monetary Policy Rules for Small, Open, Emerging Economies. *Journal of Monetary Economics* 50:1109–1146.

Lubik, T. A. and F. Schorfheide. 2007. Do Central Banks Respond to Exchange Rate Movements: A Structural Investigation. *Journal of Monetary Economics* 54:1069–1087.

Mavroeidis, S. 2004. Weak Identification of Forward-Looking Models in Monetary Economics. *Oxford Bulletin of Economics and Statistics* 66:609–635.

McKibbin W. J. 1988. Policy Analysis with the MSG2 Model. *Australian Economic Papers*, supplement, pp. 126–150.

McKibbin, W. J., and J. D. Sachs 1989. The McKibbin-Sachs Global Model: Theory and Specification. *NBER Working Paper* 3100.

Medina, J. P., and C. Soto. 2006. Model for Analysis and Simulations: A Small Open Economy DSGE for Chile. Paper presented at the Central Bank of Chile Workshop on Macroeconomic Modeling in Central Banks. http://www.bcentral.cl/conferencias-seminarios/otras-conferencias/pdf/modelling2006/soto_medina.pdf.

Murchison S., and A. Rennison. 2006. ToTEM: The Bank of Canada's New Quarterly Projection Model. *Bank of Canada Technical* Report No. 97.

Murphy, C. W. 1988. An Overview of the Murphy Model. In *Macroeconomic Modelling in Australia, Australian Economic Papers*. M. E. Burns and C.W. Murphy (eds), pp. 61–8.

Nason, J. M. and G. W. Smith. 2005. Identifying the New Keynesian Phillips Curve. *Working Paper 2005-1*, Federal Reserve Bank of Atlanta.

Nickell, S. 1985. Error Correction, Partial Adjustment and All That. *Oxford Bulletin of Economics and Statistics* 47:119–130.

Powell, A. A. and C. W. Murphy. 1995. *Inside a Modern Macroeconometric Model*. Lecture Notes on Economics and Mathematical Systems 428, Berlin: Springer.

Preston, A. J. 1978. Concepts of Structure and Model Identifiability for Econometric Systems. In *Stability and Inflation*. A. R. Bergstrom *et al.* (eds). Chichester: Wiley pp. 275–297.

Ramsey, F. P. 1928. A Mathematical Theory of Saving. *Economic Journal* 38:543–559.

Rotemberg, J. J. 1982. Monopolistic Price Adjustment and Aggregate Output. *Review of Economic Theory* 114:198–203.

Rothenberg, T. 1971. Identification in Parametric Models. *Econometrica* 39:577–591.

Sargan, J. D. 1964. Wages and Prices in the United Kingdom: A Study in Econometric Methodology. In *Econometric Analysis for National Economic Planning*. P. E. Hart, G. Mills, and J. K. Whitaker (eds). London: Butterworth. pp. 22–54.

Schmitt-Grohe, S., and M. Uribe 2003. Closing Small Open Economy Models. *Journal of International Economics* 61:163–185.

Schorfheide, F. 2000. Loss Function-Based Evaluation of DSGE Models. *Journal of Applied Econometrics* 15:645–670.

Smets, F., and R. Wouters. 2003. An Estimated Dynamic Stochastic General Equilibrium Model of the Euro Area. *Journal of the European Economic Association* 1:1123–1175.

Solow, R. M. 1956. A Contribution to the Theory of Economic Growth. *Quarterly Journal of Economics* 70:65–94.

Swan, T. W. 1956. Economic Growth and Capital Accumulation. *Economic Record* 32:334–361.

Wallis, K. 1988. Some Recent Developments in Macroeconometric Modelling in the United Kingdom. *Australian Economic Papers* 27:7–25.

Wallis, K. F. 1995. Large Scale Macroeconometric Modelling. In *Handbook of Applied Econometrics*, Volume 1: Macroeconomics. M. H. Pesaran and M. R. Wickens (eds). Oxford: Blackwell.

Wallis, K. F. and J. D. Whitley. 1991. Macro Models and Macro Policy in the 1980s. *Oxford Review of Economic Policy* 7:118–127.

Watson, M. W. 1993. Measures of Fit for Calibrated Models. *Journal of Political Economy* 101:1011–1041.

Yaari, M. E. 1965. Uncertainty Lifetime, Life Insurance and the Theory of Consumer. *Review of Economic Studies* 32:137–1350.

10

Forecasting with Interval and Histogram Data: Some Financial Applications

Javier Arroyo, Gloria González-Rivera, and Carlos Maté

CONTENTS

10.1 Introduction

In economics we customarily deal with classical data sets. When we collect information on a set of variables of interest, either in a cross-sectional or/and time series framework, our sample information is a collection of data points $\{y_i\}, i = 1 \ldots n$ or $\{y_t\}, t = 1 \ldots T$ where y_i or $y_t \in \mathbb{R}$ takes a single value in $\mathbb{R}$. In many instances, the single value is the result of an aggregation procedure, spatial or temporal, over information collected at a very disaggregated level. Some pertinent examples follow.

In financial markets the price of an asset (stocks, bonds, exchange rates, etc.) is observed at a very high frequency, i.e., tick by tick; however, there is a huge number of studies where the analysis is performed at the daily frequency using the closing price, or even at lower frequencies such as weekly or monthly. It may be claimed that tick-by-tick pricing will generate a huge amount of data from which it will be difficult to discriminate information from noise, but on the other extreme, by analyzing just closing prices we will be discarding valuable intraday information. We can think of alternative ways of collecting information, for instance, we can gather the maximum and minimum prices in a day so that the information to be analyzed will come in an interval format; or the daily interquartile prices such that the interval will run from the price at the 25% quartile to the price at the 75% quartile; or we can construct daily histograms with all the intraday prices. In these cases the data point is no longer a single value but a collection of values represented by the daily low/high interval, or the interquartile interval, or the daily histogram. The intervals or the histograms, when indexed by time, will constitute an interval time series or a histogram time series.

Another instance refers to the information collected by national statistical institutes in relation to income and population dynamics. Census surveys provide socioeconomic information on all individuals in a nation that is customarily disseminated in an aggregated format, for instance a time series of average income per capita. The objective of these national surveys is not to follow the dynamics of single individuals, which most likely will be different from one period to the next, but the dynamics of a collective. However, summarizing national information by averages, though informative, is a poor approach that throws away the internal variation provided by the disaggregated information about the single units. Once more, disseminating the data in a richer format such as intervals or histograms will provide a more complete picture of income and population dynamics. There are many other areas such as marketing, environmental sciences, quality control, medical sciences, etc. in which the information is rich enough to make the object of analysis not the single-valued variable but the interval-valued or the histogram-valued variable.

Interval- and histogram-valued data can be classified as symbolic data sets as opposed to classical data sets. Symbolic data is a proposal to deal with the massive information contained in nowadays super large data sets found across many disciplines. While the analysis of these data sets requires some summary procedure to bring them to a manageable size, the objective is to retain as much of their original knowledge as possible. An extensive review of this new field, which started in the late 1980s and early 1990s, is provided by Billard and Diday (2003, 2006), who define the complexity of symbolic data, review the current methods of analysis and state the challenges that lie ahead.

Economics and business are disciplines in which data sets are becoming consistently larger due to sophisticated information systems that collect and store huge amount of data. However, the development of new methodologies to deal with the characteristics of large data sets is moving at a slower pace. A case on time is the aforementioned high-frequency financial data and the challenges brought by it such as irregularly spaced observations with strong intraday patterns and a complex dependence structure. There are other examples in the economics literature that emphasize the richness of the data, though eventually the analysis is performed within the boundaries of classical inferential methods. For instance, the article by Zellner and Tobias (2000) provides the time series of the median and interquartile range of the industrial production growth rates of 18 countries but eventually the authors focus on the single-valued time series of the median growth rates. The article by González-Rivera, Lee, and Mishra (2008) presents a stylized time series of cross-sectional returns of the constituents of the SP500 index grouped in histograms (see Figure 10.1). However, the authors focus on the dependence structure of the single-valued time series of the time-varying cross-sectional ranks (VCR). Both of these instances could be viewed from the perspective of symbolic data: in Zellner and Tobias (2000) the data is an interval-valued time series and in González-Rivera, Lee, and Mishra (2008) is a histogram-valued time series.

There is an emergent literature in economics and statistics dealing with interval-valued data in a regression framework. Manski and Tamer (2002) examined a regression model where some regressors are interval-valued, like interval wealth and income, and some others are point-valued. Lima Neto

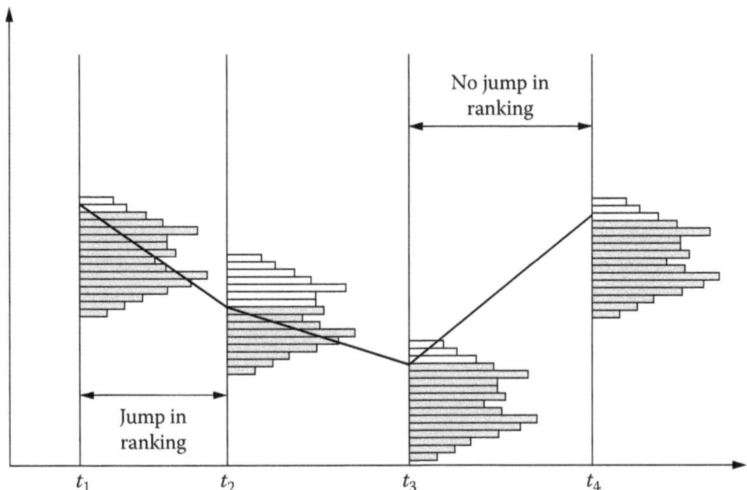

FIGURE 10.1
Stylized time series of the histograms of the cross-sectional returns of the constituents of the SP500 index. (From González-Rivera, G., T.-H. Lee, and S. Mishra. 2008. Jumps in cross-sectional rank and expected returns: a mixture model. *Journal of Applied Econometrics* 23:585–606.)

and de Carvalho (2010) proposed a constrained linear regression model for interval-valued data. Maia, de Carvalho, and Ludermir (2008) implemented ARIMA and neural networks models to forecast the center and radii of intervals. Han et al. (2008) analyzed the sterling-dollar exchange rate time series based on an interval linear model. Cheung, Cheung, and Wan (2009) analyzed the range of daily stock prices by proposing a VECM for the daily interval of high and low prices. García-Ascanio and Maté (2010) forecast monthly electricity demand with hourly interval data. A different approach to regression that treats intervals as convex compact random sets is proposed in González-Rodríguez et al. (2007) and Blanco et al. (2008). Regression models with histogram-valued data are almost nonexistent so that they offer wide opportunities for further research.

This chapter focuses on the forecasting of interval and histogram-valued data. The surveys and review articles by Diday and his coauthors focus on descriptive and multivariate methods of analysis adapted from the classical statistical methodology. To our knowledge, the development of forecasting methods for interval and histogram-valued data is in its infancy so that this chapter is a contribution to that end. We start with a preliminary section defining the structure of the data and basic descriptive statistics. There are two main sections, one for interval data and another for histogram data. In the first, we review how classical regression methods can be adapted to analyze intervals. The main insight is that the interval can be defined by its center and radius or by its minimum and maximum, so that we construct two time series to which classical methods can be applied. In this vein, we build a system, either VAR or VEC models, from which an interval forecast will be obtained. In a different approach based on the arithmetic of intervals and on notions of distances between intervals, we adapt classical filtering techniques like the exponential smoothing and nonparametric techniques like the k-Nearest Neighbors (k-NN) algorithm to produce the interval forecast. In the second main section, we deal with histogram-valued data. In this case the object of analysis is considerably more difficult to analyze and we focus exclusively on the adaptation of smoothing techniques and the k-NN. To construct a histogram forecast, we will not base our operations on the arithmetic of histograms but on the key idea of the "barycentric" histogram as the "average" measure. We should stress that no attempt has been made, either with a time series of intervals or histogram, to uncover the data generating mechanism but rather to forecast the future under the premise that it should not be very far from some average (weighted or unweighted) of the past.

10.2 Interval Data

In this section, we will define interval data and the interval random variable. As a foundation for the forthcoming analysis, we succinctly introduce the algebra of intervals. We will focus on the empirical first and second moments of the interval random variable. The main objective of this section is to discuss (1) regression analysis with interval data, and (2) the forecasting problem. A financial application will showcase the contribution of (1) and (2) to the modeling of economic and financial data. While we will not discuss the nature of interval data, we acknowledge that there are many reasons why interval data may arise. Among others, interval data are generated when the data collection process genuinely produces intervals, or when there are not exact numerical values to quantify a variable, or when there is uncertainty of any kind in the values of the variable, or when variability of a variable is the focus of analysis, or when the measurement tools produce measurement errors. Regardless of the origin, the researcher will be facing data that come with an interval format and this is the primary object of analysis.

10.2.1 Preliminaries

We start with the basic notion of an interval following Kulpa (2006). Let $(E, \leq)$ be a partially ordered set. An interval is generally defined as follows:

Definition 10.1 *An interval* $[a]$ *over the base set* $(E, \leq)$ *is an ordered pair* $[a] = [a_L, a_U]$, *where* $a_L, a_U \in E$ *are the endpoints or bounds of the interval such that* $a_L \leq a_U$.

The interval is called degenerate when $a_L = a_U$, in which case the interval reduces to a point. An interval is the set of elements bounded by the endpoints, these included, namely, $[a] = \{e \in E \mid a_L \leq e \leq a_U\}$. When the base set E is the set of real numbers $\mathbb{R}$, the intervals are subsets of the real line $\mathbb{R}$.

An equivalent representation of an interval is given by the center (midpoint) and radius (half range) of the interval, namely, $[a] = \langle a_C, a_R \rangle$, where $a_C = (a_L + a_U)/2$ and $a_R = (a_U - a_L)/2$.

10.2.1.1 Basic Interval Arithmetic

In order to proceed with our analysis we need an algebra to operate with intervals. Basic interval arithmetic (Moore 1966; Moore, Kearfott, and Cloud 2009) is based on the following principle: let $[a]$ and $[b]$ be two intervals and $\Box$ be an arithmetic operator, then $[a]\Box[b]$ is the smallest interval which contains $a \Box b$, $\forall a \in [a]$ and $\forall b \in [b]$. Interval addition, subtraction, multiplication and division are particular cases of this principle and are defined by

$$[a] + [b] = [a_L + b_L, a_U + b_U] \tag{10.1}$$

$$[a] - [b] = [a_L - b_U, a_U - b_L] \tag{10.2}$$

$$[a] \cdot [b] = [\min\{a_L \cdot b_L, a_L \cdot b_U, a_U \cdot b_L, a_U \cdot b_U\}, \max\{a_L \cdot b_L, a_L \cdot b_U, a_U \cdot b_L, a_U \cdot b_U\}] \tag{10.3}$$

$$[a]/[b] = [a] \cdot (1/[b]), \text{ with } 1/[b] = [1/b_U, 1/b_L]. \tag{10.4}$$

It is worth noting that interval arithmetic subsumes the classical one, in the sense that, if the operands are degenerate intervals, the result of interval operations will be equal to the result obtained by the single number arithmetic. In interval arithmetic, addition and multiplication satisfy the associative and commutative properties. The distributive property does not always hold, but the subdistributive property is satisfied, which is defined as

$$[a]([b]+[c]) \subseteq [a][b]+[a][c]. \tag{10.5}$$

If $[a]$ is a degenerate interval, then this property becomes the distributive property. The interval arithmetic is key for the development of regression techniques and for the adaptation of forecasting methods to interval data.

10.2.1.2 Interval Random Variable

We proceed with the definition of an interval random variable. Let $(\Omega, \mathcal{F}, P)$ be a probability space, where Ω is the set of elementary events, $\mathcal{F}$ is the σ-field of events and $P : \mathcal{F} \to [0, 1]$ the σ-additive probability measure; and define a partition of Ω into sets $A(x)$ such $A_X(x) = \{\omega \in \Omega | X(\omega) = x\}$, where $x \in [x_L, x_U]$, then:

Definition 10.2 *A mapping $X : \mathcal{F} \to [x_L, x_U] \subset \mathbb{R}$, such that for all $x \in [x_L, x_U]$ there is a set $A_X(x) \in \mathcal{F}$, is called an interval random variable.*

10.2.1.3 Descriptive Statistics

The descriptive statistics of an interval random variable are proposed by Bertrand and Goupil (2000). For an interval random variable X, suppose that we have a sample of m individuals ($i = 1, 2, \dots, m$) and for each i, an interval data point $[x]_i \equiv [x_{Li}, x_{Ui}]$. A key assumption for the forthcoming descriptive statistics is that the values in a given interval, i.e., $x_{Li} \leq x \leq x_{Ui}$, are uniformly distributed within the interval. Furthermore, we assume that each individual has the same probability $1/m$ of being observed. Then, the *empirical density function* $f_X(x)$ is a mixture of m uniform distributions

$$f_X(x) = \frac{1}{m} \sum_{i:x\in[x]_i} \frac{I(x \in [x]_i)}{\| [x]_i \|} = \frac{1}{m} \sum_{i:x\in[x]_i} \frac{1}{x_{Ui} - x_{Li}} \qquad x \in \mathbb{R}, \tag{10.6}$$

where $I(x \in [x]_i)$ is an indicator function that takes the value 1 when $x \in [x]_i$ and zero otherwise; and $\| \ [x]_i \ \|$ is the length of the interval $[x]_i$.

Based on the density function (Equation 10.6), the sample mean is obtained by solving the following integral

$$\bar{X} = \int_{-\infty}^{\infty} x f(x) dx = \frac{1}{m} \sum_{i:x \in [x]_i} \frac{1}{x_{Ui} - x_{Li}} \int_{x_{Li}}^{x_{Ui}} x dx$$
$$= \frac{1}{2m} \sum_i (x_{Ui} + x_{Li}) = \frac{1}{m} \sum_i x_{C_i}, \tag{10.7}$$

concluding that the sample mean of an interval random variable is the average of the centers of the intervals in the sample. Analogously, the sample variance is calculated by solving the integral

$$S_X^2 = \int_{-\infty}^{\infty} (x - \bar{X})^2 f(x) dx = \left(\int_{-\infty}^{\infty} x^2 f(x) dx \right) - \bar{X}^2, \tag{10.8}$$

which can be rewritten in terms of the interval bounds as

$$S_X^2 = \frac{1}{3m} \sum_i (x_{Ui}^2 + x_{Ui} x_{Li} + x_{Li}^2) - \frac{1}{4m^2} \left[\sum_i (x_{Ui} + x_{Li}) \right]^2. \tag{10.9}$$

The sample variance combines the variability of the centers as well as the variability within each interval. When the interval is degenerate, both sample moments, the mean and the variance, collapse to the sample mean and variance of the classical data.

10.2.2 The Regression Problem

Now suppose that we have two interval random variables Y and X for which we collect a sample of intervals $([x]_i, [y]_i)$ for $i = 1, 2, \ldots, m$. The interval data point i is a rectangle centered in the centers of $[x]_i$ and $[y]_i$ and whose sides are equal to the length of the respective intervals. A graphical representation of this data is provided in Figures 10.2 to 10.4. In this section, we review the analysis of a regression model with interval data. The classical regression model can be adapted to interval data by focusing on the centers of the interval, or on the maximum and minimum of the interval, or on the center and radius of the interval. The advantage of this approach is that statistical inference is readily available.

The simplest approach to estimate a regression model with interval data is provided by Billard and Diday (2000). It consists of fitting a regression line to the centers of the intervals, $y_{Ci} = \beta' x_{Ci} + \varepsilon_{Ci}$, so that the objective function to minimize is

$$\min_{\hat{\beta}} \sum_i \hat{\varepsilon}_{Ci}^2 = \sum_i (y_{Ci} - \hat{\beta}' x_{Ci})^2, \tag{10.10}$$

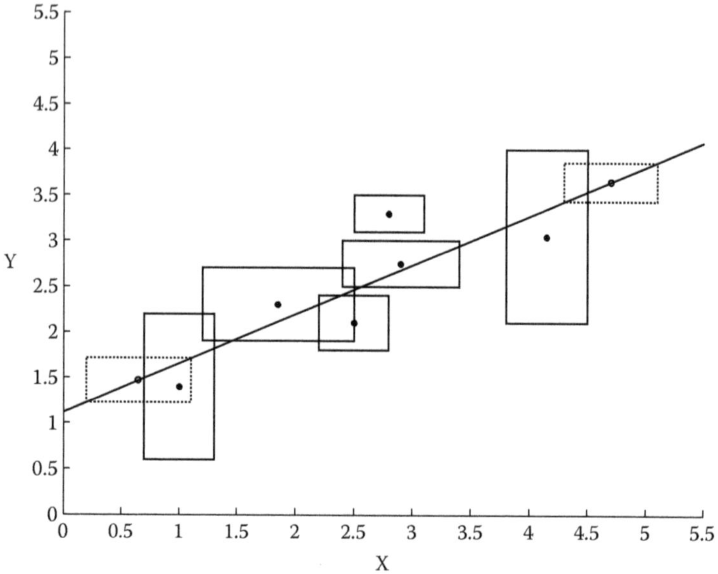

FIGURE 10.2
Fitting of a regression line to the centers of the intervals (From Billard, L., and E. Diday. 2000. Regression analysis for interval-valued data. In *Data Analysis, Classification and Related Methods: Proceedings of the 7th Conference of the IFCS, IFCS 2002*. Berlin: Springer. pp. 369–374.) The estimated rectangles according to the regression line are represented by a dashed line.

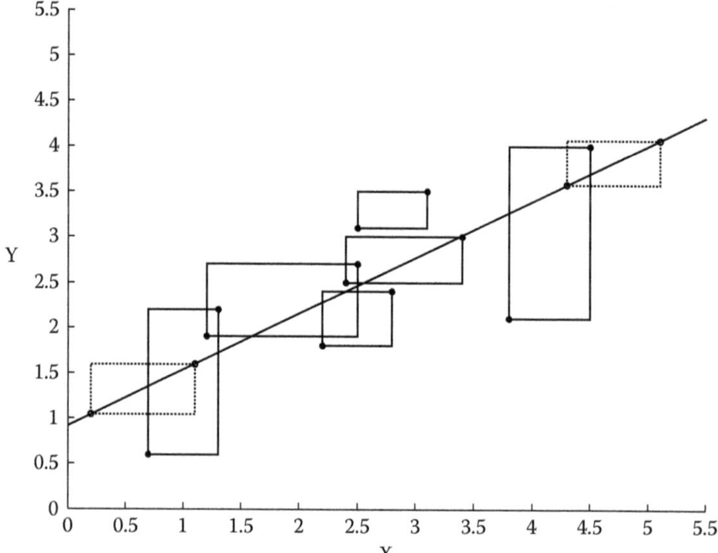

FIGURE 10.3
Regression line according to Brito (2007). The estimated rectangles from the regression line are represented by a dashed line.

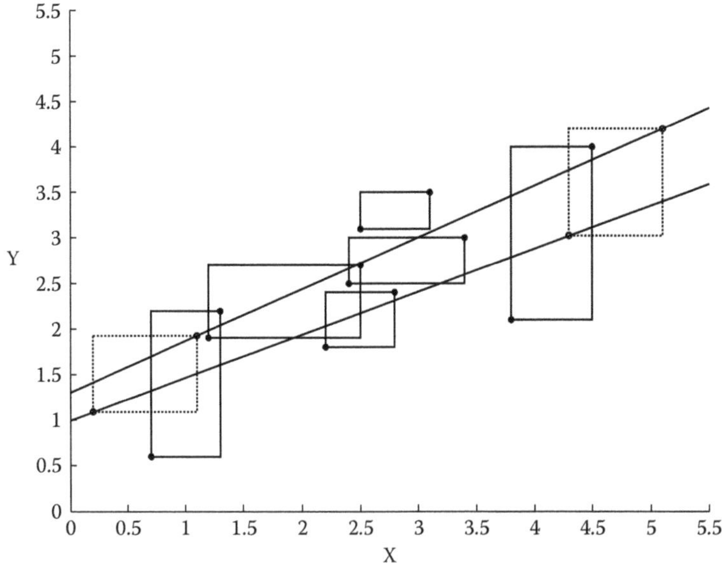

FIGURE 10.4
Regression lines fitted to the minima and maxima of the intervals (Billard, L., and E. Diday. 2002. Symbolic regression analysis. In *Classification, Clustering and Data Analysis: Proceedings of the 8th Conference of the IFCS, IFCS 2002*. Berlin: Springer. pp. 281–288.) The estimated rectangles according to the regression lines are represented by a dashed line.

the solution to this problem is the classical least squares estimator $\hat{\beta} = (X_C' X_C)^{-1} X_C' Y_C$ and standard statistical inference will apply under the standard assumptions about the error term of the regression. Though this model will provide information about the average centrality of the intervals, it disregards the range of the intervals that is an important feature of interval data.

There are several proposals aimed to incorporate the length of the interval into the analysis. Brito (2007) proposes to minimize the following objective function

$$\min_{\hat{\beta}} \sum_i \left(\hat{\varepsilon}_{Li}^2 + \hat{\varepsilon}_{Ui}^2\right) = \sum_i \left(y_{Li} - \hat{\beta}' x_{Li}\right)^2 + \sum_i \left(y_{Ui} - \hat{\beta}' x_{Ui}\right)^2, \tag{10.11}$$

which is equivalent to run two constrained (same regression coefficients) regressions on the lower bounds $y_{Li} = \beta' x_{Li} + \varepsilon_{Li}$ and the upper bounds $y_{Ui} = \beta' x_{Ui} + \varepsilon_{Ui}$ of the intervals. For the case of one regressor model, the OLS estimators have the following expression

$$\hat{\beta}_1 = \frac{\tilde{S}_{XY}}{\tilde{S}_X^2} = \frac{\frac{1}{2m}\sum_i \left[(x_{Li} - \bar{X})(y_{Li} - \bar{Y}) + (x_{Ui} - \bar{X})(y_{Ui} - \bar{Y})\right]}{\frac{1}{2m}\sum_i \left[(x_{Li} - \bar{X})^2 + (x_{Ui} - \bar{X})^2\right]} \tag{10.12}$$

$$\hat{\beta}_0 = \bar{Y} - \hat{\beta}_1 \bar{X}$$

where $\bar{X}$ and $\bar{Y}$ are given in Equation 10.7. Brito (2007) calls the numerator $\tilde{S}_{XY}$ the co-dispersion measure and the denominator $\tilde{S}^2_X$ the dispersion measure, which is different from Equation 10.9. This regression line passes through the average center $(\bar{X}, \bar{Y})$, but the slope is guided by the range of the intervals, whose effect is summarized by the sum of the covariance between the lower bounds of $[x]_i$ and $[y]_i$ and the covariance between the upper bounds of $[x]_i$ and $[y]_i$. In other words, the researcher collects a sample of points as (x_{Li}, y_{Li}) and (x_{Ui}, y_{Ui}) and fits a unique regression line to the full sample. Equivalently, we can understand Brito's proposal as a constrained system of equations

$$\underset{2m\times 1}{\begin{bmatrix} Y_L \\ Y_U \end{bmatrix}} = \underset{2m\times k}{\begin{bmatrix} X_L \\ X_U \end{bmatrix}} \underset{k\times 1}{\beta} + \underset{2m\times 1}{\begin{bmatrix} \varepsilon_L \\ \varepsilon_U \end{bmatrix}}, \tag{10.13}$$

for which the OLS estimator is

$$\hat{\beta}_{OLS} = \left[X_L' X_L + X_U' X_U\right]^{-1}\left[X_L' Y_L + X_U' Y_U\right]. \tag{10.14}$$

However, the vector ε is likely to be heteroscedastic, i.e., $\sigma^2_L \neq \sigma^2_U$

$$\Omega = \begin{pmatrix} \sigma^2_L & \sigma_{LU} \\ \sigma_{LU} & \sigma^2_U \end{pmatrix} \otimes I, \tag{10.15}$$

where $I_{m\times m}$ is the identity matrix. In this case, the GLS estimator $\hat{\beta}_{GLS} = [X'\Omega^{-1}X]^{-1}[X'\Omega^{-1}Y]$ would be more efficient than the OLS. A feasible GLS estimator will depend on the proposed model of heteroscedasticity. In the simplest heteroscedastic case, where $\sigma^2_L \neq \sigma^2_U$, the estimated $\hat{\Omega}$ will be obtained by replacing the population moments σ^2_L, σ^2_U and σ_{LU} with their sample counterparts.

An alternative proposal by Billard and Diday (2000, 2002) is to estimate two different regression lines, one for the minima and another for the maxima of the intervals with no restrictions across lines as in

$$\begin{aligned} y_{Li} &= \beta_L' x_{Li} + \varepsilon_{Li} \\ y_{Ui} &= \beta_U' x_{Ui} + \varepsilon_{Ui}. \end{aligned} \tag{10.16}$$

The estimation of the model proceeds by minimizing the following objective function

$$\min_{\hat{\beta}_L, \hat{\beta}_U} \sum_i \left(\hat{\varepsilon}^2_{Li} + \hat{\varepsilon}^2_{Ui}\right), \tag{10.17}$$

which is equivalent to perform two separate minimizations, $\min_{\hat{\beta}_L} \sum_i \hat{\varepsilon}^2_{Li}$ and $\min_{\hat{\beta}_U} \sum_i \hat{\varepsilon}^2_{Ui}$ because of the absence of cross-equation restrictions. This approach can also be written as a system of seemingly unrelated regression equations (SURE)

$$\underset{2m\times 1}{\begin{bmatrix} Y_L \\ Y_U \end{bmatrix}} = \underset{2m\times 2k}{\begin{bmatrix} X_L & 0 \\ 0 & X_U \end{bmatrix}} \underset{2k\times 1}{\begin{bmatrix} \beta_L \\ \beta_U \end{bmatrix}} + \underset{2m\times 1}{\begin{bmatrix} \varepsilon_L \\ \varepsilon_U \end{bmatrix}} \tag{10.18}$$

that is estimated by GLS, i.e., $\hat{\beta}_{GLS} = [X'\Omega^{-1}X]^{-1}[X'\Omega^{-1}Y]$. If $\Omega = I$ the GLS estimator reduces to the OLS estimator. However, given that $y_{Li} \leq y_{Ui}$ and $x_{Li} \leq x_{Ui}$, it is very likely that ε_{Li} and ε_{Ui} will be correlated and $\Omega \neq I$, thus the GLS estimator will be more efficient than the OLS. The feasible GLS will be constructed as in the previous approach. In practice, since there are not restrictions in the system, we could have some observations for which the estimated dependent variable is such that $\hat{y}_{Li} > \hat{y}_{Ui}$, which obviously contradicts the logic of interval data.

The last approach based on classical regression techniques is proposed by Lima Neto and de Carvalho (2008). It consists on running two independent regression models for the center and the radius (or range) of the intervals. Recall that $x_{Ci} = (x_{Li} + x_{Ui})/2$ and $x_{Ri} = (x_{Ui} - x_{Li})/2$. The model is

$$\begin{aligned} y_{Ci} &= \beta_C' x_{Ci} + \varepsilon_{Ci} \\ y_{Ri} &= \beta_R' x_{Ri} + \varepsilon_{Ri} \end{aligned} \tag{10.19}$$

and the objective function to minimize is

$$\min_{\hat{\beta}_C, \hat{\beta}_R} \sum_i \left(\hat{\varepsilon}_{Ci}^2 + \hat{\varepsilon}_{Ri}^2\right), \tag{10.20}$$

which, in the absence of cross-equation restrictions and with spherical disturbances, is equivalent to perform two separate minimizations, $\min_{\hat{\beta}_C} \sum_i \hat{\varepsilon}_{C_i}^2$ and $\min_{\hat{\beta}_R} \sum_i \hat{\varepsilon}_{Ri}^2$. The corresponding estimator is the classical OLS but the properties of the error term may dictate the choice of a GLS estimator, within a SURE system, as more appropriate than the OLS estimator. Other estimators as MLE or QMLE can also be implemented. However, the radius, being strictly positive, will not be normally distributed and a MLE estimator based on multivariate normality of the vector $(\varepsilon_{Ci}, \varepsilon_{Ri})'$ will be at least highly inefficient.

Figures 10.2 to 10.4 describe the graphical differences among the three regression lines proposed by Billard and Diday (2000, 2002) and Brito (2007). The proposal by Lima Neto and de Carvalho (2008) cannot be graphed in the same set of coordinates (X, Y).

10.2.3 The Prediction Problem

In this section, we define an interval-valued time series (ITS), we propose an approach to measure dissimilarities between intervals in ITS, and we implement forecasting methods for ITS based on smoothing filters and nonparametric estimators like the k-NN. Neither of these two approaches aims to specify a model for an ITS that approximates a hidden data-generating mechanism, but rather they should be viewed as automatic procedures to extract information from a noisy signal from which eventually we can extrapolate a future value.

Definition 10.3 *An interval-valued stochastic process is a collection of interval random variables that are indexed by time, i.e.,* $\{X_t\}$ *for* $t \in T \subset \mathbb{R}$, *with each* X_t *following Definition 10.2.*

An interval-valued time series is a realization of an interval-valued stochastic process and it will be equivalently denoted as $\{[x]_t\} = \{[x_{Lt}, x_{Ut}]\} = \{\langle x_{Ct}, x_{Rt}\rangle\}$ for $t = 1, 2, \ldots, T$.

10.2.3.1 Accuracy of the Forecast

It is customary in classical time series to assess the forecast as a function of the difference between the realized value and the forecast value. In ITS, one may be tempted to calculate the difference $[x]_{t+1} - [\hat{x}]_{t+1}$ but, because the interval difference bounds all the possible results when considering single real numbers in the two operands, see property (Equation 10.2), the resulting interval will have an excessive width and thus, it will not be deemed appropriate to measure the accuracy of a forecast (Arroyo, Espínola, and Maté 2010). The following example will clarify this point.

Suppose that $[x]_{t+1} = [\hat{x}]_{t+1} = [a_L, a_U], a_L < a_U$. Since the realized value is identical to the forecast, the forecast error must be zero $[x]_{t+1} - [\hat{x}]_{t+1} = [0, 0]$. If this difference is the interval difference (Equation 10.2), then it must be the case that $[A] = [a, a]$ with $a \in \mathbb{R}$, which is a contradiction with our assumption $a_L < a_U$. If $[a_L, a_U]$ is a nondegenerate interval, the result of the difference is an interval with the center in zero and with a length twice the length of the interval $[a_L, a_U]$, e.g., if $[a_L, a_U] = [1, 2]$, $[x]_t - [\hat{x}]_t = [-1, 1]$. Given these shortcomings, Arroyo and Maté (2006) propose the use of distances to quantify the dissimilarity (the forecast error) between the realized and the forecast intervals. The properties of distances, i.e., nonnegativity, symmetry, and triangle inequality, make them a suitable tool for this purpose. A distance, proposed by González et al. (2004), is defined as

$$D_K([x], [y]) = \frac{1}{\sqrt{2}}\sqrt{(x_L - y_L)^2 + (x_U - y_U)^2} = \sqrt{(x_C - y_C)^2 + (x_R - y_R)^2}, \tag{10.21}$$

which can be understood as an Euclidean-like distance considering the description of the intervals by their minimum and their maximum or, alternatively, by their center and by their radii. There is a large number of distances proposed in the literature, each with its advantages and disadvantages so that their use will depend on the needs of the researcher. In the forthcoming sections we will implement the Euclidean-type distance because of its intuitive and mathematical appeal.

Now, the assessment of a forecast will proceed by the choice of a distance measure and a loss function. Given a realized and a forecast ITS, $\{[x]_t\}$ and $\{[\hat{x}]_t\}$ with $t = 1, \ldots, T$, Arroyo, Espínola, and Maté (2010) propose the Mean Distance Error to quantify the accuracy of the forecast

$$MDE^q(\{[x]_t\}, \{[\hat{x}]_t\}) = \left(\frac{\sum_{t=1}^{T} (D^q([x]_t, [\hat{x}]_t))}{T} \right)^{\frac{1}{q}}, \tag{10.22}$$

where D is a distance such as D_K in Equation 10.21, and q is the order of the distance, such that for $q = 1$ the mean distance error is similar in spirit to the mean absolute error (MAE) loss function, and for $q = 2$ to the root mean squared error (RMSE) loss function. Other loss functions, statistical or economic/business based, can also be chosen to evaluate a forecast. The important point is that the quantification of the error should be based on a distance measure.

10.2.3.2 Smoothing Methods

Smoothing is a filtering technique that consists on averaging values of a time series, and by doing that, removing noise. These methods are easy to implement and they constitute a benchmark to evaluate the forecasting ability of more sophisticated methods (Gardner 2006). With the help of the arithmetic of intervals, it is relatively easy to adapt these smoothing procedures to ITS (Arroyo, Espínola, and Maté 2010). We begin with exponential smoothing though there is an even simpler smoothing procedure provided by just a moving average of order q.

10.2.3.2.1 Exponential Smoothing Given an ITS $\{[x_t]\}$ for $t = 1, 2, \ldots, T$, the forecast for the $t + 1$ period of a simple exponential smoothing in recursive form is written as

$$[\hat{x}]_{t+1} = \alpha[x]_t + (1 - \alpha)[\hat{x}]_t, \tag{10.23}$$

where $\alpha \in [0, 1]$. This representation weights the most recent observation and its forecast. In classic time series, the simple exponential smoothing can be equivalently represented in error correction form. However, with ITS both representations are not equivalent due to the properties of the interval arithmetic. To understand this difference, let us write the error correction representation

$$[\hat{x}]_{t+1} = [\hat{x}]_t + \alpha[e]_t, \tag{10.24}$$

where $[e]_t$ would be the interval error in t, $[e]_t = [x]_t - [\hat{x}]_t$. Due to the subdistributive property (Equation 10.5) of interval arithmetic, the relation between both expressions is the following

$$\alpha[x]_t + (1 - \alpha)[\hat{x}]_t \subseteq \alpha[x]_t - \alpha[\hat{x}]_t + [\hat{x}]_t = [\hat{x}]_t + \alpha([x]_t - [\hat{x}]_t), \tag{10.25}$$

which means that the recursive form yields tighter intervals than the error correction form. Due to this fact, the error correction form should not be considered in ITS forecasting. In addition, the error correction representation is not equivalent to the ITS moving average with exponentially decreasing weights, while the recursive form is. By backward substitution in Equation 10.23, and for t large, the simple exponential smoothing becomes

$$[\hat{x}]_{t+1} \simeq \sum_{j=1}^{t} \alpha(1-\alpha)^{j-1}[x]_{t-(j-1)}, \tag{10.26}$$

which is a moving average with exponentially decreasing weights.

Since the interval arithmetic subsumes the classical arithmetic, the smoothing methods for ITS subsume those for classic time series, so that if the intervals in the ITS are degenerated then the smoothing results will be identical to those obtained with the classical smoothing methods. When using Equation 10.23, all the components of the interval — center, radius, minimum, and maximum — are equally smoothed, i.e.,

$$\hat{x}_{\Gamma,t+1} = \alpha x_{\Gamma,t} + (1-\alpha)\hat{x}_{\Gamma,t} \quad \text{where } \Gamma \in \{L, U, C, R\}, \tag{10.27}$$

which means that, in a smoothed ITS, both the position and the width of the intervals will show less variability than in the original ITS, and that the smoothing factor will be the same for all components of the interval.

Additional smoothing procedures, like exponential smoothing with trend, or damped trend, or seasonality, can be adapted to ITS following the same principles presented in this section.

10.2.3.3 k-NN Method

The k-Nearest Neighbors (k-NN) method is a classic pattern recognition procedure that can be used for time series forecasting (Yakowitz 1987). The k-NN forecasting method in classic time series consists of two steps: identification of the k sequences in the time series that are more similar to the current one, and computation of the forecast as the weighted or unweighted average of the k-closest sequences determined in the previous step.

The adaptation of the k-NN method to forecast ITS consists of the following steps:

1. The ITS, $\{[x]_t\}$ with $t = 1, \ldots, T$, is organized as a series of d-dimensional interval-valued vectors

$$[x]_t^d = ([x]_t, [x]_{t-1}, \ldots, [x]_{t-(d-1)})', \tag{10.28}$$

   where $d \in \mathbb{N}$ is the number of lags.
2. We compute the dissimilarity between the most recent interval-valued vector $[x]_T^d = ([x]_T, [x]_{T-1}, \ldots, [x]_{T-d+1})'$ and the rest of the vectors in $\{[x]_t^d\}$. We use a distance measure to assess the dissimilarity between

vectors, i.e.,

$$D_t([x]_T^d, [x]_t^d) = \left(\frac{\sum_{i=1}^{d} \left(D^q([x]_{T-i+1}, [x]_{t-i+1}) \right)}{d} \right)^{\frac{1}{q}}, \tag{10.29}$$

where $D([x]_{T-i+1}, [x]_{t-i+1})$ is a distance such as the kernel-based distance shown in Equation 10.21, q is the order of the measure that has the same effect that in the error measure shown in Equation 10.22.

3. Once the dissimilarity measures are computed for each $[x]_t^d$, $t = T-1, T-2, \ldots, d$, we select the k closest vectors to $[x]_T^d$. These are denoted by $[x]_{T_1}^d, [x]_{T_2}^d, \ldots, [x]_{T_k}^d$.
4. Given the k closest vectors, their subsequent values, $[x]_{T_1+1}, [x]_{T_2+1} \ldots, [x]_{T_k+1}$, are averaged to obtain the final forecast

$$[\hat{x}]_{T+1} = \sum_{p=1}^{k} \omega_p \cdot [x]_{T_p+1}, \tag{10.30}$$

where $[x]_{T_p+1}$ is the consecutive interval of the sequence $[x]_{T_p}^d$, and ω_p is the weight assigned to the neighbor p, with $\omega_p \geq 0$ and $\sum_{p=1}^{k} \omega_p = 1$. Equation 10.30 is computed according to the rules of interval arithmetic. The weights are assumed to be equal for all the neighbors $\omega_p = 1/k \forall p$, or inversely proportional to the distance between the last sequence $[x]_T^d$ and the considered sequence $[x]_{T_p}^d$

$$\omega_p = \frac{\psi_p}{\sum_{l=1}^{k} \psi_l}, \tag{10.31}$$

with $\psi_p = (D_{T_p}([x]_T^d, [x]_{T_p}^d) + \xi)^{-1}$ for $p = 1, \ldots, k$. The constant $\xi = 10^{-8}$ prevents the weight to explode when the distance between two sequences is zero.

The optimal values $\hat{k}$ and $\hat{d}$, which minimize the mean distance error (Equation 10.22) in the estimation period, are obtained by conducting a two-dimensional grid search.

10.2.4 Interval-Valued Dispersion: Low/High SP500 Prices

In this section, we apply the aforementioned interval regression and prediction methods to the daily interval time series of low/high prices of the SP500 index. We will denote the interval as $[p_{L,t}, p_{U,t}]$. There is strand in the financial literature — Parkinson (1980), Garman and Klass (1980), Ball and Torous (1984), Rogers and Satchell (1991), Yang and Zhang (2000), and Alizadeh, Brandt, and Diebold (2002) among others — that deals with functions of the range of the interval, $p_U - p_L$, in order to provide an estimator of the volatility σ of asset returns. In this chapter we do not pursue this route. The object of analysis is the interval $[p_{L,t}, p_{U,t}]$ itself and our goal is the construction of the one-step-ahead forecast $[\hat{p}_{L,t+1}, \hat{p}_{U,t+1}]$. Obviously such a forecast can be an input to produce a forecast $\hat{\sigma}_{t+1}$ of volatility. One of the advantage of forecasting the low/high interval versus forecasting volatility is that the prediction error of the interval is based on observables as opposed to the prediction error for the volatility forecast for which "observed" volatility may be a problem.

The sample period goes from January 3, 2000 to September 30, 2008. We consider two sets of predictions:

1. Low volatility prediction set (year 2006): estimation period that goes from January 3, 2000 to December 30, 2005 (1508 trading days) and prediction period that goes from January 3, 2006 to December 29, 2006 (251 trading days).
2. High volatility prediction set (year 2008): estimation period that goes from January 2, 2002 to December 31, 2007 (1510 trading days) and prediction period that goes from January 2, 2008 to September 30, 2008 (189 trading days).

A plot of the first ITS $[p_{L,t}, p_{U,t}]$ is presented in Figure 10.5.

Following the classical regression approach to ITS, we are interested in the properties and time series regression models of the components of the interval, i.e., p_L, p_U, p_C, and p_R. We present the most significant and unrestricted time series models for $[p_{L,t}, p_{U,t}]$ and $\langle p_{C,t}, p_{R,t}\rangle$ in the spirit of the regression proposals of Billard and Diday (2000, 2002) and Lima Neto and de Carvalho (2008) reviewed in the previous sections. To save space we omit the univariate modeling of the components of the interval but these results are available upon request. However, we need to report that for p_L and p_U, we cannot reject a unit root, which is expected because these are price levels of the SP500, and that p_C has also a unit root because it is the sum of two unit root processes. In addition, p_L and p_U are cointegrated of order one with cointegrating vector $(1, -1)$, which implies that p_R is a stationary process given

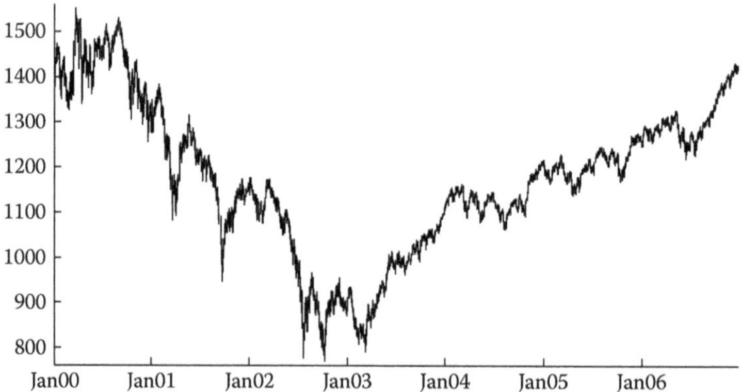

FIGURE 10.5
ITS of the weekly low/high from January 2000 to December 2006.

that $p_R = (p_U - p_L)/2$. Following standard model selection criteria and time series specification tools, the best model for $\langle \Delta p_{C,t}, p_{R,t} \rangle$ is a VAR(3) and for $[p_{L,t}, p_{U,t}]$ a VEC(3). The estimation results are presented in Tables A.1 and A.2 in the appendix.

In Table A.1, the estimation results for $\langle \Delta p_{C,t}, p_{R,t} \rangle$ in both periods are very similar. The radius $p_{R,t}$ exhibits high autoregressive dependence and it is negatively correlated with the previous change in the center of the interval $\Delta p_{C,t-1}$ so that positive surprises in the center tend to narrow down the interval. On the other hand $\Delta p_{C,t}$ has little linear dependence and it is not affected by the dynamics of the radius. There is Granger causality from the center to the radius, but not vice versa. The radius equation enjoys a relative high adjusted R-squared of about 40% while the center is basically not linearly predictable. In general terms, there is a strong similarity between the modeling of $\langle \Delta p_{C,t}, p_{R,t} \rangle$ and the most classical modeling of volatility with ARCH models for financial returns. The processes $p_{R,t}$ and the conditional variance of an asymmetric ARCH model, i.e., $\sigma^2_{t|t-1} = \alpha_0 + \alpha_1 \varepsilon^2_{t-1} + \alpha_2 \varepsilon_{t-1} + \beta \sigma^2_{t-1|t-2}$, share the autoregressive nature and the well-documented negative correlation of past innovations and volatility. The unresponsiveness of the center to the information in the dynamics of the radius is also similar to the findings in ARCH-in-mean processes where it is difficult to find significant effects of volatility on the return process.

In Table A.2, we report the estimation results for $[p_{L,t}, p_{U,t}]$ for both periods 2000–2005 and 2002–2007. In general, there is much less linear dependence in the short-run dynamics of $[p_{L,t}, p_{U,t}]$, which is expected as we are modeling financial prices. There is Granger-causality running both ways, from Δp_L to Δp_U and vice versa. Overall, the 2002-2007 period seems to be noisier (R-squared of 14%) than the 2000–2005 (R-squared of 20%–16%).

Based on the estimation results of the VAR(3) and VEC(3) models, we proceed to construct the one-step-ahead forecast of the interval $[\hat{p}_{L,t+1|t}, \hat{p}_{U,t+1|t}]$. We also implement the exponential smoothing methods and the k-NN method for ITS proposed in the above sections and compare their respective forecasts. For the smoothing procedure, the estimated value of α is $\hat{\alpha} = 0.04$ in the estimation period 2000–2005 and $\hat{\alpha} = 0.03$ in 2002–2007. We have implemented the k-NN with equal weights and with inversely proportional as in Equation 10.31. In the period 2000–2005, the numbers of neighbors is $\hat{k} = 23$ (equal weights) and $\hat{k} = 24$ (proportional weights); in 2002–2007 $\hat{k} = 18$ for the k-NN with equal weights and $\hat{k} = 24$ for proportional weights. In both estimation periods, the length of the vector is $\hat{d} = 2$ for the k-NN with equal weights and $\hat{d} = 3$ for the proportional weights. The estimation of α, k, and d has been performed by minimizing the mean distance MDE (Equation 10.22) with $q = 2$. In both methods, smoothing and k-NN, the centers of the intervals have been first-differenced to proceed with the estimation and forecasting. However, in the following comparisons, the estimated differenced centers are transformed back to present the estimates and forecasts in levels. In Table 10.1 we show the performance of the five models measured by the MDE ($q = 2$) in the estimation and prediction

TABLE 10.1

Performance of the Forecasting Methods: MDE ($q = 2$)

	Period 2000–2006		Period 2002–2008	
	Estimation	Prediction	Estimation	Prediction
Models	**2000–2005**	**2006**	**2002–2007**	**2008**
VAR(3)	9.359	6.611	7.614	15.744
VEC(3)	9.313	6.631	7.594	15.766
k-NN (eq.weights)	9.419	6.429	7.625	15.865
k-NN (prop.weights)	9.437	6.303	7.617	16.095
Smoothing	9.833	6.698	7.926	16.274
Naive	10.171	7.056	8.231	16.549

periods. We have also added a "naive" model that does not entail any estimation and whose forecast is the observation in the previous period, i.e., $[\hat{p}_{L,t+1|t}, \hat{p}_{U,t+1|t}] = [p_{L,t}, p_{U,t}]$.

For both low- and high-volatility periods the performance ranking of the six models is very similar. The worst performer is the naive model followed by the smoothing model. In 2006, the k-NN procedures are superior to the VAR(3) and VEC(3) models, but in 2008 the VAR and VEC systems perform slightly better than the k-NNs. The high-volatility year 2008 is clearly more difficult to forecast, the MDE in 2008 is twice as much as the MDE in the estimation period 2002–2007. On the contrary, in the low volatility year 2006, the MDE in the prediction period is about 30% lower than the MDE in the estimation period 2000–2005. A statistical comparison of the MDEs of the five models in relation to the naive model is provided by the Diebold and Mariano test of unconditional predictability (Diebold and Mariano 1995). The null hypothesis to test is the equality of the MDEs, i.e., $H_0 : E(D^2_{(\text{naive})} - D^2_{(\text{other})}) = 0$ versus $H_1 : E(D^2_{(\text{naive})} - D^2_{(\text{other})}) > 0$. If the null hypothesis is rejected the other model is superior to the naive model. The results of this test are presented in Table 10.2.

In 2006 all the five models are statistically superior to the benchmark naive model. In 2008 the smoothing procedure and the k-NN with proportional weights are statistically equivalent to the naive model while the remaining three models outperform the naive.

TABLE 10.2

Results of the Diebold and Mariano Test

	T-Test for $H_0 : E(D^2_{(\text{naive})} - D^2_{(\text{other})}) = 0$	
Models	**2006**	**2008**
VAR(3)	2.86	2.67
VEC(3)	2.26	2.46
k-NN(eq.weights)	3.55	2.43
k-NN(prop.weights)	4.17	1.79
Smoothing	5.05	1.15

We also perform a complementary assessment of the forecasting ability of the five models by running some regressions of the Mincer–Zarnowitz type. In the prediction periods, for the minimum p_L and the maximum p_U, we run separate regressions of the realized observations on the predicted observations as in $p_{L,t} = c + \beta \hat{p}_{L,t} + \varepsilon_t$ and $p_{U,t} = c + \beta \hat{p}_{U,t} + v_t$. Under a quadratic loss function, we should expect an unbiased forecast, i.e., $\beta = 1$ and $c = 0$. However, the processes $p_{L,t}$ and $\hat{p}_{L,t}$ are $I(1)$ and, as expected, cointegrated, so that these regressions should be performed with care. The point of interest is then to test for a cointegration vector of $(1, -1)$. To test this hypothesis using an OLS estimator with the standard asymptotic distribution, we need to consider that in the $I(1)$ process $\hat{p}_{L,t}$, i.e., $\hat{p}_{L,t} = \hat{p}_{L,t-1} + \nu_t$, the innovations ε_t and ν_t are not independent; in fact because $\hat{p}_{L,t}$ is a forecast of $p_{L,t}$ the correlation $\rho(\nu_{t+i}, \varepsilon_t) \neq 0$ for $i > 0$. To remove this correlation, the cointegrating regression will be augmented with some terms to finally estimate a regression as $p_{L,t} = c + \beta \hat{p}_{L,t} + \sum_i \gamma_i \Delta \hat{p}_{L,t+i} + e_t$ (the same argument applies to $p_{U,t}$). The hypothesis of interest is $H_0 : \beta = 1$ versus $H_1 : \beta \neq 1$. A t-statistic for this hypothesis will be asymptotically standard normal distributed. We may also need to correct the t-test if there is some serial correlation in e_t. In Table 10.3 we present the testing results.

We reject the null for the smoothing method for both prediction periods and for both $p_{L,t}$ and $p_{U,t}$ processes. Overall the prediction is similar for 2006 and 2008. The VEC(3) and the k-NN methods deliver better forecasts across the four instances considered. For those models in which we fail to reject $H_0 : \beta = 1$, we also calculate the unconditional average difference between the realized and the predicted values, i.e, $\bar{p} = \sum_t (p_t - \hat{p}_t)/T$. The magnitude of this average is in the single digits, so that for all purposes, it is insignificant given that the level of the index is in the thousands. In Figure 10.6 we show the k-NN (equal weights)-based forecast of the interval low/high of the SP500 index for November and December 2006.

TABLE 10.3

Results of the t-Test for Cointegrating Vector $(1, -1)$

	Asymptotic (Corrected) t-Test $H_0 : \beta = 1$ **versus** $H_1 : \beta \neq 1$ $p_t = c + \beta \hat{p}_t + \sum_i \gamma_i \Delta \hat{p}_{t+i} + e_t$			
	2006		**2008**	
	min: $p_{L,t}$	**max:** $p_{U,t}$	**min:** $p_{L,t}$	**max:** $p_{U,t}$
VAR(3)	3.744*	−1.472	3.024*	−2.712*
VEC(3)	1.300	0.742	2.906*	−2.106
k-NN (eq.weights)	0.639	−4.191*	1.005	−2.270
k-NN (prop.weights)	3.151*	−2.726*	1.772	−1.731
Smoothing	−3.542*	−2.544*	2.739*	−3.449*

*Rejection of the null hypothesis at the 1% significance level.

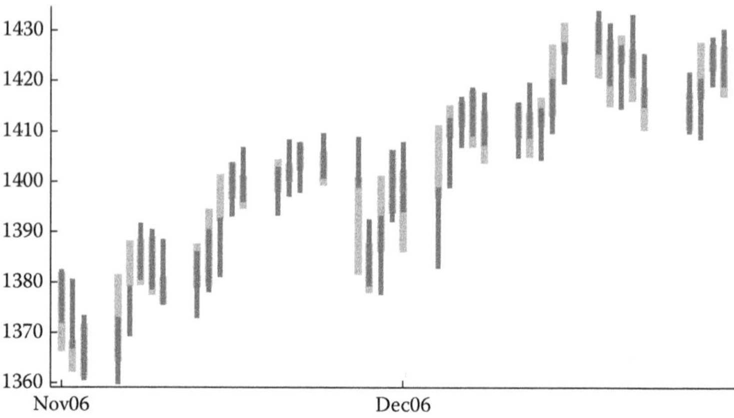

FIGURE 10.6
k-NN based forecast (black) of the low/high prices of the SP500; realized ITS (grey).

10.3 Histogram Data

In this section, our premise is that the data is presented to the researcher as a frequency distribution, which may be the result of an aggregation procedure, or the description of a population or any other grouped collective. We start by describing histogram data and some univariate descriptive statistics. Our main objective is to present the prediction problem by defining a histogram time series (HTS) and implementing smoothing techniques and nonparametric methods like the k-NN algorithm. As we have seen in the section on interval data, these two methods require the calculation of suitable averages. To this end, instead of relying on the arithmetic of histograms, we introduce the barycentric histogram that is an average of a set of histograms. The choice of appropriate distance measures is key to the calculation of the barycenter, and eventually of the forecast of a HTS.

10.3.1 Preliminaries

Given a variable of interest X, we collect information on a group of individuals or units that belong to a set S. For every element $i \in S$, we observe a datum such as

$$h_{X_i} = \{([x]_{i1}, \pi_{i1}), \ldots, ([x]_{in_i}, \pi_{in_i})\}, \quad \text{for } i \in S, \tag{10.32}$$

where π_{ij} , $j = 1, \ldots, n_i$ is a frequency that satisfies $\pi_{ij} \geq 0$ and $\sum_{j=1}^{n_i} \pi_{ij} = 1$; and $[x]_{ij} \subseteq \mathbb{R}$, $\forall i, j$, is an interval (also known as bin) defined as $[x]_{ij} \equiv [x_{Lij}, x_{Uij})$ with $-\infty < x_{Lij} \leq x_{Uij} < \infty$ and $x_{Ui\,j-1} \leq x_{Lij}$ $\forall i, j$, for $j \geq 2$. The datum h_{X_i} is a histogram and the data set will be a collection of histograms $\{h_{X_i}, i = 1, \ldots, m\}$.

As in the case of interval data, we could summarize the histogram data set by its empirical density function from which the sample mean and the sample variance can be calculated (Billard and Diday 2006). The sample mean is

$$\bar{X} = \frac{1}{2m} \sum_{i=1}^{m} \sum_{j=1}^{n_i} (x_{Uij} + x_{Lij})\pi_{ij}, \tag{10.33}$$

which is the average of the weighted centers for each interval; and the sample variance is

$$S_X^2 = \frac{1}{3m} \sum_{i=1}^{m} \sum_{j=1}^{n_i} (x_{Uij}^2 + x_{Uij}x_{Lij} + x_{Lij}^2)\pi_{ij} - \frac{1}{4m^2} \left[\sum_{i=1}^{m} \sum_{j=1}^{n_i} (x_{Uij} + x_{Lij})\pi_{ij} \right]^2,$$

which combines the variability of the centers as well as the intra-interval variability. Note that the main difference between these sample statistics and those in Equations 10.7 and 10.9 for interval data is the weight provided by the frequency $\pi_{i,j}$ associated with each interval $[x]_{i,j}$.

Next, we proceed with the definition of a histogram random variable. Let $(\Omega, \mathcal{F}, P)$ be a probability space, where Ω is the set of elementary events, $\mathcal{F}$ is the σ-field of events and $P : \mathcal{F} \to [0, 1]$ the σ-additive probability measure; and define a partition of Ω into sets $A_X(x)$ such that $A_X(x) = \{\omega \in \Omega | X(\omega) = x\}$, where $x \in \{h_{X_i}, i = 1, \ldots, m\}$.

Definition 10.4 *A mapping $h_X : \mathcal{F} \to \{h_{X_i}\}$, such that, for all $x \in \{h_{X_i}, i = 1 \ldots m\}$ there is a set $A_X(x) \in \mathcal{F}$, is called a histogram random variable.*

Then, the definition of stochastic process follows as:

Definition 10.5 *A histogram-valued stochastic process is a collection of histogram random variables that are indexed by time, i.e., $\{h_{X_t}\}$ for $t \in T \subset \mathbb{R}$, with each h_{X_t} following Definition 10.4.*

A histogram-valued time series is a realization of a histogram-valued stochastic process and it will be equivalently denoted as $\{h_{X_t}\} \equiv \{h_{X_t}, \ t = 1, 2, \ldots, T\}$.

10.3.2 The Prediction Problem

In this section, we propose a dissimilarity measure for HTS based on a distance. We present two distance measures that will play a key role in the estimation and prediction stages. They will also be instrumental to the definition of a barycentric histogram, which will be used as the average of a set of histograms. Finally, we will present the implementation of the prediction methods.

10.3.2.1 Accuracy of the Forecast

Suppose that we construct a forecast for $\{h_{X_t}\}$, which we denote as $\{\hat{h}_{X_t}\}$. It is sensible to define the forecast error as the difference $h_{X_t} - \hat{h}_{X_t}$. However, the difference operator based on histogram arithmetic (Colombo and Jaarsma 1980) does not provide information on how dissimilar the histograms h_{X_t} and $\hat{h}_{X_t}$ are. In order to avoid this problem, Arroyo and Maté (2009) propose the mean distance error (MDE), which in its most general form is defined as

$$MDE^q(\{h_{X_t}\}, \{\hat{h}_{X_t}\}) = \left(\frac{\sum_{t=1}^{T} D^q(h_{X_t}, \hat{h}_{X_t})}{T}\right)^{\frac{1}{q}}, \tag{10.34}$$

where $D(h_{X_t}, \hat{h}_{X_t})$ is a distance measure such as the Wasserstein or the Mallows distance to be defined shortly and q is the order of the measure, such that for $q = 1$ the resulting accuracy measure is similar to the MAE and for $q = 2$ to the RMSE.

Consider two density functions, $f(x)$ and $g(x)$, with their corresponding cumulative distribution functions (CDF), $F(x)$ and $G(x)$, the Wasserstein distance between $f(x)$ and $g(x)$ is defined as

$$D_W(f, g) = \int_0^1 |F^{-1}(t) - G^{-1}(t)| dt, \tag{10.35}$$

and the Mallows as

$$D_M(f, g) = \sqrt{\int_0^1 (F^{-1}(t) - G^{-1}(t))^2 dt}, \tag{10.36}$$

where $F^{-1}(t)$ and $G^{-1}(t)$ with $t \in [0, 1]$ are the inverse CDFs of $f(x)$ and $g(x)$, respectively. The dissimilarity between two functions is essentially measured by how far apart their t-quantiles are, i.e., $F^{-1}(t) - G^{-1}(t)$. In the case of Wasserstein, the distance is defined in the L_1 norm and in the Mallows in the L_2 norm. When considering Equation 10.34, $D(h_{X_t}, \hat{h}_{X_t})$ will be calculated by implementing the Wasserstein or Mallows distance. By using the definition of the CDF of a histogram in Billard and Diday (2006), the Wasserstein and Mallows distances between two histograms h_X and h_Y can be written analytically as functions of the centers and radii of the histogram bins, i.e.,

$$D_W(h_X, h_Y) = \sum_{j=1}^{n} \pi_j |x_{Cj} - y_{Cj}| \tag{10.37}$$

$$D_M^2(h_X, h_Y) = \sum_{j=1}^{n} \pi_j \left[(x_{Cj} - y_{Cj})^2 + \frac{1}{3}(x_{Rj} - y_{Rj})^2\right]. \tag{10.38}$$

10.3.2.2 The Barycentric Histogram

Given a set of K histograms h_{X_k} with $k = 1, \ldots, K$, the barycentric histogram h_{X_B} is the histogram that minimizes the distances between itself and all the K histograms in the set. The optimization problem is

$$\min_{h_{X_B}} \sum_{k=1}^{K} \left[D^r(h_{X_k}, h_{X_B})\right]^{1/r}, \tag{10.39}$$

where $D(h_{X_k}, h_{X_B})$ is a distance measure. The concept is introduced by Irpino and Verde (2006) to define the prototype of a cluster of histogram data. As Verde and Irpino (2007) show, the choice of the distance determine the properties of the barycenter.

When the chosen distance is Mallows, for $r = 2$, the optimal barycentric histogram $h^*_{X_B}$ has the following center/radius characteristics. Once the k histograms are rewritten in terms of n^* bins, for each bin $j = 1, \ldots, n^*$, the barycentric center x^*_{Cj} is the mean of the centers of the corresponding bin in each histogram and the barycentric radius x^*_{Rj} is the mean of the radii of the corresponding bin in each of the K histograms,

$$x^*_{Cj} = \frac{\sum_{k=1}^{K} x_{Ckj}}{K} \tag{10.40}$$

$$x^*_{Rj} = \frac{\sum_{k=1}^{K} x_{Rkj}}{K}. \tag{10.41}$$

When the distance is Wasserstein, for $r = 1$ and for each bin $j = 1, \ldots, n^*$, the barycentric center x^*_{Cj} is the median of the centers of the corresponding bin in each of the K histograms,

$$x^*_{Cj} = \text{median}(x_{Ckj}) \quad \text{for } k = 1, \ldots, K \tag{10.42}$$

and the radius x^*_{Rj} is the corresponding radius of the bin where the median x^*_{Cj} falls among the K histograms. For more details on the optimization problem, please see Arroyo and Maté (2009).

10.3.2.3 Exponential Smoothing

The exponential smoothing method can be adapted to histogram time series by replacing averages with the barycentric histogram, as it was shown in Arroyo and Maté (2008).

Let $\{h_{X_t}\}\, t = 1, \ldots, T$ be a histogram time series, the exponentially smoothed forecast is given by the following equation

$$\hat{h}_{X_{t+1}} = \alpha h_{X_t} + (1 - \alpha)\hat{h}_{X_t}, \tag{10.43}$$

where $\alpha \in [0, 1]$. Since the right-hand side is a weighted average of histograms, we can use the barycenter approach so that the forecast is the solution to the following optimization exercise

$$\hat{h}_{X_{t+1}} \equiv \arg\min_{\hat{h}_{X_{t+1}}} \left(\alpha D^2(\hat{h}_{X_{t+1}}, h_{X_t}) + (1-\alpha) D^2(\hat{h}_{X_{t+1}}, \hat{h}_{X_t})\right)^{1/2}, \quad (10.44)$$

where $D(\cdot, \cdot)$ is the Mallows distance. The use of the Wasserstein distance is not suitable in this case because of the properties of the median, which will ignore the weighting scheme (with the exception of $\alpha = 0.5$) so intrinsically essential to the smoothing technique. For further developments of this issue see Arroyo, González-Rivera, Maté and Muñoz-San Roque (2010).

For t large, the recursive form (Equation 10.43) can be easily rewritten as a moving average

$$\hat{h}_{X_{t+1}} \simeq \sum_{j=1}^{t} \alpha(1-\alpha)^{j-1} h_{X_{t-(j-1)}}, \quad (10.45)$$

which in turn can also be expressed as the following optimizations problem

$$\hat{h}_{X_{t+1}} \equiv \arg\min_{\hat{h}_{X_{t+1}}} \left[\sum_{j=1}^{t} \alpha(1-\alpha)^{j-1} D^2(\hat{h}_{X_{t+1}}, h_{X_{t-(j-1)}})\right]^{1/2}, \quad (10.46)$$

with $D(\cdot, \cdot)$ as the Mallows distance. The Equations 10.44 and 10.46 are equivalent.

Figure 10.7 shows an example of the exponential smoothing using Equation 10.44 for the histograms $h_{X_t} = \{([19, 20), 0.1), ([20, 21), 0.2), ([21, 22], 0.7)\}$ and $\hat{h}_{X_t} = \{([0, 3), 0.35), ([3, 6), 0.3), ([6, 9], 0.35)\}$ with $\alpha = 0.9$ and $\alpha = 0.1$. In both cases, the resulting histogram averages the location, the support, and the shape of both histograms h_{X_t} and $\hat{h}_{X_t}$ in a suitable way.

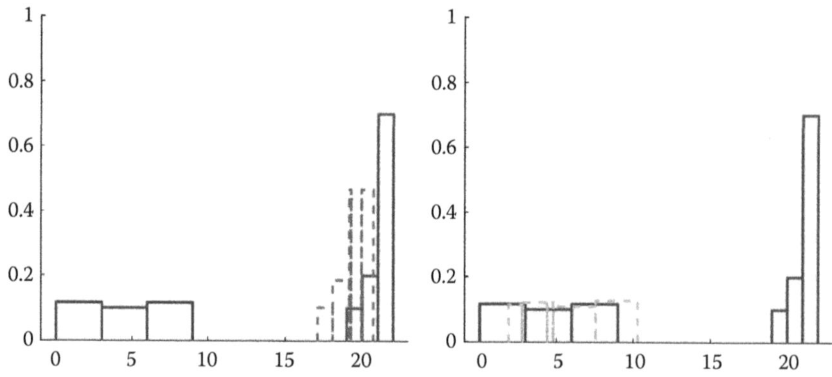

FIGURE 10.7
Exponential smoothing of histograms using the recursive formulation with $\alpha = 0.9$ (left) and $\alpha = 0.1$ (right). In each part of the figure, the barycenter is the dash-lined histogram.

10.3.2.4 k-NN Method

The adaptation of the k-NN method to forecast HTS was proposed by Arroyo and Maté (2009). The method consists of similar steps to those described in the interval section:

1. The HTS, $\{h_{X_t}\}$ with $t = 1, \ldots, T$, is organized as a series of d-dimensional histogram-valued vectors $\{h^d_{X_t}\}$ where

$$h^d_{X_t} = (h_{X_t}, h_{X_{t-1}}, \ldots, h_{X_{t-(d-1)}})', \tag{10.47}$$

where $d \in \mathbb{N}$ is the number of lags and $t = d, \ldots, T$.
2. We compute the dissimilarity between the most recent histogram-valued vector $h^d_{X_T} = (h_{X_T}, h_{X_{T-1}}, \ldots, h_{X_{T-(d-1)}})'$ and the rest of the vectors in $\{h^d_{X_t}\}$ by implementing the following distance measure

$$D_t(h^d_{X_T}, h^d_{X_t}) = \left(\frac{\sum_{i=1}^d \left(D^q(h_{X_{T-i+1}}, h_{X_{t-i+1}})\right)}{d}\right)^{\frac{1}{q}}, \tag{10.48}$$

where $D^q(h_{X_{T-i+1}}, h_{X_{t-i+1}})$ is the Mallows or the Wasserstein distance of order q.
3. Once the dissimilarity measures are computed for each $h^d_{X_t}$, $t = T - 1, T-2, \ldots, d$, we select the k closest vectors to $h^d_{X_T}$. These are denoted by $h^d_{X_{T_1}}, h^d_{X_{T_2}}, \ldots, h^d_{X_{T_k}}$.
4. Given the k closest vectors, their subsequent values, $h_{X_{T_1+1}}, h_{X_{T_2+1}}, \ldots, h_{X_{T_k+1}}$, are averaged by means of the barycenter approach to obtain the final forecast $\hat{h}_{X_{T+1}}$ as in

$$\hat{h}_{X_{T+1}} \equiv \arg\min_{\hat{h}_{X_{T+1}}} \left[\sum_{p=1}^{k} \omega_p D^r(\hat{h}_{X_{T+1}}, h_{X_{T_p+1}})\right]^{1/r}, \tag{10.49}$$

where $D(\hat{h}_{X_{T+1}}, h_{X_{T_p+1}})$ is the Mallows distance with $r = 2$ or the Wasserstein distance with $r = 1$, $h_{X_{T_p+1}}$ is the consecutive histogram in the sequence $h^d_{X_{T_p}}$, and ω_p is the weight assigned to the neighbor p, with $\omega_p \geq 0$ and $\sum_{p=1}^k \omega_p = 1$. As in the case of the interval-valued data, the weights may be assumed to be equal for all the neighbors $\omega_p = 1/k \ \forall p$, or inversely proportional to the distance between the last sequence $h^d_{X_T}$ and the considered sequence $h^d_{X_{T_p}}$.

The optimal values, $\hat{k}$ and $\hat{d}$, which minimize the mean distance error (Equation 10.34) in the estimation period, are obtained by conducting a two-dimensional grid search.

10.3.3 Histogram Forecast for SP500 Returns

In this section, we implement the exponential smoothing and the k-NN methods to forecast the one-step-ahead histogram of the returns to the constituents of the SP500 index. We collect the weekly returns of the 500 firms in the index from 2002 to 2005. We divide the sample into an estimation period of 156 weeks running from January 2002 to December 2004, and a prediction period of 52 weeks that goes from January 2005 to December 2005. The histogram data set consists of 208 weekly equiprobable histograms. Each histogram has four bins, each one containing 25% of the firms' returns.

For the smoothing procedure, the estimated value of α is $\hat{\alpha} = 0.13$. We have implemented the k-NN with equal weights and with inversely proportional as in Equation 10.31 using the Mallows and Wasserstein distances. With the Mallows distance, the estimated numbers of neighbors is $\hat{k} = 11$ and the length of the vector is $\hat{d} = 9$ for both weighting schemes. With the Wasserstein distance, $\hat{k} = 12, \hat{d} = 9$ (equal weights), and $\hat{k} = 17$, $\hat{d} = 8$ (proportional weights). The estimation of α, k, and d has been performed by minimizing the Mallows MDE with $q = 1$, except for the Wasserstein-based k-NN which used the Wasserstein MDE with $q = 1$. In Table 10.4, we show the performance of the five models measured by the Mallows-based MDE ($q = 1$) in the estimation and prediction periods. We have also added a "naive" model that does not entail any estimation and for which the one-step-ahead forecast is the observation in the previous period, i.e., $\hat{h}_{X_{t+1|t}} = h_{X_t}$.

In the estimation and prediction period, the naive model is clearly outperformed by the rest of the five models. In the estimation period, the five models exhibit similar performance with a MDE of 4.9 approximately. In the prediction period, the exponential smoothing and the Wasserstein-based k-NN seem to be superior to the Mallows-based k-NN. We should note that the MDEs in the prediction period are about 11% lower than the MDEs in the estimation period.

For the prediction year 2005, we provide a statistical comparison of the MDEs of the five models in relation to the naive model by implementing the Diebold and Mariano test of unconditional predictability (Diebold and Mariano 1995). The null hypothesis to test is the equality of the MDEs, i.e., $H_0 : E(D_{(\text{naive})} - D_{(\text{other})}) = 0$ versus $H_1 : E(D_{(\text{naive})} - D_{(\text{other})}) > 0$. If the null

TABLE 10.4

Performance of the Forecasting Methods: MDE ($q = 1$)

Models	Estimation 2002–2004	Prediction 2005
Mall. k-NN (eq.weights)	4.988	4.481
Mall. k-NN (prop.weights)	4.981	4.475
Wass. k-NN (eq.weights)	4.888	4.33
Wass. k-NN (prop.weights)	4.882	4.269
Exp. Smoothing	4.976	4.344
Naive	6.567	5.609

TABLE 10.5

Results of the Diebold and Mariano Test

	t-Test for $H_0 : E(D_{(\text{naive})} - D_{(\text{other})}) = 0$
Models	**2005 Prediction Year**
Mall. k-NN(eq.weights)	2.32
Mall. k-NN(prop.weights)	2.69
Wass. k-NN(eq.weights)	2.29
Wass. k-NN(prop.weights)	2.29
Exp. smoothing	3.08

hypothesis is rejected, the "other" model is superior to the naive model. The results of this test are presented in Table 10.5.

In 2005, all the five models are statistically superior to the benchmark naive model, though the rejection of the null is stronger for the exponential smoothing and the Mallows-based k-NN models with proportional weights.

In Figure 10.8, we present the 2005 one-step-ahead histogram forecast obtained with the exponential smoothing procedure and we compare it to the realized value. For each time period, we draw two histograms: the realized histogram (the right one) and the forecast histogram (the left one). Overall the forecast follows very closely the realized value except for those observations that have extreme returns. The fit can be further appreciated when we zoom in the central 50% mass of the histograms (Figure 10.9).

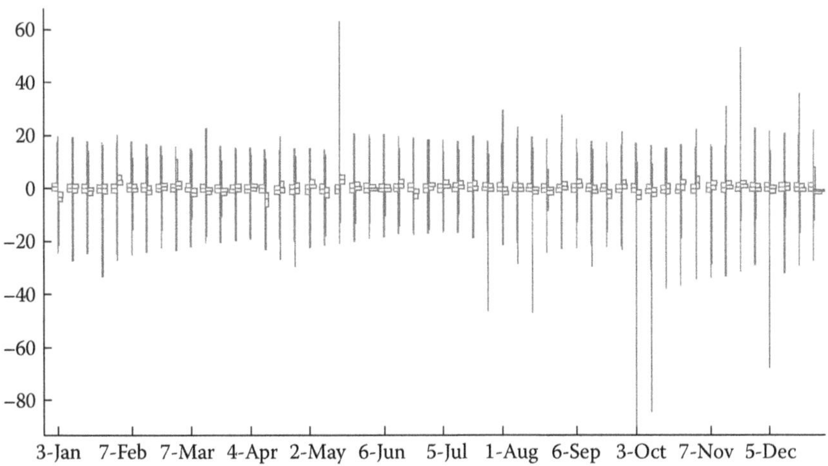

FIGURE 10.8
2005 realized histograms (the right ones) and exponential smoothed one-step-ahead histogram forecasts (the left ones) for the HTS of SP500 returns. Weekly data.

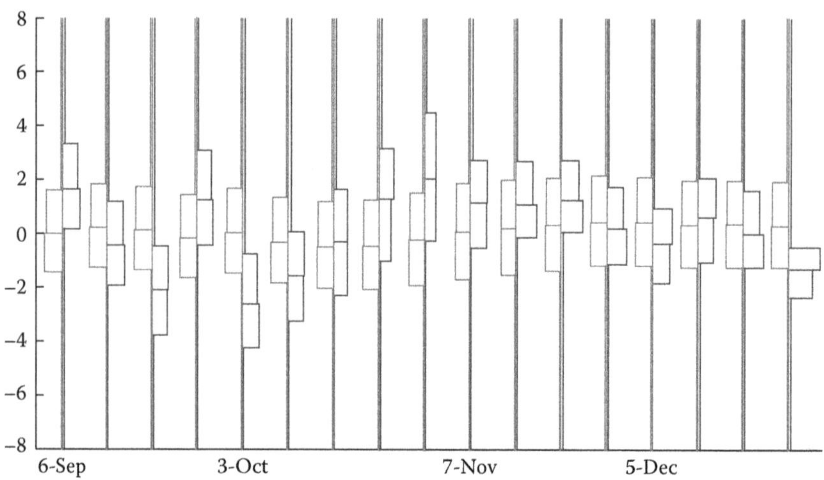

FIGURE 10.9
Zoom of Figure 10.8 from September to December 2005.

10.4 Summary and Conclusions

Large databases prompt the need for new methods of processing information. In this article we have introduced the analysis of interval-valued and histogram-valued data sets as an alternative to classical single-valued data sets and we have shown the promise of this approach to deal with economic and financial data.

With interval data, most of the current efforts have been directed to the adaptation of classical regression models as the interval is decomposed into two single-valued variables, either the center/radius or the min/max. The advantage of this decomposition is that classical inferential methods are available. Methodologies that analyze the interval per se fall into the realm of random sets theory and though there is some important research on regression analysis with random sets, inferential procedures are almost nonexistent. Being our current focus is the prediction problem, we have explored two different venues to produce a forecast with interval time series (ITS). First, we have implemented the classical regression approach to the analysis of ITS, and secondly we have proposed the adaptation to ITS of filtering techniques, such as smoothing, and nonparametric methods, such as the k-NN, to ITS. The latter venue requires the use of interval arithmetic to construct the appropriate averages and the introduction of distance measures to assess the dissimilarity between intervals and to quantify the prediction error. We have implemented these ideas with the SP500 index. We modeled the center/radius time series and the low/high time series of what we called interval-valued dispersion of the SP500 index and compared their one-step-ahead forecasts to those of a smoothing procedure and k-NN methods. A VEC model for the low/high series and the k-NN methods have the best forecasting performance.

With histogram data, the analysis becomes more complex. Regression analysis with histograms is in its infancy and the venues for further developments are large. We have focused exclusively in the prediction problem with smoothing methods and nonparametric methods. A key concept for the implementation of these two procedures is the introduction of the barycentric histogram that is a device that works as an average (weighted or unweighted) of a set of histograms. As with ITS, the introduction of the appropriate distances to judge dissimilarities among histograms and to assess forecast errors are fundamental ingredients in the analysis. The collection over time of cross-sectional returns of the firms in the SP500 index provides a nice histogram time series (HTS), on which we have implemented the aforementioned methods to eventually produce the one-step-ahead histogram forecast. Simple smoothing techniques seem to work remarkably well.

There are still many unexplored areas in ITS and HTS. A very important question is the search for a model. This will require the understanding of the notion of dependence in ITS and HTS. A first step in this direction is provided by González-Rivera and Arroyo (2010) who construct autocorrelation functions for HTS and ITS. From an econometric point of view, model building requires further research on identification, estimation, testing, and model selection procedures. Economic and financial questions will benefit greatly from this new approach to the analysis of large data sets.

10.5 Acknowledgment

We thank the referees and the editors for useful and constructive comments. Arroyo acknowledges support from the Spanish Council for Science and Innovation (grant TIN2008-06464-C03-01) and from the *Programa de Creación y Consolidación de Grupos de Investigación UCM-Banco Santander*. González-Rivera acknowledges the financial support provided by the University Scholar Award. Maté acknowledges the financial support provided by the project "Forecasting models from symbolic data (PRESIM)" from Universidad Pontificia Comillas.

Appendix

Estimation Results for ITS SP500 Index

TABLE A.1

Estimation of the VAR(3) Model for the Differenced Center and Radius Time Series

Estimation Sample 2000–2005			Estimation Sample 2002–2007		
VAR	**D(Cen)**	**Rad**	**VAR**	**D(Cen)**	**Rad**
D(Cen(−1))	0.33218	−0.09764	**D(Cen(−1))**	0.279225	−0.074092
	0.0262	0.00997		0.02619	0.00978
	[12.6803]	[−9.79410]		[10.6611]	[−7.57934]
D(Cen(−2))	−0.181348	−0.001809	**D(Cen(−2))**	−0.092471	−0.010534
	0.02742	0.01043		0.02713	0.01012
	[−6.61378]	[−0.17332]		[−3.40879]	[−1.04037]
D(Cen(−3))	0.050564	0.00429	**D(Cen(−3))**	0.006178	−0.013364
	0.02616	0.00996		0.02629	0.00981
	[1.93281]	[0.43091]		[0.23500]	[−1.36214]
Rad(−1)	0.066659	0.150616	**Rad(−1)**	−0.00284	0.152907
	0.06593	0.02509		0.06731	0.02512
	[1.01103]	[6.00287]		[−0.04219]	[6.08652]
Rad(-2)	−0.049629	0.313259	**Rad(-2)**	0.046537	0.27345
	0.06319	0.02405		0.0649	0.02422
	[−0.78541]	[13.0270]		[0.71705]	[11.2886]
Rad(-3)	0.129442	0.285272	**Rad(-3)**	−0.01386	0.276629
	0.0648	0.02466		0.06635	0.02477
	[1.99747]	[11.5678]		[−0.20888]	[11.1699]
C	−1.319847	2.088036	C	−0.045805	2.074405
	0.60607	0.23064		0.5355	0.19987
	[−2.17772]	[9.05315]		[−0.08554]	[10.3788]

TABLE A.2

Estimation of the VEC(3) Model for Low/High Time Series

Estimation Sample 2000–2005			Estimation Sample 2002–2007		
Error Correction:	**D(Low)**	**D(High)**	**Error Correction:**	**D(Low)**	**D(High)**
CointEq1	−0.438646	0.007023	**CointEq1**	−0.124897	0.121926
	0.05364	0.04758		0.04103	0.03692
	[−8.17770]	[0.14761]		[−3.04419]	[3.30283]
D(Low(-1))	0.112549	0.515586	**D(Low(-1))**	−0.165406	0.425054
	0.05429	0.04816		0.0489	0.044
	[2.07293]	[10.7050]		[−3.38238]	[9.66024]
D(Low(-2))	−0.093605	0.193326	**D(Low(-2))**	−0.314249	0.130253
	0.0505	0.0448		0.04863	0.04375
	[−1.85344]	[4.31532]		[−6.46233]	[2.97698]
D(Low(-3))	0.026446	0.112943	**D(Low(-3))**	−0.15041	0.061275
	0.0396	0.03512		0.0399	0.0359
	[0.66790]	[3.21547]		[−3.76992]	[1.70691]

TABLE A.2 (Continued)

Estimation of the VEC(3) Model for Low/High Time Series

Estimation Sample 2000–2005			Estimation Sample 2002–2007		
Error Correction:	**D(Low)**	**D(High)**	**Error Correction:**	**D(Low)**	**D(High)**
D(High(-1))	0.313542	−0.287591	**D(High(-1))**	0.524179	−0.221533
	0.05905	0.05238		0.05188	0.04668
	[5.30959]	[−5.49018]		[10.1046]	[−4.74625]
D(High(-2))	−0.073453	−0.382411	**D(High(-2))**	0.248088	−0.239401
	0.05604	0.04971		0.05323	0.04789
	[−1.31078]	[−7.69307]		[4.66085]	[−4.99871]
D(High(-3))	0.04646	−0.065429	**D(High(-3))**	0.182654	−0.073329
	0.04356	0.03864		0.04262	0.03835
	[1.06663]	[−1.69337]		[4.28593]	[−1.91234]
C	−0.064365	−0.118124	**Cointegrating Eq:**	**CointEq1**	
	0.28906	0.25642	**Low(-1)**	1	
	[−0.22267]	[−0.46068]	**High(-1)**	−1.002284	
Cointegrating Eq:	**Co-intEq1**			0.00318	
Low(-1)	1			[−315.618]	
High(-1)	−1.001255		C	16.82467	
	0.00268			3.81466	
	[−373.870]			[4.41053]	
@TREND(1)	−0.012818				
	0.00105				
	[−12.1737]				
C	27.97538				

References

Alizadeh, S., M. Brandt, and F. Diebold. 2002. Range-based estimation of stochastic volatility models. *Journal of Finance* 57(3):1047–1091.

Arroyo, J., R. Espínola, and C. Maté. 2010. Different approaches to forecast interval time series: a comparison in finance. *Computational Economics* Forthcoming.

Arroyo, J., G. González–Rivera, C. Maté, and A. Muñoz–San Roque. 2010. Smoothing methods for histogram-valued time series. An application to value-at-risk. Working paper, Department of Computer Science and Artificial Intelligence, Universidad Complutense de Madrid, Madrid. Spain.

Arroyo, J., and C. Maté. 2006. Introducing interval time series: accuracy measures. In *COMPSTAT 2006, Proceedings in Computational Statistics*. Heidelberg: Physica-Verlag, pp. 1139–1146.

Arroyo, J., and C. Maté. 2008. Forecasting time series of observed distributions with smoothing methods based on the barycentric histogram. In *Computational Intelligence in Decision and Control. Proceedings of the 8th International FLINS Conference*, pp. 61–66. Singapore: World Scientific.

Arroyo, J., and C. Maté. 2009. Forecasting histogram time series with k-nearest neighbours methods. *International Journal of Forecasting* 25(1):192–207.

Ball, C., and W. Torous. 1984. The maximim likelihood estimation of security price volatility: theory, evidence, and an application to option pricing. *Journal of Business* 57(1):97–112.

Bertrand, P., and F. Goupil. 2000. Descriptive statistics for symbolic data, *Analysis of Symbolic Data. Exploratory Methods for Extracting Statistical Information from Complex Data*. Berlin: Springer. pp. 103–124.

Billard, L., and E. Diday. 2000. Regression analysis for interval-valued data. In *Data Analysis, Classification and Related Methods: Proceedings of the 7th Conference of the IFCS, IFCS 2002*. Berlin: Springer. pp. 369–374.

Billard, L., and E. Diday. 2002. Symbolic regression analysis. In *Classification, Clustering and Data Analysis: Proceedings of the 8th Conference of the IFCS, IFCS 2002*. Berlin: Springer. pp. 281–288.

Billard, L., and E. Diday. 2003. From the statistics of data to the statistics of knowledge: symbolic data analysis. *Journal of the American Statistical Association* 98(462):470–487.

Billard, L., and E. Diday. 2006. *Symbolic Data Analysis: Conceptual Statistics and Data Mining*. 1st ed. Chichester: Wiley & Sons.

Blanco, Á., A. Colubi, N. Corral, and G. González-Rodríguez. 2008. On a linear independence test for interval-valued random sets. In *Soft Methods for Handling Variability and Imprecision. Advances in Soft Computing*. Berlin: Springer. pp. 111–117.

Brito, P. 2007. Modelling and analysing interval data. In *Proceedings of the 30th Annual Conference of GfKl*. Berlin: Springer. pp. 197–208.

Cheung, Y.-L., Y.-W. Cheung, and A. T. K. Wan. 2009. A high-low model of daily stock price ranges. *Journal of Forecasting* 28(2):103–119.

Colombo, A., and R. Jaarsma. 1980. A powerful numerical method to combine random variables. *IEEE Transactions on Reliability* 29(2):126–129.

Diebold, F. X., and R. S. Mariano. 1995. Comparing predictive accuracy. *Journal of Business and Economic Statistics* 13(3):253–263.

García-Ascanio, C., and C. Maté. 2010. Electric power demand forecasting using interval time series: a comparison between VAR and iMLP. *Energy Policy* 38(2):715–725.

Gardner, E. S. 2006. Exponential smoothing: the state of the art. Part 2. *International Journal of Forecasting* 22(4): 637–666.

Garman, M. B., and M. J. Klass. 1980. On the estimation of security price volatilities from historical data. *Journal of Business* 53(1):67–78.

González, L., F. Velasco, C. Angulo, J. A. Ortega, and F. Ruiz. 2004. Sobrenúcleos, distancias y similitudes entre intervalos. *Inteligencia Artificial, Revista Iberoamericana de IA* 8(23):111–117.

González-Rivera, G., and J. Arroyo. 2010. Time series modeling of histogram-valued data. The daily histogram time series of SP500 intradaily returns. *International Journal of Forecasting*. Forthcoming.

González-Rivera, G., T.-H. Lee, and S. Mishra. 2008. Jumps in cross-sectional rank and expected returns: a mixture model. *Journal of Applied Econometrics* 23:585–606.

González-Rodríguez, G., A. Blanco, N. Corral, and A. Colubi. 2007. Least squares estimation of linear regression models for convex compact random sets. *Advances in Data Analysis and Classification* 1:67–81.

Han, A., Y. Hong, K. Lai, and S. Wang. 2008. Interval time series analysis with an application to the sterling-dollar exchange rate. *Journal of Systems Science and Complexity* 21(4):550–565.

Irpino, A., and R. Verde. 2006. A new Wasserstein based distance for the hierarchical clustering of histogram symbolic data. In *Data Science and Classification, Proceedings of the IFCS 2006*. Berlin: Springer. pp. 185–192.

Kulpa, Z. 2006. A diagrammatic approach to investigate interval relations. *Journal of Visual Languages and Computing* 17(5):466–502.

Lima Neto, E. A., and F. d. A. T. de Carvalho. 2008. Centre and range method for fitting a linear regression model to symbolic interval data. *Computational Statistics and Data Analysis* 52:1500–1515.

Lima Neto, E. d. A., and F. d. A. T. de Carvalho. 2010. Constrained linear regression models for symbolic interval-valued variables. *Computational Statistics & Data Analysis* 54(2):333–347.

Maia, A. L. S., F. d. A. de Carvalho, and T. B. Ludermir. 2008. Forecasting models for interval-valued time series. *Neurocomputing* 71(16–18):3344–3352.

Manski, C., and E. Tamer. 2002. Inference on regressions with interval data on a regressor or outcome. *Econometrica* 70(2):519–546.

Moore, R. E. 1966. *Interval Analysis*. Englewood Cliffs, NJ: Prentice Hall.

Moore, R. E., R. B. Kearfott, and M. J. Cloud (eds.) 2009. *Introduction to Interval Analysis*. Philadelphia, PA: SIAM Press.

Parkinson, M. 1980. The extreme value method for estimating the variance of the rate of return. *The Journal of Business* 53(1):61.

Rogers, L., and S. Satchell. 1991. Estimation variance from high, low, and closing prices. *Annals of Applied Probability* 1(4):504–512.

Verde, R., and A. Irpino. 2007. Dynamic clustering of histogram data: using the right metric. In *Selected Contributions in Data Analysis and Classification*. Berlin: Springer. pp. 123–134.

Yakowitz, S. 1987. Nearest-neighbour methods for time series analysis. *Journal of Time Series Analysis* 8(2):235–247.

Yang, D., and Q. Zhang. 2000. Drift independent volatility estimation based on high, low, open, and close prices. *Journal of Business* 73(3):477–492.

Zellner, A., and J. Tobias. 2000. A note on aggregation, disaggregation and forecasting performance. *Journal of Forecasting* 19:457–469.

11

Predictability of Asset Returns and the Efficient Market Hypothesis

M. Hashem Pesaran

CONTENTS

11.1 Introduction

Economists have long been fascinated by the nature and sources of variations in the stock market. By the early 1970s a consensus had emerged among financial economists suggesting that stock prices could be well approximated by a random walk model and that changes in stock returns were basically unpredictable. Fama (1970) provides an early, definitive statement of this position. Historically, the "random walk" theory of stock prices was preceded by theories relating movements in the financial markets to the business cycle. A prominent example is the interest shown by Keynes in the variation in stock returns over the business cycle.

The efficient market hypothesis (EMH) evolved in the 1960s from the random walk theory of asset prices advanced by Samuelson (1965). Samuelson showed that in an informationally efficient market price changes must be unforecastable. Kendall (1953), Cowles (1960), Osborne (1959), Osborne (1962), and many others had already provided statistical evidence on the random nature of equity price changes. Samuelson's contribution was, however, instrumental in providing academic respectability for the hypothesis, despite the fact that the random walk model had been around for many years; having been originally discovered by Louis Bachelier, a French statistician, back in 1900.

Although a number of studies found some statistical evidence against the random walk hypothesis, these were dismissed as economically unimportant (could not generate profitable trading rules in the presence of transaction costs) and statistically suspect (could be due to data mining). For example, Fama (1965), concluded that "... there is no evidence of important dependence from either an investment or a statistical point of view." Despite its apparent empirical success, the random walk model was still a statistical statement and not a coherent theory of asset prices. For example, it need not hold in markets populated by risk averse traders, even under market efficiency.

There now exist many different versions of the EMH, and one of the aims of this chapter is to provide a simple framework where alternative versions of the EMH can be articulated and discussed. We begin with an overview of the statistical properties of asset returns at different frequencies (daily, weekly, and monthly), and consider the evidence on return predictability, risk aversion, and market efficiency. We then focus on the theoretical foundation of the EMH, and show that market efficiency could coexist with heterogeneous beliefs and individual "irrationality," so long as individual errors are cross sectionally weakly dependent in the sense defined by Chudik, Pesaran, and Tosetti (2010). But at times of market euphoria or gloom these individual errors are likely to become cross sectionally strongly dependent and the collective outcome could display significant departures from market efficiency. Market efficiency could be the norm, but most likely it will be punctuated by episodes of bubbles and crashes. To test for such episodes we argue in favour of compiling survey data on individual expectations of price changes that are combined with information on whether such expectations are compatible with market equilibrium. A trader who believes that asset prices are too high (low) might still expect further price rises (falls). Periods of bubbles and crashes could result if there are sufficiently large numbers of such traders that are prepared to act on the basis of their beliefs. The chapter also considers if periods of market inefficiency can be exploited for profit. We conclude with some general statements on new research directions.

We begin with some basic concepts and set out how returns are computed over different horizons and assets, and discuss some of the known stylized facts about returns by means of simple statistical models.

11.2 Prices and Returns

11.2.1 Single Period Returns

Let P_t be the price of a security at date t. The absolute price change over the period $t-1$ to t is given by $P_t - P_{t-1}$, the relative price change by

$$R_t = (P_t - P_{t-1})/P_{t-1}$$

the gross return (excluding dividends) on security by

$$1 + R_t = P_t/P_{t-1}$$

and the log price change by

$$r_t = \Delta \ln(P_t) = \ln(1 + R_t)$$

It is easily seen that for *small relative price changes* the log-price change and the relative price change are almost identical.

In the case of daily observations when dividends are negligible, $100 \cdot R_t$ measures the *percent return* on the security, and $100 \cdot r_t$ is the continuously compounded return. R_t is also known as discretely compounded return. The continuously compounded return, r_t, is particularly convenient in the case of temporal aggregation (multi-period returns; see Subsection 11.2.2), while the discretely compounded returns are convenient for use in cross-sectional aggregation, namely, aggregation of returns across different instruments in a portfolio. For example, for a portfolio composed of N instruments with weights $w_{i,t-1}$, ($\sum_{i=1}^{N} w_{i,t-1} = 1$, $w_{i,t-1} \geq 0$) we have

$$R_{pt} = \sum_{i=1}^{N} w_{i,t-1} R_{it}, \quad \text{(percent return)}$$

$$r_{pt} = \ln\left(\sum_{i=1}^{N} w_{i.t-1} e^{r_{it}}\right), \quad \text{(continuously compounded)}$$

Often r_{pt} is approximated by $\sum_{i=1}^{N} w_{i,t-1} r_{it}$.

When dividends are paid out we have

$$\begin{aligned} R_t &= (P_t - P_{t-1})/P_{t-1} + D_t/P_{t-1} \\ &\approx \Delta \ln(P_t) + D_t/P_{t-1} \end{aligned}$$

where D_t is the dividend paid out during the holding period.

11.2.2 Multi-Period Returns

Single-period price changes (returns) can be used to compute multi-period price changes or returns. Denote the return over the most recent h periods by $R_t(h)$ then (abstracting from dividends)

$$R_t(h) = \frac{P_t - P_{t-h}}{P_{t-h}}$$

or

$$1 + R_t(h) = P_t/P_{t-h}$$

and

$$r_t(h) = \ln(P_t/P_{t-h}) = r_t + r_{t-1} + \cdots + r_{t-h+1}$$

where r_{t-i}, $i = 0, 1, 2, \ldots, h-1$ are the single-period returns. For example, weekly returns are defined by $r_t(5) = r_t + r_{t-1} + \cdots + r_{t-4}$. Similarly, since there are 25 business days in one month, then the 1-month return can be computed as the sum of the last 25 1-day returns, or $r_t(25)$.

11.2.3 Overlapping Returns

Note that multi-period returns have overlapping daily observations. In the case of weekly returns, $r_t(5)$ and $r_{t-1}(5)$ have the four daily returns, $r_{t-1} + r_{t-2} + r_{t-3} + r_{t-4}$ in common. As a result the multi-period returns will be serially correlated even if the underlying daily returns are not serially correlated. One way of avoiding the overlap problem would be to sample the multi-period returns h periods apart. But this is likely to be inefficient as it does not make use of all available observations. A more appropriate strategy would be to use the overlapping returns but allow for the fact that this will induce serial correlations. For further details see Pesaran, Pick, and Timmermann (2010).

11.3 Statistical Models of Returns

A simple model of returns (or log-price changes) is given by

$$\begin{aligned} r_{t+1} &= \Delta \ln(P_{t+1}) = p_{t+1} - p_t \\ &= \mu_t + \sigma_t \varepsilon_{t+1}, \qquad t = 1, 2, \ldots, T \end{aligned} \tag{11.1}$$

where μ_t and σ_t^2 are the conditional mean and the conditional variance of returns (with respect to the information set Ω_t available at time t) and ε_{t+1} represents the unpredictable component of return. Two popular distributions for ε_{t+1} are

$$\varepsilon_{t+1} \mid \Omega_t \sim IID\ \mathcal{Z}$$

$$\varepsilon_{t+1} \mid \Omega_t \sim \left(\sqrt{\tfrac{v-2}{v}}\right) IID\ \mathcal{T}_v$$

where $\mathcal{Z} \sim N(0,1)$ stands for a standard normal distribution, and $\mathcal{T}_v$ stands for Student's t with v degrees of freedom. Unlike the normal distribution that has moments of all orders, $\mathcal{T}_v$ only has moments of order $v-1$ and smaller. For the Student's t to have a variance, for example, we need $v > 2$.

Since $r_{t+1} = \ln(1 + R_{t+1})$, where $R_{t+1} = (P_{t+1} - P_t)/P_t$, it then follows that under $\varepsilon_{t+1} \mid \Omega_t \sim IID\ \mathcal{Z}$, the price level, P_{t+1} conditional on Ω_t will be lognormally distributed. Note that $\Omega_t = (P_t, P_{t-1}, \dots)$ and $\Omega_t = (r_t, r_{t-1}, \dots)$ convey the same information and are equivalent. Hence, $P_{t+1} = P_t \exp(r_{t+1})$, and we have[1]

$$\begin{aligned} E(P_{t+1} \mid \Omega_t) &= P_t E(\exp(r_{t+1}) \mid \Omega_t) \\ &= P_t \exp\left(\mu_t + \frac{1}{2}\sigma_t^2\right). \end{aligned}$$

Similarly,

$$\mathrm{Var}(P_{t+1} \mid \Omega_t) = P_t^2 \exp(2\mu_t + \sigma_t^2)\left[\exp(\sigma_t^2) - 1\right].$$

In practice, it is much more convenient to work with log returns, r_{t+1}, rather than asset prices.

The probability density functions of $\mathcal{Z}$ and $\mathcal{T}_v$ are given by

$$f(\mathcal{Z}) = (2\pi)^{-1/2} \exp\left[\frac{-\mathcal{Z}^2}{2}\right], \quad -\infty < \mathcal{Z} < \infty \tag{11.2}$$

and

$$f(\mathcal{T}_v) = \frac{1}{\sqrt{v}B(v/2, 1/2)}\left[1 + \frac{\mathcal{T}_v^2}{v}\right]^{-(v+1)/2} \tag{11.3}$$

where $-\infty < \mathcal{T}_v < \infty$, and $B(v/2, 1/2)$ is the beta function defined by

$$B(\alpha, \beta) = \frac{\Gamma(\alpha)\Gamma(\beta)}{\Gamma(\alpha+\beta)}, \quad \Gamma(\alpha) = \int_0^\infty u^{\alpha-1} e^{-u} du.$$

[1] Using properties of the moment generating function of normal variates, if $x \sim N(\mu_x, \sigma_x^2)$ then, $E[\exp(x)] = \exp(\mu_x + .5\sigma_x^2)$.

It is easily seen that

$$E\left(\mathcal{T}_v\right) = 0, \text{ and } \operatorname{Var}\left(\mathcal{T}_v\right) = \frac{v}{v-2}.$$

A large part of financial econometrics is concerned with alternative ways of modeling the conditional mean (mean returns), μ_t, the conditional variance (asset return volatility), σ_t, and the cumulative probability distribution of the errors, ε_{t+1}. A number of issues need to be addressed in order to choose an adequate model. In particular:

- Is the distribution of returns normal?
- Is the distribution of returns constant over time?
- Are returns statistically independent over time?
- Are squares or absolute values of returns independently distributed over time?
- What are the cross correlation of returns on different instruments?

The above modeling issues can be readily extended to the case where we are concerned with a vector of asset returns, $\mathbf{r}_t = (r_{1t}, r_{2t}, \ldots r_{mt})'$. In this case we also need to model the pair-wise conditional correlations of asset returns, namely,

$$\operatorname{Corr}(r_{it}, r_{jt} \mid \Omega_t) = \frac{\operatorname{Cov}(r_{it}, r_{jt} \mid \Omega_t)}{\sqrt{\operatorname{Var}(r_{it} \mid \Omega_t) \operatorname{Var}(r_{jt} \mid \Omega_t)}}.$$

Typically the conditional variances and correlations are modeled using exponential smoothing procedures or the multivariate generalized autoregressive conditional heteroscedastic models developed in the econometric literature.

11.3.1 Percentiles, Critical Values, and Value at Risk

Suppose a random variable r (say daily returns on an instrument) has the probability density function $f(r)$. Then the pth percentile of the distribution of r, denoted by C_p, is defined as that value of return such that p percent of the returns fall below it. Mathematically we have

$$p = \Pr(r < C_p) = \int_{-\infty}^{C_p} f(r)dr.$$

In the literature on risk management C_p is used to compute "value at risk" or *VaR* for short. For $p = 1\%$, C_p associated with the one-sided critical value of the normal distribution is given by -2.33σ, where σ is the standard deviation of returns.

In hypothesis testing C_p is known as the critical value of the test associated with a (one-sided) test of size p. In the case of two-sided tests of size p, the associated critical value is computed as $C_{p/2}$.

11.3.2 Measures of Departure from Normality

The normal probability density function for r_{t+1} conditional on the information at time t, Ω_t, is given by

$$f(r_{t+1}) = (2\pi\sigma_t^2)^{-1/2} \exp\left[-\frac{1}{2\sigma_t^2}(r_{t+1} - \mu_t)^2\right]$$

with $\mu_t = E(r_{t+1} \mid \Omega_t)$ and $\sigma_t^2 = E\left[(r_{t+1} - \mu_t)^2 \mid \Omega_t\right]$ being the conditional mean and variance. If the return process is stationary, unconditionally we also have $\mu = E(r_{t+1})$, and $\sigma^2 = E[(r_{t+1} - \mu_t)^2]$.

Skewness and tail-fatness measures are defined by

$$\text{Skewness} = \sqrt{b_1} = m_3/m_2^{3/2}$$

$$\text{Kurtosis} = b_2 = m_4/m_2^2$$

where

$$m_j = \frac{\sum_{t=1}^{T}(r_t - \bar{r})^j}{T}, \qquad j = 2, 3, 4.$$

For a normal distribution $\sqrt{b_1} \approx 0$, and $b_2 \approx 3$. In particular

$$\hat{\mu} = \bar{r} = \sum_{t=1}^{T} r_t/T, \quad \hat{\sigma} = \sqrt{\frac{\sum_{t=1}^{T}(r_t - \bar{r})^2}{T-1}}$$

The Jarque–Bera's (1980) test statistic for departure from normality is given by

$$JB = T\left\{\frac{1}{6}b_1 + \frac{1}{24}(b_2 - 3)^2\right\}.$$

Under the joint null hypothesis that $b_1 = 0$ and $b_2 = 3$, the JB statistic is asymptotically distributed (as $T \to \infty$) as a chi-squared with 2 degrees of freedom, χ_2^2. Therefore, a value of JB in excess of 5.99 will be statistically significant at the 95 percent confidence level, and the null hypothesis of normality will be rejected.

11.4 Empirical Evidence: Statistical Properties of Returns

Table 11.1 gives a number of statistics for daily returns (×100) on four main equity index futures, namely, S&P 500 (SP), FTSE 100 (FTSE), German DAX (DAX), and Nikkei 225 (NK), over the period January 3, 2000 to August 31, 2009 (for a total of 2519 observations).

The kurtosis coefficients are particularly large for all the four equity futures and exceed the benchmark value of 3 for the normal distribution. There is

TABLE 11.1

Descriptive Statistics for Daily Returns on SP 500, FTSE 100, German DAX, and Nikkei 225

Variables	SP	FTSE	DAX	NK
Maximum	14.11	10.05	12.83	20.70
Minimum	−9.88	−9.24	−8.89	−13.07
Mean ($\bar{r}$)	−0.01	−0.01	−0.01	−0.01
S. D. ($\hat{\sigma}$)	1.39	1.33	1.65	1.68
Skewness ($\sqrt{b_1}$)	0.35	0.06	0.24	0.16
Kurtosis (b_2)	14.30	9.70	8.50	17.80
JB statistic	13453.6	4713.1	3199.2	23000.8

some evidence of positive skewness, but it is of second order importance as compared to the magnitude of excess kurtosis coefficient given by, $b_2 - 3$. The large values of excess kurtosis are reflected in the huge values of the *JB* statistics reported in Equation 11.1. Also under the assumption that returns are normally distributed, we would have expected the maximum and minimum of daily returns to fall (with 99% confidence) in the region of $\pm 2.33 \times$ S. D., which is ± 3.24 for SP500, as compared to the observed values of −9.88 and 14.11. See also Figure 11.1.

The departure from normality is particularly pronounced over the past decade where markets have been subject to two important episodes of financial crises: the collapse of markets in 2000 after the dot-com bubble and the stock market crash of 2008 after the 2007 credit crunch. (see Figure 11.2).

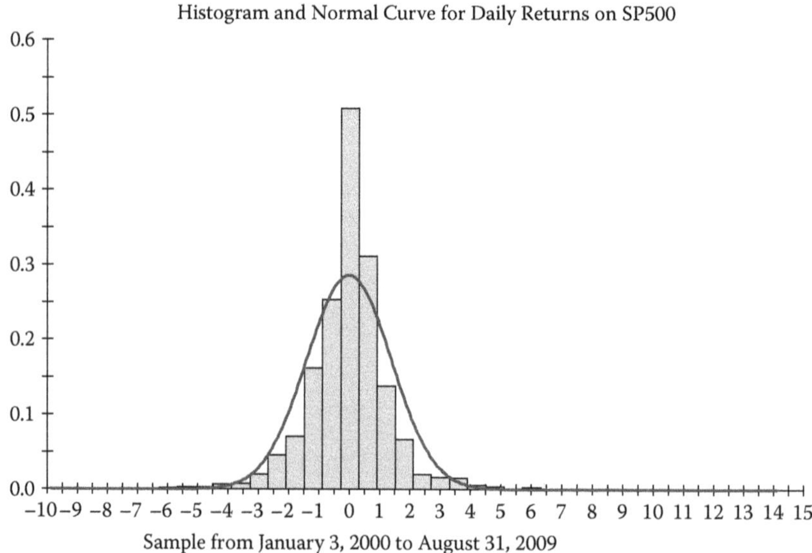

FIGURE 11.1
Histogram and Normal curve for daily returns on SP500 (over the period January 3, 2009 to August 31, 2009).

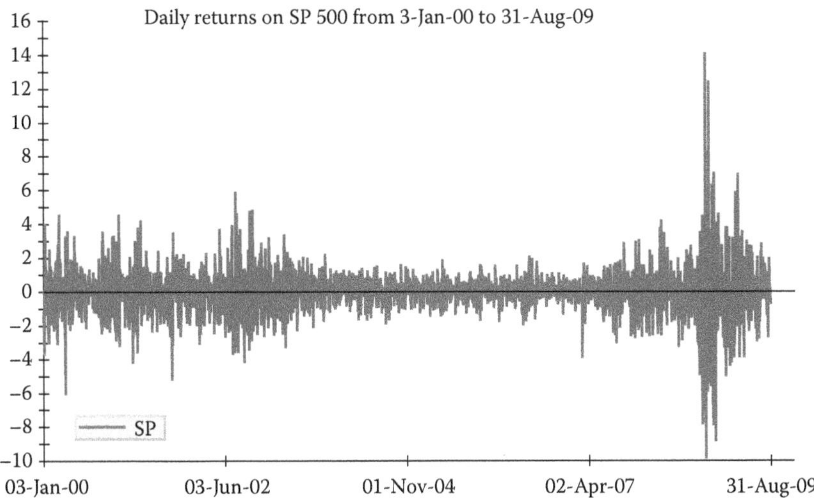

FIGURE 11.2
Daily returns on SP500 (over the period January 3, 2000 to August 31, 2009).

However, the evidence of departure from normality can be seen in daily returns even before 2000. For example, over the period January 3, 1994 to December 31, 1999 (1565 daily observations) kurtosis coefficient of returns on SP500 was 9.5 which is still well above the benchmark value of 3. The recent financial crisis has accentuated the situation but can not be viewed as the cause of the observed excess kurtosis of equity returns.

Similar results are also obtained if we consider weekly returns. The kurtosis coefficients estimated using weekly returns over the period January 2000 to the end of August 2009 (504 weeks) were 12.4, 15.07, 8.9, and 15.2 for SP500, FTSE, DAX, and Nikkei, respectively. These are somewhat lower than the estimates obtained using daily observations for SP500 and Nikkei, but are quite a bit higher for FTSE. For DAX daily and weekly observations yield a very similar estimate of the kurtosis coefficient.

For currencies the kurtosis coefficient of returns (measured in terms of U.S. dollar) varies from 4.5 for euro to 13.8 for the Australian dollar. The estimates computed using daily observations over the period January 3, 2000, to August 31, 2009 are summarized in Table 11.2. The currencies considered are the British pound (GBP), euro (EU), Japanese yen (JPY), Swiss franc (CHF), Canadian dollar (CAD), and Australian dollar (AD), all measured in terms of U.S. dollar.

The returns on government bonds are generally less fat-tailed than the returns on equities and currencies. But their distribution still shows a significant degree of departure from normality.

Table 11.3 reports descriptive statistics on daily returns on the main four government bond futures: U.S. T-Note 10Y (BU), Europe Euro Bund 10Y (BE), Japan Government Bond 10Y (BJ), and UK Long Gilts 8.75-13Y (BG) over the period 03 Jan., 00 to 31 Aug., 09.

TABLE 11.2

Descriptive Statistics for Daily Returns on British Pound, Euro, Japanese Yen, Swiss Franc, Canadian Dollar, and Australian Dollar

Variables	JPY	EU	GBP	CHF	CAD	AD
Maximum	4.53	3.17	3.41	4.58	5.25	6.21
Minimum	−3.93	−3.01	−5.04	−3.03	−3.71	−9.50
Mean ($\bar{r}$)	−0.006	0.016	0.007	0.012	0.013	0.022
S. D. ($\hat{\sigma}$)	−0.65	0.65	0.60	0.70	0.59	0.90
Skewness ($\sqrt{b_1}$)	−0.28	0.01	−0.35	0.12	0.09	−0.76
Kurtosis (b_2)	5.99	4.50	7.20	4.90	9.10	13.80

It is clear that for all the three asset classes there are significant departures from normality which needs to be taken into account when analyzing financial time series.

11.4.1 Other Stylized Facts about Asset Returns

Asset returns are typically uncorrelated over time, are difficult to predict, and, as we have seen, tend to have distributions that are fat-tailed. In contrast the absolute or squares of asset returns (that measure risk), namely $|r_t|$ or r_t^2, are serially correlated and tend to be predictable. It is interesting to note that r_t can be written as

$$r_t = \text{sign}(r_t)\,|r_t|$$

where $\text{sign}(r_t) = +1$ if $r_t > 0$ and $\text{sign}(r_t) = -1$ if $r_t \leq 0$. Since $|r_t|$ is predictable, it is, therefore, the nonpredictability of $\text{sign}(r_t)$, or the direction of the market, which lies behind the difficulty of predicting returns.

The extent to which returns are predictable depends on the forecast horizon, the degree of market volatility, and the state of the business cycle. Predictability tends to rise during crisis periods. Similar considerations also apply to the degree of fat-tailedness of the underlying distribution and the cross correlations of asset returns. The return distributions become less fat-tailed as the horizon is increased, and cross correlations of asset returns become more predictable with the horizon. Cross correlation of returns also tends to increase

TABLE 11.3

Descriptive Statistics for Daily Returns on U.S. T-Note 10Y, Europe Euro Bund 10Y, Japan Government Bond 10Y, and UK Long Gilts 8.75-13Y

Variables	BU	BE	BG	BJ
Maximum	3.63	1.48	2.43	1.53
Minimum	−2.40	−1.54	−1.85	−1.41
Mean ($\bar{r}$)	0.00	0.01	0.01	0.01
S. D. ($\hat{\sigma}$)	0.43	0.32	0.35	0.24
Skewness ($\sqrt{b_1}$)	−0.004	−0.18	0.02	−0.18
Kurtosis (b_2)	6.67	4.49	6.02	6.38

with market volatility. The analysis of time variations in the cross correlation of asset returns is beyond the scope of this chapter. However, the interested reader might wish to consult Pesaran and Pesaran (2010) where multivariate conditional volatility models are fitted to weekly returns on equities, bonds, and currencies.

In the case of daily returns, equity returns tend to be negatively serially correlated. During normal times they are small and only marginally significant statistically, but become relatively large and attain a high level of statistical significance during crisis periods. These properties are illustrated in the following empirical application.

The first and second order serial correlation coefficients of daily returns on SP500 over the period January 3, 2000 to August 31, 2007, are −0.015 (0.0224) and −0.0458 (0.0224), respectively, but increase to −0.068 (0.0199) and −0.092 (0.0200) once the sample is extended to the end of August 2009 which covers the 2008 global financial crisis.[2] Similar patterns are also observed for other equity indices. For currencies the evidence is more mixed. In the case of major currencies such as euro and yen, there is little evidence of serial correlation in returns and this outcome does not seem much affected by whether one considers normal or crisis periods. For other currencies there is some evidence of negative serial correlation, particularly at times of crisis. For example, over the period January 3, 2000 to August 31, 2009 the first-order serial correlation of daily returns on Australian dollar amounts to −0.056 (0.0199), but becomes statistically insignificant if we exclude the crisis period. There is also very little evidence of serial correlation in daily returns on the four major government bonds that we have been considering. This outcome does not depend on whether the crisis period is included in the sample. Irrespective of whether the underlying returns are serially correlated, their absolute values (or their squares) are highly serially correlated, often over many periods. For example, over the January 3, 2000 to August 31, 2009 period the first and second order serial correlation coefficients of absolute return on SP500 are 0.2644 (0.0199), 0.3644 (0.0204); for euro they are 0.0483 (0.0199) and 0.1125 (0.0200); and for U.S. 10Y bond they are 0.0991 (0.0199) and 0.1317 (0.0201). The serial correlation in absolute returns tends to decay very slowly and continues to be statistically significant event after 120 trading days. See Figure 11.3.

It is also interesting to note that there is little correlation between r_t and $|r_t|$. Based on the full sample ending in August 2009, this correlation is −0.0003 for SP500, 0.025 for euro, and 0.009 for the U.S. 10Y bond.

11.4.2 Monthly Stock Market Returns

Many of the regularities and patterns documented for returns using daily or weekly observations can also be seen in monthly observations, once a sufficiently long period is considered. For the U.S. stock market long historical monthly data on prices and dividends are compiled by Shiller and can be

[2] The figures in parentheses are standard errors.

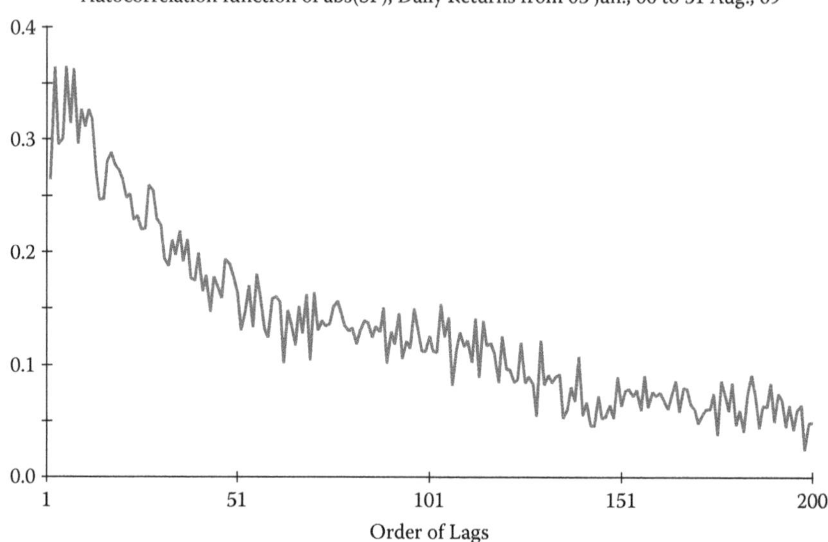

FIGURE 11.3
Autocorrelation function of the absolute values of SP500 (over the period January 3, 2009 to August 31, 2009)

downloaded from his homepage.[3] An earlier version of this data set has been analyzed in Shiller (2005). Monthly returns on SP500 (inclusive of dividends) is computed as

$$\text{RSP}_t = 100\left(\frac{\text{SP}_t - \text{SP}_{t-1} + \text{SPDIV}_t}{\text{SP}_{t-1}}\right)$$

where SP_t is the monthly spot price index of SP500 and SPDIV_t denotes the associated dividends on the SP500 index. Over the period 1871m1 to 2009m9 (a total of 1664 monthly observations) the coefficient of skewness and kurtosis of RSP amounted to 1.07% and 23.5%, respectively. The excess kurtosis coefficient of 20.5 is much higher than the figure of 11.3 obtained for the daily observations on SP over the period January 3, 2000 to August 31, 2009. Also as before the skewness coefficient is relatively small. However, the monthly returns show a much higher degree of serial correlation and a lower degree of volatility as compared to daily or weekly returns. The correlation coefficients of RSP are 0.346 (0.0245) and 0.077 (0.027), and the serial correlation coefficients continue to be statistically significant up to the lag order of 12 months. Also, the pattern of serial correlations in absolute monthly returns, $|\text{RSP}_t|$, is not that different from that of the serial correlation in RSP_t, which suggests a lower degree of return volatility (as compared to the volatility of daily or weekly returns) once the effects of mean returns are taken into account.

[3] See http://www.econ.yale.edu/~shiller/data.htm.

Similar, but less pronounced, results are obtained if we exclude the 1929 stock market crash and focus on the post World War II period. The coefficients of skewness and kurtosis of monthly returns over the period 1948m1 to 2009m9 (741 observations) are –0.49 and 5.2, respectively. The first- and second-order serial correlation coefficients of returns are 0.361 (0.0367) and 0.165 (0.041), respectively. The main difference between these subsample estimates and those obtained for the full sample is the much lower estimate for the kurtosis coefficient. But even the lower post-1948 estimates suggest a significant degree of fat-tailedness in the monthly returns.

11.5 Stock Return Regressions

Consider the linear excess return regression

$$R_{t+1} - r_t^f = a + b_1 x_{1t} + b_2 x_{2t} + \cdots + b_k x_{kt} + \varepsilon_{t+1} \quad (11.4)$$

where R_{t+1} is the one-period holding return on an stock index, such as FTSE or Dow Jones, defined by

$$R_{t+1} = (P_{t+1} + D_{t+1} - P_t)/P_t. \quad (11.5)$$

P_t is the stock price at the end of the period and D_{t+1} is the dividend paid out over the period t to $t+1$, and x_{it} , $i = 1, 2, \ldots, k$ are the factors/variables thought to be important in predicting stock returns. Finally, r_t^f is the return on the government bond with one-period to maturity (the period to maturity of the bond should be exactly the same as the holding period of the stock). $R_{t+1} - r_t^f$ is known as the excess return (return on stocks in excess of the return on the safe asset). Note also that r_t^f would be known to the investor/trader at the end of period t, before the price of stocks, P_{t+1}, is revealed at the end of period $t+1$.

Examples of possible stock market predictors are past changes in macroeconomic variables such as interest rates, inflation, dividend yield (D_t/P_{t-1}), price earnings ratio, output growth, and term premium (the difference in yield of a high-grade and a low-grade bond such as AAA rated minus BAA rated bonds).

For individual stocks the relevant stock market regression is the capital asset pricing model (CAPM), augmented with potential predictors:

$$R_{i,t+1} = a_i + b_{1i} x_{1t} + b_{2i} x_{2t} + \cdots + b_{ki} x_{kt} + \beta_i R_{t+1} + \varepsilon_{i,t+1} \quad (11.6)$$

where $R_{i,t+1}$ is the holding period return on asset i (shares of firm i), defined similarly as R_{t+1}. The asset-specific regressions (Equation 11.6) could also include firm specific predictors, such as R_{it} or its higher order lags, book-to-market value or size of firm i. Under market efficiency, as characterized by CAPM,

$$a_i = 0, b_{1i} = b_{2i} = \cdots = b_{ki} = 0$$

and only the "betas," β_i, will be significantly different from zero. Under CAPM, the value of β_i captures the risk of holding the share i with respect to the market.

11.6 Market Efficiency and Stock Market Predictability

It is often argued that if stock markets are efficient then it should not be possible to predict stock returns, namely, that none of the variables in the stock market regression (Equation 11.4) should be statistically significant. Some writers have even gone so far as to equate stock market efficiency with the non-predictability property. But this line of argument is not satisfactory and does not help in furthering our understanding of how markets operate. The concept of market efficiency needs to be defined separately from predictability. In fact, it is easily seen that stock market returns will be nonpredictable only if market efficiency is combined with risk neutrality.

11.6.1 Risk Neutral Investors

Suppose there exists a risk free asset such as a government bond with a known payout. In such a case an investor with an initial capital of $\$A_t$ is faced with two options:

Option 1: Hold the risk-free asset and receive

$$\$(1 + r_t^f) A_t$$

at the end of the next period.

Option 2: Switch to stocks by purchasing A_t/P_t shares, hold them for one period and expect to receive

$$\$ (A_t/P_t) (P_{t+1} + D_{t+1})$$

at the end of period $t + 1$.

A risk-neutral investor will be indifferent between the certainty of $\$(1 + r_t^f) A_t$, and his/her expectations of the uncertain payout of option 2. Namely, for such a risk neutral investor

$$(1 + r_t^f) A_t = E\left[(A_t/P_t)(P_{t+1} + D_{t+1}) \,|\Omega_t\right] \tag{11.7}$$

where Ω_t is the investor's information at the end of period t. This relationship is called the "arbitrage condition." Using Equation 11.5 we now have

$$P_{t+1} + D_{t+1} = P_t(1 + R_{t+1})$$

and the above arbitrage condition can be simplified to

$$E\left[(1 + R_{t+1})\,|\Omega_t\right] = \left(1 + r_t^f\right)$$

or

$$E\left(R_{t+1} - r_t^f\,|\Omega_t\right) = 0. \tag{11.8}$$

This result establishes that if the investor forms his/her expectations of future stock (index) returns taking account of all market information efficiently, then the excess return, $R_{t+1} - r_t^f$, should not be predictable using any of the market information that is available at the end of period t. Notice that r_t^f is known at time t and is therefore included in Ω_t. Hence, under the joint hypothesis of market efficiency and risk neutrality we must also have $E\left(R_{t+1}\,|\Omega_t\right) = r_t^f$.

The above set up can also be used to derive conditions under which asset prices can be characterized as a random walk model. Suppose, the risk free rate, r_t^f, in addition to being known at time t, is also constant over time and given by r^f. Then using Equation 11.7 we can also write

$$P_t = \left(\frac{1}{1+r}\right) E\left[(P_{t+1} + D_{t+1})\,|\Omega_t\right]$$

or

$$P_t = \left(\frac{1}{1+r^f}\right)\left[E\left(P_{t+1}\,|\Omega_t\right) + E\left(D_{t+1}\,|\Omega_t\right)\right].$$

Under the rational expectations hypothesis and assuming that the "transversality condition"

$$\lim_{j\to\infty}\left(\frac{1}{1+r^f}\right)^j E\left(P_{t+j}\,|\Omega_t\right) = 0$$

holds we have the familiar result

$$P_t = \sum_{j=1}^{\infty}\left(\frac{1}{1+r^f}\right)^j E\left(D_{t+j}\,|\Omega_t\right) \tag{11.9}$$

that equates the level of stock price to the present discounted stream of the dividends expected to occur to the asset over the infinite future. The transversality condition rules out rational speculative bubbles and is satisfied if the asset prices are not expected to rise faster than the exponential decay rate determined by the discount factor, $0 < 1/(1 + r^f) < 1$. It is now easily seen that if D_t follows a random walk so will P_t. For example, suppose

$$D_t = D_{t-1} + \varepsilon_t \tag{11.10}$$

where ε_t is a white noise process. Then

$$E\left(D_{t+j}\,|\Omega_t\right) = D_t$$

and

$$P_t = \frac{D_t}{r^f} \tag{11.11}$$

Therefore, we also have

$$P_t = P_{t-1} + u_t \tag{11.12}$$

where $u_t = \varepsilon_t / r^f$.

The random walk property holds even if $r^f = 0$, since in such a case it would be reasonable to expect no dividends are also paid out, namely $D_t = 0$. In this case the arbitrage condition becomes

$$E\left(P_{t+1} \left| \Omega_t \right.\right) = P_t \tag{11.13}$$

which is satisfied by the random walk model but is in fact more general than the random walk model. An asset price that satisfies Equation 11.13 is a martingale process. Random walk processes with zero drift are martingale processes but not all martingale processes are random walks. For example, the price process

$$P_{t+1} = P_t + \lambda \left\{ (\Delta P_{t+1})^2 - E\left[(\Delta P_{t+1})^2 \left| \Omega_t \right. \right] \right\} + \varepsilon_t$$

where ε_t is a white noise process, is a martingale process with respect to the information set Ω_t, but it is clearly not a random walk process, unless $\lambda = 0$.

Other modifications of the random walk theory is obtained if it is assumed that dividends follow a geometric random walk which is more realistic than the linear dividend model assumed in Equation 11.10. In this case

$$D_{t+1} = D_t \exp(\mu_d + \sigma_d \nu_{t+1}) \tag{11.14}$$

where μ_d and σ_d are the mean and standard deviation of the growth rate of the dividends. If it is further assumed that $\nu_{t+1} \left| \Omega_t \right.$ is $N(0, 1)$, we have

$$E\left(D_{t+j} \left| \Omega_t \right.\right) = D_t \exp\left(j\mu_d + \frac{1}{2} j\sigma_d^2 \right)$$

Using this result in Equation 11.9 now yields [assuming that $(1 + r^f)^{-1} \exp$ $(\mu_d + \frac{1}{2}\sigma_d^2) < 1$]

$$P_t = \frac{D_t}{\rho} \tag{11.15}$$

where

$$\rho = (1 + r^f) \exp\left(-\mu_d - \frac{1}{2}\sigma_d^2 \right) - 1$$

The condition $(1+r^f)^{-1}\exp\left(\mu_d + \frac{1}{2}\sigma_d^2\right) < 1$ ensures that the infinite sum in Equation 11.9 is convergent and $\rho > 0$. Under this set up $\ln(P_t) = \ln(D_t) - \ln(\rho)$, and

$$\ln(P_t) = \ln(P_{t-1}) + \mu_d + \sigma_d \nu_t \tag{11.16}$$

which establishes that in this case it is log prices that follow the random walk model. This is a special case of the statistical model of return, (Equation 11.1), discussed in Section 11.3, where $\mu_t = \mu_d$, and $\sigma_t = \sigma_d$.

There are, however, three different types of empirical evidence that shed doubt on the empirical validity of the present value model under risk neutrality.

1. The model predicts a constant price-dividend ratio for a large class of the dividend processes. Two prominent examples, the linear and the geometric random walk models, (Equations 11.10 and 11.14) are discussed earlier. For more general dividend processes the price-dividend ratio, $\rho_t = P_t/D_t$, could be time-varying, but it must be mean-reverting, in the sense that shocks to prices and dividends must eventually cancel out. In reality, the price-dividend ratio varies considerably over time, shows a high degree of persistence, and in general it is not possible to reject the hypothesis that the processes for ρ_t or $\ln(\rho_t)$ contain a unit root. For the Shiller data discussed in Subsection 11.4.2 the autocorrelation coefficient of the log dividend to price ratio computed over the period 1871m1 to 2009m9 is 0.994 (0.024) and falls very gradually as its order is increased and amounts to 0.879 (0.111) at the lag order 12.
2. We have already established that under risk neutrality excess returns must not be predictable. See Equation 11.8. Yet there is ample evidence of excess return predictability at least in periods of high market volatility. For example, it is possible to explain 15% of the variations in monthly excess returns on SP500 over the period 1872m2 to 2009m9 by running a linear regression of the excess return on a constant and its 12 lagged values — namely, by a univariate $AR(12)$ process. This figure rises to 19% if we exclude the 1929 stock market crash and focus on the post-1948 period. See also the references cited in Subsection 11.7.1.
3. To derive the geometric random walk model of asset prices (Equation 11.16) from the present value model under risk neutrality, we have assumed that innovations to the dividend process are normally distributed. This implies that innovations to asset returns must also be normally distributed. But the empirical evidence discussed in Section 11.4 clearly shows that innovations to asset returns tend to be fat-tailed, and often significantly depart from normality. This anomaly between the theory and the evidence is also difficult to reconcile. Under the present value model prices will have fat-tailed innovations only if the dividends that drive asset prices are also fat-tailed. But under the geometric

random walk model for dividends (Equation 11.14), $E\left(D_{t+j}\,|\Omega_t\right)$ need not exist if the dividend innovations, ν_t, are fat-tailed. One important example arises when ν_t has the Student t distribution as defined by Equation 11.3. For the derivation of the present value expression in this case we need $E(\exp(\sigma_d \nu_{t+j}))$, which is the moment generating function of ν_{t+j} evaluated at σ_d. But the Student t distribution does not have a moment generating function, and hence the present value formula cannot be computed when innovations to the dividends are t distributed.

11.6.2 Risk Averse Investors

In addition to the above documented empirical shortcomings, it is also important to note that risk neutrality is a behavioral assumption and need not hold even if all market information is processed efficiently by all the market participants. A more reasonable way to proceed is to allow some or all of the investors to be risk averse. In this more general case the certain pay out, $(1+r_t^f)A_t$, and the expectations of the uncertain pay out, $E\left[(A_t/P_t)\left(P_{t+1}+D_{t+1}\right)|\Omega_t\right]$, will not be the same and differ by a (possibly) time-varying risk premium which could also vary with the level of the initial capital, A_t. More specifically, we have

$$E\left[(A_t/P_t)\left(P_{t+1}+D_{t+1}\right)|\Omega_t\right]=\left(1+r_t^f\right)A_t+\lambda_t A_t$$

where λ_t is the premium per \$ of invested capital required (expected) by the investor. It is now easily seen that

$$E\left(R_{t+1}-r_t^f\,|\Omega_t\right)=\lambda_t$$

and it is no longer necessarily true that under market efficiency excess returns are nonpredictable. The extent to which excess returns can be predicted will depend on the existence of a historically stable relationship between the risk premium, λ_t, and the macro and business cycle indicators such as changes in interest rates, dividends, and various business cycle indicators.

In the context of the consumption capital asset pricing model λ_t is determined by the *ex ante* correlation of excess returns and changes in the marginal utility of consumption. In the case of a representative consumer with the single period utility function, $u(c_t)$, the first-order intertemporal optimization condition (the Euler equation) is given by

$$E\left[\left(R_{t+1}-r_t^f\right)\frac{u'(c_{t+1})}{u'(c_t)}\,|\Omega_t\right]=0 \qquad (11.17)$$

where c_t denotes the consumer's real consumption in period t. Using the above condition it is now easily seen that[4]

$$\lambda_t = -\frac{\mathrm{Cov}\left[R_{t+1}, \frac{u'(c_{t+1})}{u'(c_t)} \,|\Omega_t\right]}{E\left[\frac{u'(c_{t+1})}{u'(c_t)} \,|\Omega_t\right]} = -\frac{\mathrm{Cov}\left[R_{t+1}, u'(c_{t+1}) \,|\Omega_t\right]}{E\left[u'(c_{t+1}) \,|\Omega_t\right]}$$

For a power utility function, $u(c_t) = (c_t^{1-\gamma} - 1)/(1-\gamma)$, and $u'(c_{t+1})/u'(c_t) = \exp(-\gamma\Delta\ln(c_{t+1}))$, where $\gamma > 0$ is the coefficient of relative risk aversion. In this case λ_t is given by

$$\lambda_t = \frac{-\mathrm{Cov}\left[R_{t+1}, \exp(-\gamma\Delta\ln(c_{t+1})) \,|\Omega_t\right]}{E\left[\exp(-\gamma\Delta\ln(c_{t+1})) \,|\Omega_t\right]}. \tag{11.18}$$

This result shows that the risk premium depends on the covariance of asset returns with the marginal utility of consumption. The premium demanded by the investor to hold the stock is higher if the return on the asset co-varies positively with consumption. The extent of this co-variation depends on the magnitude of the risk aversion coefficient γ. For plausible values of γ (in the range 1–3) and historically observed values of the consumption growth, we would expect λ_t to be relatively small, below 1% per annum. However, using annual observations over relatively long periods one obtains a much larger estimate for λ_t. This was first pointed out by Mehra and Prescott (1985) who found that in the 90 years from 1889 to 1978 the average estimate of λ_t in fact amounted to 6.18% per annum, which could only be reconciled with the theory if one was prepared to consider an implausibly large value for the relative risk aversion coefficient (in the regions of 30 or 40). The large discrepancy between the historical estimate of λ_t based on $R_{t+1} - r_t^f$, and the theory-consistent estimate of λ_t based on Equation 11.18, is known as the "equity premium puzzle." There have been many attempts in the literature to resolve the puzzle by modifications to the utility function, attitudes toward risk, allowing for the possibility of rare events, and the heterogeneity in asset holdings and preferences across consumers. For reviews see Kocherlakota (2003) and Mehra and Prescott (2003).

But even if the mean discrepancy between $E(R_{t+1} - r_t^f \,|\Omega_t)$ and λ_t as given by Equation 11.18 is resolved, the differences in the higher moments of historically and theory-based risk premia are likely to be important empirical issues of concern. It seems difficult to reconcile the high volatility of excess returns with the low volatility of consumption growth that are observed historically.

[4] Let $X_{t+1} = R_{t+1} - r_t^f$ and $Y_{t+1} = u'(c_{t+1})/u'(c_t)$, and write the Euler equation (Equation 11.17) as

$$E\left[X_{t+1}Y_{t+1} \,|\Omega_t\right] = 0 = \mathrm{Cov}[X_{t+1}Y_{t+1}|\Omega_t] + E\left[X_{t+1} \,|\Omega_t\right] E\left[Y_{t+1} \,|\Omega_t\right].$$

Then the required results follow immediately, also noting that r_t^f is known at time t and hence has a zero correlation with $u'(c_{t+1})/u'(c_t)$.

11.7 Return Predictability and Alternative Versions of the Efficient Market Hypothesis

In his 1970 review, Fama distinguishes between three different forms of the EMH:

1. The *weak* form asserts that all price information is fully reflected in asset prices, in the sense that current price changes cannot be predicted from past prices. This weak form was also introduced in an unpublished paper by Roberts (1967).
2. The *semi-strong* form that requires asset price changes to fully reflect all publicly available information and not only past prices.
3. The *strong* form that postulates that prices fully reflect information even if some investor or group of investors have monopolistic access to some information.

Fama regarded the strong form version of the EMH as a benchmark against which the other forms of market efficiencies are to be judged. With respect to the weak form version he concludes that the test results strongly support the hypothesis, and considered the various departures documented as economically unimportant. He reached a similar conclusion with respect to the semi-strong version of the hypothesis although, as he noted, the empirical evidence available at the time was rather limited and far less comprehensive as compared to the evidence on the weak version.

The three forms of the EMH present different degrees whereby public and private information are revealed in transaction prices. It is difficult to reconcile all three versions to the mainstream asset pricing theory, and as we shall see below a closer connection is needed between market efficiency and the specification of the model economy that underlies it.

11.7.1 Dynamic Stochastic Equilibrium Formulations and the Joint Hypothesis Problem

Evidence on the semi-strong form of the EMH was revisited by Fama in a second review of the *Efficient Capital Markets* published in 1991. By then it was clear that the distinction between the weak and the semi-strong forms of the EMH was redundant. The random walk model could not be maintained either, in view of more recent studies, in particular that of Lo and MacKinlay (1988).

A large number of studies in the finance literature had confirmed that stock returns over different horizons (days, weeks, and months) can be predicted to some degree by means of interest rates, dividend yields and a variety of macroeconomic variables exhibiting clear business cycle variations. A number of studies also showed that returns tend to be more predictable the longer the forecast horizon. While the vast majority of these studies had looked at the US stock market, an emerging literature has also considered the UK stock market.

US studies include Balvers, Cosimano, and MacDonald (1990), Breen, Glosten, and Jagannathan (1989), Campbell (1987), Fama and French (1989), and subsequently by Ferson and Harvey (1993), Kandel and Stambaugh (1996), Pesaran and Timmermann (1994), and Pesaran and Timmermann (1995). See Granger (1992) for a survey of the methods and results in the literature. UK studies after 1991 included Clare, Thomas, and Wickens (1994), Clare, Psaradakis, and Thomas (1995), Black and Fraser (1995), and Pesaran and Timmermann (2000).

Theoretical advances over Samuelson's seminal paper by Leroy (1973), Rubinstein (1976), and Lucas (1978) also made it clear that in the case of risk averse investors tests of predictability of excess returns could not on their own confirm or falsify the EMH. The neoclassical theory cast the EMH in the context of dynamic stochastic (general) equilibrium models and showed that excess returns weighted by marginal utility could be predictable. Only under risk neutrality, where marginal utility was constant, did the equilibrium condition imply the nonpredictability of excess returns.

As Fama (1991) noted in his second review, the test of the EMH involved a joint hypothesis — market efficiency and the underlying equilibrium asset pricing model. He concluded that "Thus, market efficiency *per se* is not testable" (see p. 1575). This did not, however, mean that market efficiency was not a useful concept. Almost all areas of empirical economics are subject to the joint hypotheses problem.

11.7.2 Information and Processing Costs and the EMH

The EMH, in the sense of asset "prices fully reflect all available information" was also criticized by Grossman and Stiglitz (1980) who pointed out that there must be "sufficient profit opportunities, i.e. inefficiencies, to compensate investors for the cost of trading and information-gathering."

Only in the extreme and unrealistic case where all information and trading costs are zero would one expect prices to fully reflect all available information. But if information is in fact costless it would be known even before market prices are established.

As Fama recognized a weaker and economically more sensible version of the efficiency hypothesis would be needed, namely, "prices reflect information to the point where the marginal benefits of acting on information (the profits to be made) do not exceed the marginal costs." This in turn makes the task of testing the market efficiency even more complicated and would require equilibrium asset pricing models that allowed for information and trading costs in markets with many different traders and with nonconvergent beliefs.

In view of these difficulties some advocates of the EMH have opted for a trade-based notion, and define markets as efficient if it would not be possible for the investors "... to earn above-average returns without accepting above-average risks" (Malkiel 2003, see p. 60). This notion can take account of information and transaction costs and does not involve testing joint hypotheses. But this is far removed from the basic idea of markets as efficient allocators of capital investment across countries, industries, and firms.

Beating the market as a test of market efficiency also poses new challenges. Whilst it is certainly possible to construct trading strategies (inclusive of transaction costs) with Sharpe ratios that exceed those of the market portfolios *ex post*, such evidence is unlikely to be convincing to the advocates of the EMH. It could be argued that they are carried out with the benefit of hindsight, and are unlikely to be repeated in real time. In this connections the following considerations would need to be born in mind:

1. Data mining/data snooping (Pesaran and Timmermann 2005).
2. Structural change and model instability (choice of observation window).
3. The positive relationship that seems to exist between transaction costs and predictability.
4. Market volatility and learning.
5. The "Beat the market" test is not that helpful either in shedding light on the nature and the extent of market inefficiencies. A more structural approach would be desirable.

11.8 Theoretical Foundations of the EMH

At the core of the EMH lies the following three basic premises:

1. *Investor rationality*: It is assumed that investors are rational, in the sense that they correctly update their beliefs when new information is available.
2. *Arbitrage*: Individual investment decisions satisfy the arbitrage condition, and trade decisions are made guided by the calculus of the subjective expected utility theory à la Savage.
3. *Collective rationality*: Differences in beliefs across investors cancel out in the market.

To illustrate how these premises interact, suppose that at the start of period (day, week, month) t there are N_t traders (investors) that are involved in acts of arbitrage between a stock and a safe (risk-free) asset. Denote the one-period holding returns on these two assets by R_{t+1} and r_t^f, respectively. Following a similar line of argument as in Subsection 11.6.2, the arbitrage condition for trader i is given by

$$\hat{E}_i\left(R_{t+1} - r_t^f \,|\Omega_{it}\right) = \lambda_{it} + \delta_{it}$$

where $\hat{E}_i\left(R_{t+1} - r_t^f \,|\Omega_{it}\right)$ is his/her subjective expectations of the excess return, $R_{t+1} - r_t^f$ taken with respect to the information set

$$\Omega_{it} = \Psi_{it} \cup \Phi_t$$

where Φ_t is the component of the information that is publicly available, $\lambda_{it} > 0$ represents trader's risk premium, and $\delta_{it} > 0$ is her/his information and trading costs per unit of funds invested. In the absence of information and trading costs, λ_{it} can be characterized in terms of the trader's utility function, $u_i(c_{it})$, where c_t is his/her real consumption expenditure during the period t to $t+1$, and is given by

$$\lambda_{it} = \hat{E}_i\left(R_{t+1} - r_t^f \,|\Omega_{it}\right) = \frac{-\hat{\text{C}}\text{ov}_i(m_{i,t+1}, R_{t+1}|\Omega_{it})}{\hat{E}_i(m_{i,t+1}|\Omega_{it})}$$

where $\hat{\text{C}}\text{ov}_i\,(.\,|\Omega_{it})$ is the subjective covariance operator condition on the trader's information set, Ω_{it}, $m_{i,t+1} = \beta_i u_i'(c_{i,t+1})/u_i'(c_{it})$, which is known as the "stochastic discount factor," $u_i'(.)$ is the first derivative of the utility function, and β_i is his/her discount factor.

The expected returns could differ across traders due to the differences in their perceived conditional probability distribution function of $R_{t+1} - r_t^f$, the differences in their information sets, Ω_{it}, the differences in their risk preferences, and/or endowments. Under the rational expectations hypothesis

$$\hat{E}_i\left(R_{t+1} - r_t^f \,|\Omega_{it}\right) = E\left(R_{t+1} - r_t^f \,|\Omega_{it}\right)$$

where $E\left(R_{t+1} - r_t^f \,|\Omega_{it}\right)$ is the "true" or "objective" conditional expectation. Furthermore, in this case

$$E\left[\hat{E}_i\left(R_{t+1} - r_t^f \,|\Omega_{it}\right)|\Phi_t\,\right] = E\left[E\left(R_{t+1} - r_t^f \,|\Omega_{it}\right)|\Phi_t\right]$$

and since $\Phi_t \subset \Omega_{it}$ we have

$$E\left[\hat{E}_i\left(R_{t+1} - r_t^f \,|\Omega_{it}\right)|\Phi_t\,\right] = E\left(R_{t+1} - r_t^f \,\,|\Phi_t\right)$$

Therefore, under the REH, taking expectations of the individual arbitrage conditions with respect to the public information set yields

$$E\left(R_{t+1} - r_t^f \,\,|\Phi_t\,\right) = E\,(\lambda_{it} + \delta_{it}\,|\Phi_t\,)$$

which also implies that $E\,(\lambda_{it} + \delta_{it}\,|\Phi_t\,)$ *must* be the same across all i, or

$$E\left(R_{t+1} - r_t^f \,|\Phi_t\right) = E\,(\lambda_{it} + \delta_{it}\,|\Phi_t\,) = \rho_t, \text{ for all } i$$

where ρ_t is an average market measure of the combined risk premia and transaction costs. The REH combined with perfect arbitrage ensures that different *traders* have the same expectations of $\lambda_{it} + \delta_{it}$. Rationality and market discipline override individual differences in tastes, information processing abilities, and other transaction related costs and renders the familiar representative agent arbitrage condition:

$$E\left(R_{t+1} - r_t^f \,|\Phi_t\right) = \rho_t. \tag{11.19}$$

This is clearly compatible with trader-specific λ_{it} and δ_{it}, so long as

$$\lambda_{it} = \lambda_t + \varepsilon_{it}, \; E(\varepsilon_{it} | \Phi_t) = 0$$
$$\delta_{it} = \delta_t + \upsilon_{it}, \; E(\upsilon_{it} | \Phi_t) = 0$$

where ε_{it} and υ_{it} are distributed with mean zero independently of Φ_t, and λ_t and δ_t are known functions of the publicly available information.

Under this setting the extent to which excess returns can be predicted will depend on the existence of a historically stable relationship between the risk premium, λ_t, and the macro and business cycle indicators such as changes in interest rates, dividends, and various other indicators.

The rational expectations hypothesis is rather extreme, and is unlikely to hold at all times in all markets. Even if one assumes that in financial markets learning takes place reasonably fast, there will still be periods of turmoil where market participants will be searching in the dark, trying and experimenting with different models of $R_{t+1} - r_t^f$ often with marked departures from the common rational outcomes, given by $E(R_{t+1} - r_t^f | \Phi_t)$.

Herding and correlated behavior across some of the traders could also lead to further departures from the equilibrium RE solution. In fact the objective probability distribution of $R_{t+1} - r_t^f$ might itself be affected by market transactions based on subjective estimates $\hat{E}_i(R_{t+1} - r_t^f | \Omega_{it})$.

Market inefficiencies provide further sources of stock market predictability by introducing a wedge between a "correct" *ex ante* measure $E(R_{t+1} - r_t^f | \Phi_t)$, and its average estimate by market participants, which we write as

$$\sum_{i=1}^{N_t} w_{it} \hat{E}_i(R_{t+1} - r_t^f | \Omega_{it})$$

where w_{it} is the market share of the ith trader.

Let

$$\bar{\xi}_{wt} = \sum_{i=1}^{N_t} w_{it} \hat{E}_i(R_{t+1} - r_t^f | \Omega_{it}) - E(R_{t+1} - r_t^f | \Phi_t)$$

and note that it can also be written as (since $\sum_{i=1}^{N_t} w_{it} = 1$)

$$\bar{\xi}_{wt} = \sum_{i=1}^{N_t} w_{it} \xi_{it} \tag{11.20}$$

where

$$\xi_{it} = \hat{E}_i(R_{t+1} - r_t^f | \Omega_{it}) - E(R_{t+1} - r_t^f | \Phi_t). \tag{11.21}$$

ξ_{it} measures the degree to which individual expectations differ from the correct (but unobservable) expectations, $E(R_{t+1} - r_t^f | \Phi_t)$. A nonzero ξ_{it} could arise from individual irrationality, but not necessarily so. Rational individuals faced with an uncertain environment, costly information and limitations on computing power could rationally arrive at their expectations of future price changes that with hindsight differ from the correct ones.[5] A nonzero ξ_{it} could also arise due to disparity of information across traders (including information asymmetries), and heterogeneous priors due to model uncertainty or irrationality. Nevertheless, despite such individual deviations, $\bar{\xi}_{wt}$ which measures the extent of market or collective inefficiency, could be quite negligible. When N_t is sufficiently large, individual "irrationality" can cancel out at the level of the market, so long as $\xi_{it}, i = 1, 2, \ldots, N_t$ are not cross sectionally strongly dependent, and no single trader dominates the market, in the sense that $w_{it} = O(N_t^{-1})$ at any time.[6] Under these conditions at each point in time, t, the average expected excess returns across the individual traders converges in quadratic means to the expected excess return of a representative trader, namely, we have

$$\sum_{i=1}^{N_t} w_{it} \hat{E}_i \left(R_{t+1} - r_t^f | \Omega_{it} \right) \overset{q.m.}{\rightarrow} E\left(R_{t+1} - r_t^f | \Phi_t \right), \text{ as } N_t \rightarrow \infty.$$

In such periods the representative agent paradigm would be applicable and predictability of excess return will be governed solely by changes in business cycle conditions and other publicly available information.[7]

However, in periods where traders' individual expectations become strongly correlated (say, as the result of herding or common over-reactions to distressing news) $\bar{\xi}_{wt}$ need not be negligible even in thick markets with many traders; and market inefficiencies and profitable opportunities could prevail. Markets could also display inefficiencies without exploitable profitable opportunities if $\bar{\xi}_{wt}$ is nonzero but there is no stable predictable relationship between $\bar{\xi}_{wt}$ and business cycle or other variables that are observed publicly.

The evolution and composition of $\bar{\xi}_{wt}$ can also help in shedding light on possible bubbles or crashes developing in asset markets. Bubbles tend to develop in the aftermath of technological innovations that are commonly acknowledged to be important, but with uncertain outcomes. The emerging common beliefs about the potential advantages of the new technology and the difficulties individual agents face in learning how to respond to the new investment opportunities can further increase the gap between average market expectations of excess returns and the associated objective rational

---

[5] This is in line with the premise of the recent paper by Angeletos, Lorenzoni, and Pavan (2010) who maintain the axiom of rationality, but allow for dispersed information and the possibility of information spillovers in the financial markets to explain market inefficiencies.

[6] Concepts of weak and strong cross-section dependence are defined and discussed in Chudik, Pesaran, and Tosetti (2010).

[7] The heterogeneity of expectations across traders can also help in explaining large trading volume observed in the financial markets; a feature that has proved difficult to explain in representative agent asset pricing models. But see Scheinkman and Xiong (2003) who relate the occurrence of bubbles and crashes to changes in trading volume.

expectations outcome. Similar circumstances can also prevail during a crash phase of the bubble when traders tend to move in tandem trying to reduce their risk exposures all at the same time. Therefore, one would expect that during bubbles and crashes the individual errors, ξ_{it}, to become more correlated, such that the average errors, $\bar{\xi}_{wt}$, are no longer negligible. In contrast, at times of market calm the individual errors are likely to be weakly correlated, with the representative agent rational expectations model being a reasonable approximation.

More formally note that since r_t^f and P_t are known at time t, then

$$\xi_{it} = \hat{E}_i\left(\frac{P_{t+1} + D_{t+1}}{P_t} \,|\Omega_{it}\right) - E\left(\frac{P_{t+1} + D_{t+1}}{P_t} \,|\Phi_t\right).$$

Also, to simplify the exposition, assume that the length of the period t is sufficiently small so that dividends are of secondary importance and

$$\xi_{it} \approx \hat{E}_i\left(\Delta \ln(P_{t+1}) \,|\Omega_{it}\right) - f_t,$$

where $f_t = E\left(\Delta \ln(P_{t+1}) \,|\Phi_t\right)$ is the unobserved price change expectations. Individual deviations, ξ_{it}, could then become strongly correlated if individual expectations $\hat{E}_i\left(\Delta \ln(P_{t+1}) \,|\Omega_{it}\right)$ differ systematically from f_t. For example, suppose that

$$\hat{E}_i\left(\Delta \ln(P_{t+1}) \,|\Omega_{it}\right) = \theta_{it}\Delta \ln(P_t),$$

but $f_t = 0$, namely, in the absence of heterogeneous expectations $\Delta \ln(P_{t+1})$ would have been unpredictable with a zero mean. Then it is easily seen that $\bar{\xi}_{wt} = \bar{\theta}_{wt}\Delta \ln(P_t)$, where $\bar{\theta}_{wt} = \Sigma_{i=1}^{N_t} w_{it}\theta_{it}$. It is clear that $\bar{\xi}_{wt}$ need not converge to zero if in period t the majority of market participants believe future price changes are positively related to past price changes, so that $\lim_{N_t\to\infty} \bar{\theta}_{wt} > 0$. In this simple example price bubbles or crashes occur when $\bar{\theta}_{wt}$ becomes positive over a relatively long period.

It should be clear from the above discussion that testing for price bubbles requires disaggregated time series information on individual beliefs and unobserved price change expectations, f_t. Analysis of aggregate time series observations can provide historical information about price reversals and some of their proximate causes. But such information is unlikely to provide conclusive evidence of bubble formation and its subsequent collapse. Survey data on traders' individual beliefs combined with suitable market proxies for f_t are likely to be more effective in empirical analysis of price bubbles.

An individual investor could be asked to respond to the following two questions regarding the current and future price of a given asset:

1. Do you believe the current price is (a) just right (in the sense that the price is in line with market fundamentals), (b) is above the fundamental price, or (c) is below the fundamental price?
2. Do you expect the market price next period to (a) stay about the level it is currently, (b) fall, or (c) rise?

In cases where the market is equilibrating we would expect a close association between the proportion of respondents who select 1a and 2a, 1b and 2b, and 1c and 2c. But in periods of bubbles (crashes) one would expect a large proportion of respondents who select 1b (1c) to also select 2c (2b).

In situations where the equilibrating process is well established and commonly understood, the second question is redundant. For example, if an individual states that the room temperature is too high, it will be understood that he/she would prefer less heating. The same is not applicable to financial markets and hence responses to both questions are needed for a better understanding of the operations of the markets and their evolution over time.

11.9 Exploiting Profitable Opportunities in Practice

In financial markets the EMH is respected but not worshipped. It is recognized that markets are likely to be efficient most of the time but not all the time. Inefficiencies could arise particularly during periods of important institutional and technological changes. It is not possible to know when and where market inefficiencies arise in advance — but it is believed that they will arise from time to time. Market traders love volatility as it signals news and change with profit possibilities to exploit. Identification of exploitable predictability tend to be fully diversified across markets for bonds, equities, and foreign exchange. Misalignments across markets for different assets and in different countries often present the most important opportunities. Examples include statistical arbitrage and global macro arbitrage trading rules.

Predictability and market liquidity are often closely correlated; less liquid markets are likely to be more predictable. Market predictability and liquidity need to be jointly considered in developing profitable trading strategies. Return forecasting models used in practice tend to be recursive and adaptive along the lines developed in Pesaran and Timmermann (1995) and recently reviewed in Pesaran and Timmermann (2005). The recursive modeling (RM) approach is also in line with the more recent developments in behavioral finance. The RM approach aims at minimizing the effect of hindsight and data snooping (a problem that afflicts all *ex post* return regressions), and is explicitly designed to take account of the potential instability of the return regressions over time. For example, Pesaran and Timmermann (1995) find that the switching trading rule manages to beat the market only during periods of high volatility where learning might be incomplete and markets inefficient.

Pesaran and Timmermann (2005) provide a review of the recursive modeling approach, its use in derivation of trading rules, and discuss a number of practical issues in their implementation such as the choice of the universe of factors over which to search, choice of the estimation window, how to take account of measurement and model uncertainty, how to cross validate the RM, and how and when to introduce model innovations.

The RM approach still faces many challenges ahead. As Pesaran and Timmermann (2005) conclude:

> Automated systems reduce, but do not eliminate, the need for discretion in real time decision making. There are many ways that automated systems can be designed and implemented. The space of models over which to search is huge and is likely to expand over time. Different approximation techniques such as genetic algorithms, simulated annealing and MCMC algorithms can be used. There are also many theoretically valid model selection or model averaging procedures. The challenge facing real time econometrics is to provide insight into many of these choices that researchers face in the development of automated systems.

Return forecasts need to be incorporated in sound risk management systems. For this purpose point forecasts are not sufficient and joint probability forecast densities of a large number of interrelated asset returns will be required. Transaction and slippage costs need to be allowed for in the derivation of trade rules. Slippage arises when long (short) orders, optimally derived based on currently observed prices, are placed in rising (falling) markets. Slippage can be substantial, and is in addition to the usual transactions costs.

Familiar risk measures such as the Sharpe ratio and the VaR are routinely used to monitor and valuate the potential of trading systems. But due to cash constraint (for margin calls, etc.) it is large drawdowns that are most feared. Prominent recent examples are the downfall of long-term capital which experienced substantial drawdowns in 1998 following the Russian financial crisis, and the collapse of Lehman Brothers during the global financial crisis of 2008.

Successful traders might not be (and usually are not) better in forecasting returns than many others in the market. What they have is a sense of "big" opportunities when they are confident of making a "kill."

11.10 New Research Directions

We have identified two important sources of return predictability and possible profitable opportunities. One relates to the familiar business cycle effects and involves modeling ρ_t, defined by Equation 11.19, in terms of the publicly available information, Φ_{t-1}. The second relates to the average deviations of individual traders' expectations from the "correct" unknown expectations, as measured by $\bar{\xi}_{wt}$ and defined by Equation 11.20. As noted earlier this component could vary considerably over time and need not be related to business cycle factors. It tends to be large during periods of financial crisis when the correlation of mispricing across traders rises, and negligible during periods of market calm when correlations are low. Over the past three decades much of the research in finance and macroeconomics has focused on modeling of ρ_t, and by comparison little attention has been paid to $\bar{\xi}_{wt}$. This is clearly an important area for future research. Our discussions also point to a number of related areas for further research. There are clearly

- Limits to rational expectations (for an early treatment see (Pesaran 1987), also see the recent paper on "Survey Expectations" by Pesaran and Weale (2006).
- Limits to arbitrage due to liquidity requirements and institutional constraints.
- Herding and correlated behavior with noise traders entering markets during bull periods and deserting during bear periods.

Behavioral finance, complexity theory, and the adaptive markets hypothesis recently advocated by Lo (2004) all try, in one way or another, to address the above sources of the departures from the EMH. Some of the recent developments in behavioral finance are reviewed in Baberis and Thaler (2003).

Farmer and Lo (1999) focus on the recent research that views the financial markets from a biological perspective and, specifically, within an evolutionary framework in which markets, instruments, institutions, and investors interact and evolve dynamically according to the "law" of economic selection. Under this view, financial agents compete and adapt, but they do not necessarily do so in an optimal fashion.

Special care should also be exercised in evaluation of return predictability and trading rules. To minimize the effects of hindsight in such an analysis, recursive modeling techniques discussed in Pesaran and Timmermann (1995), Pesaran and Timmermann (2000), and Pesaran and Timmermann (2005) seem much more appropriate than the return regressions on a fixed set of regressors/factors that are estimated *ex post* on historical data.

11.11 Acknowledgment

I am grateful to Elisa Tosetti for valuable help with the preparation of this paper, and to Alex Chudik, Bill Janeway, Ron Smith, Ansgar Walther, and Aman Ullah for helpful comments. Part of this paper is based on my presentation at the CFS symposium "Market Efficiency Today" held in Frankfurt/Main on October 6, 2005 in honor of Eugene F. Fama.

References

Angeletos, G., G. Lorenzoni, and A. Pavan. 2010. Beauty contests and irrational exuberance: A neoclassical approach. Working Paper 15883. Cambridge, MA: National Bureau of Economic Research.

Baberis, N. and R. Thaler. 2003. A survey of behavioral finance. In G. M. Constantinides, M. Harris, and R. Stultz (eds.), *Handbook of Behavioral Economics of Finance*. Amsterdam: Elsevier Science. pp. 1052–1090.

Balvers, R. J., T. F. Cosimano, and B. MacDonald. 1990. Predicting stock returns in an efficient market. *The Journal of Finance* 45:1109–1128.

Black, A., and P. Fraser. 1995. UK stock returns: Predictability and business conditions. *The Manchester School Supplement* 63:85–102.

Breen, W., L. R. Glosten, and R. Jagannathan. 1989. Economic significance of predictable variations in stock index returns. *Journal of Finance* 44:1177–1189.

Campbell, J. Y. 1987. Stock returns and the term structure. *Journal of Financial Economics* 18:373–399.

Chudik, A., M. H. Pesaran, and E. Tosetti. 2010. Weak and strong cross section dependence and estimation of large panels. European Central Bank: ECB Working Paper No. 1100.

Clare, A. D., Z. Psaradakis, and S. H. Thomas. 1995. An analysis of seasonality in the UK equity market. *Economic Journal* 105:398–409.

Clare, A. D., S. H. Thomas, and M. R. Wickens. 1994. Is the gilt-equity yield ratio useful for predicting UK stock return? *Economic Journal* 104:303–315.

Cowles, A. 1960. A revision of previous conclusions regarding stock price behavior. *Econometrica* 28:909–915.

Fama, E. F. 1965. The behavior of stock market prices. *Journal of Business* 38:34–105.

Fama, E. F. 1970. Efficient capital markets: A review of theory and empirical work. *The Journal of Finance* 25:383–417.

Fama, E. F. 1991. Efficient capital markets: II. *The Journal of Finance* 46:1575–1617.

Fama, E. F. and K. R. French. 1989. Business conditions and expected returns on stocks and bonds. *Journal of Financial Economics* 25:23–49.

Farmer, D., and A. Lo. 1999. Frontiers of finance; evolution and efficient markets. In *Proceedings of the National Academy of Sciences*. 96:9991–9992.

Ferson, W. E., and C. R. Harvey. 1993. The risk and predictability of international equity returns. *Review of Financial Studies* 6:527–566.

Granger, C. W. J. 1992. Forecasting stock market prices: Lessons for forecasters. *International Journal of Forecasting* 8:3–13.

Grossman, S., and J. Stiglitz. 1980. On the impossibility of informationally efficient markets. *American Economic Review* 70:393–408.

Jarque, C. M. and A. K. Bera. 1980. Efficient tests for normality, homoscedasticity and serial independence of regression residuals. *Economics Letters* 6:255–259.

Kandel, S., and R. F. Stambaugh. 1996. On the predictability of stock returns: An asset-allocation perspective. *Journal of Finance* 51:385–424.

Kendall, M. 1953. The analysis of economic time series - part 1: Prices. *Journal of the Royal Statistical Society* 96:11–25.

Kocherlakota, N. R. 2003. The equity premium: It's still a puzzle. *Journal of Economic Literature* 34:42–71.

Leroy, S. 1973. Risk aversion and the martingale property of stock returns. *International Economic Review* 14:436–446.

Lo, A. 2004. The adaptive markets hypothesis: Market efficiency from an evolutionary perspective. *Journal of Portfolio Management* 30:15–29.

Lo, A., and C. MacKinlay. 1988. Stock market prices do not follow random walks: evidence from a simple specification test. *Review of Financial Studies* 1:41–66.

Lucas, R. E. 1978. Asset prices in an exchange economy. *Econometrica* 46:1429–1446.

Malkiel, B. G. 2003. The efficient market hypothesis and its critics. *Journal of Economic Perspectives* 17:59–82.

Mehra, R., and E. Prescott. 1985. The equity premium: A puzzle. *Journal of Monetary Economics* 15:146–161.

Mehra, R., and E. C. Prescott. 2003. The equity premium puzzle in retrospect. In M. H. G. M. Constantinides and R. Stulz (eds). *Handbook of the Economics of Finance*. Amsterdam: North Holland. pp. 889–938.

Osborne, M. 1959. Brownian motion in the stock market. *Operations Research* 7:145–173.

Osborne, M. 1962. Periodic structures in the brownian motion of stock prices. *Operations Research* 10:345–379.

Pesaran, B., and M. H. Pesaran. 2010. Conditional volatility and correlations of weekly returns and the VaR analysis of 2008 stock market crash. *Economic Modelling*, forthcoming.

Pesaran, M. H. 1987. *The Limits to Rational Expectations*. Oxford, U.K.: Basil Blackwell. Reprinted with corrections 1989.

Pesaran, M. H., A. Pick, and A. Timmermann. 2010. Variable selection, estimation and inference for multi-period forecasting problems. Netherlands Central Bank: DNB Working Paper 250.

Pesaran, M. H., and A. Timmermann. 1994. Forecasting stock returns: An examination of stock market trading in the presence of transaction costs. *Journal of Forecasting* 13:335–367.

Pesaran, M. H., and A. Timmermann. 1995. The robustness and economic significance of predictability of stock returns. *Journal of Finance* 50:1201–1228.

Pesaran, M. H., and A. Timmermann. 2000. A recursive modelling approach to predicting uk stock returns. *The Economic Journal* 110:159–191.

Pesaran, M. H., and A. Timmermann. 2005. Real time econometrics. *Econometric Theory* 21:212–231.

Pesaran, M. H., and M. Weale. 2006. Survey expectations. *Handbook of Economic Forecasting*. Amsterdam: North-Holland.

Roberts, H. 1967. Statistical versus clinical prediction in the stock market. Unpublished manuscript, Center for Research in Security Prices, University of Chicago.

Rubinstein, M. 1976. The valuation of uncertain income streams and the pricing of options. *Bell Journal of Economics* 7:407–425.

Samuelson, P. 1965. Proof that properly anticipated prices fluctuate randomly. *Industrial Management Review Spring* 6:41–49.

Scheinkman, J. A., and W. Xiong. 2003. Overconfidence and speculative bubbles. *Journal of Political Economy* 111:1183–1219.

Shiller, R. J. 2005. *Irrational Exuberance.* (2nd ed.). Princeton NJ: Princeton University Press.

12

A Factor Analysis of Bond Risk Premia

Sydney C. Ludvigson and Serena Ng

CONTENTS

12.1 Introduction

The expectations theory of the term structure posits that variables in the information set at time t should have no predictive power for excess bond returns. Consider the predictive regression

$$r_{t+h} = a + b'Z_t + e_{th}$$

where r_{t+h} is excess returns for holding period h, and Z_t is a set of predictors. Conventional tests often reject the null hypothesis that the parameter vector b is zero. Some suggest that over-rejections may arise if r is stationary and the variables Z are highly persistent, making inference highly distorted in finite samples. For this reason, researchers often use finite sample corrections or the bootstrap to conduct inference. However, it is often the case that robust inference still points to a rejection of the null hypothesis.

For a long time, the Zs found to have predictive power are often financial variables such as default premium, term premium, dividend price ratio, and measures of stock market variability and liquidity. Cochrane and Piazzesi (2005) find that a linear combination of five forward spreads explains between 30% and 35% of the variation in next year's excess returns on bonds with maturities ranging from 2 to 5 years. Yet theory suggests that predictive power for excess bond returns should come from macroeconomic variables. Campbell (1999) and Wachter (2006) suggest that bond and equity risk premia should covary with a slow-moving habit driven by shocks to aggregate consumption. Brandt and Wang (2003) argue that risk premia are driven by shocks to inflation as well as aggregate consumption; notably, both are macroeconomic shocks.

In an effort to reconcile theory and evidence, recent work has sought to establish and better understand the relation between excess returns and macroeconomic variables. Piazzesi and Swanson (2004) find that the growth of nonfarm payroll employment is a strong predictor of excess returns on federal funds futures contracts. Ang and Piazzesi (2003) use a no-arbitrage factor model of the term structure of interest rates that also allows for time-varying risk premia and finds that the pricing kernel is driven by a few observed macroeconomic variables and unobserved yield factors. Kozicki and Tinsley (2005) use affine models to link the term structure to perceptions of monetary policy. Duffie (2008) finds that an "expectations" factor unrelated to the level and the slope has strong predictive power for short-term interest rates and excess returns, and that this expectations factor has a strong inverse relation with industrial production. Notably, these studies have focused on the relation between expected excess bond returns, risk premia, and a few selected macroeconomic variables. The evidence falls short of documenting a direct relation between expected excess bond returns (bond risk premia) and the macroeconomy.

In Ludvigson and Ng (2007), we used a new approach. We used a small number of estimated (static) factors instead of a handful of observed predictors in the predictive regressions, where the factors are estimated from a large panel of macroeconomic data using the method of asymptotic principal components (PCA). Such a predictive regression is a special case of what is known as a "factor augmented regression" (FAR).[1] The factors enable us to substantially reduce the dimension of the predictor set while still being able to use the information underlying the variables in the panel. Furthermore, our latent factors are estimated without imposing a no-arbitrage condition or any parametric structure. Thus, our testing framework is nonstructural, both from an economic and a statistical point of view. We find that latent factors associated

[1] See Bai and Ng (2008) for a survey on this literature.

with real economic activity have significant predictive power for excess bond returns even in the presence of financial predictors such as forward rates and yield spreads. Furthermore, we find that bond returns and yield risk premia are more countercyclical when these risk premia are constructed to exploit information in the factors.

This chapter investigates the robustness of our earlier findings with special attention paid to how the factors are estimated. We first reestimate the FAR on a panel of 131 series over a longer sample. As in our previous work, these (static) factors, denoted $\hat{f}_t$, are estimated by PCA. We then consider an alternative set of factor estimates, denoted $\hat{g}_t$, that differ from the PCA estimates in two important ways. First, we use a priori information to organize the 131 series into 8 blocks. Second, we estimate a dynamic factor model for each of the eight blocks using a Bayesian procedure.

Compared with our previous work, we now use information in the large macroeconomic panel in a different way, and we estimate dynamic factors using a Bayesian method. It is thus useful to explain the motivation for doing so. The factors estimated from large panels of data are often criticized for being difficult to interpret, and organizing the data in blocks (such as output and price) provides a natural way to name the factors estimated from a block of data. At this point, we could have used PCA to estimate one static factor for each block. We could also have estimated dynamic factors using dynamic principal components, which is frequency-domain based. Whichever principal components estimator we choose, the estimates will not be precise as the number of series in each block is no longer 131 but a much smaller number. Bayesian estimation is more appropriate for the newly organized panels of data and Bayesian estimation yields a direct assessment of sampling variability. Using an estimator that is not principal components based also allows us to more thoroughly assess whether the FAR estimates are sensitive to how the factors are estimated. This issue, to our knowledge, has not been investigated in the literature. Notably, the factors that explain most of the variation in the large macroeconomic panel of data need not be the same as the factors most important for predicting excess bond returns. Thus for each of the two sets of factor estimates, namely, $\hat{f}_t$ and $\hat{g}_t$, we consider a systematic search of the relevant predictors, including an out-of-sample criterion to guard against overfitting the predictive regression with too many factors. We also assess the stability of the relation between excess bond returns and the factors over the sample.

An appeal of FAR is that when N and T are large and $\sqrt{T}/N$ tends to zero, the estimated factors in the FAR can be treated as though they are the true but latent factors. There is no need to account for sampling error incurred when the factors are estimated. Numerous papers have studied the properties of the (static and dynamic) principal components estimators in a forecasting context.[2] To date, little is known about the properties of the FAR estimates when $\sqrt{T}/N$ is not negligible. We show that principal components estimation may induce a

[2] See, for example, Boivin and Ng (2005).

bias in the parameter estimates of the predictive regression and suggest how a bias correction can be constructed. For our application, this bias is very small.

Our main finding is that macro factors have strong predictive power for excess bond returns and that this result holds up regardless of which method is used to estimate the factors. The reason is that both methods are capable of isolating the factor for real activity, which contributes significantly to variations in excess bond returns. However, the prior information that permits us to easily give names to the factors also constrains how information in the large panel is used. Thus, as far as predictability is concerned, the factors estimated from the large panel tend to be better predictors than the factors estimated from the eight blocks of data, for the same total number of series used in estimation. Recursive estimation of the predictive regressions finds that the macroeconomic factors are statistically significant throughout the entire sample, even though the degree of predictability varies over the 45 years considered. While the estimated bond and yield risk premia without the macro factors are acyclical, these premia are countercyclical when the estimated factors are used to forecast excess returns. This implies that investors must be compensated for risks associated with recessions.

Our empirical work is based on a macroeconomic panel that extends the one used in Stock and Watson (2005), which has since been used in a number of factor analyses.[3] The original data set consists of monthly observations for 132 macroeconomic time series from 1959:1 to 2003:12. We extend their data to 2007:12 and our panel consists of 131 series. Our empirical work uses data from 1964:1 to 2007:12.

12.2 Predictive Regressions

For $t = 1, \ldots T$, let $rx_{t+1}^{(n)}$ denote the continuously compounded (log) excess return on an n-year discount bond in period $t+1$. Excess returns are defined as $rx_{t+1}^{(n)} \equiv r_{t+1}^{(n)} - y_t^{(1)}$, where $r_{t+1}^{(n)}$ is the log holding period return from buying an n-year bond at time t and selling it as an $n-1$ year bond at time $t+1$, and $y_t^{(1)}$ is the log yield on the one-year bond. That is, if $p_t^{(n)}$ is log price of n-year discount bond at time t, then the log yield is $y_t^{(n)} \equiv -(1/n)p_t^{(n)}$.

A standard approach to assessing whether excess bond returns are predictable is to select a set of K predetermined conditioning variables at time t, given by the $K \times 1$ vector Z_t, and then estimate

$$rx_{t+1}^{(n)} = \beta' Z_t + \epsilon_{t+1} \tag{12.1}$$

by least squares. For example, Z_t could include the individual forward rates studied in Fama and Bliss (1987), the single forward factor studied in Cochrane and Piazzesi (2005), or other predictor variables based on a few macroeconomic series. Such a procedure may be restrictive when the number of eligible

[3] See, for example, Bai and Ng (2006b) and DeMol, Giannone, and Reichlin (2006).

predictors is quite large. In particular, suppose we observe a $T \times N$ panel of macroeconomic data with elements $x_t = (x_{1t}, x_{2t}, \ldots x_{Nt})'$, $t = 1, ..., T$, where the cross-sectional dimension, N, is large, and possibly larger than the number of time periods, T. The set of eligible predictors consists of the union of x_t and Z_t. With standard econometric tools, it is not obvious how a researcher could use the information contained in the panel because unless we have a way of ordering the importance of the N series in forming conditional expectations (as in an autoregression), there are potentially 2^N possible combinations to consider. The regression

$$rx_{t+1}^{(n)} = \gamma' x_t + \beta' Z_t + \epsilon_{t+1} \tag{12.2}$$

quickly runs into degrees-of-freedom problems as the dimension of x_t increases, and estimation is not even feasible when $N + K > T$.

The approach we consider is to posit that x_{it} has a factor structure so that if these factors were observed, we would have replaced Equation 12.2 by the following (infeasible) "factor augmented regression"

$$rx_{t+1}^{(n)} = \alpha' F_t + \beta' Z_t + \epsilon_{t+1}, \tag{12.3}$$

where F_t is a set of k factors whose dimension is much smaller than that of x_t but has good predictive power for $rx_{t+1}^{(n)}$. Equation 12.1 is nested within the factor-augmented regression, making Equation 12.3 a convenient framework to assess the importance of x_{it} via F_t, even in the presence of Z_t. The Z_t that we will use as benchmark is the forward rate factor used in Cochrane and Piazzesi (2005). This variable, hereafter referred to as *CP*, is a simple average of the one-year yield and four forward rates. These authors find that the predictive power of forward rates, yield spreads, and yield factors are subsumed in CP_t. To implement the regression given by Equation 12.3, we need to resolve two problems. First, F_t is latent and we must estimate it from data. Second, we need to isolate those factors with predictive power for our variable of interest, $rx_{t+1}^{(n)}$.

12.3 Estimation of Latent Factors

The first problem is dealt with by replacing F_t with an estimated value $\widehat{F}_t$ that is close to F_t in some well-defined sense, and this involves making precise a model from which F_t can be estimated. We will estimate two factor models, one static and one dynamic, using data retrieved from the Global Insight database and the Conference Board. The data are collected to incorporate as many series as that used in Stock and Watson (2005). However, one series (ao048) is no longer available on a monthly basis after 2003. Accordingly, our new data set consists of 131 series from 1959:1 to 2007:12, though our empirical analysis starts in 1964:1 because of availability of the bond yield data. As in the original Stock and Watson data, some series need to be transformed to be stationary. In general, real variables are expressed in growth rates, first differences are used for nominal interest rates, and second log differences are used for prices. The data description is given in appendix. This data can be downloaded from our Web site http://www.econ.nyu.edu/user/ludvigsons/Data&ReplicationFiles.zip.

12.3.1 Static Factors

Let N be the number of cross-section units and T be the number of time series observations. For $i = 1, \ldots N, \quad t = 1, \ldots T$, a *static* factor model is defined as

$$x_{it} = \lambda_i' f_t + e_{it}. \tag{12.4}$$

In factor analysis, e_{it} is referred to as the idiosyncratic error and λ_i are the factor loadings. This is a vector of weights that unit i put on the corresponding r (static) common factors f_t. In finance, x_{it} is the return for asset i in period t, f_t is a vector of systematic risk, λ_i is the exposure to the risk factors, and e_{it} is the idiosyncratic returns. Although the model specifies a static relationship between x_{it} and f_t, f_t itself can be a dynamic vector process that evolves according to

$$A(L) f_t = u_t,$$

where $A(L)$ is a polynomial (possibly of infinite order) in the lag operator. The idiosyncratic error e_{it} can also be a dynamic process, and e_{it} can also be cross-sectionally correlated.

We estimate f_t using the method of asymptotic principal components (PCA) originally developed by Connor and Korajzcyk (1986) for a small T large N environment. Letting "hats" denote estimated values, the $T \times r$ matrix $\hat{f}$ is $\sqrt{T}$ times the r eigenvectors corresponding to the r largest eigenvalues of the $T \times T$ matrix $xx'/(TN)$ in decreasing order with $\hat{f}'\hat{f} = I_r$. The normalization is necessary as the matrix of factor loadings Λ and f are not separately identifiable. The normalization also yields $\hat{\Lambda} = x'\hat{f}/T$. Intuitively, for each t, $\hat{f}_t$ is a linear combinations of each element of the $N \times 1$ vector $x_t = (x_{1t}, \ldots, x_{Nt})'$, where the linear combination is chosen optimally to minimize the sum of squared residuals $x_t - \Lambda f_t$. Bai and Ng (2002) and Stock and Watson (2002a) showed that the space spanned by f_t can be consistently estimated by $\hat{f}_t$ defined as above when $N, T \to \infty$. The number of static factors in x_t can be determined by the panel information criteria developed in Bai and Ng (2002). For the panel of 131 series under investigation, the IC_2 criterion finds eight factors over the full sample of 576 observations (with the maximum number of factors set to 20).

A common criticism of the method of principal components estimator is that the factors can be difficult to interpret. Our interpretation of the factors is based on the marginal R^2s, obtained by regressing each of the 131 series on the eight factors, one at a time. Because the factors are mutually uncorrelated, the marginal R^2 is also the explanatory power of the factor in question holding other factors fixed. Extending the sample to include three more years of data did not change our interpretation of the factors. Figures 12.1 through 12.8 show the marginal R-square statistics from regressing the series number given on the x-axis onto the estimated factor named in the heading. As in Ludvigson and Ng (2007), $\hat{f}_1$ is a real activity factor that loads heavily on employment and output data. The second factor loads heavily on interest rate spreads, while the third and fourth factors load on prices. Factor 5 loads on interest rates (much more strongly than the interest rate spreads). Factor 6 loads predominantly on the housing variables while factor 7 loads on measures of the money supply. Factor 8 loads on variables relating to the stock market. Thus, loosely speaking, factors 5–8 are more strongly related to money, credit, and finance.

While knowing that there are eight factors in the macroeconomic panel is useful information in its own right, of interest here are not the N variables $x_t = (x_{1t}, \ldots, x_{Nt})'$, but the scalar variable $rx_{t+1}^{(n)}$ which is not in x_t. Factors that are pervasive for the large panel of data need not be important for predicting $rx_{t+1}^{(n)}$. For this reason, we make a distinction between $F_t \subset f_t$ and f_t. The predictive regression of interest is

$$rx_{t+1}^{(n)} = \alpha_F' \hat{F}_t + \beta_F' Z_t + \epsilon_{t+1}, \tag{12.5}$$

which has a vector of generated regressors, $\hat{F}_t$.

Consistency of $\hat{\alpha}_F$ follows from the fact that the difference between $\hat{f}_t$ and the space spanned by f_t vanishes at rate $\min[N, T]$, a result established in Bai and Ng (2002).[4] Bai and Ng (2006a) showed that if $\sqrt{T}/N \to 0$ as $N, T \to \infty$, the sampling uncertainty from first-step estimation is negligible. The practical implication is that standard errors can be computed for the estimates of α_F as though the true F_t were used in the regression. This is in contrast to the case when $\hat{F}_t$ is estimated from a first-step regression with a finite number of predictors. As shown in Pagan (1984), the standard errors for $\hat{\alpha}_F$ in such a case are incorrect unless they are adjusted for the estimation error incurred in the first step of F_t.

12.3.2 Dynamic Factors

An advantage of the method of principal components is that it can handle a large panel of data at little computation cost, one reason being that little structure is imposed on the estimation. To be convinced that factor augmented regressions are useful in analyzing economic issues of interest, we need to show that estimates of the FAR are robust to the choice of the estimator *and* to the specification of the factor model. To this end, we consider an alternative way of estimating the factors with two fundamental differences.

[4] It is useful to remark that the convergence rate established in Stock and Watson (2002a) is too slow to permit consistent estimation of the parameters in Equation 12.5.

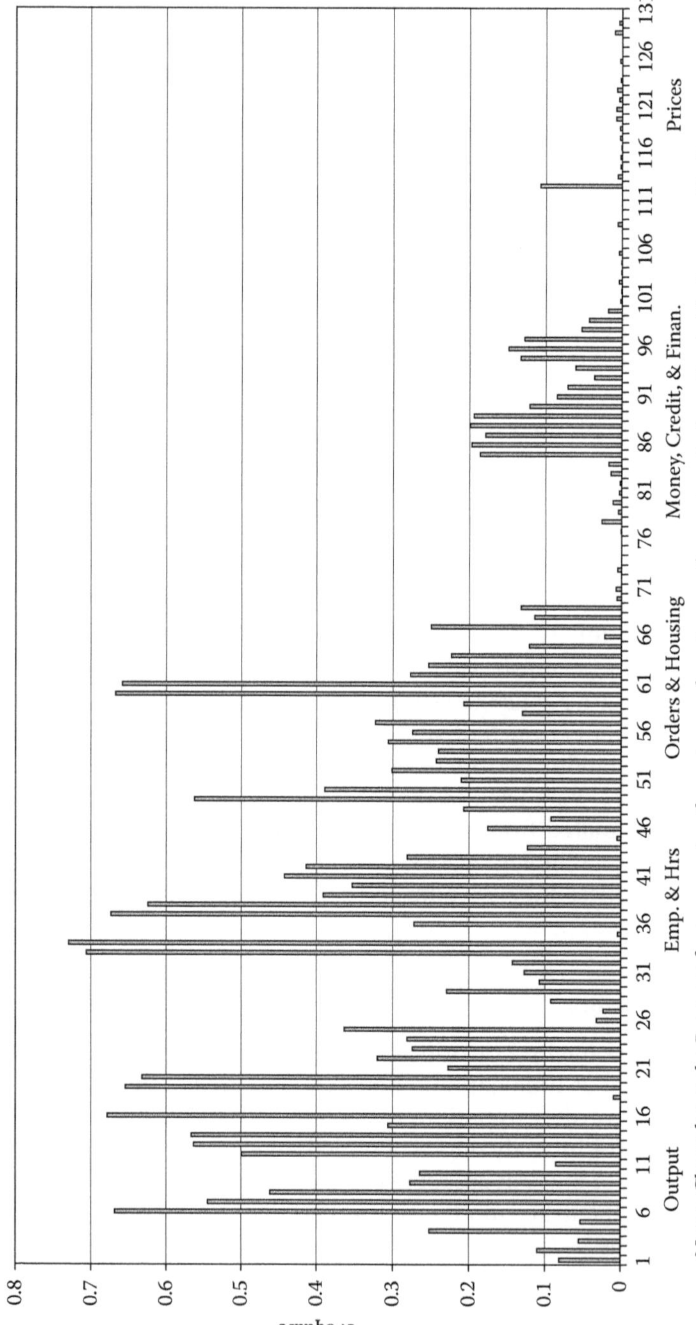

Notes: Chart shows the R-square from regressing the series number given on the x-axis onto the estimated factor named in the heading. See the appendix for a description of the numbered series. The factors are estimated using data from 1964:1-2007:12.

FIGURE 12.1
Marginal R-squares for F_1.

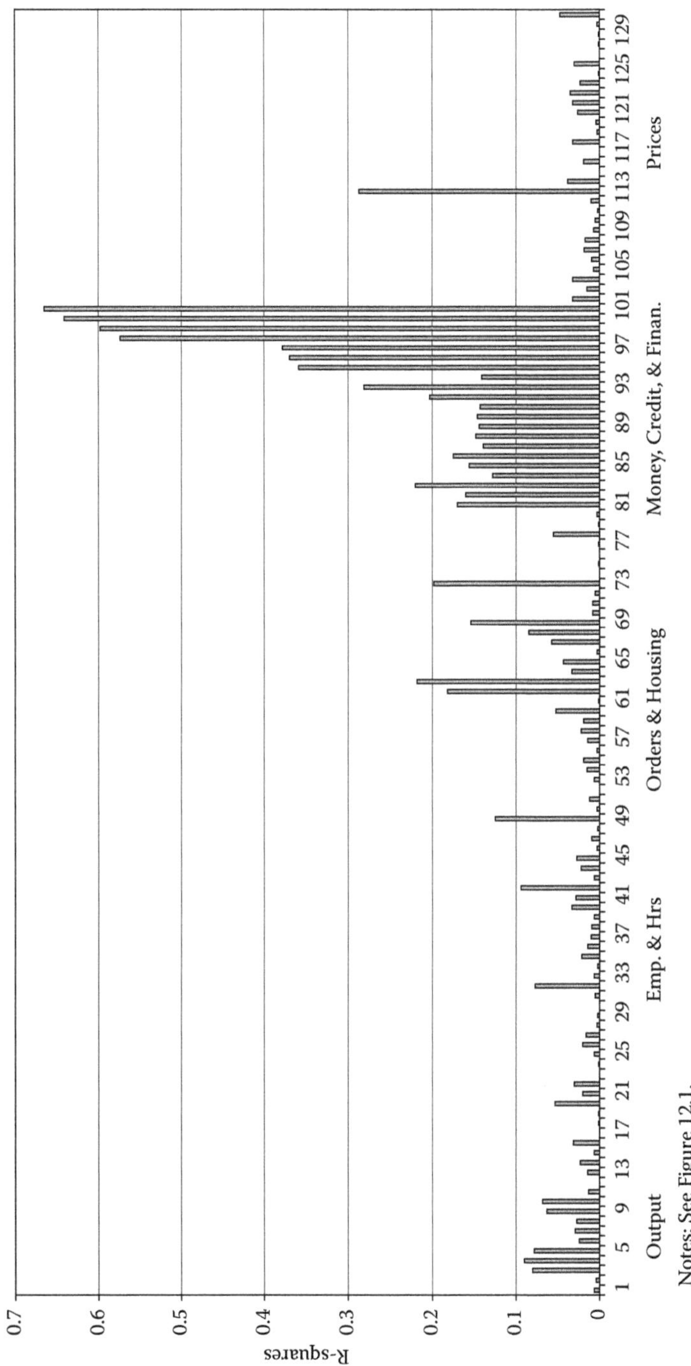

FIGURE 12.2
Marginal R-squares for F_2.

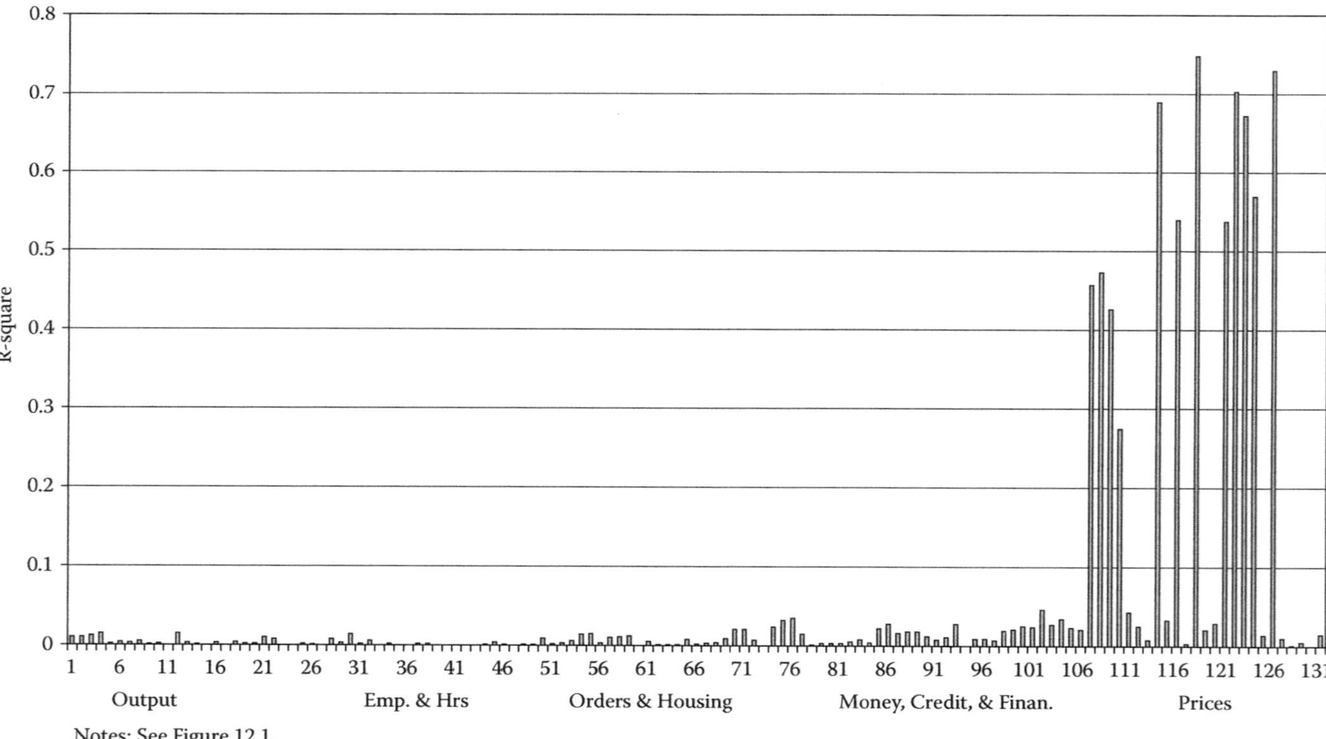

Notes: See Figure 12.1.

FIGURE 12.3
Marginal R-squares for F_3.

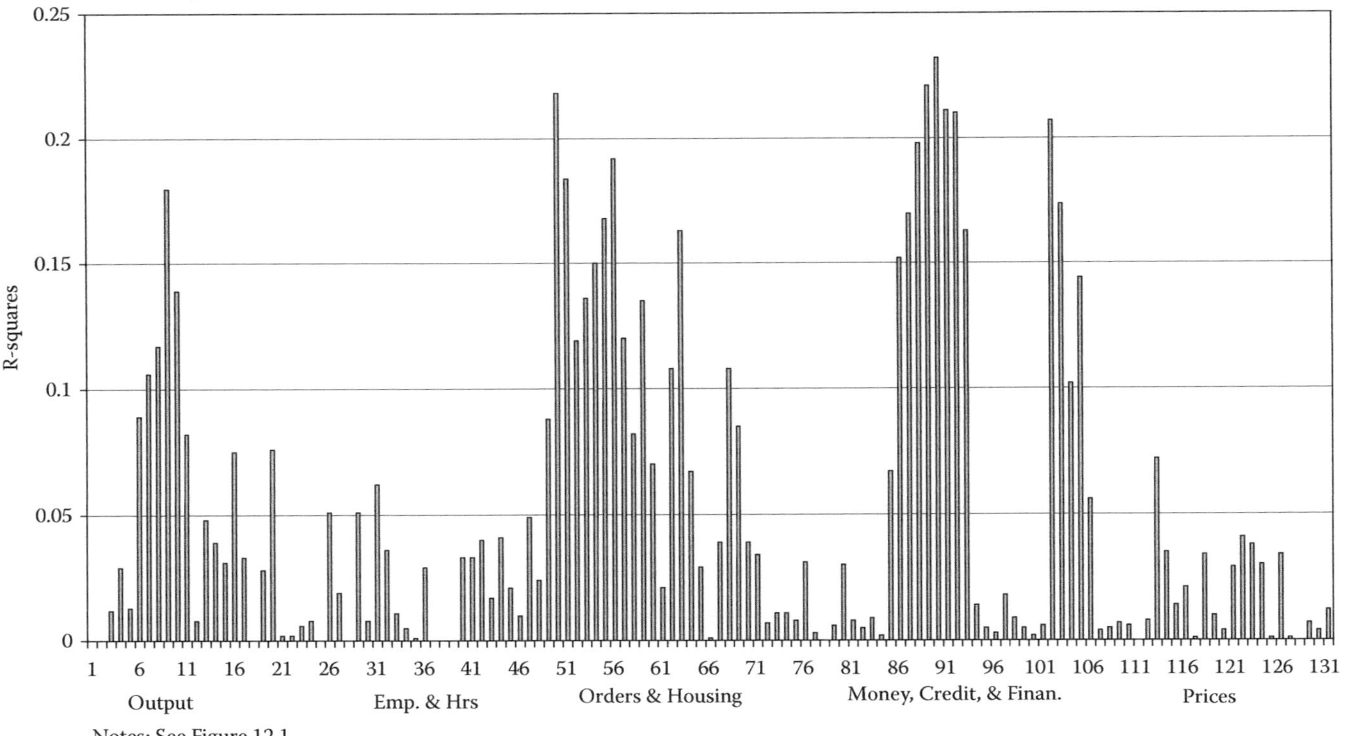

FIGURE 12.4
Marginal R-squares for F_4.

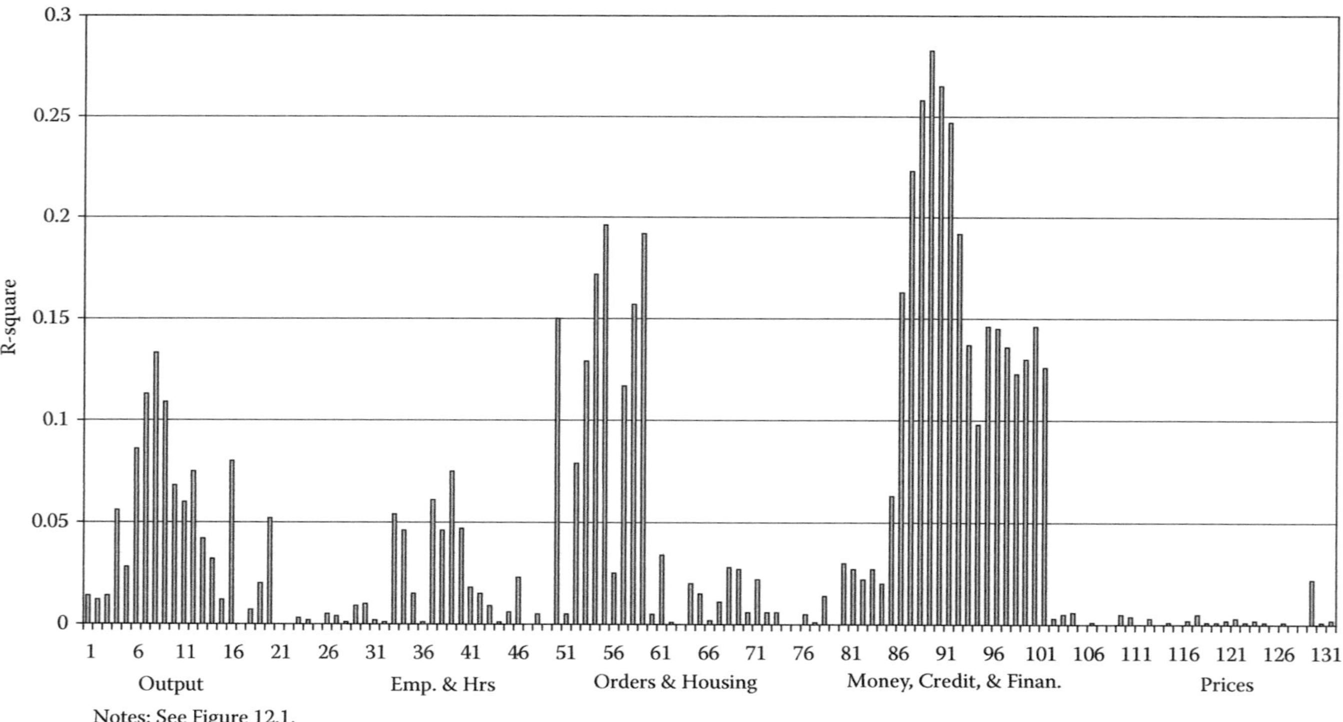

Notes: See Figure 12.1.

FIGURE 12.5
Marginal R-squares for F_5.

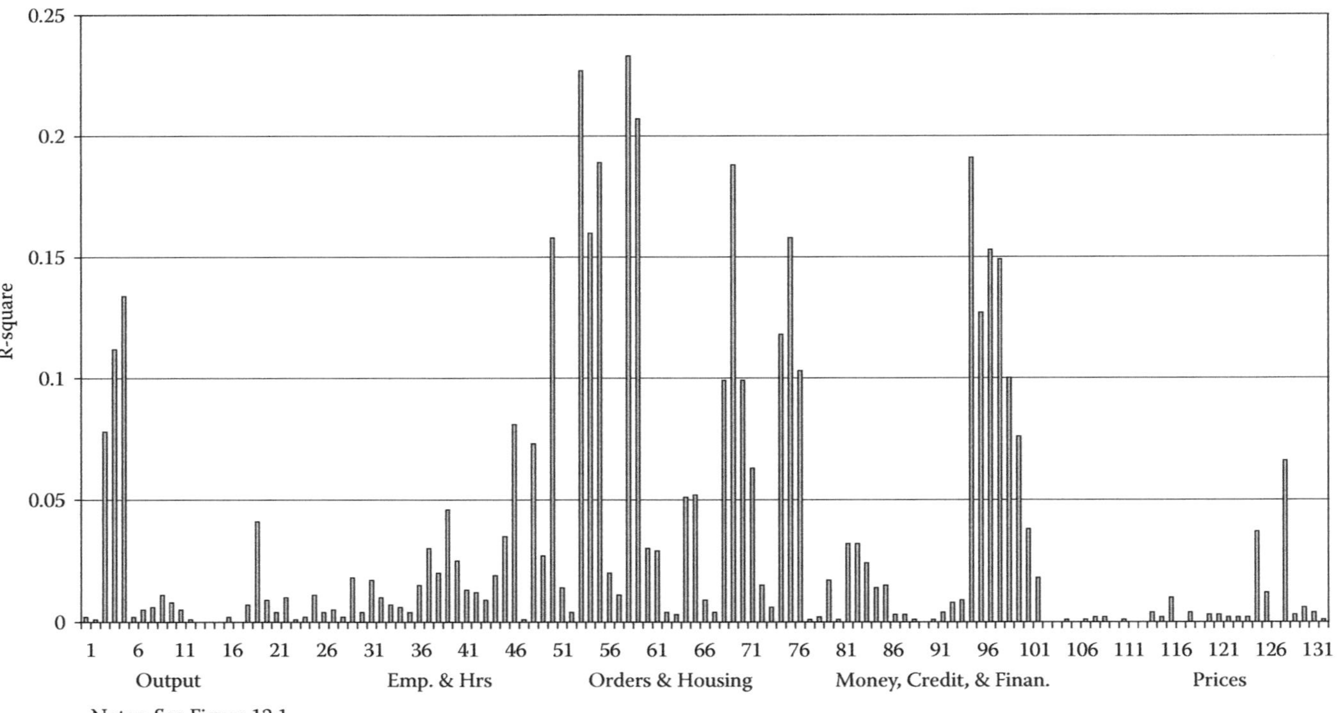

Notes: See Figure 12.1.

FIGURE 12.6
Marginal R-squares for F_6.

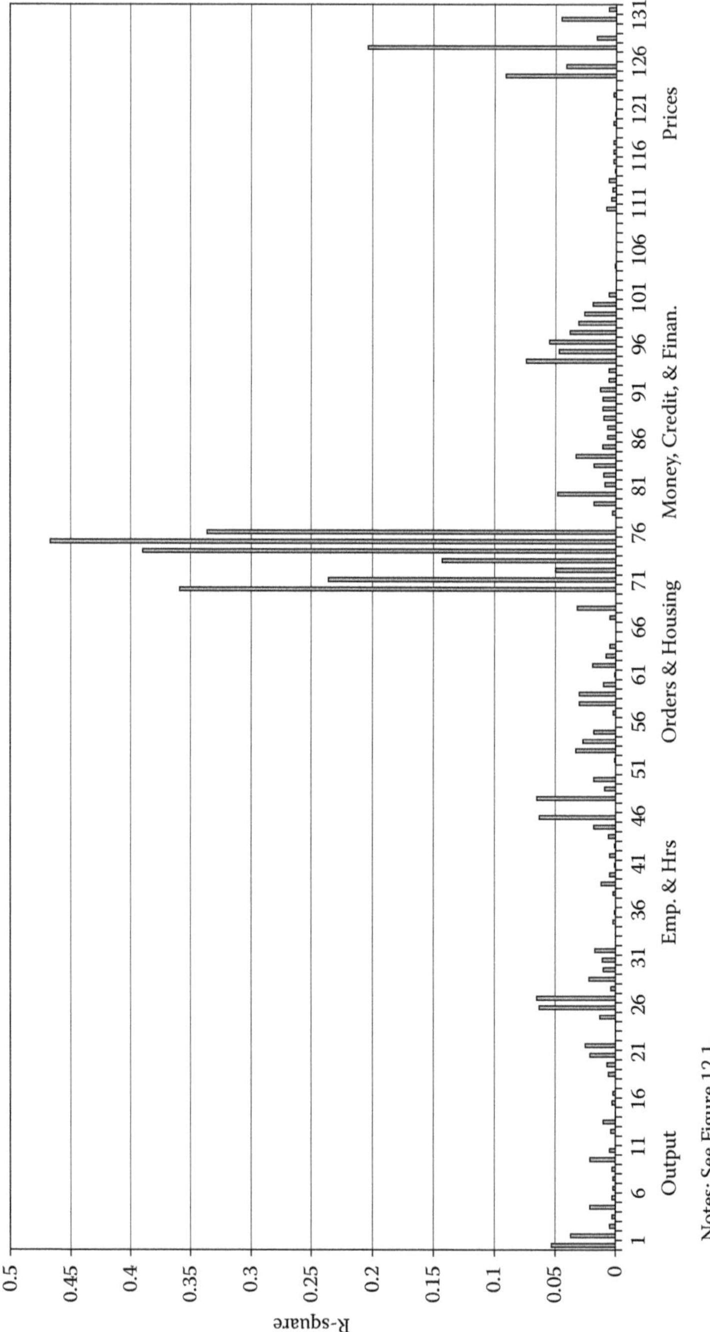

Notes: See Figure 12.1.

FIGURE 12.7
Marginal R-squares for F_7.

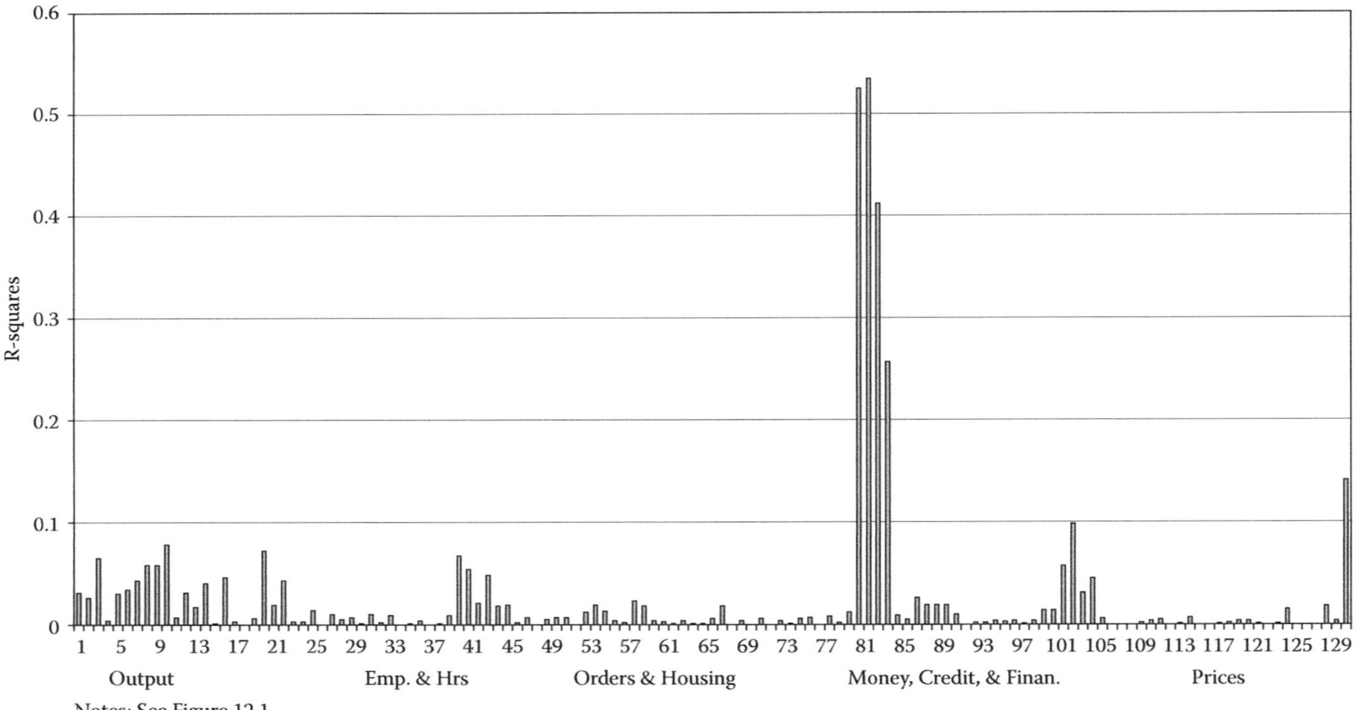

Notes: See Figure 12.1.

FIGURE 12.8
Marginal R-squares for F_8.

First, we use prior information to organize the data into eight blocks. These are (1) output, (2) labor market, (3) housing sector, (4) orders and inventories, (5) money and credit, (6) bond and forex, (7) prices, and (8) stock market. The largest block is the labor market which has 30 series, while the smallest group is the stock market block, which only has four series. The advantage of estimating the factors (which will now be denoted g_t) from blocks of data is that the factor estimates are easy to interpret.

Second, we estimate a dynamic factor model specified as

$$x_{it} = \beta_i'(L)g_t + e_{xit}, \tag{12.6}$$

where $\beta_i(L) = (1 - \lambda_{i1}L - \ldots - \lambda_{is}L^s)$ is a vector of dynamic factor loadings of order s and g_t is a vector of q "dynamic factors" evolving as

$$\psi_g(L)g_t = \epsilon_{gt},$$

where $\psi_g(L)$ is a polynomial in L of order p_G, ϵ_{gt} are i.i.d. errors. Furthermore, the idiosyncratic component e_{xit} is an autoregressive process of order p_X so that

$$\psi_x(L)e_{xit} = \epsilon_{xit}.$$

This is the factor framework used in Stock and Watson (1989) to estimate the coincident indicator with $N = 4$ variables. Here, our N can be as large as 30.

The dimension of g_t, (which also equals the dimension of ϵ_t), is referred to as the number of dynamic factors. The main distinction between the static and the dynamic model is best understood using a simple example. The model $x_{it} = \beta_{i0}g_t + \beta_{i1}g_{t-1} + e_{it}$ is the same as $x_{it} = \lambda_{i1}f_{1t} + \lambda_{i2}f_{2t}$ with $f_{1t} = g_t$ and $f_{2t} = g_{t-1}$. Here, the number of factors in the static model is two but there is only one factor in the dynamic model. Essentially, the static model does not take into account that f_t and f_{t-1} are dynamically linked. Forni et al. (2005) showed that when N and T are both large, the space spanned by g_t can also be consistently estimated using the method of dynamic principal components originally developed in Brillinger (1981). Boivin and Ng (2005) find that static and dynamic principal components have similar forecast precision, but that static principal components are much easier to compute. It is an open question whether to use the static or the dynamic factors in predictive regressions though the majority of factor augmented regressions use the static factor estimates. Our results will shed some light on this issue.

We estimate a dynamic factor model for each of the eight blocks. Given the definition of the blocks, it is natural to refer to g_{1t} as an output factor, g_{7t} as a price factor, and so on. However, as some blocks have a small number of series, the (static or dynamic) principal components estimator which assumes that N and T are both large will give imprecise estimates. We therefore use the Bayesian method of Monte Carlo Markov Chain (MCMC). MCMC samples a chain that has the posterior density of the parameters as its stationary distribution. The posterior mean computed from draws of the chain are then unbiased for g_t. For factor models, Kose, Otrok, and Whiteman (2003) use an algorithm that involves inversion of N matrices that are of dimension $T \times T$, which can be computationally demanding. The algorithms used in Aguilar and West (2000), Geweke and Zhou (1996), and Lopes and West (2004) are extensions of the MCMC method developed in Carter and Kohn (1994) and Fruhwirth-Schnatter (1994). Our method is similar and follows the implementation in Kim and Nelson (2000) of the Stock–Watson coincident indicator closely. Specifically, we first put the dynamic factor model into a state-space framework. We assume $p_X = p_G = 1$ and $s_g = 2$ for every block. For $i = 1, \ldots N_b$ (the number of series in block b), let x_{ibt} be the observation for unit i of block b at time t. Given that $p_X = 1$, the measurement equation is

$$(1 - \psi_{bi} L) x_{bit} = (1 - \psi_{bi} L)(\beta_{bi0} + \beta_{bi1} L + \beta_{bi2} L^2) g_{bt} + \epsilon_{Xbit}$$

or more compactly,

$$x^*_{bit} = \beta^*_i(L) g_{bt} + \epsilon_{Xbit}.$$

Given that $p_G = 1$, the transition equation is

$$g_{bt} = \psi_{gb} g_{bt-1} + \epsilon_{gbt}.$$

We assume $\epsilon_{Xbit} \sim N(0, \sigma^2_{Xbi})$ and $\epsilon_{gb} \sim N(0, \sigma^2_{gb})$. We use principal components to initialize g_{bt}. The parameters $\beta_b = (\beta_{b1}, \ldots, \beta_{b,Nb})$, $\psi_{Xb} = \psi_{Xb1}, \ldots,$ $\psi_{Xb,Nb}$ are initialized to zero. Furthermore, $\sigma_{Xb} = (\sigma_{Xb1}, \ldots, \sigma_{Xb,N_b})$, ψ_{gb}, and σ^2_{gb} are initialized to random draws from the uniform distribution. For $b = 1, \ldots, 8$ blocks, Gibbs sampling can now be implemented by successive iteration of the following steps:

1. Draw $g_b = (g_{b1}, \ldots g_{bT})'$ conditional on β_b, ψ_{Xb}, σ_{Xb} and the $T \times N_b$ data matrix x_b.
2. Draw ψ_{gb} and σ^2_{gb} conditional on g_b.
3. For each $i = 1, \ldots N_b$, draw β_{bi}, ψ_{Xbi} and σ^2_{Xbi} conditional on g_b and x_b.

We assume normal priors for $\beta_{bi} = (\beta_{i0}, \beta_{i1}, \beta_{i2})$, ψ_{Xbi} and ψ_{gb}. Given conjugacy, β_{bi}, ψ_{Xbi}, ψ_{gb}, are simply draws from the normal distributions whose posterior means and variances are straightforward to compute. Similarly, σ^2_{gb} and σ^2_{Xbi} are draws from the inverse chi-square distribution. Because the model is linear and Gaussian, we can run the Kalman filter forward to obtain the conditional mean $g_{bT|T}$ and conditional variance $P_{bT|T}$. We then draw g_{bT} from its conditional distribution, which is normal, and proceed backwards to generate draws $g_{bt|T}$ for $t = T - 1, \ldots, 1$ using the Kalman filter. For identification, the loading on the first series in each block is set to 1. We take 12,000 draws and discard the first 2000. The posterior means are computed from every 10th draw after the burn-in period. The $\hat{g}_t$s used in subsequent analysis are the means of these 1000 draws.

As in the case of static factors, not every g_{bt} need to have predictive power for excess bond returns. Let $G_t \subset g_t = (g_{1t}, \ldots g_{8t})$ be those that do. The analog to Equation 12.5 using dynamic factors is

$$rx^{(n)}_{t+1} = \alpha'_G \hat{G}_t + \beta'_G Z_t + \epsilon_{t+1}, \tag{12.7}$$

TABLE 12.1

First Order Autocorrelation Coefficients

	$\hat{f}_t$	t	$\hat{g}_t$	t
1	0.767	20.589	−0.361	−6.298
2	0.748	18.085	0.823	22.157
3	−0.239	−2.852	0.877	32.267
4	0.456	7.594	0.660	14.385
5	0.362	6.819	−0.344	−1.635
6	0.422	4.232	0.448	4.552
7	−0.112	−0.672	0.050	0.609
8	0.225	4.526	0.157	2.794

We have now obtained two sets of factor estimates using two distinct methodologies. We can turn to an assessment of whether the estimates of the predictive regression are sensitive to how the factors are estimated.

12.3.3 Comparison of $\hat{f}_t$ and $\hat{g}_t$

Table 12.1 reports the first order autocorrelation coefficients for f_t and g_t. Both sets of factors exhibit persistence, with $\hat{f}_{1t}$ being the most correlated of the eight $\hat{f}_t$, and $\hat{g}_{3t}$ being the most serially correlated amongst the $\hat{g}_t$. Table 12.2 reports the contemporaneous correlations between $\hat{f}$ and $\hat{g}$. The real activity factor $\hat{f}_1$ is highly correlated with the $\hat{g}_t$ estimated from output, labor, and manufacturing blocks. $\hat{f}_2$, $\hat{f}_4$, and $\hat{f}_5$ are correlated with many of the $\hat{g}$, but the correlations with the bond/exchange rate seem strongest. $\hat{f}_3$ is predominantly a price factor, while $\hat{f}_8$ is a stock market factor. $\hat{f}_7$ is most correlated with $\hat{g}_5$, which is a money market factor. $\hat{f}_8$ is highly correlated with $\hat{g}_8$, which is estimated from stock market data.

The contemporaneous correlations reported in Table 12.2 do not give a full picture of the correlation between $\hat{f}_t$ and $\hat{g}_t$ for two reasons. First, the $\hat{g}_t$ are not mutually uncorrelated, and second, they do not account for correlations that might occur at lags. To provide a sense of the dynamic correlation between $\hat{f}$

TABLE 12.2

Correlation between $\hat{f}_t$ and g_t

	$\hat{g}_1$ Output	$\hat{g}_2$ Labor	$\hat{g}_3$ Housing	$\hat{g}_4$ Mfg.	$\hat{g}_5$ Money	$\hat{g}_6$ Finance	$\hat{g}_7$ Prices	$\hat{g}_8$ Stocks
$\hat{f}_1$	0.601	0.903	0.551	0.766	−0.067	0.489	0.126	−0.092
$\hat{f}_2$	0.181	−0.120	0.376	0.269	0.095	−0.462	−0.227	0.449
$\hat{f}_3$	0.037	0.027	−0.150	−0.010	−0.148	0.144	−0.800	−0.067
$\hat{f}_4$	−0.303	0.118	0.253	−0.128	0.185	−0.417	−0.194	0.092
$\hat{f}_5$	0.306	0.179	−0.365	0.026	0.046	−0.474	−0.009	0.183
$\hat{f}_6$	0.103	−0.140	0.321	0.179	−0.398	0.008	0.050	0.177
$\hat{f}_7$	0.064	−0.023	0.125	0.004	0.743	0.088	−0.078	0.100
$\hat{f}_8$	−0.241	0.073	−0.023	0.111	−0.057	0.119	−0.052	0.689

TABLE 12.3

Long Run Correlation between $\hat{f}_t$ and $\hat{g}_t$

	$\hat{g}_1$ **Output**	$\hat{g}_2$ **Labor**	$\hat{g}_3$ **Housing**	$\hat{g}_4$ **Mfg.**	$\hat{g}_5$ **Money**	$\hat{g}_6$ **Finance**	$\hat{g}_7$ **Prices**	$\hat{g}_8$ **Stocks**	R^2
$\hat{f}_1$	0.447	0.536	0.215	0.066	−0.008	0.140	−0.002	−0.038	0.953
$\hat{f}_2$	0.548	−0.466	0.296	0.299	0.031	−0.536	−0.135	0.266	0.689
$\hat{f}_3$	0.100	0.026	−0.152	−0.036	−0.007	0.211	−0.390	−0.026	0.935
$\hat{f}_4$	−0.925	0.699	0.491	−0.242	0.004	−0.444	−0.077	−0.064	0.723
$\hat{f}_5$	0.682	0.417	−0.624	−0.135	−0.000	−0.488	0.018	0.146	0.790
$\hat{f}_6$	0.070	−0.357	0.467	−0.098	−0.294	0.144	0.061	0.100	0.490
$\hat{f}_7$	0.226	−0.252	0.136	−0.095	0.540	0.325	−0.080	0.180	0.692
$\hat{f}_8$	−0.986	0.447	−0.224	0.167	0.025	0.313	−0.049	0.905	0.797

Reported are estimates of $A_{r.0}$, obtained from the regression: $\hat{f}_{rt} = A_{r.0}\hat{g}_t + \sum_{i=1}^{p-1} A_{r.i}\Delta g_{t-i} + e_t$ with $p = 4$.

and $\hat{g}_t$, we first standardize $\hat{f}_t$ and $\hat{g}_t$ to have unit variance. We then consider the regression

$$\hat{f}_{rt} = a + A_{r.0}\hat{g}_t + \sum_{i=1}^{p-1} A_{r.i}\Delta\hat{g}_{t-i} + e_{it},$$

where for $r = 1, \ldots, 8$ and $i = 0, \ldots, p-1$, $A_{r.i}$ is a 8×1 vector of coefficients summarizing the dynamic relation between $\hat{f}_{rt}$ and lags of $\hat{g}_t$. The coefficient vector $A_{r.0}$ summarizes the long-run relation between $\hat{g}_t$ and $\hat{f}_t$. Table 12.3 reports results for $p = 4$, along with the R^2 of the regression. Except for $\hat{f}_6$, the current value and lags of $\hat{g}_t$ explain the principal components quite well. While it is clear that $\hat{f}_1$ is a real activity factor, the remaining $\hat{f}$s tend to load on variables from different categories. Tables 12.2 and 12.3 reveal that $\hat{g}_t$ and $\hat{f}_t$ reduce the dimensionality of information in the panel of data in different ways. Evidently, the $\hat{f}_t$s are weighted averages of the $\hat{g}_t$s and their lags. This can be important in understanding the results to follow.

12.4 Predictive Regressions

Let $\hat{H}_t \subset \hat{h}_t$, where $\hat{h}_t$ is either $\hat{f}_t$ or $\hat{g}_t$. Our predictive regression can generically be written as

$$rx_{t+1}^{(n)} = \alpha'\hat{H}_t + \beta' CP_t + \epsilon_{t+1}. \tag{12.8}$$

Equation 12.8 allows us to assess whether $\widehat{H}_t$ has predictive power for excess bond returns, conditional on the information in CP_t. In order to assess whether macro factors $\widehat{H}_t$ have unconditional predictive power for future returns, we also consider the restricted regression

$$rx_{t+1}^{(n)} = \alpha'\widehat{H}_t + \epsilon_{t+1}. \tag{12.9}$$

Since $\hat{F}_t$ and $\hat{G}_t$ are both linear combinations of $x_t = (x_{1t}, \ldots x_{Nt})'$, say $F_t = q_F' x_t$ and $G_t = q_G' x_t$, we can also write Equation 12.8 as

$$rx_{t+1}^{(n)} = \alpha^{*\prime} x_t + \beta' CP_t + \epsilon_{t+1}$$

where $\alpha^{*\prime} = \alpha_F' q_F'$ or $\alpha_G' q_G'$. The conventional regression Equation 12.1 puts a weight of zero on all but a handful of x_{it}. When $\hat{H}_t = \hat{F}_t$, q_F is related to the k eigenvectors of $xx'/(NT)$ that will not, in general, be numerically equal to zero. When $\hat{H}_t = \hat{G}_t$, q_G and thus α^* will have many zeros since each column of $\hat{G}_t$ is estimated using a subset of x_t. Viewed in this light, a factor augmented regression with PCA down-weights unimportant regressors. A FAR estimated using blocks of data sets put some but not all coefficients on x_t equal to zero. A conventional regression is most restrictive as it constrains almost the entire α^* vector to zero.

As discussed earlier, factors that are pervasive in the panel of data x_{it} need not have predictive power for $rx_{t+1}^{(n)}$, which is our variable of interest. In Ludvigson and Ng (2007), $\hat{H}_t = \hat{F}_t$ was determined using a method similar to that used in Stock and Watson (2002b). We form different subsets of $\hat{f}_t$, and/or functions of $\hat{f}_t$ (such as $\hat{f}_{1t}^2$). For each candidate set of factors, $\widehat{F}_t$, we regress $rx_{t+1}^{(n)}$ on $\widehat{F}_t$ and CP_t and evaluate the corresponding in-sample BIC and $\bar{R}^2$. The in-sample BIC for a model with k regressors is defined as

$$\text{BIC}_{in}(k) = \hat{\sigma}_k^2 + k\frac{\log T}{T},$$

where $\hat{\sigma}_k^2$ is the variance of the regression estimated over the entire sample. To limit the number of specifications we search over, we first evaluate r univariate regressions of returns on each of the r factors. Then, for only those factors found to be significant in the r univariate regressions, we evaluate whether the squared and the cubed terms help reduce the BIC criterion further. We do not consider other polynomial terms, or polynomial terms of factors not important in the regressions on linear terms.

In this chapter, we again use the BIC to find the preferred set of factors, but we perform a systematic and therefore much larger search. Instead of relying on results from preliminary univariate regressions to guide us to the final model, we directly search over a large number of models with different numbers of regressors. We want to allow excess bond returns to be possibly nonlinear in the eight factors and hence include the squared terms as candidate regressors. If we additionally include all the cubic terms, and given that we have eight factors and CP to consider, we would have over thirteen million (2^{27}) potential models. As a compromise, we limit our candidate regressor set to eighteen variables: $(\hat{f}_{1t}, \ldots, \hat{f}_{8t}; \hat{f}_{1t}^2, \ldots, \hat{f}_{8t}^2; \hat{f}_{1t}^3, CP_t)$. We also restrict the maximum number of predictors to eight. This leads to an evaluation of 106,762 models.[5]

[5] This is obtained by considering $C_{18,j}$ for $j = 1, \ldots, 8$, where $C_{n,k}$ denotes choosing k out of n potential predictors.

The purpose of this extensive search is to assess the potential impact on the forecasting analysis of fishing over large numbers of possible predictor factors. As we show, the factors chosen by the larger, more systematic, search are the same as those chosen by the limited search procedure used in Ludvigson and Ng (2007). This suggests that data mining does not in practice unduly influence the findings in this application, since we find that the same few key factors always emerge as important predictor variables regardless of how extensive the search is.

It is well known that variables found to have predictive power in-sample do not necessarily have predictability out of sample. As discussed in Hansen (2008), in-sample overfitting generally leads to a poor out-of-sample fit. One is less likely to produce spurious results based on an out-of-sample criterion because a complex (large) model is less likely to be chosen in an out-of-sample comparison with simple models when both models nests the true model. Thus, when a complex model is found to outperform a simple model out of sample, it is stronger evidence in favor of the complex model. To this end, we also find the best among 106,762 models as the minimizer of the out-of-sample BIC. Specifically, we split the sample at $t = T/2$. Each model is estimated using the first $T/2$ observations. For $t = T/2+1, \dots, T$, the values of predictors in the second half of the sample are multiplied into the parameters estimated using the first half of the sample to obtain the fit, denoted $\hat{r}x_{t+12}$. Let $\tilde{e}_t = rx_{t+12} - \hat{r}x_{t+12}$ and $\tilde{\sigma}_k^2 = \frac{1}{T/2}\sum_{t=T/2+1}^{T} \tilde{e}_t^2$ be the out-of-sample error variance corresponding to model j. The out-of-sample BIC is defined as

$$\mathrm{BIC}_{\mathrm{out}}(j) = \log \tilde{\sigma}_j^2 + \frac{\dim_j \log(T/2)}{T/2},$$

where dim_j is the size of model j. By using an out-of-sample BIC selection criterion, we guard against the possibility of spurious overfitting. Regressors with good predictive power only over a subsample will not likely be chosen. As the predictor set may differ depending on whether the CP factor is included (i.e., whether we consider Equations 12.8 and 12.9), the two variable selection procedures are repeated with CP excluded from the potential predictor set. Using the predictors selected by the in- and the out-of-sample BIC, we reestimate the predictive regression over the entire sample. In the next section, we show that the predictors found by this elaborate search are the same handful of predictors found in Ludvigson and Ng (2007) and that these handful of macroeconomic factors have robust significant predictive power for excess bond returns beyond the CP factor.

We also consider as predictor a linear combination of $\hat{h}_t$ along the lines of Cochrane and Piazzesi (2005). This variable, denoted $\hat{H}8_t$ is defined as $\hat{\gamma}'\hat{h}_t^+$ where $\hat{\gamma}$ is obtained from the following regression:

$$\frac{1}{4}\sum_{n=2}^{5} rx_{t+1}^n = \gamma_0 + \gamma'\hat{h}_t^+, \tag{12.10}$$

with $\hat{h}_t^+ = (\hat{h}_{1t}, \ldots, \hat{h}_{8t}, \hat{h}_{1t}^3)$. The estimates are as follows:

	$h_t = \hat{f}_t$		$h_t = \hat{g}_t$	
	$\hat{\gamma}$	$t_{\hat{\gamma}}$	$\hat{\gamma}$	$t_{\hat{\gamma}}$
h_1	−1.681	−4.983	0.053	0.343
h_2	0.863	3.009	−1.343	−2.593
h_3	−0.018	−0.203	−0.699	−1.891
h_4	−0.626	−2.167	0.628	1.351
h_5	−0.264	−1.463	−0.001	−0.012
h_6	−0.720	−2.437	−0.149	−0.691
h_7	−0.426	−2.140	−0.018	−0.210
h_8	0.665	3.890	−0.418	−2.122
h_1^3	0.115	3.767	0.049	1.733
cons	0.900	2.131	0.764	1.518
$\bar{R}^2$	0.261		0.104	

Notice that we could also have replaced $\hat{h}_t$ in the above regression with $\hat{H}_t$, where $\hat{H}_t$ comprises predictors selected by either the in- or the out-of-sample BIC. However, $\hat{H}8_t$ is a factor-based predictor that is arguable less vulnerable to the effects of data mining because it is simply a linear combination of all the estimated factors.

Tables 12.4 to 12.7 report results for maturities of 2, 3, 4, and 5 years. The first four columns of each table are based on the static factors (i.e., $\hat{H}_t = \hat{F}_t$), while columns 5 to 8 are based on the dynamic factors (i.e., $\hat{H}_t = \hat{G}_t$). Of these, columns 1, 2, 5, and 6 include the CP variable, while columns 3, 4, 7, and 8 do not include the CP. Columns 9 and 10 report results using $\hat{F}8$ with and without CP and columns 11 and 12 do the same with $\hat{G}8$ in place. Our benchmark is a regression that has the CP variable as the sole predictor. This is reported in last column, i.e., column 13.

12.4.1 Two-Year Returns

As can be seen from Table 12.4, the CP alone explains 0.309 of the variance in the 2-year excess bound returns. The variable $\hat{F}_8$ alone explains 0.279 (column 10), while $\hat{G}_8$ alone explains only 0.153 of the variation (column 12). Adding $\hat{F}8$ to the regression with the CP factor (column 9) increases $\bar{R}^2$ to 0.419, and adding $\hat{G}8$ (column 11) to CP yields an $\bar{R}^2$ of 0.401. The macroeconomic factors thus have nontrivial predictive power above and beyond the CP factor.

We next turn to regressions when both the factors and CP are included. In Ludvigson and Ng (2007), the static factors $\hat{f}_{1t}, \hat{f}_{2t}, \hat{f}_{3t}, \hat{f}_{4t}, \hat{f}_{8t}$, and CP are found to have the best predictive power for excess returns. The in-sample BIC still finds the same predictors to be important, but adds $\hat{f}_{6t}$ and $\hat{f}_{5t}^2$ to the predictor list. It is, however, noteworthy that some variables selected by the BIC have individual t statistics that are not significant. The resulting model has an $\bar{R}^2$ of 0.460 (column 1). The out-of-sample BIC selects smaller models and finds $\hat{f}_1, \hat{f}_8, \hat{f}_5^2, \hat{f}_1^3$, and the CP to be important regressors (column 2).

TABLE 12.4

Regressions $rx_{t+1}^{(2)} = a + \alpha' \hat{H}_t + \beta' CP_t + \epsilon_{t+1}$

	$\hat{H} = \hat{F}$				$\hat{H} = \hat{G}$				$\hat{H} = \hat{F}$		$\hat{H} = \hat{G}$		
	In	Out	In	Out	In	Out	In	Out					
$\hat{H}$	1	2	3	4	5	6	7	8	9	10	11	12	13
$\hat{H}_1$	−0.761	−0.793	−0.935	−0.931	-	-	0.147	0.170	-	-	-	-	-
tstat	−5.387	−4.848	−5.748	−5.449	-	-	2.947	2.623	-	-	-	-	-
$\hat{H}_2$	-	-	0.325	0.326	−0.494	−0.627	−0.699	−0.646	-	-	-	-	-
tstat	-	-	2.663	2.520	−3.151	−3.623	−2.905	−3.062	-	-	-	-	-
$\hat{H}_3$	-	-	-	-	−0.492	-	−0.532	−0.487	-	-	-	-	-
tstat	-	-	-	-	−4.813	-	−2.889	−3.012	-	-	-	-	-
$\hat{H}_4$	−0.291	-	−0.399	−0.399	-	-	0.186	-	-	-	-	-	-
tstat	−2.716	-	−3.103	−2.974	-	-	1.039	-	-	-	-	-	-
$\hat{H}_6$	−0.151	-	−0.281	−0.280	0.137	-	−0.163	-	-	-	-	-	-
tstat	−1.322	-	−1.949	−1.795	1.679	-	−1.594	-	-	-	-	-	-
$\hat{H}_7$	−0.128	-	−0.143	−0.144	-	-	-	-	-	-	-	-	-
tstat	−1.577	-	−1.517	−1.365	-	-	-	-	-	-	-	-	-
$\hat{H}_8$	0.240	0.241	0.302	-	−0.136	-	−0.164	-	-	-	-	-	-
tstat	2.981	3.297	3.575	-	−1.562	-	−1.997	-	-	-	-	-	-
$\hat{H}_2^2$	-	-	-	-	-	−0.100	-	-	-	-	-	-	-
tstat	-	-	-	-	-	−2.147	-	-	-	-	-	-	-
$\hat{H}_4^2$	-	-	-	-	−0.074	-	−0.121	−0.118	-	-	-	-	-
tstat	-	-	-	-	−3.165	-	−3.167	−3.076	-	-	-	-	-
$\hat{H}_5^2$	−0.080	−0.110	-	-	-	-	-	-	-	-	-	-	-
tstat	−2.468	−2.925	-	-	-	-	-	-	-	-	-	-	-
$\hat{H}_6^2$	-	-	-	-	−0.086	−0.083	−0.084	−0.080	-	-	-	-	-

(continued)

TABLE 12.4 (Continued)

Regressions $rx^{(2)}_{t+1} = a + \alpha'\hat{H}_t + \beta' CP_t + \epsilon_{t+1}$

	$\hat{H} = \hat{F}$				$\hat{H} = \hat{G}$								
	In	Out	In	Out	In	Out	In	Out	$\hat{H} = \hat{F}$		$\hat{H} = \hat{G}$		
$\hat{H}$	1	2	3	4	5	6	7	8	9	10	11	12	13
tstat	-	-	-	-	−6.245	−6.804	−3.642	−3.176	-	-	-	-	-
$\hat{H}^3_1$	0.044	0.047	0.057	0.056	0.019	-	-	-	-	-	-	-	-
tstat	2.912	2.887	3.081	3.338	2.254	-	-	-	-	-	-	-	-
CP	0.385	0.411	-	-	0.452	0.433	-	-	0.336	-	0.413	-	0.455
tstat	5.647	6.981	-	-	7.488	7.738	-	-	4.437	-	6.434	-	8.836
$\hat{H}8$	-	-	-	-	-	-	-	-	0.332	0.482	0.427	0.544	-
tstat	-	-	-	-	-	-	-	-	4.336	7.212	3.880	3.493	-
$\bar{R}^2$	0.460	0.430	0.283	0.258	0.477	0.407	0.200	0.192	0.419	0.279	0.401	0.153	0.309

Note: The table reports estimates from OLS regressions of excess bond returns on the lagged variables named in column 1. The dependent variable rx^n_{t+1} is the excess log return on the n year Treasury bond. $\widehat{H}_t$ denotes a set of regressors formed from consisting of functions of $\hat{f}_t$ or $\hat{g}_t$ where $\hat{f}_t$ is a set of eight factors estimated by the method of principal components, and $\hat{g}_t$ is a vector of eight dynamic factors estimated by Bayesian factors. The panel of data used in estimation consists of 131 individual series over the period 1964:1 to 2007:12. $\hat{H}8_t$ is the single factor constructed as a linear combination of the eight estimated factors and $\hat{f}^3_1$. CP_t is the Cochrane and Piazzesi (2005) factor that is a linear combination of five forward spreads. Newey and West (1987) corrected t-statistics have lag order 18 months and are reported in parentheses. A constant is always included in the regression even though its estimate is not reported in the table.

TABLE 12.5

Regressions $rx^{(3)}_{t+1} = a + \alpha'\hat{H}_t + \beta' C P_t + \epsilon_{t+1}$

	$\hat{H} = \hat{F}$				$\hat{H} = \hat{G}$				$\hat{H} = \hat{F}$		$\hat{H} = \hat{G}$		
	In	Out	In	Out	In	Out	In	Out					
$\hat{H}$	1	2	3	4	5	6	7	8	9	10	11	12	13
$\hat{H}_1$	−1.232	−1.280	−1.624	−1.592	-	-	-	-	-	-	-	-	-
tstat	−5.079	−4.581	−5.553	−5.479	-	-	-	-	-	-	-	-	-
$\hat{H}_2$	−0.028	-	0.694	0.703	−0.782	−1.094	−1.259	−1.056	-	-	-	-	-
tstat	−0.147	-	2.851	2.982	−2.805	−3.773	−2.983	−3.092	-	-	-	-	-
$\hat{H}_3$	-	-	-	-	−0.807	-	−0.843	−0.734	-	-	-	-	-
tstat	-	-	-	-	−4.297	-	−2.667	−2.548	-	-	-	-	-
$\hat{H}_4$	−0.423	-	−0.588	−0.592	-	-	0.421	-	-	-	-	-	-
tstat	−2.193	-	−2.518	−2.496	-	-	1.225	-	-	-	-	-	-
$\hat{H}_6$	−0.433	-	−0.598	−0.590	-	-	−0.356	-	-	-	-	-	-
tstat	−1.890	-	−2.294	−2.269	-	-	−2.006	-	-	-	-	-	-
$\hat{H}_7$	−0.338	-	−0.360	−0.342	-	-	-	-	-	-	-	-	-
tstat	−2.138	-	−2.109	−1.989	-	-	-	-	-	-	-	-	-
$\hat{H}_8$	0.389	0.428	0.550	0.553	−0.308	-	−0.329	-	-	-	-	-	-
tstat	2.593	3.190	3.718	3.738	−2.018	-	−2.143	-	-	-	-	-	-
$\hat{H}_1^2$	-	-	0.156	-	-	-	-	-	-	-	-	-	-
tstat	-	-	0.854	-	-	-	-	-	-	-	-	-	-
$\hat{H}_2^2$	-	-	-	-	-	−0.208	-	-	-	-	-	-	-
tstat	-	-	-	-	-	−2.668	-	-	-	-	-	-	-
$\hat{H}_3^2$	0.111	-	-	-	-	-	-	-	-	-	-	-	-
tstat	1.999	-	-	-	-	-	-	-	-	-	-	-	-

(*continued*)

TABLE 12.5 (Continued)

Regressions $rx^{(3)}_{t+1} = a + \alpha' \hat{H}_t + \beta' CP_t + \epsilon_{t+1}$

	$\hat{H} = \hat{F}$				$\hat{H} = \hat{G}$				$\hat{H} = \hat{F}$		$\hat{H} = \hat{G}$		
	In	Out	In	Out	In	Out	In	Out					
$\hat{H}$	1	2	3	4	5	6	7	8	9	10	11	12	13
$\hat{H}^2_4$	-	-	-	-	−0.190	-	−0.250	−0.275	-	-	-	-	-
tstat	-	-	-	-	−3.925	-	−3.005	−3.622	-	-	-	-	-
$\hat{H}^2_5$	-	−0.161	-	-	-	-	-	-	-	-	-	-	-
tstat	-	−2.179	-	-	-	-	-	-	-	-	-	-	-
$\hat{H}^2_6$	-	-	-	-	−0.152	−0.147	−0.140	−0.127	-	-	-	-	-
tstat	-	-	-	-	−7.130	−6.883	−3.307	−2.551	-	-	-	-	-
$\hat{H}^2_7$	-	-	-	-	0.089	-	-	-	-	-	-	-	-
tstat	-	-	-	-	2.687	-	-	-	-	-	-	-	-
$\hat{H}^3_1$	0.095	0.086	0.141	0.106	0.032	-	0.031	-	-	-	-	-	-
tstat	3.235	3.204	2.922	3.445	2.233	-	1.942	-	-	-	-	-	-
CP	0.760	0.784	-	-	0.847	0.821	-	-	0.644	-	0.786	-	0.856
tstat	5.329	6.885	-	-	7.516	7.770	-	-	4.661	-	6.381	-	8.301
$\hat{H}8$	-	-	-	-	-	-	-	-	0.588	0.877	0.710	0.931	-
tstat	-	-	-	-	-	-	-	-	4.494	7.133	3.624	3.256	-
$\bar{R}^2$	0.455	0.424	0.268	0.267	0.475	0.418	0.182	0.167	0.432	0.277	0.404	0.135	0.328

TABLE 12.6

Regressions $rx_{t+1}^{(4)} = a + \alpha' \hat{H}_t + \beta' CP_t + \epsilon_{t+1}$

	$\hat{H} = \hat{F}$				$\hat{H} = \hat{G}$								
	In	Out	In	Out	In	Out	In	Out	$\hat{H} = \hat{F}$		$\hat{H} = \hat{G}$		
$\hat{H}$	1	2	3	4	5	6	7	8	9	10	11	12	13
$\hat{H}_1$	−1.521	−1.521	−2.011	−2.050	-	-	-	-	-	-	-	-	-
tstat	−5.138	−4.149	−5.013	−5.290	-	-	-	-	-	-	-	-	-
$\hat{H}_2$	-	-	1.069	1.069	−0.952	−1.342	−1.619	−1.601	-	-	-	-	-
tstat	-	-	3.028	3.095	−2.680	−3.754	−2.812	−2.848	-	-	-	-	-
$\hat{H}_3$	-	-	-	-	−1.036	-	−1.080	−1.078	-	-	-	-	-
tstat	-	-	-	-	−4.127	-	−2.486	−2.401	-	-	-	-	-
$\hat{H}_4$	−0.436	-	−0.689	−0.681	-	-	0.590	0.452	-	-	-	-	-
tstat	−1.595	-	−1.957	−1.978	-	-	1.221	0.927	-	-	-	-	-
$\hat{H}_5$	-	-	−0.321	-	-	-	-	-	-	-	-	-	-
tstat	-	-	−1.475	-	-	-	-	-	-	-	-	-	-
$\hat{H}_6$	−0.668	-	−0.889	−0.889	-	-	−0.605	-	-	-	-	-	-
tstat	−2.160	-	−2.522	−2.449	-	-	−2.333	-	-	-	-	-	-
$\hat{H}_7$	−0.534	–	−0.535	−0.541	-	-	-	-	-	-	-	-	-
tstat	−2.401	-	−2.222	−2.209	-	-	-	-	-	-	-	-	-
$\hat{H}_8$	0.578	0.636	0.820	0.822	−0.474	-	−0.521	-	-	-	-	-	-
tstat	2.820	3.365	3.935	3.914	−2.344	-	−2.277	-	-	-	-	-	-
$\hat{H}_1^2$	-	−0.146	-	-	-	-	-	-	-	-	-	-	-
tstat	-	−0.770	-	-	-	-	-	-	-	-	-	-	-
$\hat{H}_2^2$	-	-	-	-	-	−0.284	-	-	-	-	-	-	-
tstat	-	-	-	-	-	−2.934	-	-	-	-	-	-	-

(continued)

TABLE 12.6 (Continued)

Regressions $rx_{t+1}^{(4)} = a + \alpha' \hat{H}_t + \beta' CP_t + \epsilon_{t+1}$

	$\hat{H} = \hat{F}$				$\hat{H} = \hat{G}$				$\hat{H} = \hat{F}$		$\hat{H} = \hat{G}$		
	In	Out	In	Out	In	Out	In	Out					
$\hat{H}$	1	2	3	4	5	6	7	8	9	10	11	12	13
$\hat{H}_3^2$	0.177	-	-	-	-	-	-	-	-	-	-	-	-
tstat	2.527	-	-	-	-	-	-	-	-	-	-	-	-
$\hat{H}_4^2$	-	-	-	-	−0.262	-	−0.354	−0.367	-	-	-	-	-
tstat	-	-	-	-	−3.692	-	−2.976	−3.552	-	-	-	-	-
$\hat{H}_5^2$	-	−0.228	-	-	-	-	-	-	-	-	-	-	-
tstat	-	−2.309	-	-	-	-	-	-	-	-	-	-	-
$\hat{H}_6^2$	-	-	-	-	−0.231	−0.227	−0.219	−0.189	-	-	-	-	-
tstat	-	-	-	-	−6.923	−9.811	−4.375	−3.248	-	-	-	-	-
$\hat{H}_7^2$	-	-	-	-	0.148	0.104	-	-	-	-	-	-	-
tstat	-	-	-	-	3.258	2.233	-	-	-	-	-	-	-
$\hat{H}_1^3$	0.131	0.081	0.142	0.148	0.037	-	0.036	-	-	-	-	-	-
tstat	3.436	1.483	3.938	3.602	1.964	-	1.599	-	-	-	-	-	-
CP	1.115	1.158	-	-	1.238	1.219	-	-	0.955	-	1.150	-	1.235
tstat	6.077	7.028	-	-	7.821	8.197	-	-	4.765	-	6.417	-	8.224
$\hat{H}8$	-	-	-	-	-	-	-	-	0.777	1.204	0.864	1.188	-
tstat	-	-	-	-	-	-	-	-	4.474	7.247	3.388	3.061	-
$\bar{R}^2$	0.473	0.441	0.263	0.260	0.496	0.445	0.171	0.155	0.452	0.273	0.416	0.114	0.357

TABLE 12.7

Regressions $rx_{t+1}^{(5)} = a + \alpha' \hat{H}_t + \beta' CP_t + \epsilon_{t+1}$

	$\hat{H} = \hat{F}$				$\hat{H} = \hat{G}$				$\hat{H} = \hat{F}$		$\hat{H} = \hat{G}$		
	In	Out	In	Out	In	Out	In	Out					
$\hat{H}$	1	2	3	4	5	6	7	8	9	10	11	12	13
$\hat{H}_1$	−1.653	−1.373	−2.214	−2.277	0.308	-	0.326	-	-	-	-	-	-
tstat	−4.723	−3.686	−4.503	−4.819	1.701	-	2.049	-	-	-	-	-	-
$\hat{H}_2$	-	-	1.355	1.355	−1.145	−1.573	−1.928	−1.609	-	-	-	-	-
tstat	-	-	3.111	3.195	−2.653	−3.691	−2.760	−2.994	-	-	-	-	-
$\hat{H}_3$	-	-	-	-	−1.161	-	−1.199	−1.003	-	-	-	-	-
tstat	-	-	-	-	−3.615	-	−2.224	−2.021	-	-	-	-	-
$\hat{H}_4$	−0.516	-	−0.818	−0.805	-	-	0.654	-	-	-	-	-	-
tstat	−1.478	-	−1.861	−1.881	-	-	1.128	-	-	-	-	-	-
$\hat{H}_5$	-	-	−0.523	-	-	-	-	-	-	-	-	-	-
tstat	-	-	−1.969	-	-	-	-	-	-	-	-	-	-
$\hat{H}_6$	−0.856	-	−1.120	−1.120	-	-	−0.678	-	-	-	-	-	-
tstat	−2.150	-	−2.566	−2.462	-	-	−2.049	-	-	-	-	-	-
$\hat{H}_7$	−0.686	-	−0.685	−0.694	-	-	-	-	-	-	-	-	-
tstat	−2.479	-	−2.321	−2.299	-	-	-	-	-	-	-	-	-
$\hat{H}_8$	0.702	0.725	0.985	0.988	−0.563	-	−0.608	-	-	-	-	-	-
tstat	2.756	3.292	3.956	3.907	−2.217	-	−2.156	-	-	-	-	-	-
$\hat{H}_1^2$	-	−0.563	-	-	-	-	-	-	-	-	-	-	-

(continued)

TABLE 12.7 (Continued)

Regressions $rx_{t+1}^{(5)} = a + \alpha' \hat{H}_t + \beta' CP_t + \epsilon_{t+1}$

	$\hat{H} = \hat{F}$				$\hat{H} = \hat{G}$				$\hat{H} = \hat{F}$		$\hat{H} = \hat{G}$		
	In	Out	In	Out	In	Out	In	Out					
$\hat{H}$	1	2	3	4	5	6	7	8	9	10	11	12	13
tstat	-	−3.037	-	-	-	-	-	-	-	-	-	-	-
$\hat{H}_2^2$	-	-	-	-	-	−0.339	-	-	-	-	-	-	-
tstat	-	-	-	-	-	−2.955	-	-	-	-	-	-	-
$\hat{H}_3^2$	0.204	-	-	-	-	-	-	-	-	-	-	-	-
tstat	2.327	-	-	-	-	-	-	-	-	-	-	-	-
$\hat{H}_4^2$	-	-	-	-	−0.357	-	−0.465	−0.466	-	-	-	-	-
tstat	-	-	-	-	−4.429	-	−3.497	−3.684	-	-	-	-	-
$\hat{H}_6^2$	-	-	-	-	−0.269	−0.279	−0.253	−0.234	-	-	-	-	-
tstat	-	-	-	-	−6.235	−9.685	−4.407	−3.596	-	-	-	-	-
$\hat{H}_7^2$	-	-	-	-	0.179	-	-	-	-	-	-	-	-
tstat	-	-	-	-	3.221	-	-	-	-	-	-	-	-
$\hat{H}_1^3$	0.150	-	0.160	0.170	-	-	-	-	-	-	-	-	-
tstat	3.310	-	3.893	3.440	-	-	-	-	-	-	-	-	-
CP	1.316	1.394	-	-	1.457	1.413	-	-	1.115	-	1.359	-	1.453
tstat	5.603	6.985	-	-	7.237	7.409	-	-	4.370	-	5.969	-	7.576
$\hat{H}8$	-	-	-	-	-	-	-	-	0.938	1.437	0.955	1.338	-
tstat	-	-	-	-	-	-	-	-	4.542	7.281	3.078	2.854	-
$\bar{R}^2$	0.435	0.392	0.251	0.245	0.453	0.408	0.152	0.135	0.422	0.259	0.377	0.097	0.330

Among the dynamic factors, $\hat{g}_2$ (labor market), $\hat{g}_8$ (stock market), $\hat{g}_6^2$ (bonds and foreign exchange) along with CP are selected by both BIC procedures as predictors (columns 5 and 6). Interestingly, the output factor $\hat{g}_1$ is not significant when the CP is included. The out-of-sample BIC has an $\bar{R}^2$ of 0.407, showing that there is a substantial amount of variation in the 2-year excess bond returns that can be predicted by macroeconomic factors. The in-sample BIC additionally selects $\hat{g}_{3t}$, $\hat{g}_{6t}$ and some higher-order terms with an $\bar{R}^2$ of 0.477. Thus, predictive regressions using $\hat{f}_t$ and $\hat{g}_t$ both find a factor relating to real activity ($\hat{f}_{1t}$ or $\hat{g}_{1t}$) and one relating to the stock market ($\hat{f}_{8t}$ or $\hat{g}_{18}$) to have significant predictive power for 2-year excess bond returns.

Results when the regressions do not include the CP variable are in columns 3, 4, 7, and 8. Evidently, $\hat{f}_2$ is now important according to both the in- and out-of-sample BIC, showing that the main effect of CP is to render $\hat{f}_2$ redundant. Furthermore, the out-of-sample BIC now selects a model that is only marginally more parsimonious than that selected by the in-sample BIC. The regressions with $\hat{F}$ alone have an $\bar{R}^2$ of 0.283 and 0.258, respectively, slightly less than what is obtained with CP as the only regressor.

Regressions based on the dynamic factors are qualitatively similar. The factors $\hat{g}_1$, $\hat{g}_3$, and $\hat{g}_4$, found not to be important when CP is included are now selected as relevant predictors when CP is dropped. Without CP, the dynamic factors selected by the in-sample BIC explain 0.2 of the 1-year-ahead variation in excess bond returns, while the more parsimonious model selected by the out-of-sample BIC has an $\bar{R}^2$ of 0.192. These numbers are lower than what we obtain in columns (3) and (4) using $\hat{F}_t$ as predictors.

It is important to stress that we consider the two sets of factor estimates not to perform a horse race of whether the PCA or the Bayesian estimator is better. The purpose instead is to show that macroeconomic factors have predictive power for excess bond returns irrespective of the way we estimate the factors. Although the precise degree of predictability depends on how the factors are estimated, a clear picture emerges. At least 20% of the variation in excess bound returns can be predicted by macroeconomic factors even in the presence of the CP factor.

12.4.2 Longer Maturity Returns and Overview

Tables 12.5 to 12.7 report results for returns with maturity of 3, 4, and 5 years. Most of the static factors found to be useful in predicting $rx_{t+1}^{(2)}$ by the in-sample BIC remain useful in predicting the longer maturity returns. These predictors include $\hat{f}_{1t}$, $\hat{f}_{4t}$, $\hat{f}_{6t}$, $\hat{f}_{7t}$, $\hat{f}_{8t}$, $\hat{f}_{1t}^3$, and CP. Of these, $\hat{f}_{1t}$, $\hat{f}_{8t}$, and CP are also selected by the out-of-sample BIC procedure. The nonlinear term $\hat{f}_{1t}^3$ is an important predictor in equations for all maturity returns except the 5 years. The factors add at least 10 basis points to the $\bar{R}^2$ with CP as the sole predictor.

The dynamic factors found important in explaining 2-year excess return are generally also relevant in regressions for longer maturity excess returns. The in-sample BIC finds $\hat{g}_{2t}$, $\hat{g}_{3t}$, $\hat{g}_{8t}$, $\hat{g}_{4t}^2$, $\hat{g}_{6t}^2$ along with the CP to be important

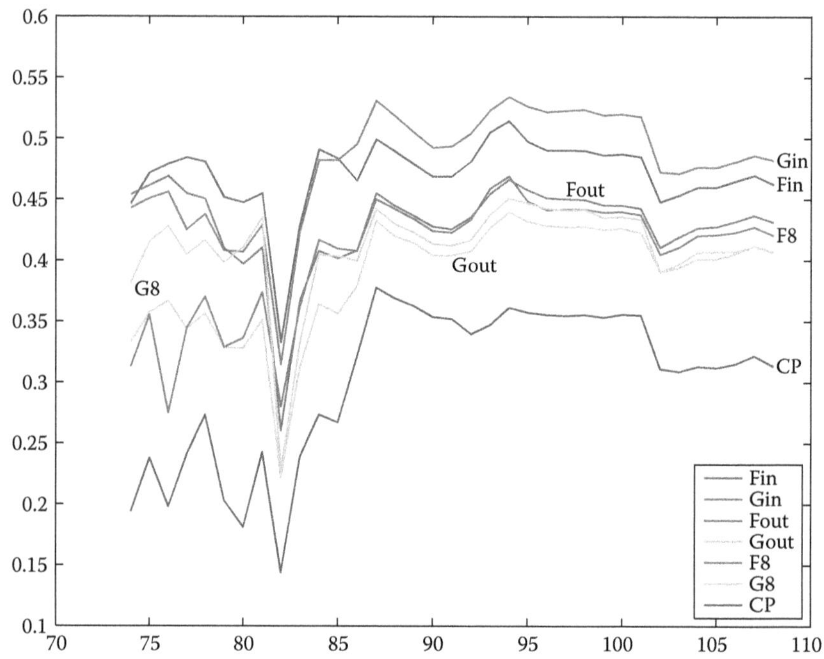

FIGURE 12.9

Adjusted R-squares, with CP. Fin and Gin are the $\bar{R}^2$ from rolling estimation of Equation 12.8, with predictors selected by the in-sample BIC. Fout and Gout use predictors selected by the out-of-sample BIC. F8 and G8 use a linear combination of eight factors as predictors, where the weights are based on Equation 12.10.

in regressions of all maturities. The output factor is again not significant in regressions with 3- and 4-year maturities. It is marginally significant in the 5-year maturity, but has the wrong sign. While $\hat{g}_8$ was relevant in the 2-year regression, it is not an important predictor in the regressions for longer maturity returns. The out-of-sample BIC finds dynamic factors from the labor market ($\hat{g}_{2t}$), the bond and foreign exchange markets ($\hat{g}_{6t}$). Together, these factors have incremental predictive power for excess bond returns over CP, improving the $\bar{R}^2$ by slightly less than 10 basis points.

The relevance of macroeconomic variables in explaining excess bond returns is reinforced by the results in columns 10 and 12, which show that a simple linear combination of the eight factors still adds substantial predictive power beyond the CP factor. This result is robust across all four maturities considered, noting that the coefficient estimate on $\hat{H}8$ increases with the holding period without changing the statistical significance of the coefficient.

To see if the predictability varies over the sample, we also consider rolling regressions. Starting with the first regression that spans the sample 1964:1 to 1974:12, we add 12 monthly observations each time and record the $\bar{R}^2$. Figure 12.9 shows the $\bar{R}^2$ for regressions with CP included. Apart from a notable drop around the 1983 recession, $\bar{R}^2$ is fairly constant. Figure 12.10

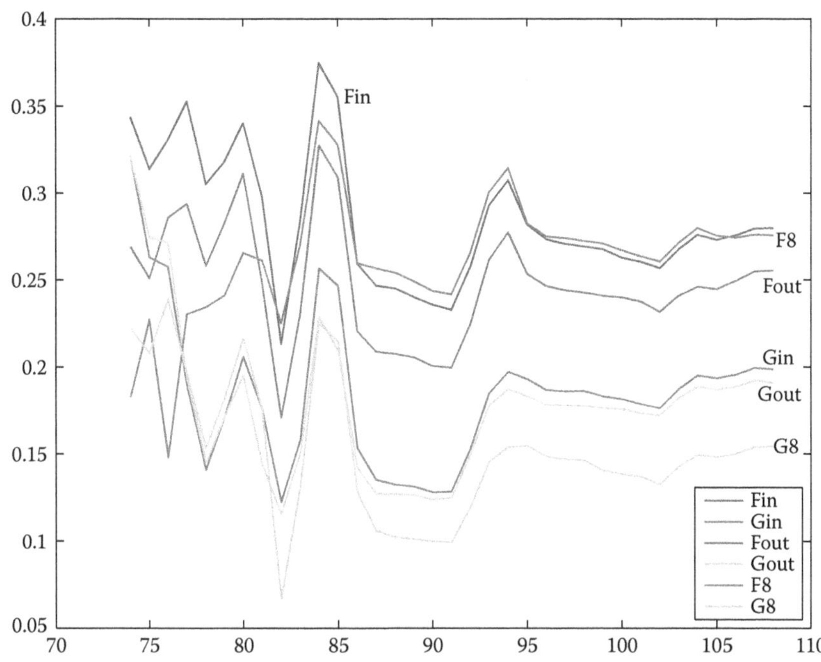

FIGURE 12.10
Adjusted R-squares, without CP. Fin and Gin are the $\bar{R}^2$ from rolling estimation of (8), with predictors selected by the in-sample BIC. Fout and Gout use predictors selected by the out-of-sample BIC. F8 and G8 use a linear combination of eight factors as predictors, where the weights are based on (10).

depicts the $\bar{R}^2$ for regressions without CP. Notice that the $\bar{R}^2$ that corresponds to $\hat{F}8_t$ tends to be 15 basis points higher than $\hat{G}8_t$. As noted earlier, each of the eight $\hat{f}_t$ is itself a combination of the current and lags of the eight $\hat{g}_t$. This underscores the point that imposing a structure on the data to facilitate interpretation of the factors comes at the cost of not letting the data find the best predictive combination possible.

The results reveal that the estimated factors consistently have stronger predictive power for one- and multi-year ahead excess bond returns. The most parsimonious specification has just two variables — $\hat{H}8$ and CP_t — explaining over 40% of the variation in rx^n_{t+1} of every maturity. A closer look reveals that the real activity factor $\hat{f}_{1t}$ is the strongest factor predictor, both numerically and statistically. As $\hat{g}_{1t}$ tends not to be selected as predictor, this suggests that the part of $\hat{f}_{1t}$ that has predictive power for excess bond returns is derived from real activity other than output. However, the dynamic factors $\hat{g}_{2t}$ (labor market) and $\hat{g}_{3t}$ (housing) have strong predictive power. Indeed, $\hat{f}_{1t}$ is highly correlated with $\hat{g}_{2t}$ and the coefficients for these predictors tend to be negative. This means that excess bond returns of every maturity are countercyclical, especially with the labor market. This result is in accord with the models of Campbell (1999) and Wachter (2006), which posit that forecasts of excess returns should be countercyclical because risk aversion is low in good times and high in recessions. We will subsequently show that yield risk premia, which are based on forecasts of excess returns, are also countercyclical.

12.5 Inference Issues

The results thus far assume that N and T are large and that $\sqrt{T}/N$ tends to zero. In this section, we first consider the implication for factor augmented regressions when $\sqrt{T}/N$ may not be small as is assumed. We then examine the finite sample inference issues.

12.5.1 Asymptotic Bias

If excess bond returns truly depend on macroeconomic factors, then consistent estimates of the factors should be better predictors than the observed variables because these are contaminated measures of real activity.[6] An appealing feature of PCA is that if $\sqrt{T}/N \to 0$ as $N, T \to \infty$, then $\hat{F}_t$ can be treated in the predictive regression as though it were F_t. To see why this is the case, consider again the infeasible predictive regression, dropping the observed predictors W_t for simplicity. We have

$$\begin{aligned} rx^n_{t+1} &= \alpha_F^{+\prime} F_t + \epsilon_{t+1} \\ &= \alpha_F' \hat{F}_t + \alpha_F'(HF_t - \hat{F}_t) + \epsilon_{t+1}, \end{aligned}$$

where $\alpha_F = \alpha H^{-1}$, and H is a $r \times r$ matrix defined in Bai and Ng (2006a). Let $S_{\hat{F}\hat{F}} = T^{-1}\sum_{t=1}^T \hat{F}_t \hat{F}_t'$. Then

$$\sqrt{T}(\hat{\alpha}_F - \alpha_F) = \hat{S}_{\hat{F}\hat{F}}^{-1}\left(\frac{1}{\sqrt{T}}\sum_{t=1}^{T} \hat{F}_t \epsilon_{t+1}\right) + S_{\hat{F}\hat{F}}^{-1}\left(\frac{1}{\sqrt{T}}\sum_{t=1}^{T} \hat{F}_t(HF_t - \hat{F}_t)\right)\alpha_F. \tag{12.11}$$

But $T^{-1}\hat{F}'(FH' - \hat{F}) = O_p(\min[N, T]^{-1})$, a result that follows from Bai (2003). Thus if $\sqrt{T}/N \to 0$, the second term is negligible. It follows that

$$\sqrt{T}(\hat{\alpha}_F - \alpha_F) \xrightarrow{d} N(0, \text{Avar}(\hat{\alpha}_F)),$$

where

$$\text{Avar}(\hat{\alpha}_F) = \text{plim}\ S_{\hat{F}\hat{F}}^{-1}\widehat{\text{Avar}}(g_t) S_{\hat{F}\hat{F}}^{-1},$$

$\widehat{\text{Avar}}(g_t)$ is an estimate of the asymptotic variance of $g_{t+1} = \hat{\epsilon}_{t+1}\hat{F}_t$.

[6] Moench (2008) finds that factors estimated from a large panel of macroeconomic data explain the short rate better than output and inflation.

Consider now the case when $\sqrt{T}$ is comparable to N. Although the first term on the right-hand side of Equation 12.11 is mean zero, the second term is a $O_p(1)$ random variable that may not be mean zero. This generates a bias in the asymptotic distribution for $\hat{\alpha}_F$.

PROPOSITION 12.1 *Suppose assumptions A–E of Bai and Ng (2006a) hold and let $\hat{F}_t \subset \hat{f}_t$, where $\hat{f}_t$ are the principal component estimates of f_t, $x_{it} = \lambda_{it}' f_t + e_{it}$. Let $\hat{\alpha}_F$ be obtained from least squares estimation of the FAR $y_{t+h} = \alpha_F' \hat{F}_t + e_{t+h}$. An estimate of the bias in $\hat{\alpha}_F$ is*

$$\hat{B}_1 \approx -S_{\hat{F}\hat{F}}^{-1}\left(\frac{1}{NT}\sum_{t=1}^{T}\widehat{\text{Avar}_t}(\hat{F}_t)\right)\hat{\alpha}_F,$$

where $\text{Avar}_t(F_t) = V^{-1}\Gamma_t V^{-1}$, V is a $r \times r$ diagonal matrix of the eigenvalues of $(N \cdot T)^{-1}xx'$, and $\Gamma_t = \lim_{N\to\infty} N^{-1}\sum_{i=1}^{N}\sum_{j=1}^{N} E(\lambda_i\lambda_j' e_{it}e_{jt})$. Let $\hat{\alpha}_F^B = \hat{\alpha}_F - \hat{B}_1$ be the biased corrected estimate. Then

$$\sqrt{T}(\hat{\alpha}_F^B - \alpha_F)\xrightarrow{d} N\Big(0, \text{Avar}(\hat{\alpha}_F)\Big).$$

The asymptotic variance for the bias corrected estimator is the same as $\hat{\alpha}_F$. Proposition 1 makes use of the fact that

$$\begin{aligned}\frac{1}{T}\sum_{t=1}^{T}\hat{F}_t(HF_t - \hat{F}_t)' &= \frac{1}{T}\sum_{t=1}^{T}(\hat{F}_t - HF_t)(HF_t - \hat{F}_t)' + HF_t(HF_t - \hat{F}_t)\\ &= -E\left[\frac{1}{T}\sum_{t=1}^{T}(\hat{F}_t - HF_t)(HF_t - \hat{F}_t)'\right] + o_p(1)\\ &= -\frac{1}{NT}\sum_{t=1}^{T}\text{Avar}(\hat{F}_t) + o_p(1).\end{aligned}$$

The estimation of $\text{Avar}_t(\hat{F}_t)$ was discussed in Bai and Ng (2006a). If $E(e_{it}^2) = \sigma^2$ for all i and t, $\text{Avar}_t(F_t)$ is the same for all t. Although Γ_t will depend on t if e_{it} is heteroskedastic, a consistent estimate of Γ_t can be obtained for each t when the errors are not cross-sectionally correlated, i.e., $E(e_{it}e_{jt}) = 0$. Alternatively, if $E(e_{it}e_{jt}) = \sigma_{ij} \neq 0$ for some or all t, panel data permit an estimate of $\text{Avar}(\hat{F}_t)$ that does not depend on t even when the e_{it} are cross-sectionally correlated. This estimator of Γ_t, referred to as CS-HAC in Bai and Ng (2006a), will be used later.

As this result on bias is new, we consider a small Monte Carlo experiment to gauge the magnitude of the bias as N and T changes. We consider a model with $r = 1$ and 2 factors. We assume $\lambda_i \sim N(0, 1)$ and $F_t \sim N(0, 1)$. These are only simulated once. Samples of $x_{it} = \lambda_i F_t + e_{it}$ and $y_t = \alpha' F_t + \epsilon_t$ are obtained by simulating $e_{it} \sim \sigma N(0, 1)$ and $\epsilon_t \sim N(0, 1)$ for $i = 1, \ldots N$, $t = 1, \ldots T$. We let $\alpha = 1$ when $r = 1$ and $\alpha = (1, 2)$ when $r = 2$. We consider three values of σ.

The smaller σ is, the more informative are the data for the factors. The results are as follows:

Estimated Bias for $\hat{\alpha}_1$
DGP: $y_t = F_t'\alpha + \epsilon_t, \quad x_{it} = \lambda_i F_t + e_{it}$

	$\sigma = 1$				$\sigma = 4$			
$r = 1$	$T = 50$	100	200	500	50	100	200	500
$N = 50$	−0.025	−0.020	−0.022	−0.019	−0.171	−0.156	−0.210	−0.242
100	−0.009	−0.009	−0.009	−0.012	−0.107	−0.107	−0.115	−0.138
200	−0.004	−0.004	−0.005	−0.004	−0.058	−0.058	−0.068	−0.071
500	−0.002	−0.002	−0.002	−0.002	−0.024	−0.030	−0.031	−0.034
$r = 2$								
50	0.014	−0.035	0.026	0.017	0.002	−0.244	−0.077	−0.124
100	−0.020	0.003	−0.018	−0.020	0.116	−0.170	−0.056	−0.158
200	−0.010	0.001	0.007	−0.009	−0.104	−0.036	0.077	−0.092
500	−0.005	0.002	−0.004	0.001	−0.047	−0.043	0.028	0.031

As the true value of α is one, the entries can also be interpreted as percent bias. For large N and T, the bias is quite small and ignoring the sampling error in $\hat{F}_t$ should be inconsequential. Bias is smaller when $T/N = c$ than when $N/T = c$ for the same $c > 1$, confirming that the factors are more precisely estimated when there are more cross-section units to wash out the idiosyncratic noise. However, when σ is large and the data are uninformative about the factors, the bias can be well over 10% and as large as 20%. In such cases, the bias is also increasing in the number of estimated factors.

12.5.2 Bias When the Predictors Are Functions of $\hat{f}_t$

Our predictive regression has two additional complications. First, some of our predictors are powers of the estimated factors. Second, $\hat{F}8_t$ is a linear combination of a subset of $\hat{f}_t$ and $\hat{f}_{1t}^3$, which is a nonlinear function of $\hat{f}_{1t}$. To see how to handle the first problem, consider the case of the scalar predictor, $\hat{m}_t = m(\hat{f}_{1t})$ and let $m_t = m(Hf_{1t})$ where m takes its argument to the power b. The factor augmented regression becomes

$$y_t = \alpha_F' \hat{m}_t + \alpha_F'(m_t - \hat{m}_t) + \epsilon_t,$$

where $\alpha_F = \alpha H^{-b}$. The required bias correction is now of the form

$$B_2 = S_{\hat{m}\hat{m}'}^{-1} \left(\frac{1}{T} \sum_{t=1}^{T} \hat{m} \left(m_t - \hat{m}_t \right)' \alpha_F \right).$$

But since m is continuous in $\hat{f}_{1t}$,

$$m(\hat{f}_{1t}) = m(f_t) + m_{f_{1,t}}(\hat{f}_{1t} - Hf_{1t}),$$

where $m_{\hat{f},t} = \frac{\partial \hat{m}_t}{\partial f_{1t}}|_{f_{1t}=Hf_{1t}}$. We have

$$\hat{m}_t - m_t = b(Hf_{1t})^{b-1}(\hat{f}_{1t} - Hf_{1t}) = O_p(\min[N, T]^{-1}).$$

Given the foregoing result, it is then straightforward to show that

$$T^{-1}\sum_{t=1}^{T}\hat{m}_t(m_t - \hat{m}_t)' = \left[T^{-1}\sum_{t=1}^{T} m_{f_{1},t}\text{Avar}(\hat{f}_{1t})m'_{f_{1},t}\right] + o_p(1).$$

Extending the argument to the case when m_t is a vector leads to the bias correction

$$\hat{\mathcal{B}}_2 = -S^{-1}_{\hat{m}\hat{m}'}\left(T^{-1}\sum_{t=1}^{T} m_{\hat{F},t}\text{Avar}(\hat{F}_{1t})m'_{\hat{F},t}\right)\alpha_F.$$

Finally, consider the predictive regression

$$y_t = \alpha'_F\hat{M}_t + \epsilon_t,$$

where $\hat{M}_t = \hat{\gamma}_0 + \hat{\gamma}'\hat{m}_t$. The bias can be estimated by

$$\hat{\mathcal{B}}_3 = \hat{\gamma}'\hat{\mathcal{B}}_2\hat{\gamma}.$$

In our application, $\hat{\gamma}$ is obtained from estimation of Equation 12.10.

While in theory, these bias corrections are required only when $\sqrt{T}/N$ does not tend to zero, in finite samples, the bias correction might be desirable even when $\sqrt{T}/N$ is small. We calculate the biased corrected estimates for two specifications of the predictive regressions. The first is when the predictors are selected by the in-sample BIC (column 1 of Tables 12.4 to 12.7). As this tends to lead to a larger model, the bias is likely more important. The second is when $\hat{F}8_t$ is used as predictor (column 9 of Tables 12.4 to 12.7), which is the most parsimonious of our specifications. Note that the observed predictor CP is not associated with first-step estimation error. As such, this predictor does not contribute to bias.

Reported in Table 12.8 are results using the CS-HAC, which allows the idiosyncratic errors to be cross-sectionally correlated. Results when the errors are heteroskedastic but cross-sectionally uncorrelated are similar. The results indicate that the bias is quite small. For the present application, the effect of the bias correction is to increase the absolute magnitude of the coefficient estimates in the predictive regressions. The t-statistics (not reported) are correspondingly larger. The finding that the macroeconomic factors have predictive power for excess bond returns is not sensitive to the assumption underlying the asymptotically validity of the FAR estimates.

12.5.3 Bootstrap Inference

According to asymptotic theory, heteroskedasticity and autocorrelation consistent standard errors that are asymptotically $N(0, 1)$ can be used to obtain

TABLE 12.8

Biased Corrected Estimates: $rx_{t+1}^{(n)} = a + \alpha'\hat{F}_t + \beta' CP_t + \epsilon_{t+1}$

$\hat{F}$	$n = 2$		$n = 3$		$n = 4$		$n = 5$	
$\hat{H}_1$	−0.761	-	−1.232	-	−1.521	-	−1.653	-
$\tilde{\alpha}$	−0.785	-	−1.277	-	−1.576	-	−1.724	-
bias	0.024	-	0.045	-	0.054	-	0.072	-
$\hat{H}_2$	-	-	−0.028	-	-	-	-	-
$\tilde{\alpha}$	-	-	−0.059	-	-	-	-	-
bias	-	-	0.032	-	-	-	-	-
$\hat{H}_4$	−0.291	-	−0.423	-	−0.436	-	−0.516	-
$\tilde{\alpha}$	−0.307	-	−0.454	-	−0.472	-	−0.564	-
bias	0.016	-	0.031	-	0.036	-	0.048	-
$\hat{H}_6$	−0.151	-	−0.433	-	−0.668	-	−0.856	-
$\tilde{\alpha}$	−0.168	-	−0.468	-	−0.710	-	−0.912	-
bias	0.018	-	0.035	-	0.042	-	0.055	-
$\hat{H}_7$	−0.128	-	−0.338	-	−0.534	-	−0.686	-
$\tilde{\alpha}$	−0.145	-	−0.372	-	−0.573	-	−0.737	-
bias	0.017	-	0.034	-	0.039	-	0.051	-
$\hat{H}_8$	0.240	-	0.389	-	0.578	-	0.702	-
$\tilde{\alpha}$	0.225	-	0.355	-	0.542	-	0.654	-
bias	0.016	-	0.033	-	0.036	-	0.048	-
$\hat{H}_3^2$	-	-	0.111	-	0.177	-	0.204	-
$\tilde{\alpha}$	-	-	0.114	-	0.181	-	0.209	-
bias	-	-	−0.004	-	−0.004	-	−0.006	-
$\hat{H}_5^2$	−0.080	-	-	-	-	-	-	-
$\tilde{\alpha}$	−0.078	-	-	-	-	-	-	-
bias	−0.003	-	-	-	-	-	-	-
$\hat{H}_1^3$	0.044	-	0.095	-	0.131	-	0.150	-
$\tilde{\alpha}$	0.045	-	0.096	-	0.133	-	0.153	-
bias	−0.001	-	−0.002	-	−0.002	-	−0.003	-
CP	0.385	0.336	0.760	0.644	1.115	0.955	1.316	1.115
$\tilde{\alpha}$	0.381	0.343	0.760	0.660	1.108	0.980	1.306	1.147
bias	0.004	−0.007	–	−0.016	0.007	−0.026	0.010	−0.032
$\hat{H}8$	-	0.332	-	0.588	-	0.777	-	0.938
$\tilde{\alpha}$	-	0.342	-	0.607	-	0.802	-	0.972
bias	-	−0.010	-	−0.019	-	−0.025	-	−0.035

Note: The bias unadjusted estimates are reported in columns 1 and 9 of Tables 12.4 to 12.7, respectively.

TABLE 12.9

Bootstrap Estimates When $\hat{H}_t = \hat{F}_t$: Regression $rx_{t+1}^{(n)} = \alpha'\hat{F}_t + \beta' CP_t + \epsilon_{t+1}$

	$\hat{\alpha}$	bias	Bootstrap 95% CI	Bootstrap 99% CI	Bootstrap under the Null 95% CI	Bootstrap under the Null 99% CI
				$n = 2$		
$\hat{H}_1$	−0.761	0.012	(−1.143 −0.343)	(−1.071 −0.399)	(−0.021 −0.015)	(−0.021 −0.016)
$\hat{H}_4$	−0.291	−0.006	(−0.554 −0.031)	(−0.508 −0.073)	(−0.003 0.003)	(−0.002 0.003)
$\hat{H}_6$	−0.151	−0.002	(−0.467 0.166)	(−0.408 0.100)	(−0.015 0.016)	(−0.015 0.016)
$\hat{H}_7$	−0.128	−0.004	(−0.285 0.027)	(−0.258 −0.010)	(−0.008 0.011)	(−0.007 0.009)
$\hat{H}_8$	0.240	0.004	(0.054 0.425)	(0.088 0.404)	(−0.011 0.010)	(−0.010 0.008)
$\hat{H}_5^2$	−0.080	0.003	(−0.187 0.040)	(−0.170 0.015)	(−0.010 −0.003)	(−0.009 −0.003)
$\hat{H}_1^3$	0.044	−0.001	(0.010 0.076)	(0.016 0.071)	(−0.000 0.000)	(−0.000 0.000)
CP	0.385	−0.003	(0.262 0.516)	(0.276 0.490)	(0.003 0.009)	(0.003 0.008)
$\bar{R}^2$	0.460		(0.237 0.523)	(0.261 0.500)	(0.019 0.045)	(0.021 0.042)
				$n = 3$		
$\hat{H}_1$	−1.232	0.027	(−1.914 −0.506)	(−1.797 −0.655)	(−0.021 −0.015)	(−0.021 −0.016)
$\hat{H}_2$	−0.028	−0.017	(−0.574 0.505)	(−0.486 0.426)	(−0.001 0.005)	(−0.000 0.005)
$\hat{H}_4$	−0.423	−0.004	(−0.881 0.030)	(−0.811 −0.050)	(−0.003 0.003)	(−0.003 0.003)
$\hat{H}_6$	−0.433	0.012	(−0.969 0.093)	(−0.870 0.024)	(−0.014 0.015)	(−0.013 0.014)
$\hat{H}_7$	−0.338	−0.002	(−0.585 −0.094)	(−0.549 −0.140)	(−0.009 0.010)	(−0.007 0.009)
$\hat{H}_8$	0.389	−0.002	(0.082 0.669)	(0.140 0.632)	(−0.009 0.008)	(−0.008 0.007)
$\hat{H}_3^2$	0.111	−0.003	(−0.046 0.250)	(−0.006 0.221)	(0.000 0.002)	(0.000 0.002)
$\hat{H}_1^3$	0.095	−0.002	(0.034 0.145)	(0.046 0.136)	(0.000 0.001)	(0.000 0.001)
CP	0.760	−0.001	(0.546 0.980)	(0.582 0.935)	(0.003 0.009)	(0.003 0.008)
$\bar{R}^2$	0.455		(0.280 0.559)	(0.303 0.533)	(0.013 0.035)	(0.014 0.032)

(*continued*)

TABLE 12.9 (Continued)

Bootstrap Estimates When $\hat{H}_t = \hat{F}_t$: Regression $rx_{t+1}^{(n)} = \alpha' \hat{F}_t + \beta' CP_t + \epsilon_{t+1}$

	$\hat{\alpha}$	bias	Bootstrap 95% CI	Bootstrap 99% CI	Bootstrap under the Null 95% CI	Bootstrap under the Null 99% CI
				$n = 4$		
$\hat{H}_1$	−1.521	0.047	(−2.488 −0.480)	(−2.323 −0.617)	(−0.021 −0.015)	(−0.021 −0.016)
$\hat{H}_4$	−0.436	0.001	(−1.048 0.178)	(−0.958 0.090)	(−0.004 0.003)	(−0.003 0.003)
$\hat{H}_6$	−0.668	−0.002	(−1.410 0.131)	(−1.297 0.002)	(−0.014 0.015)	(−0.013 0.014)
$\hat{H}_7$	−0.534	0.004	(−0.942 −0.178)	(−0.849 −0.230)	(−0.009 0.010)	(−0.007 0.008)
$\hat{H}_8$	0.578	0.004	(0.119 1.022)	(0.206 0.957)	(−0.010 0.009)	(−0.009 0.007)
$\hat{H}_3^2$	0.177	−0.001	(−0.031 0.375)	(0.002 0.339)	(0.000 0.002)	(0.000 0.002)
$\hat{H}_1^3$	0.131	−0.003	(0.055 0.206)	(0.068 0.189)	(0.000 0.001)	(0.000 0.001)
CP	1.115	−0.006	(0.820 1.401)	(0.861 1.348)	(0.003 0.009)	(0.003 0.009)
$\bar{R}^2$	0.473		(0.277 0.567)	(0.303 0.545)	(0.014 0.036)	(0.016 0.034)
			$n = 5$			
$\hat{H}_1$	−1.653	0.026	(−2.832 −0.429)	(−2.648 −0.606)	(−0.021 −0.015)	(−0.021 −0.016)
$\hat{H}_4$	−0.516	−0.004	(−1.306 0.321)	(−1.190 0.169)	(−0.003 0.003)	(−0.003 0.003)
$\hat{H}_6$	−0.856	0.011	(−1.870 0.190)	(−1.666 0.012)	(−0.014 0.014)	(−0.013 0.014)
$\hat{H}_7$	−0.686	0.012	(−1.182 −0.119)	(−1.071 −0.244)	(−0.007 0.010)	(−0.006 0.009)
$\hat{H}_8$	0.702	−0.004	(0.139 1.286)	(0.224 1.160)	(−0.009 0.008)	(−0.009 0.007)
$\hat{H}_3^2$	0.204	0.000	(−0.059 0.491)	(−0.017 0.419)	(0.000 0.002)	(0.001 0.002)
$\hat{H}_1^3$	0.150	−0.001	(0.051 0.242)	(0.069 0.232)	(0.000 0.001)	(0.000 0.001)
CP	1.316	−0.009	(0.896 1.723)	(0.945 1.663)	(0.003 0.009)	(0.003 0.008)
$\bar{R}^2$	0.435		(0.225 0.518)	(0.251 0.488)	(0.015 0.036)	(0.016 0.033)

robust t-statistics for the in-sample regressions. Moreover, provided $\sqrt{T}/N$ goes to zero as the sample increases, the $\widehat{F}_t$ can be treated as observed regressors, and the usual t-statistics are valid (Bai and Ng 2006a). To guard against inadequacy of the asymptotic approximation in finite samples, we consider bootstrap inference in this section.

To proceed with a bootstrap analysis, we need to generate bootstrap samples of $rx_{t+1}^{(n)}$, and thus the exogenous predictors Z_t (here just CP_t), as well as of the estimated factors $\widehat{F}_t$. Bootstrap samples of $rx_{t+1}^{(n)}$ are obtained in two ways: first by imposing the null hypothesis of no predictability, and second, under the alternative that excess returns are forecastable by the factors and conditioning variables studied above. The use of monthly bond price data to construct continuously compounded annual returns induces an MA(12) error structure in the annual log returns. Thus, under the null hypothesis that the expectations hypothesis is true, annual compound returns are forecastable up to an MA(12) error structure, but are not forecastable by other predictor variables or additional moving average terms.

Bootstrap sampling that captures the serial dependence of the data is straightforward when, as in this case, there is a parametric model for the dependence under the null hypothesis. In this event, the bootstrap may be accomplished by drawing random samples from the empirical distribution of the residuals of a $\sqrt{T}$ consistent, asymptotically normal estimator of the parametric model, in our application a twelfth-order moving average process. We use this approach to form bootstrap samples of excess returns under the null. Under the alternative, excess returns still have the MA(12) error structure induced by the use of overlapping data, but estimated factors $\widehat{F}_t$ are presumed to contain additional predictive power for excess returns above and beyond that implied by the moving average error structure.

To create bootstrapped samples of the factors, we re-sample the $T \times N$ panel of data, x_{it}. For each i, we assume that the idiosyncratic errors e_{it} and the errors u_t in the factor process are AR(1) processes. Least squares estimation of $\widehat{e}_{it} = \rho_i \widehat{e}_{it-1} + v_{it}$ yields the estimates $\widehat{\rho}_i$ and $\widehat{v}_{it}$, $t = 2, \ldots, T$, recalling that $\widehat{e}_{it} = x_{it} - \widehat{\lambda}_i' \widehat{f}_t$. These errors are then re-centered. To generate a new panel of data, for each i, $\widehat{v}_{it}$ is re-sampled (while preserving the cross-section correlation structure) to yield bootstrap samples of $\widehat{e}_{it}$. In turn, bootstrap values of x_{it} are constructed by adding the bootstrap estimates of the idiosyncratic errors, $\widehat{e}_{it}$, to $\widehat{\lambda}_i' \widehat{F}_t$. Applying the method of principal components to the bootstrapped data yields a new set of estimated factors. Together with bootstrap samples of CP_t created under the assumption that it is an AR(1), we have a complete set of bootstrap regressors in the predictive regression.

Each regression using the bootstrapped data gives new estimates of the regression coefficients. This is repeated B times. Bootstrap confidence intervals for the parameter estimates and $\bar{R}^2$ statistics are calculated from $B = 10,000$ replications. We compute 90th and 95th percentiles of $\hat{\beta}_F$ and $\hat{\alpha}_F$, as well as the bootstrap estimate of the bias. This also allows us to compare the adequacy of our calculations for asymptotic bias considered in the previous subsection. The exercise is repeated for 2-, 3-, 4-, and 5-year excess bond returns.

To conserve space, the results in Table 12.9 are reported only for the largest model (corresponding to column 1 of Tables 12.4 to 12.7). The results based on bootstrap inference are consistent with asymptotic inference. In particular, the magnitude of predictability found in the historical data is too large to be accounted for by sampling error of the size we currently have. The coefficients on the predictors and factors are statistically different from zero at the 95% level and are well outside the 95% confidence interval under the null of no predictability. The bootstrap estimate of the bias on coefficients associated with the estimated factors are small, and the $\bar{R}^2$ are similar in magnitude to what was reported in Tables 12.4 to 12.7.

12.5.4 Posterior Inference

In Tables 12.4 to 12.7, we have used the posterior mean of G_t in the predictive regression computed from 1000 draws (taken from a total of 25,000 draws) from the posterior distribution of G_t. The $\hat{\alpha}$ do not reflect sampling uncertainty about G_t. To have a complete account of sampling variability, we estimate the predictive regressions for each of the 1000 draws of G_t. This gives us the posterior distribution for α as well as the corresponding t-statistic.

Reported in Table 12.10 are the posterior mean of α_G along with the 5% and 95% percentage points of the t-statistic. The point estimates reported in Tables 12.4 to 12.7 are very close to the posterior means. Sampling variability from having to estimate the dynamic factors has little effect on the estimates of the factor augmented regressions.

So far we find that macroeconomic factors have nontrivial predictive power for bond excess returns and that the sampling error induced by $\hat{F}_t$ or $\hat{G}_t$ in the predictive regressions are numerically small. Multiple factors contribute to the predictability of excess returns, so it is not possible to infer the cyclicality of return risk premia by observing the signs of the individual coefficients on factors in forecasting regressions of excess returns. But Tables 12.4 to 12.7 provide a summary measure of how the factors are related to future excess returns by showing that excess bond returns are high when the linear combinations of all factors, $\hat{F}8_t$ and $\hat{G}8_t$, are high. Figures 12.11 and 12.12 show that $\hat{F}8_t$ and $\hat{G}8_t$ are in turn high when real activity (as measured by industrial production growth) is low. The results therefore imply that excess returns are forecast to be high when economic activity is slow or contracting. That is, return risk premia are countercyclical. This is confirmed by the top panels of Figures 12.13 and 12.14, which plot return risk premia along with industrial production growth. The bottom panels of these figures show that the factors contribute significantly to the countercyclicality of risk-premia. Indeed, when factors are excluded (but CP_t is included), risk-premia are a-cyclical. Of economic interest is whether yield risk-premia are also countercyclical. We now turn to such an analysis.

TABLE 12.10

Posterior Mean: $rx_{t+1}^{(n)} = a + \alpha'\hat{G}_t + \beta' CP_t + \epsilon_{t+1}$

$\hat{F}$	$n = 2$		$n = 3$		$n = 4$		$n = 5$	
$\hat{H}_1$	-	-	-	-	-	-	0.288	-
$t_{.05}$	-	-	-	-	-	-	1.275	-
$t_{.95}$	-	-	-	-	-	-	1.912	-
$\hat{H}_2$	−0.506	-	−0.801	-	−0.976	-	−1.159	-
$t_{.05}$	−3.676	-	−3.239	-	−3.140	-	−3.099	-
$t_{.95}$	−2.942	-	−2.622	-	−2.477	-	−2.397	-
$\hat{H}_3$	−0.456	-	−0.746	-	−0.959	-	−1.074	-
$t_{.05}$	−5.335	-	−4.749	-	−4.616	-	−3.302	-
$t_{.95}$	−4.050	-	−3.637	-	−3.482	-	−3.374	-
$\hat{H}_6$	0.139	-	-	-	-	-	-	-
$t_{.05}$	1.819	-	-	-	-	-	-	-
$t_{.95}$	1.712	-	-	-	-	-	-	-
$\hat{H}_8$	−0.139	-	−0.309	-	−0.473	-	−0.561	-
$t_{.05}$	−1.872	-	−2.366	-	−2.622	-	−2.523	-
$t_{.95}$	−1.332	-	−1.732	-	−1.994	-	−1.863	-
$\hat{H}_4^2$	−0.070	-	−0.183	-	−0.253	-	−0.348	-
$t_{.05}$	−2.395	-	−2.982	-	−2.920	-	−3.713	-
$t_{.95}$	−2.787	-	−3.319	-	−3.089	-	−3.681	-
$\hat{H}_6^2$	−0.086	-	−0.154	-	−0.235	-	−0.274	-
$t_{.05}$	−5.427	-	−6.109	-	−6.109	-	−5.559	-
$t_{.95}$	−6.629	-	−7.223	-	−6.838	-	−6.138	-
$\hat{H}_7^2$	-	-	0.087	-	0.146	-	0.178	-
$t_{.05}$	-	-	2.408	-	2.866	-	2.852	-
$t_{.95}$	-	-	2.404	-	3.006	-	2.914	-
$\hat{H}_1^3$	0.019	-	0.032	-	0.037	-	-	-
$t_{.05}$	2.092	-	2.090	-	1.836	-	-	-
$t_{.95}$	2.346	-	2.357	-	2.095	-	-	-
CP	0.452	0.416	0.845	0.790	1.236	1.155	1.456	1.365
$t_{.05}$	7.200	6.334	7.285	6.300	7.568	6.348	7.012	5.900
$t_{.95}$	7.566	6.919	7.641	6.770	7.926	6.760	7.331	6.262
$\hat{H}8$	-	0.428	-	0.712	-	0.867	-	0.959
$t_{.05}$	-	3.330	-	3.096	-	2.888	-	2.610
$t_{.95}$	-	4.316	-	4.033	-	3.803	-	3.489
$\bar{R}^2_{0.95}$	0.471	0.399	0.469	0.403	0.489	0.415	0.448	0.377
$\bar{R}^2_{0.05}$	0.469	0.397	0.467	0.401	0.488	0.413	0.446	0.375

Note: Reported are the mean estimates when a predictive regression is run for each draw of G_t. Estimates when the regressors are the posterior mean of the G_t are reported in columns 5 and 10 of Tables 12.4 to 12.7, respectively.

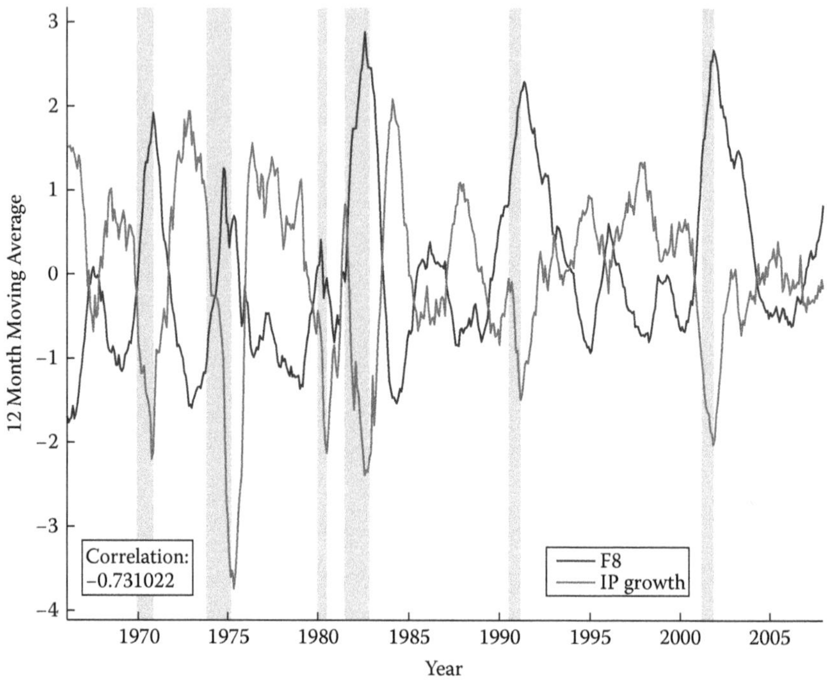

FIGURE 12.11
F8 and IP Growth

12.6 Countercyclical Yield Risk Premia

The *yield risk premium* or *term premium* should not be confused with the term spread, which is simply the difference in yields between the n-period bond and the one-period bond. Instead, the yield risk premium is a component of the the n-period yield:

$$y_t^{(n)} = \underbrace{\frac{1}{n}E_t\big(y_t^{(1)} + y_{t+1}^{(1)} + \cdots + y_{t+n-1}^{(1)}\big)}_{\text{expectations component}} + \underbrace{\varkappa_t^{(n)}}_{\text{yield risk premium}}. \tag{12.12}$$

Under the expectations hypothesis, the yield risk premium, $\varkappa_t^{(n)}$, is assumed constant.

It is straightforward to show that the yield risk premium is identically equal to the average of expected future return risk premia of declining maturity:

$$\varkappa_t^{(n)} = \frac{1}{n}\big[E_t\big(rx_{t+1}^{(n)}\big) + E_t\big(rx_{t+2}^{(n-1)}\big) + \cdots + E_t\big(rx_{t+n-1}^{(2)}\big)\big]. \tag{12.13}$$

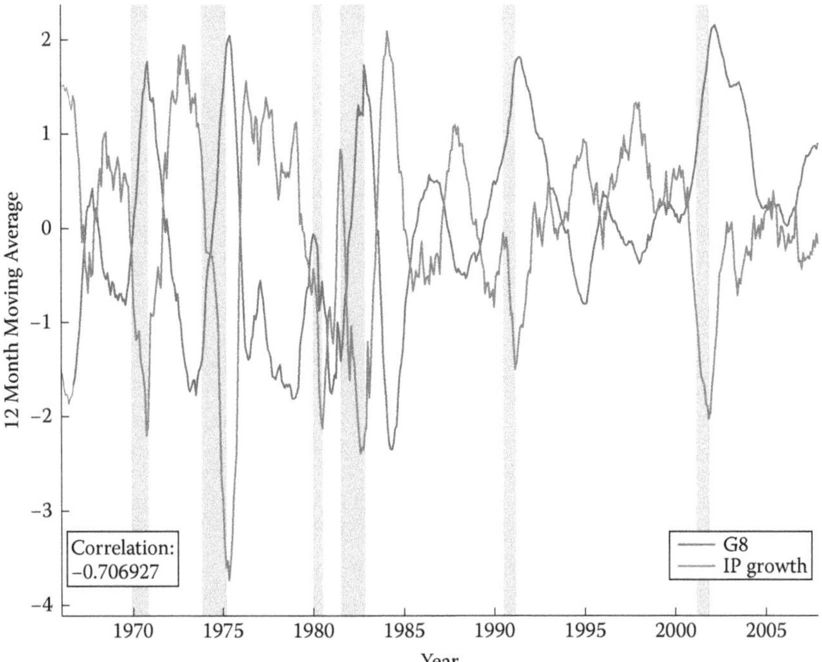

FIGURE 12.12
G8 and IP Growth.

To form an estimate of the risk premium component in yields, $\varkappa_t^{(n)}$, we need estimates of the multistep ahead forecasts that appear on the right-hand side of Equation 12.13. Denote estimated variables with "hats." Then

$$\widehat{\varkappa}_t^{(n)} = \frac{1}{n}\left[\widehat{E}_t\left(rx_{t+1}^{(n)}\right) + \widehat{E}_t\left(rx_{t+2}^{(n-1)}\right) + \cdots + \widehat{E}_t\left(rx_{t+n-1}^{(2)}\right)\right], \tag{12.14}$$

where $\widehat{E}_t(\cdot)$ denotes an estimate of the conditional expectation $E_t(\cdot)$ formed by a linear projection. As estimates of the conditional expectations are simply linear forecasts of excess returns, multiple steps ahead our earlier results for the FAR have direct implications for risk premia in yields.

To generate multistep ahead forecasts we estimate a monthly pth-order vector autoregression (VAR). The idea behind the VAR is that multistep ahead forecasts may be obtained by iterating one-step ahead linear projections from the VAR. The VAR vector contains observations on excess returns, the Cochrane–Piazzesi factor, CP_t and $\hat{H}_t$, where $\hat{H}_t$ are the estimated factors (or a linear combination of them). Let

$$Z_t^U \equiv \left[rx_t^{(5)}, rx_t^{(4)}, \ldots, rx_t^{(2)}, CP_t, \hat{H}8_t\right]'$$

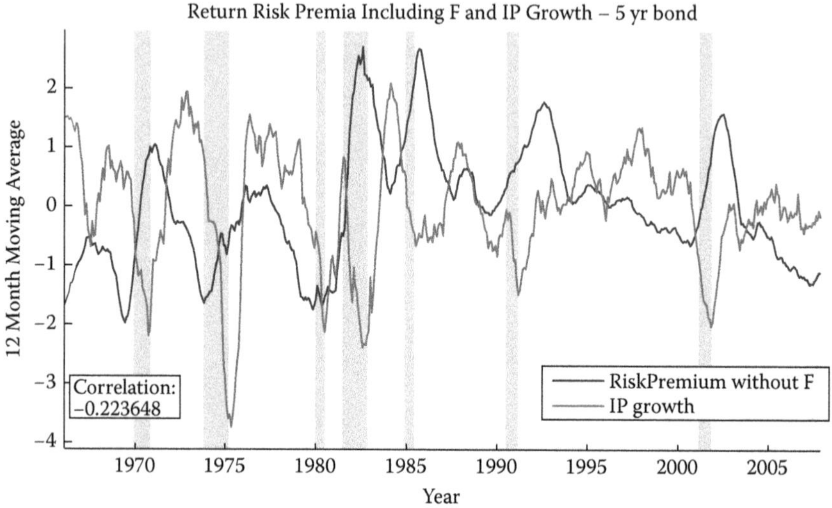

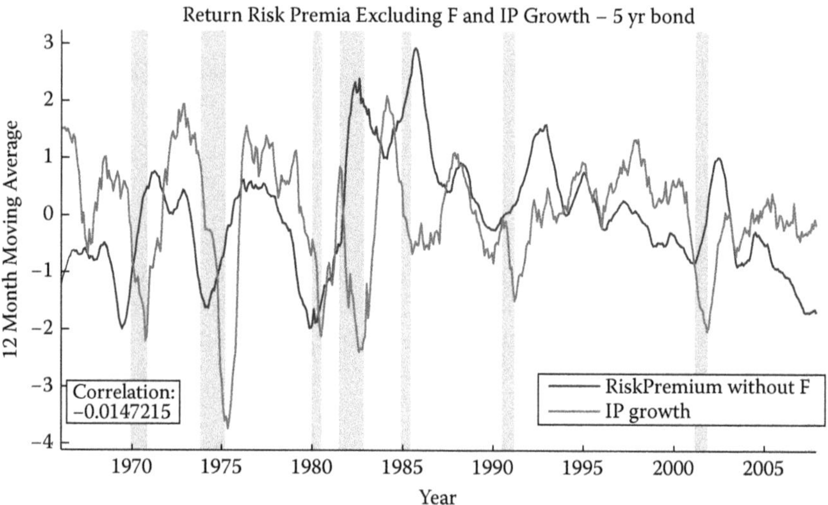

FIGURE 12.13
Return Risk Premia.

where $\hat{H}8$ is either $\hat{F}8$ or $\hat{G}8$. For comparison, we will also form bond forecasts with a restricted VAR that excludes the estimated factors, but still includes CP_t as a predictor variable:

$$Z_t^R \equiv \left[rx_t^{(5)}, rx_t^{(4)}, \ldots, rx_t^{(2)}, CP_t\right]'.$$

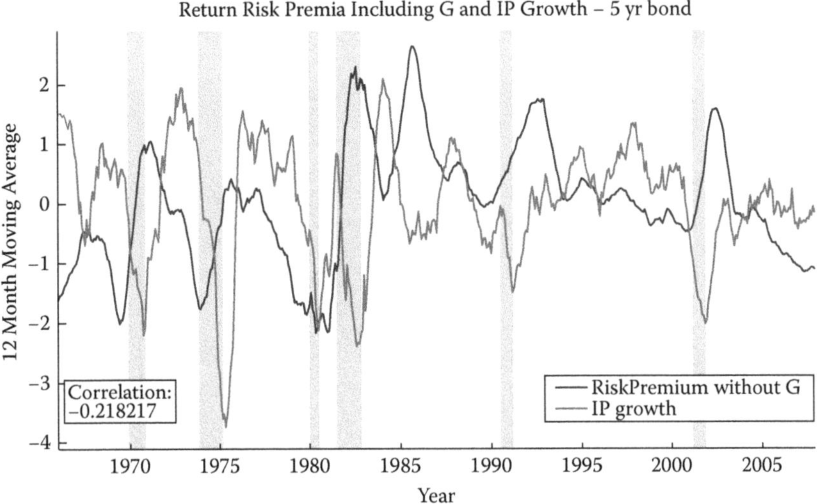

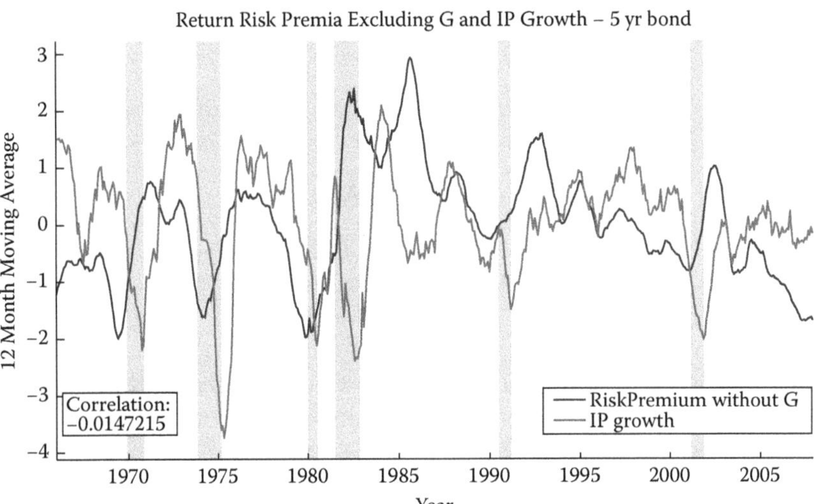

FIGURE 12.14
Return Risk Premia.

We use a monthly VAR with $p = 12$ lags, where, for notational convenience, we write the VAR in terms of mean deviations[7]:

$$Z_{t+1/12} - \boldsymbol{\mu} = \boldsymbol{\Phi}_1 (Z_t - \boldsymbol{\mu}) + \boldsymbol{\Phi}_2(Z_{t-1/12} - \boldsymbol{\mu}) + \cdots + \boldsymbol{\Phi}_p(Z_{t-11/12} - \boldsymbol{\mu}) + \varepsilon_{t+1/12}. \tag{12.15}$$

[7] This is only for notational convenience. The estimation will include the means.

Let k denote the number of variables in Z_t. Then Equation 12.15 can be expressed as a $VAR(1)$:

$$\boldsymbol{\xi}_{t+1/12} = \mathbf{A}\boldsymbol{\xi}_t + \mathbf{v}_{t+1/12}, \tag{12.16}$$

where,

$$\underset{(kp\times 1)}{\boldsymbol{\xi}_{t+1/12}} \equiv \begin{bmatrix} Z_t - \boldsymbol{\mu} \\ Z_{t-1/12} - \boldsymbol{\mu} \\ \cdot \\ \cdot \\ \cdot \\ Z_{t-11/12} - \boldsymbol{\mu} \end{bmatrix} \qquad \underset{(kp\times 1)}{\mathbf{v}_t} \equiv \begin{bmatrix} \varepsilon_{t+1/12} \\ \mathbf{0} \\ \cdot \\ \cdot \\ \cdot \\ \mathbf{0} \end{bmatrix}$$

$$\underset{(kp\times kp)}{\mathbf{A}} = \begin{bmatrix} \boldsymbol{\Phi}_1 & \boldsymbol{\Phi}_2 & \boldsymbol{\Phi}_3 & \cdot & \cdot & \boldsymbol{\Phi}_{p-1} & \boldsymbol{\Phi}_p \\ \mathbf{I}_n & \mathbf{0} & \mathbf{0} & \cdot & \cdot & \mathbf{0} & \mathbf{0} \\ \mathbf{0} & \mathbf{I}_n & \mathbf{0} & \cdot & \cdot & \mathbf{0} & \mathbf{0} \\ \cdot & \cdot & \cdot & \cdot & \cdot & \cdot & \cdot \\ \cdot & \cdot & \cdot & \cdot & \cdot & \cdot & \cdot \\ \cdot & \cdot & \cdot & \cdot & \cdot & \cdot & \cdot \\ \mathbf{0} & \mathbf{0} & \mathbf{0} & \cdot & \cdot & \mathbf{I}_n & \mathbf{0} \end{bmatrix}.$$

Multistep ahead forecasts are straightforward to compute using the first-order VAR:

$$E_t \boldsymbol{\xi}_{t+j/12} = \mathbf{A}^j \boldsymbol{\xi}_t.$$

When $j = 12$, the monthly VAR produces forecasts of 1-year ahead variables, $E_t\boldsymbol{\xi}_{t+1} = \mathbf{A}^{12}\boldsymbol{\xi}_t$; when $j = 24$, it computes 2-year ahead forecasts, and so on. Define a vector ej that picks out the jth element of $\boldsymbol{\xi}_t$, i.e., $e1'\boldsymbol{\xi}_t \equiv rx_t^{(5)}$. In the notation above, we have $e1_{(kp\times 1)} = [1, 0, 0, \ldots, 0]'$, $e2_{(kp\times 1)} = [0, 1, 0, \ldots, 0]'$, analogously for $e3$ and $e4$. Thus, given estimates of the VAR parameters $\mathbf{A}$, we may form estimates of the conditional expectations on the right-hand side of Equation 12.14 using the VAR forecasts of return risk premia. For example, the estimate of the expectation of the 5-year bond, 1 year ahead, is given by $\widehat{E}_t(rx_{t+1}^{(5)}) = e1'\mathbf{A}^{12}\boldsymbol{\xi}_t$; the estimate of the expectation of the 4-year bond, 2 years ahead, is given by $\widehat{E}_t(rx_{t+2}^{(4)}) = e2'\mathbf{A}^{24}\boldsymbol{\xi}_t$, and so on.

Letting $\hat{H}_t = \hat{F}5_t$ where $\hat{F}5_t$ is a linear combination of $\hat{f}_{1t}$, $\hat{f}_{1t}^3$, $\hat{f}_{3t}$, $\hat{f}_{4t}$, and $\hat{f}_{8t}$. we showed in Ludvigson and Ng (2007) that both yield and return risk premia are more countercyclical and reach greater values in recessions than in the absence of $\hat{H}_t$. Here, we verify that this result holds up for different choices of $\hat{H}_t$. To this end, we let $\hat{H}_t$ be the static and dynamic factors selected by the out-of-sample BIC. These two predictor sets embody information in fewer factors than the ones implied by the in-sample BIC, $\hat{H}8$, or $F5_t$ used in Ludvigson and Ng (2007). The point is to show that a few macroeconomic factors are enough to generate an important difference in the properties of risk premia. Specifically, without $\hat{F}_t$ in Z_t^u, the correlation between the estimated return risk premium and IP growth is -0.014. With $\hat{F}_t$ in Z_t^u, the correlation is −0.223. These correlations are −0.045 and −0.376 for yield risk premia. With $\hat{G}_t$ in Z_t^U, the correlation of IP growth with return and yield risk premium are −0.218 and −0.286, respectively. Return and yield risk premia are thus more countercyclical when the factors are used to forecast excess returns.

Figure 12.15 shows the 12-month moving average of risk-premium component of the 5-year bond yield. As we can see, yield risk premia were particularly high in the 1982–1983 recession, as well as shortly after the 2001 recession. Figure 12.16 shows the yield risk premia estimated with and without using $\hat{F}_t$ to forecast excess returns, while Figure 12.17 shows a similar picture with and without $\hat{G}_t$. The difference between the risk premia estimated with and without the factors is largest around recessions. For example, the yield risk premium on the 5-year bond estimated using the information contained in $\hat{F}_t$ or $\hat{G}_t$ was over 2% in the 2001 recession, but it was slightly below 1% without $\hat{G}_t$. The return risk premia (not reported) show a similar pattern.

When the economy is contracting, the countercyclical nature of the risk factors contributes to a steepening of the yield curve even as future short-term rates fall. Conversely, when the economy is expanding, the factors contribute to a flattening of the yield curve even as expectations of future short-term rates rise. This implies that information in the factors is ignored. Too much variation in the long-term yields is attributed to the expectations component in recessions. Information in the macro factors are thus important in accurate decomposition of risk premia, especially in recessions.

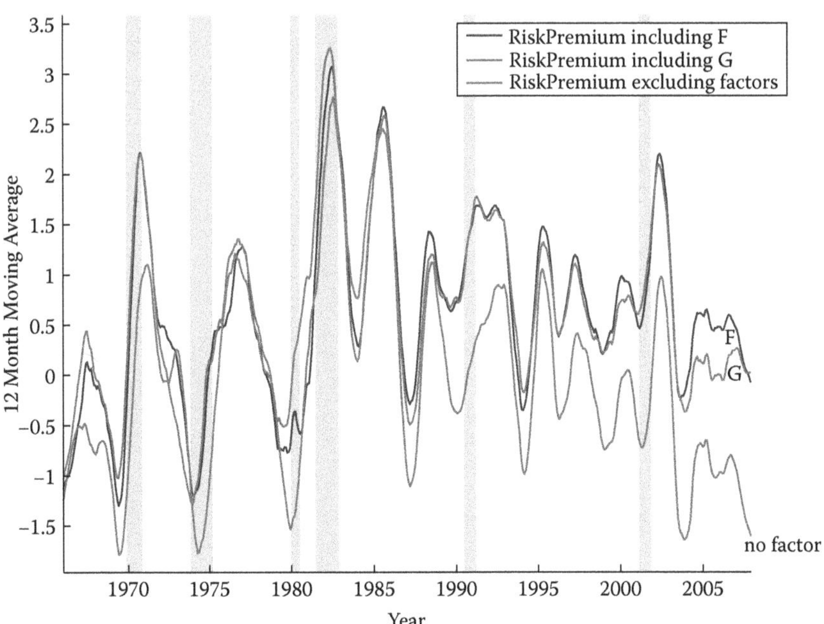

FIGURE 12.15
Yield Risk Premium with and without factors −5 yr bond.

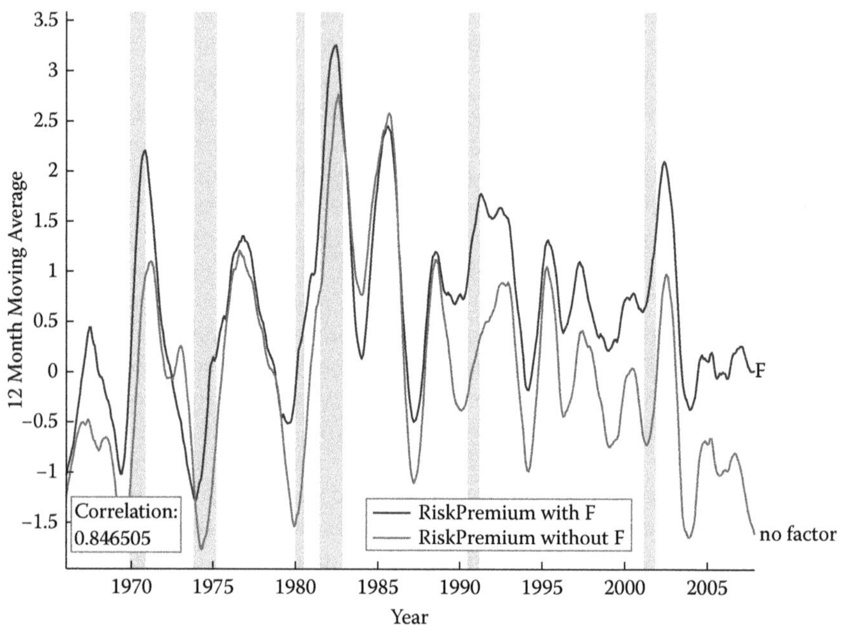

FIGURE 12.16
Yield Risk Premia Including and Excluding F −5 yr bond.

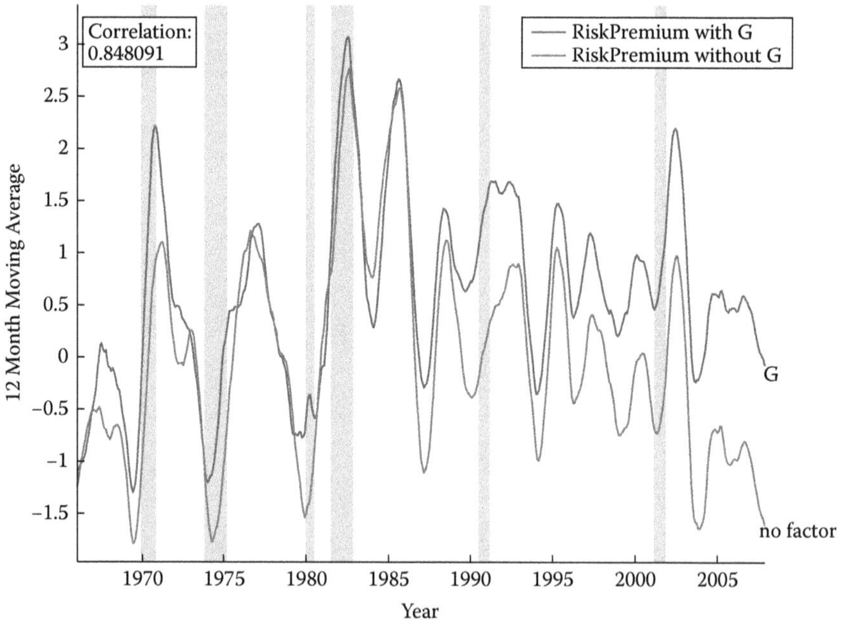

FIGURE 12.17
Yield Risk Premia Including and Excluding G −5 yr bond.

12.7 Conclusion

There is a good deal of evidence that excess bond returns are predictable by financial variables. Yet, macroeconomic theory postulates that it is real variables relating to macroeconomic activity that should forecast bond returns. This chapter presents robust evidence in support of the theory. Macroeconomic factors, especially the real activity factor, has strong predictive power for excess bond returns even in the presence of financial predictors. Our analysis consists of estimating two sets of factors and a comprehensive specification search. We also account for sampling uncertainty that might arise from estimation of the factors. While the estimated risk premia without using the macro factors to forecast excess returns are acyclical, both bond returns and yield risk premia are countercyclical when the factors are used. The evidence indicate that investors seek compensation for macroeconomic risks associated with recessions.

12.8 Acknowledgment

We thank Jushan Bai for helpful suggestions and Matt Smith for excellent research assistance. We also thank the Conference Board for providing us with some of the data. Financial support from the National Science Foundation (Grant No. 0617858 to Ludvigson and SES-0549978 to Ng) is gratefully acknowledged. Ludvigson also acknowledges financial support from the Alfred P. Sloan Foundation and the CV Starr Center at NYU. Any errors or omissions are the responsibility of the authors.

Data Appendix

This appendix lists the short name of each series, its mnemonic (the series label used in the source database), the transformation applied to the series, and a brief data description. All series are from the Global Insights Basic Economics Database, unless the source is listed (in parentheses) as TCB (The Conference Board's Indicators Database) or AC (author's calculation based on Global Insights or TCB data). In the transformation column, ln denotes logarithm, Δ ln and Δ^2 ln denote the first and second difference of the logarithm, lv denotes the level of the series, and Δ lv denotes the first difference of the series. The data are available from 1959:01 to 1997:12.

Group 1: Output and Income

No.	Gp	Short Name	Mnemonic	Tran	Descripton
1	1	PI	ypr	Δln	Personal Income (AR, Bil. Chain 2000 $) (TCB)
6	1	IP: total	ips10	Δln	Industrial Production Index–Total Index
7	1	IP: products	ips11	Δln	Industrial Production Index–Products, Total
8	1	IP: final prod	ips299	Δln	Industrial Production Index–Final Products
9	1	IP: cons gds	ips12	Δln	Industrial Production Index–Consumer Goods
10	1	IP: cons dble	ips13	Δln	Industrial Production Index–Durable Consumer Goods
11	1	IP: cons nondble	ips18	Δln	Industrial Production Index–Nondurable Consumer Goods
12	1	IP: bus eqpt	ips25	Δln	Industrial Production Index–Business Equipment
13	1	IP: matls	ips32	Δln	Industrial Production Index–Materials
14	1	IP: dble matls	ips34	Δln	Industrial Production Index–Durable Goods Materials
15	1	IP: nondble matls	ips38	Δln	Industrial Production Index–Nondurable Goods Materials
16	1	IP: mfg	ips43	Δln	Industrial Production Index–Manufacturing (Sic)
17	1	IP: res util	ips307	Δln	Industrial Production Index–Residential Utilities
18	1	IP: fuels	ips306	Δln	Industrial Production Index–Fuels
19	1	NAPM prodn	pmp	lv	Napm Production Index (Percent)
20	1	Cap util	utl11	Δlv	Capacity Utilization (SIC-Mfg) (TCB)

Group 2: Labor Market

No.	Gp	Short Name	Mnemonic	Tran	Descripton
21	2	Help wanted indx	lhel	Δlv	Index Of Help-Wanted Advertising In Newspapers (1967=100;Sa)
22	2	Help wanted/emp	lhelx	Δlv	Employment: Ratio; Help-Wanted Ads:No. Unemployed Clf
23	2	Emp CPS total	lhem	Δln	Civilian Labor Force: Employed, Total (Thous.,Sa)
24	2	Emp CPS nonag	lhnag	Δln	Civilian Labor Force: Employed, Nonagric.Industries (Thous.,Sa)
25	2	U: all	lhur	Δlv	Unemployment Rate: All Workers, 16 Years &
26	2	U: mean duration	lhu680	Δlv	Unemploy.By Duration: Average(Mean)Duration In Weeks (Sa)
27	2	U < 5 wks	lhu5	Δln	Unemploy.By Duration: Persons Unempl.Less Than 5 Wks (Thous.,Sa)

No.	Gp	Short Name	Mnemonic	Tran	Descripton
28	2	U 5–14 wks	lhu14	Δln	Unemploy.By Duration: Persons Unempl.5 To 14 Wks (Thous.,Sa)
29	2	U 15 + wks	lhu15	Δln	Unemploy.By Duration: Persons Unempl.15 Wks + (Thous.,Sa)
30	2	U 15–26 wks	lhu26	Δln	Unemploy.By Duration: Persons Unempl.15 To 26 Wks (Thous.,Sa)
31	2	U 27+ wks	lhu27	Δln	Unemploy.By Duration: Persons Unempl.27 Wks + (Thous,Sa)
32	2	UI claims	claimuii	Δln	Average Weekly Initial Claims, Unemploy. Insurance (Thous.) (TCB)
33	2	Emp: total	ces002	Δln	Employees On Nonfarm Payrolls: Total Private
34	2	Emp: gds prod	ces003	Δln	Employees On Nonfarm Payrolls–Goods-Producing
35	2	Emp: mining	ces006	Δln	Employees On Nonfarm Payrolls–Mining
36	2	Emp: const	ces011	Δln	Employees On Nonfarm Payrolls–Construction
37	2	Emp: mfg	ces015	Δln	Employees On Nonfarm Payrolls–Manufacturing
38	2	Emp: dble gds	ces017	Δln	Employees On Nonfarm Payrolls–Durable Goods
39	2	Emp: nondbles	ces033	Δln	Employees On Nonfarm Payrolls–Nondurable Goods
40	2	Emp: services	ces046	Δln	Employees On Nonfarm Payrolls–Service-Providing
41	2	Emp: TTU	ces048	Δln	Employees On Nonfarm Payrolls–Trade, Transportation, And Utilities
42	2	Emp: wholesale	ces049	Δln	Employees On Nonfarm Payrolls–Wholesale Trade.
43	2	Emp: retail	ces053	Δln	Employees On Nonfarm Payrolls–Retail Trade
44	2	Emp: FIRE	ces088	Δln	Employees On Nonfarm Payrolls–Financial Activities
45	2	Emp: Govt	ces140	Δln	Employees On Nonfarm Payrolls–Government
(46)	2	Emp-hrs nonag	a0m048	Δln	Employee Hours In Nonag. Establishments (AR, Bil. Hours) (TCB)
47	2	Avg hrs	ces151	lv	Avg Weekly Hrs of Prod or Nonsup Workers On Private Nonfarm Payrolls–Goods-Producing
48	2	Overtime: mfg	ces155	Δlv	Avg Weekly Hrs of Prod or Nonsup Workers On Private Nonfarm Payrolls–Mfg Overtime Hours
49	2	Avg hrs: mfg	aom001	lv	Average Weekly Hours, Mfg. (Hours) (TCB)
50	2	NAPM empl	pmemp	lv	Napm Employment Index (Percent)

No.	Gp	Short Name	Mnemonic	Tran	Descripton
129	2	AHE: goods	ces275	$\Delta^2 ln$	Avg Hourly Earnings of Prod or Nonsup Workers On Private Nonfarm Payrolls–Goods-Producing
130	2	AHE: const	ces277	$\Delta^2 ln$	Avg Hourly Earnings of Prod or Nonsup Workers On Private Nonfarm Payrolls–Construction
131	2	AHE: mfg	ces278	$\Delta^2 ln$	Avg Hourly Earnings of Prod or Nonsup Workers On Private Nonfarm Payrolls–Manufacturing

Group 3: Housing

No.	Gp	Short Name	Mnemonic	Tran	Descripton
51	3	Starts: nonfarm	hsfr	ln	Housing Starts:Nonfarm(1947–58);Total Farm & Nonfarm(1959–)(Thous.,Saar)
52	3	Starts: NE	hsne	ln	Housing Starts:Northeast (Thous.U.)S.A.
53	3	Starts: MW	hsmw	ln	Housing Starts:Midwest(Thous.U.)S.A.
54	3	Starts: South	hssou	ln	Housing Starts:South (Thous.U.)S.A.
55	3	Starts: West	hswst	ln	Housing Starts:West (Thous.U.)S.A.
56	3	BP: total	hsbr	ln	Housing Authorized: Total New Priv Housing Units (Thous.,Saar)
57	3	BP: NE	hsbne*	ln	Houses Authorized By Build. Permits:Northeast(Thou.U.)S.A
58	3	BP: MW	hsbmw*	ln	Houses Authorized By Build. Permits:Midwest(Thou.U.)S.A.
59	3	BP: South	hsbsou*	ln	Houses Authorized By Build. Permits:South(Thou.U.)S.A.
60	3	BP: West	hsbwst*	ln	Houses Authorized By Build. Permits:West(Thou.U.)S.A.

Group 4: Consumption, Orders and Inventories

61	4	PMI	pmi	lv	Purchasing Managers' Index (Sa)
62	4	NAPM new ordrs	pmno	lv	Napm New Orders Index (Percent)
63	4	NAPM vendor del	pmdel	lv	Napm Vendor Deliveries Index (Percent)
64	4	NAPM Invent	pmnv	lv	Napm Inventories Index (Percent)
65	4	Orders: cons gds	a1m008	Δln	Mfrs' New Orders, Consumer Goods And Materials (Mil. $) (TCB)
66	4	Orders: dble gds	a0m007	Δln	Mfrs' New Orders, Durable Goods Industries (Bil. Chain 2000 $) (TCB)
67	4	Orders: cap gds	a0m027	Δln	Mfrs' New Orders, Nondefense Capital Goods (Mil. Chain 1982 $) (TCB)
68	4	Unf orders: dble	a1m092	Δln	Mfrs' Unfilled Orders, Durable Goods Indus. (Bil. Chain 2000 $) (TCB)

69	4	M&T invent	a0m070	Δln	Manufacturing And Trade Inventories (Bil. Chain 2000 $) (TCB)
70	4	M&T invent/sales	a0m077	Δlv	Ratio, Mfg. And Trade Inventories To Sales (Based On Chain 2000 $) (TCB)
3	4	Consumption	cons-r	Δln	Real Personal Consumption Expenditures (AC) (Bill $) pi031/gmdc
4	4	M&T sales	mtq	Δln	Manufacturing And Trade Sales (Mil. Chain 1996 $) (TCB)
5	4	Retail sales	a0m059	Δln	Sales Of Retail Stores (Mil. Chain 2000 $) (TCB)
132	4	Consumer expect	hhsntn	Δlv	U. Of Mich. Index Of Consumer Expectations(Bcd-83)

Group 5: Money and Credit

No.	Gp	Short Name	Mnemonic	Tran	Descripton
71	5	M1	fm1	$\Delta^2 ln$	Money Stock: M1(Curr,Trav.Cks,Dem Dep,Other Ck'able Dep)(Bil$,Sa)
72	5	M2	fm2	$\Delta^2 ln$	Money Stock:M2(M1+O'nite Rps,Euro$,G/P&B/D & Mmmfs&Sav& Sm Time Dep(Bil$,Sa)
73	5	Currency	fmscu	$\Delta^2 ln$	Money Stock: Currency held by the public (Bil$,Sa)
74	5	M2 (real)	fm2-r	Δln	Money Supply: Real M2, fm2/gmdc (AC)
75	5	MB	fmfba	$\Delta^2 ln$	Monetary Base, Adj For Reserve Requirement Changes(Mil$,Sa)
76	5	Reserves tot	fmrra	$\Delta^2 ln$	Depository Inst Reserves:Total, Adj For Reserve Req Chgs(Mil$,Sa)
77	5	Reserves nonbor	fmrnba	$\Delta^2 ln$	Depository Inst Reserves:Nonborrowed,Adj Res Req Chgs(Mil$,Sa)
78	5	C&I loans	fclnbw	$\Delta^2 ln$	Commercial & Industrial Loans Outstanding + NonFin Comm. Paper (Mil$, SA) (Bci)
79	5	C&I loans	fclbmc	lv	Wkly Rp Lg Com'l Banks:Net Change Com'l & Indus Loans(Bil$,Saar)
80	5	Cons credit	ccinrv	$\Delta^2 ln$	Consumer Credit Outstanding–Nonrevolving(G19)
81	5	Inst cred/PI	ccipy	Δlv	Ratio, Consumer Installment Credit To Personal Income (Pct.) (TCB)

Group 6: Bond and Exchange Rates

86	6	Fed Funds	fyff	Δlv	Interest Rate: Federal Funds (Effective) (% Per Annum,Nsa)
87	6	Comm paper	cp90	Δlv	Commercial Paper Rate
88	6	3 mo T-bill	fygm3	Δlv	Interest Rate: U.S.Treasury Bills, Sec Mkt,3-Mo.(% Per Ann,Nsa)
89	6	6 mo T-bill	fygm6	Δlv	Interest Rate: U.S.Treasury Bills, Sec Mkt,6-Mo.(% Per Ann,Nsa)
90	6	1 yr T-bond	fygt1	Δlv	Interest Rate: U.S.Treasury Const Maturities,1-Yr.(% Per Ann,Nsa)
91	6	5 yr T-bond	fygt5	Δlv	Interest Rate: U.S.Treasury Const Maturities,5-Yr.(% Per Ann,Nsa)
92	6	10 yr T-bond	fygt10	Δlv	Interest Rate: U.S.Treasury Const Maturities,10-Yr.(% Per Ann,Nsa)
93	6	Aaa bond	fyaaac	Δlv	Bond Yield: Moody's Aaa Corporate (% Per Annum)
94	6	Baa bond	fybaac	Δlv	Bond Yield: Moody's Baa Corporate (% Per Annum)
95	6	CP-FF spread	scp90F	lv	cp90-fyff (AC)
96	6	3 mo-FF spread	sfygm3	lv	fygm3-fyff (AC)
97	6	6 mo-FF spread	sfygm6	lv	fygm6-fyff (AC)
98	6	1 yr-FF spread	sfygt1	lv	fygt1-fyff (AC)
99	6	5 yr-FF spread	sfygt5	lv	fygt5-fyff (AC)
100	6	10 yr-FF spread	sfygt10	lv	fygt10-fyff (AC)
101	6	Aaa-FF spread	sfyaaac	lv	fyaaac-fyff (AC)
102	6	Baa-FF spread	sfybaac	lv	fybaac-fyff (AC)
103	6	Ex rate: avg	exrus	Δln	United States;Effective Exchange Rate(Merm)(Index No.)
104	6	Ex rate: Switz	exrsw	Δln	Foreign Exchange Rate: Switzerland (Swiss Franc Per U.S.$)
105	6	Ex rate: Japan	exrjan	Δln	Foreign Exchange Rate: Japan (Yen Per U.S.$)
106	6	Ex rate: UK	exruk	Δln	Foreign Exchange Rate: United Kingdom (Cents Per Pound)
107	6	EX rate: Canada	exrcan	Δln	Foreign Exchange Rate: Canada (Canadian $ Per U.S.$)

Group 7: Prices

108	7	PPI: fin gds	pwfsa	$\Delta^2 ln$	Producer Price Index: Finished Goods (82=100,Sa)
109	7	PPI: cons gds	pwfcsa	$\Delta^2 ln$	Producer Price Index: Finished Consumer Goods (82=100,Sa)
110	7	PPI: int materials	pwimsa	$\Delta^2 ln$	Producer Price Index:Intermed Mat.Supplies & Components(82=100,Sa)
111	7	PPI: crude matls	pwcmsa	$\Delta^2 ln$	Producer Price Index: Crude Materials (82=100,Sa)
112	7	Spot market price	psccom	$\Delta^2 ln$	Spot market price index: bls & crb: all commodities(1967=100)

113	7	PPI: nonferrous materials	pw102	$\Delta^2 ln$	Producer Price Index: Nonferrous Materials (1982=100, Nsa)
114	7	NAPM com price	pmcp	lv	Napm Commodity Prices Index (Percent)
115	7	CPI-U: all	punew	$\Delta^2 ln$	Cpi-U: All Items (82–84=100,Sa)
116	7	CPI-U: apparel	pu83	$\Delta^2 ln$	Cpi-U: Apparel & Upkeep (82–84=100,Sa)
117	7	CPI-U: transp	pu84	$\Delta^2 ln$	Cpi-U: Transportation (82–84=100,Sa)
118	7	CPI-U: medical	pu85	$\Delta^2 ln$	Cpi-U: Medical Care (82–84=100,Sa)
119	7	CPI-U: comm.	puc	$\Delta^2 ln$	Cpi-U: Commodities (82–84=100,Sa)
120	7	CPI-U: dbles	pucd	$\Delta^2 ln$	Cpi-U: Durables (82–84=100,Sa)
121	7	CPI-U: services	pus	$\Delta^2 ln$	Cpi-U: Services (82–84=100,Sa)
122	7	CPI-U: ex food	puxf	$\Delta^2 ln$	Cpi-U: All Items Less Food (82–84=100,Sa)
123	7	CPI-U: ex shelter	puxhs	$\Delta^2 ln$	Cpi-U: All Items Less Shelter (82–84=100,Sa)
124	7	CPI-U: ex med	puxm	$\Delta^2 ln$	Cpi-U: All Items Less Midical Care (82–84=100,Sa)
125	7	PCE defl	gmdc	$\Delta^2 ln$	Pce, Impl Pr Defl:Pce (2000=100) (AC) (BEA)
126	7	PCE defl: dlbes	gmdcd	$\Delta^2 ln$	Pce, Impl Pr Defl:Pce; Durables (2000=100) (AC) (BEA)
127	7	PCE defl: nondble	gmdcn	$\Delta^2 ln$	Pce, Impl Pr Defl:Pce; Nondurables (2000=100) (AC) (BEA)
128	7	PCE defl: service	gmdcs	$\Delta^2 ln$	Pce, Impl Pr Defl:Pce; Services (2000=100) (AC) (BEA)

Group 8: Stock Market

No.	Gp	Short Name	Mnemonic	Tran	Descripton
82	8	S&P 500	fspcom	Δln	S&P's Common Stock Price Index: Composite (1941–43=10)
83	8	S&P: indust	fspin	Δln	S&P's Common Stock Price Index: & Industrials (1941–43=10)
84	8	S&P div yield	fsdxp	Δlv	S&P's Composite Common Stock: Dividend Yield (% Per Annum)
85	8	S&P PE ratio	fspxe	Δln	S&P's Composite Common Stock: &Price-Earnings Ratio (%,Nsa)

References

Aguilar, G., and M. West. 2000. Bayesian Dynamic Factor Models and Portfolio Allocation. *Journal of Business and Economic Statistics* 18:338–357.

Ang, A., and M. Piazzesi. 2003. A No-Arbitrage Vector Autoregression of Term Structure Dynamics with Macroeconomic and Latent Variables. *Journal of Monetary Economics* 50:745–787.

Bai, J. 2003. Inferential Theory for Factor Models of Large Dimensions. *Econometrica* 71(1):135–172.

Bai, J., and S. Ng. 2002. Determining the Number of Factors in Approximate Factor Models. *Econometrica* 70(1):191–221.

——— 2006a. Confidence Intervals for Diffusion Index Forecasts and Inference with Factor-Augmented Regressions. *Econometrica* 74(4):1133–1150.

——— 2006b. Forecasting Economic Time Series Using Targeted Predictors. *Journal of Econometrics*, forthcoming.

——— 2008. Large Dimensional Factor Analysis. *Foundations and Trends in Econometrics* 3(2):89–163.

Boivin, J., and S. Ng. 2005. Undertanding and Comparing Factor Based Forecasts. *International Journal of Central Banking* 1(3):117–152.

Brandt, M. W., and K. Q. Wang. 2003. Time-Varying Risk Aversion and Unexpected Inflation. *Journal of Monetary Economics* 50:1457–1498.

Brillinger, D. 1981. *Time Series: Data Analysis and Theory*. San Francisco: Wiley.

Campbell, J. Y. J. H. C. 1999. By Force of Habit: A Consumption-Based Explanation of Aggregate Stock Market Behavior. *Journal of Political Economy* 107:205–251.

Carter, C. K., and R. Kohn. 1994. On Gibbs Sampling for State Space Models. *Biometrika* 81(3):541–533.

Cochrane, J. H., and M. Piazzesi. 2005. Bond Risk Premia. *The American Economic Review* 95(1):138–160.

Connor, G., and R. Korajzcyk. 1986. Performance Measurement with the Arbitrage Pricing Theory: A New Framework for Analysis. *Journal of Financial Economics* 15:373–394.

DeMol, C., D. Giannone, and L. Reichlin. 2006. Forecasting Using a Large Number of Predictors: Is Bayesian Regression a Valid Alternative to Principal Components. *ECB Working Paper 700.*

Duffie, G. 2008. Information in (and not in) the Term Structure. Mimeo, Johns Hopkins University, Baltimore, MD.

Fama, E. F., and R. H. Bliss. 1987. The Information in Long-Maturity Forward Rates. *American Economic Review* 77(4):680–692.

Forni, M., M. Hallin, M. Lippi, and L. Reichlin. 2005. The Generalized Dynamic Factor Model, One Sided Estimation and Forecasting. *Journal of the American Statistical Association* 100:830–840.

Fruhwirth-Schnatter, S. 1994. Data Augmentation and Dynamic Linear Models. *Journal of Time Series Analysis* 15:183–202.

Geweke, J., and G. Zhou. 1996. Measuring the Pricing Error of the Arbitrage Pricing Theory. *Review of Financial Studies* 9(2):557–87.

Hansen, P. 2008. In-Sample and Out-of-Sample Fit: Their Joint Distribution and Its Implications for Model Selection. Stanford University, Stanford, CA.

Kim, C., and C. Nelson. 2000. *State Space Models with Regime Switching*. Cambridge, MA: MIT Press.

Kose, A., C. Otrok, and C. Whiteman. 2003. International Business Cycles: World Region and Country Specific Factors. *American Economic Review* 93(4):1216–1239.

Kozicki, S., and P. Tinsley. 2005. Term Structure Transmission of Monetary Policy. Working Paper 05–06, Fed. Reserve Bank of Kansas City, Kansas City, MO.

Lopes, H., and M. West. 2004. Bayesian Model Assessment in Factor Analysis. *Statistical Sinica* 14:41–87.

Ludvigson, S., and S. Ng. 2007. Macro Factors in Bond Risk Premia. *Review of Fiancial Studies*, forthcoming.

Moench, E. 2008. Forecasting the Yield Curve in a Data-Rich Environment: A No-Arbitrage Factor-Augmented VAR Approach. *Journal of Econometrics* 46:26–43.

Pagan, A. 1984. Econometric Issues in the Analysis of Regressions with Generated Regressors. *International Economic Review* 25:221–247.

Piazzesi, M., and E. Swanson. 2004. Futures Prices as Risk-Adjusted Forecasts of Monetary Policy. *NBER Working Paper No. 10547.*

Stock, J. H., and M. Watson. 1989. New Indexes of Coincident and Leading Economic Indications. In *NBER Macroeconomics Annual 1989*. O. J. Blanchard and S. Fischer (ed.). Cambridge, MA: MIT Press.

——— 2002a. Forecasting Using Principal Components from a Large Number of Predictors. *Journal of the American Statistical Association* 97:1167–1179.

——— 2002b. Macroeconomic Forecasting Using Diffusion Indexes. *Journal of Business and Economic Statistics* 20(2):147–162.

——— 2005. Implications of Dynamic Factor Models for VAR Analysis. NBER WP 11467.

Wachter, J. 2006. A Consumption Based Model of the Term Structure of Interest Rates. *Journal of Financial Economics* 79:365–399.

13

Dynamic Panel Data Models

Cheng Hsiao

CONTENTS

13.1 Introduction

Panel data, by blending inter-individual differences and intra-individual dynamics, have greater capacity for capturing the complexity of human behavior than data sets with only a temporal or a cross-sectional dimension (e.g., Hsiao 2003, 2007). However, typical panels focus on individual outcomes. Factors affecting individual outcomes are numerous. It is rare that the conditional density of the outcomes, y_{it}, conditional on certain variables, $\underset{\sim}{x}_{it}$, is independently, identically distributed across individual i and over time, t. To capture the effects of those omitted factors, empirical researchers often assume that, in addition to the effects of observed $\underset{\sim}{x}_{it}$, there exist unobserved individual-specific effects α_i and time-specific effects λ_t. These unobserved individual-specific and/or time-specific effects, α_i and λ_t, are supposed to capture the impacts of those omitted variables that vary across individuals but stay constant over time and the impact of those variables that vary over time but are the same for all individuals at a given time. They can be either treated as fixed constants or random variables, respectively called fixed effects (FE) or random effects (RE) model. The advantage of the FE modeling is that there is no need to postulate the relationship between the unobserved effects and the conditioning variables. The disadvantage is that it introduces the classical "incidental parameter" problems if either the time series dimension T or cross-sectional dimension, N, is finite (e.g., Neyman and Scott 1948). The advantage of the random effects modeling is that the number of unknown parameters stay constant as N and/or T increases. The disadvantage is that the relationships between the effects and the observed conditional variables have to be postulated, say, the conditional distribution of the effects given the observed factors (e.g., Hsiao 2007).

The unobserved heterogeneity across individuals and over time that are not captured by the included conditional variables could either be modeled additively or multiplicatively. Furthermore, many people believe that "all interesting economic behavior is inherently dynamic, dynamic models are the only relevant models" (e.g., Nerlove 2002). However, the estimation of dynamic models with specific effects is a great deal more difficult than the estimation of nondynamic models because the estimation of structural parameters (those parameters that are the same across i and over t) is not independent of the estimation of incidental parameters. For dynamic models there is also an issue of how to model "initial observations."

We set up the basic models in Section 13.2. Since for models involving incidental parameters the conditions for law of large numbers and central limit theorems to hold are violated, estimators based on the likelihood principle or methods of moments are no longer consistent. Section 13.3 shows the inconsistency of the maximum likelihood estimator (MLE) or covariance estimator (CV) of structural parameters in the presence of incidental parameters. Section 13.4 discusses the issues of initial values.

A general principle to obtain consistent estimators for structural parameters for models involving incidental parameters is to transform the original models into models that no longer involve incidental parameters; in Sections 13.5 and 13.6 we illustrate the implementation of this principle for the likelihood and method of moments approach by considering a simple dynamic panel data model with additive individual-specific effects. Section 13.7 discusses the estimation of dynamic models with both individual- and time-specific additive effects. Section 13.8 discusses the estimation with multiplicative individual- and time-specific effects. Section 13.9 proposes a test of additive versus multiplicative effects. Concluding remarks are in Section 13.10.

13.2 The Basic Models

We consider a dynamic model of the form[1]

$$y_{it} = \rho y_{i,t-1} + \underset{\sim}{\beta}' \underset{\sim}{x}_{it} + v_{it}, \quad |\rho| < 1, i = 1, \ldots, N, t = 1, \ldots, T, \tag{13.1}$$

and the initial values y_{io} are observable. For ease of exposition, we assume $\underset{\sim}{x}_{it}$ is a $K \times 1$ vector of strictly exogenous variables and the error term either takes the form

$$v_{it} = \alpha_i + \lambda_t + \epsilon_{it} \tag{13.2}$$

or

$$v_{it} = \alpha_i \lambda_t + \epsilon_{it}, \tag{13.3}$$

where ϵ_{it} is independently, identically distributed with mean 0 and variance σ_ϵ^2, and the individual- and time-specific effects α_i and λ_t can be either fixed or random. When α_i and λ_t are fixed constants, we impose the normalization condition $\sum_{i=1}^N \alpha_i = 0$, $\sum_{t=1}^T \lambda_t = 0$ and assume $\lim \frac{1}{N} \sum_{i=1}^N \alpha_i^2$ and $\lim \frac{1}{T} \sum_{t=1}^T \lambda_t^2$ are finite positive constants. When α_i and λ_t are random, we assume that

$$\begin{aligned}
E\alpha_i &= E\lambda_t = E\epsilon_{it} = 0, \\
E\alpha_i \underset{\sim}{x}_{it} &= E\lambda_t \underset{\sim}{x}_{is} = E\underset{\sim}{x}_{it}\epsilon_{it} = \underset{\sim}{0}, \\
E\alpha_i\lambda_t &= E\lambda_t\epsilon_{is} = E\alpha_i\epsilon_{it} = 0, \\
E\alpha_i\alpha_j &= \begin{cases} \sigma_\alpha^2, & \text{if } i = j, \\ 0, & \text{otherwise.} \end{cases} \\
E\lambda_t\lambda_s &= \begin{cases} \sigma_\lambda^2, & \text{if } t = s, \\ 0, & \text{otherwise.} \end{cases} \\
E\epsilon_{it}\epsilon_{js} &= \begin{cases} \sigma_\epsilon^2, & \text{if } i = j \text{ and } t = s, \\ 0, & \text{otherwise.} \end{cases}
\end{aligned} \tag{13.4}$$

The presence of unknown α_i introduces serial correlation that does not die out as T increases. The presence of λ_t introduces correlation across individuals that does not die out as N increases.

---

[1] When T is finite, there is no need to restrict $|\rho| < 1$ to obtain the asymptotic normality results. However, for ease of exposition we shall assume $|\rho| < 1$.

13.3 The Maximum Likelihood Estimator (MLE) (or Covariance Estimators (CV)) in the Presence of Incidental Parameters

Under the assumption that ϵ_{it} is independent normal and fixed y_{i0} the MLE of the FE model Equation 13.1 and Equation 13.2 is equal to

$$\begin{pmatrix}\hat{\rho}\\ \hat{\underset{\sim}{\beta}}\end{pmatrix} = \left[\sum_{i=1}^{N}\sum_{t=1}^{T}\begin{pmatrix} y_{i,t-1}^{*2}, & y_{i,t-1}^{*}\underset{\sim}{x}_{it}^{*'} \\ \underset{\sim}{x}_{it}^{*}y_{i,t-1}^{*}, & \underset{\sim}{x}_{it}^{*}\underset{\sim}{x}_{it}^{*'}\end{pmatrix}\right]^{-1}\left[\sum_{i=1}^{N}\sum_{t=1}^{T}\begin{pmatrix} y_{i,t-1}^{*} \\ \underset{\sim}{x}_{it}^{*}\end{pmatrix} y_{it}^{*}\right], \tag{13.5}$$

$$\hat{\alpha}_i = \bar{y}_i - \hat{\rho}\bar{y}_{i,-1} - \hat{\underset{\sim}{\beta}}'\bar{\underset{\sim}{x}}_i, \quad i = 1, \ldots, N, \tag{13.6}$$

$$\hat{\lambda}_t = \bar{y}_t - \hat{\rho}\bar{y}_{t-1} - \hat{\underset{\sim}{\beta}}'\bar{\underset{\sim}{x}}_t, \quad t = 1, \ldots, T, \tag{13.7}$$

where $y_{it}^* = (y_{it} - \bar{y}_i - \bar{y}_t + \bar{y})$, $\bar{y}_i = \frac{1}{T}\sum_{t=1}^{T} y_{it}$, $\bar{y}_t = \frac{1}{N}\sum_{i=1}^{N} y_{it}$, $\bar{y} = \frac{1}{NT}\sum_{i=1}^{N}\sum_{t=1}^{T} y_{it}$, and similarly for $\bar{y}_{t-1}$, $\bar{y}_{i,-1}$, $\bar{\underset{\sim}{x}}_i$, $\bar{\underset{\sim}{x}}_t$, $\underset{\sim}{x}_{it}^*$, v_{it}^*, $\bar{v}_i$, $\bar{v}_t$, and $\bar{v}$. The FE MLE of $(\rho, \underset{\sim}{\beta})$ is also called the covariance estimator because it is equivalent to first applying covariance transformation to sweep out α_i and λ_t,

$$y_{it}^* = \rho y_{i,t-1}^* + \underset{\sim}{\beta}'\underset{\sim}{x}_{it}^* + v_{it}^*, \tag{13.8}$$

then apply the least squares estimator to Equation 13.8. When T is finite, there are only finite number of y_{it} that contain information about α_i and α_i increases with N, the MLE is inconsistent no matter how large N is because α_i becomes incidental parameter. To illustrate this, there is no loss of generality to just consider the simple case of $\underset{\sim}{\beta} = 0$, so Equation 13.1 becomes

$$y_{it} = \rho y_{i,t-1} + v_{it}. \tag{13.9}$$

The MLE of ρ under the assumption that y_{i0} are fixed is equal to

$$\hat{\rho}_{cv} = \frac{\sum_{i=1}^{N}\sum_{t=1}^{T} y_{i,t-1}^{*}y_{it}^{*}}{\sum_{i=1}^{N}\sum_{t=1}^{T} y_{i,t-1}^{*2}} \tag{13.10}$$

The probability limit of $\hat{\rho}_{cv}$ is equal to (Hahn and Moon 2006; Hsiao and Tahmiscioglu 2008)

$$\begin{aligned}\text{plim}_{N\to\infty}(\hat{\rho}_{cv} - \rho) &= -\frac{1+\rho}{T-1}\left(1 - \frac{1}{T}\frac{1-\rho^T}{1-\rho}\right)\\ &\left\{1 - \frac{2\rho}{(1-\rho)(T-1)}\left[1 - \frac{1-\rho^T}{T(1-\rho)}\right]\right\}^{-1}.\end{aligned} \tag{13.11}$$

This estimator is biased to the order of $(1/T)$ and the bias is identical independent of whether α_i and λ_t are fixed or random and is identical to the case when λ_t are 0 for all t. (e.g., Anderson and Hsiao 1981, 1982; Hahn and Kuersteiner 2002; Hahn and Moon 2006; Hsiao and Tahmiscioglu 2008). When $T \longrightarrow \infty$, the MLE of the FE model is consistent. However, if both N and T go to infinity and lim $\left(\frac{N}{T}\right) = c > 0$, Hahn and Moon (2006) have shown that $\sqrt{NT}(\hat{\rho}_{cv} - \rho)$ is asymptotically normally distributed with mean $-\sqrt{c}(1+\rho)$ and variance $1-\rho^2$. In other words, the usual t-statistic based on $\hat{\rho}_{cv}$ could be subject to severe size distortion.

13.4 Issues of Initial Observations

One way to get around incidental parameters problem is to assume α_i and λ_t random and satisfying Equation 13.4, then the system

$$\underset{\sim}{y}_i = \underset{\sim}{y}_{i,-1}\rho + X_i\underset{\sim}{\beta} + \underset{\sim}{v}_i, \quad i = 1, \ldots, N, \tag{13.12}$$

where $\underset{\sim}{y}_i' = (y_{i1}, \ldots, y_{iT})$, $\underset{\sim}{y}_{i,-1}' = (y_{i0}, \ldots, y_{i,T-1})$, $\underset{\sim}{v}_i' = (v_{i1}, \ldots, v_{iT})$ and X_i is the $T \times K$ matrix of $(\underset{\sim}{x}_{it}')$, has

$$\begin{aligned} E\underset{\sim}{v}_i &= \underset{\sim}{0}, \\ E\underset{\sim}{v}_i\underset{\sim}{v}_i' &= \sigma_\epsilon^2 I_T + \sigma_\alpha^2 \underset{\sim}{e}_T\underset{\sim}{e}_T' + \sigma_\lambda^2 I_T, \\ E\underset{\sim}{v}_i\underset{\sim}{v}_j' &= \sigma_\lambda^2 I_T \end{aligned} \tag{13.13}$$

where I_T is T rowed identity matrix and $\underset{\sim}{e}_T$ is a $T \times 1$ vector of 1's. If $(\epsilon_{it}, \alpha_i, \lambda_t)$ are normally distributed, and y_{i0} are fixed constants, the likelihood function is

$$2\pi^{-\frac{NT}{2}}|\Omega|^{-\frac{1}{2}}\exp\left\{-\frac{1}{2}(\underset{\sim}{y} - \underset{\sim}{y}_{-1}\rho - X\underset{\sim}{\beta})'\Omega^{-1}(\underset{\sim}{y} - \underset{\sim}{y}_{-1}\rho - X\underset{\sim}{\beta})\right\} \tag{13.14}$$

where $\underset{\sim}{y} = (\underset{\sim}{y}_1', \ldots, \underset{\sim}{y}_N')'$, $\underset{\sim}{y}_{-1} = (\underset{\sim}{y}_{1,-1}', \ldots, \underset{\sim}{y}_{N,-1}')'$, $X = (X_1', \ldots, X_N')'$,

$$\Omega = \sigma_\epsilon^2 I_{NT} + \sigma_\alpha^2 I_N \otimes \underset{\sim}{e}_T\underset{\sim}{e}_T' + \sigma_\lambda^2 \underset{\sim}{e}_N\underset{\sim}{e}_N' \otimes I_T \tag{13.15}$$

and $\otimes$ denotes the kroecker product. The likelihood function no longer involves incidental parameters and the MLE is consistent and asymptotically normally distributed either N or T or both tend to infity. Given σ_ϵ^2, σ_α^2 and σ_λ^2, the MLE is identical to the generalized least squares estimator (GLS)

$$\begin{pmatrix} \tilde{\rho} \\ \tilde{\underset{\sim}{\beta}} \end{pmatrix} = \left[\begin{pmatrix} \underset{\sim}{y}_{-1}' \\ X' \end{pmatrix} \Omega^{-1}(\underset{\sim}{y}_{-1}, X)\right]^{-1} \left[\begin{pmatrix} \underset{\sim}{y}_{-1}' \\ X' \end{pmatrix} \Omega^{-1}\underset{\sim}{y}\right]. \tag{13.16}$$

However, most panels contain only finite T time series observations. The starting dates of data collection need not correspond to the starting dates of the data generating process. There is no reason to believe that the data generating process of y_{i0} to be different from the data generating process of y_{it}. If y_{i0} and y_{it} are generated from the same process, then $Ey_{i0}v_{it} = Ey_{i0}\alpha_i = 0$ implied by fixed y_{i0} assumption cannot hold.

For ease of notation, we shall assume in this section that $\lambda_t \equiv 0 \quad \forall t$. Continuous substitution of Equation 13.1 yields

$$y_{it} = \underset{\sim}{\beta}' \sum_{j=0}^{t-1} \underset{\sim}{x}_{i,t-j}\rho^j + \rho^t y_{i0} + \frac{1-\rho^t}{1-\rho}\alpha_i + \sum_{j=0}^{t-1} \epsilon_{i,t-j}\rho^j, \tag{13.17}$$

and

$$y_{i0} = \theta_{i0} + v_{i0}, \tag{13.18}$$

where $\theta_{i0} = \underset{\sim}{\beta}' \sum_{j=0}^{m} \underset{\sim}{x}_{i,-j}\rho^j$, $v_{i0} = \frac{1-\rho^m}{1-\rho}\alpha_i + \sum_{j=0}^{m} \epsilon_{i,-j}\rho^j$, assuming the process started at period $-m$. Then

$$Ev_{i0}v_{it} = \frac{1-\rho^m}{1-\rho}\sigma_\alpha^2 = c^* \neq 0. \tag{13.19}$$

Therefore, conditional on y_{i0} (or v_{i0}),

$$\underset{\sim}{y}_i = \underset{\sim}{y}_{i,-1}\rho + x_i'\underset{\sim}{\beta} + \underset{\sim}{e}_T(y_{i0} - \theta_{i0})c^* + \underset{\sim}{v}_i^*, i = 1, \ldots, N, \tag{13.20}$$

where $\underset{\sim}{v}_i^* = \underset{\sim}{v}_i - \underset{\sim}{e}_T(y_{i0} - \theta_{i0})c^*$. When T is large, the correlation between y_{it} and y_{i0} will approach zero as can be seen from Equation 13.17. When $|\rho| < 1$, asymptotically there is no difference between Equations 13.12 and 13.20, thus between treating y_{i0} fixed or y_{i0} random. However, in finite T, $y_{i,t-1}$ and y_{i0} are correlated from Equation 13.17. Regressing y_{it} on $y_{i,t-1}$ and $\underset{\sim}{x}_{it}$ is subject to omitted variable $((y_{i0} - \theta_{i0}))$ bias no matter how large N is.

To obtain consistent estimators of ρ and $\underset{\sim}{\beta}$, one should either apply GLS to Equation 13.20 (namely, the conditional system Equation 13.20 conditional on y_{i0}), or to complete the system by maximizing the joint likelihood function of $(y_{i0}, y_{i1}, \ldots, y_{iT})$,

$$\begin{aligned} y_{i0} &= \theta_{i0} + v_{i0}, \\ \underset{\sim}{y}_i &= \underset{\sim}{y}_{i,-1}\rho + X\underset{\sim}{\beta} + \underset{\sim}{v}_i, \quad i = 1, \ldots, N. \end{aligned} \tag{13.21}$$

However, θ_{i0} depends on $\underset{\sim}{x}_{i,-j}$ which are unobservable. Treating θ_{i0} as unknown parameters again will subject the system Equation 13.21 to incidental parameters when T is finite and N is large.

To get around the incidental parameters issues, Bhargava and Sargan (1983) show that if $\underset{\sim}{x}_{it}$ is generated by a homogeneous process

$$\underset{\sim}{x}_{it} = \underset{\sim}{a} + \sum_{j=0} B_j \underset{\sim}{\eta}_{i,t-j}, \sum |B_j| < \infty, \tag{13.22}$$

where $\eta_{i,t-j}$ are i.i.d. random variables with mean zero and constant variance Σ_η, then[2]

$$E(\underset{\sim}{\theta}_{i0} \mid \underset{\sim}{x}_i) = \underset{\sim}{\pi}'\underset{\sim}{x}_i, i = 1, \ldots, N, \tag{13.23}$$

[2] For ease of notation, we have merged the intercept term into $\underset{\sim}{x}_i$.

where $\underset{\sim}{x}_i = (\underset{\sim}{x}'_{i1}, \ldots, \underset{\sim}{x}'_{iT})$. Substituting Equation 13.23 into Equation 13.21 yields

$$y_{i0} = \underset{\sim}{\pi}'\underset{\sim}{x}_i + v^*_{i0},$$
$$\underset{\sim}{y}_i = \underset{\sim}{y}_{i,-1}\rho + X_i\underset{\sim}{\beta} + \underset{\sim}{v}_i, \quad i = 1, \ldots, N. \tag{13.24}$$

System Equation 13.24 no longer involves incidental parameters. Therefore, the MLE or GLS of Equation 13.24,[3]

$$\hat{\underset{\sim}{\delta}} = \left(\sum_{i=1}^{N} \tilde{X}'_i \tilde{V}^{-1} \tilde{X}_i\right)^{-1} \left(\sum_{i=1}^{N} \tilde{X}'_i \tilde{V}^{-1} \tilde{\underset{\sim}{y}}_i\right), \tag{13.25}$$

is consistent and asymptotically normally distributed either N or T or both tend to infinity with covariance matrix

$$\text{Cov}\,(\hat{\underset{\sim}{\delta}}_{\text{GLS}}) = \left(\sum_{i=1}^{N} \tilde{X}'_i \tilde{V}^{-1} \tilde{X}_i\right)^{-1} \tag{13.26}$$

where $\hat{\underset{\sim}{\delta}} = (\underset{\sim}{\pi}', \rho, \underset{\sim}{\beta}')$, $\tilde{\underset{\sim}{y}}'_i = (y_{i0}, \underset{\sim}{y}'_i)$, and

$$\tilde{X}_i = \begin{pmatrix} \underset{\sim}{x}'_i & 0 & \underset{\sim}{0}' \\ \underset{\sim}{0} & y_{i0} & \underset{\sim}{x}'_{i1} \\ \underset{\sim}{0} & y_{i1} & \underset{\sim}{x}'_{i2} \\ \cdots & \cdot & \cdots \\ \cdots & y_{i,T-1} & \underset{\sim}{x}'_{iT} \end{pmatrix}. \tag{13.27}$$

13.5 Method of Moments Estimator for Dynamic Models with Individual-Specific Effects Only

We illustrate the basic idea of generalized methods of moments and the likelihood principle for dynamic model with individual-specific effects only (i.e., $\lambda_t \equiv 0, \forall t$) in this section and the next, then discuss the estimator of models involving both additive α_i and λ_t in Section 13.7.

Taking the first difference of Equation 13.1 under the assumption of $\lambda_t = 0$ yields

$$\Delta y_{it} = \rho \Delta y_{i,t-1} + \underset{\sim}{\beta}' \Delta \underset{\sim}{x}_{it} + \Delta \epsilon_{it},$$
$$i = 1, \ldots, N,$$
$$t = 2, \ldots, T, \tag{13.28}$$

[3] Alternatively, one may apply the conditional MLE or GLS to Equation 13.20 (e.g., Blundell and Bond 1998).

where $\Delta = (1 - L)$, L denotes the lag operator so $\Delta y_{it} = y_{it} - y_{i,t-1}$. Equation 13.28 no longer involves α_i. However, $E(\Delta y_{i,t-1}\Delta\epsilon_{it}) \neq 0$. Regressing Δy_{it} on $(\Delta y_{i,t-1}, \Delta \underset{\sim}{x}'_{it})$ yields inconsistent estimators for ρ and $\underset{\sim}{\beta}$. On the other hand, $E(y_{i,t-2}\Delta\epsilon_{it}) = 0$. Therefore, $y_{i,t-2}$ can be used as instrument for $\Delta y_{i,t-1}$. However, $y_{i,t-2}$ is not the only instrument for $\Delta y_{i,t-1}$. As noted by Amemiya and MaCurdy (1986), Ahn and Schmidt (1995), Arellano and Bond (1991), Arellano and Bover (1995), Breusch, Mizon, and Schmidt (1989), etc. that all $y_{i,t-2-j}, j = 0, 1, \ldots, t-2$, and all $\underset{\sim}{x}_{it}$ satisfy the condition $E(y_{i,t-2-j}\Delta\epsilon_{it}) = 0$, $E(\underset{\sim}{x}_i \Delta\underset{\sim}{\epsilon}_{it}) = \underset{\sim}{0}$. Let $\underset{\sim}{q}_{it} = (y_{i0}, y_{i1}, \ldots, y_{i,t-2}, \underset{\sim}{x}'_i)'$, we have the moment conditions

$$E(\underset{\sim}{q}_{it}\Delta\underset{\sim}{\epsilon}_{it}) = 0, \quad t = 2, \ldots, T. \tag{13.29}$$

Stacking the $(T-1)$ first difference equation of Equation 13.28 in matrix form, we have

$$\Delta\underset{\sim}{y}_i = \Delta\underset{\sim}{y}_{i,-1}\rho + \Delta X_i\underset{\sim}{\beta} + \Delta\underset{\sim}{\epsilon}_i, \quad i = 1, \ldots, N, \tag{13.30}$$

where $\Delta\underset{\sim}{y}_i$, $\Delta\underset{\sim}{y}_{i,-1}$ and $\Delta\underset{\sim}{\epsilon}_i$ are $(T-1)\times 1$ vectors of $(\Delta y_{i2}, \ldots, \Delta y_{iT})'$, $(\Delta y_{i1}, \ldots, \Delta y_{i,T-1})'$ and $(\Delta\epsilon_{i2}, \ldots, \Delta\epsilon_{iT})'$, respectively, and ΔX_i is the $(T-1)\times K$ stacked matrix $(\Delta\underset{\sim}{x}_{i2}, \ldots, \Delta\underset{\sim}{x}_{iT})'$. Let

$$W_i = \begin{bmatrix} \underset{\sim}{q}_{i2} & \underset{\sim}{0} & \cdots & \underset{\sim}{0} \\ \underset{\sim}{0} & \underset{\sim}{q}_{i3} & \cdots & \underset{\sim}{0} \\ & \vdots & & . \\ & . & \cdots & . \\ \underset{\sim}{0} & . & \underset{\sim}{0}' & \underset{\sim}{q}_{iT} \end{bmatrix} \tag{13.31}$$

be the $(T-1)[(T-1)K + \frac{T}{2}] \times (T-1)$ block diagonal matrix. Then, we have the orthogonality conditions,

$$EW_i\Delta\underset{\sim}{\epsilon}_i = \underset{\sim}{0}. \tag{13.32}$$

Under the assumption that $(\underset{\sim}{y}'_i, \underset{\sim}{x}'_i)$ are independently, identically distributed across i, we may approximate the population moments (Equation 13.32) by the sample moments $\frac{1}{N}\sum_{i=1}^{N} W_i(\Delta\underset{\sim}{y}_i - \Delta\underset{\sim}{y}_{i,-1}\rho - \Delta x_i\underset{\sim}{\beta})$. Since there are in general more moments conditions than the number of unknowns, an efficient moment estimator is to apply the generalized lease squares principle to Equation 13.32 [Generalized methods of moments (GMM)],

$$\underset{\rho,\,\underset{\sim}{\beta}}{\text{Min}} \left(\frac{1}{N}\sum_{i=1}^{N}\Delta\underset{\sim}{\epsilon}'_i W_i\right)\Psi^{-1}\left(\frac{1}{N}\sum_{i=1}^{N} W_i\Delta\underset{\sim}{\epsilon}_i\right), \tag{13.33}$$

where $\Psi = E[\frac{1}{N^2}\sum_{i=1}^{N} W_i \Delta \underset{\sim}{\epsilon}_i \Delta \underset{\sim}{\epsilon}_i' W_i]$. Under the assumption that $\underset{\sim}{\epsilon}_i$ is i.i.d., Ψ may be approximated by $\frac{\sigma_\epsilon^2}{N^2}\sum_{i=1}^{N} W_i A W_i'$, where

$$\underset{(T-1)\times(T-1)}{A=} \begin{bmatrix} 2 & -1 & 0 & \cdots & \cdot & 0 \\ -1 & 2 & -1 & \cdots & \cdot & 0 \\ 0 & -1 & 2 & \cdots & \cdot & \cdot \\ \cdot & \cdot & \cdot & \cdots & \cdot & \cdot \\ \cdot & \cdot & \cdot & \cdots & \cdot & -1 \\ 0 & \cdot & \cdot & \cdots & -1 & 2 \end{bmatrix}. \tag{13.34}$$

Thus, the Arellano and Bover (1995) GMM estimator takes the form

$$\begin{aligned} \hat{\underset{\sim}{\theta}}_{\text{GMM}} = \begin{pmatrix} \hat{\rho} \\ \hat{\underset{\sim}{\beta}} \end{pmatrix}_{\text{GMM}} &= \left\{ \left[\sum_{i=1}^{N} \begin{pmatrix} \Delta \underset{\sim}{y}_{i,-1}' \\ \Delta X_i' \end{pmatrix} W_i' \right] \left[\sum_{i=1}^{N} W_i A W_i' \right]^{-1} \right. \\ &\quad \left. \left[\sum_{i=1}^{N} W_i (\Delta \underset{\sim}{y}_{i,-1}, \Delta X_i) \right] \right\}^{-1} \cdot \left\{ \left[\sum_{i=1}^{N} \begin{pmatrix} \Delta \underset{\sim}{y}_{i,-1}' \\ \Delta X_i' \end{pmatrix} W_i' \right] \right. \\ &\quad \left. \left[\sum_{i=1}^{N} W_i A W_i' \right]^{-1} \left[\sum_{i=1}^{N} W_i \Delta \underset{\sim}{y}_i \right] \right\}, \end{aligned} \tag{13.35}$$

with asymptotic covariance matrix

$$\text{Cov}\,(\hat{\underset{\sim}{\theta}}_{\text{GMM}}) = \left\{ \left[\sum_{i=1}^{N} \begin{pmatrix} \Delta \underset{\sim}{y}_{i,-1}' \\ \Delta X_i' \end{pmatrix} W_i' \right] \left[\sum_{i=1}^{N} W_i A W_i' \right]^{-1} \left[\sum_{i=1}^{N} W_i (\Delta \underset{\sim}{y}_i, \Delta X_i) \right] \right\}^{-1}. \tag{13.36}$$

Remark 13.1 The GMM estimator is consistent and asymptotically normally distributed whether α_i is treated as a fixed constant or a random variable because the first difference of Equation 13.1 eliminates α_i from the transformed model (Equation 13.28). However, GMM cannot estimate the coefficients of time-invariant variables, say gender, because first differencing also eliminates such variables in Equation 13.28 but the likelihood approach can if α_i is indeed random and uncorrelated with $\underset{\sim}{x}_i$.

Remark 13.2 When α_i is random and satisfies Equation 13.3, the likelihood approach uses the level equation (Equation 13.1) while the GMM approach uses the first difference equation (Equation 13.28). In general, the variation across individuals is much larger than the variation over time of an individual. Moreover, the likelihood approach uses T equations of (Equation 13.1) but the GMM uses $(T-1)$ equations of (Equation 13.28). Therefore, the likelihood approach is more efficient than the GMM approach (for detail, see Hsiao, Pesaran, and Tahmiscioglu 2002).

Remark 13.3 In implementing the likelihood approach we invoke the normality assumption. However, Equation 13.25 remains consistent and asymptotically normally distributed even v_{it} is not normally distributed. One may view estimators of the type of Equation 13.25 as a quasi-MLE (QMLE).

Remark 13.4 Although we make no assumption about the initial value distribution of y_{i0} in implementing the GMM approach, the assumption that $(\underset{\sim}{y}_i', \underset{\sim}{x}_i')$ are i.i.d. across i actually implies Equation 13.22 which is invoked to get around the incidental parameters problem in the likelihood approach. In other words, the conditions for implementing the likelihood approach are no more restrictive than the GMM approach. Moreover, as shown by Hayakawa (2009), the efficiency of GMM actually depends on the distribution of y_{i0} and σ_α^2.

Remark 13.5 The likelihood approach uses the moment conditions $E\left[\tilde{X}_i' V^{-1}\binom{v_{i0}^*}{\underset{\sim}{v}_i}\right] = \underset{\sim}{0}$, which stay fixed as N and T increases. The number of moment conditions for GMM increases at order T^2 as T increases. In finite sample, the procedure of equating sample moments to population moments can lead to severe bias in GMM as demonstrated in a Monte Carlo by Ziliak (1997). Moreover, if ρ is close to one, the correlations between Δy_{it} and $y_{i,t-j}$ for $j \geq 2$ could be weak and lead to weak instrumental variables problem as demonstrated in the Monte Carlos by Binder, Hsiao, and Pesaran (2005) and Hsiao, Pesaran, and Tahmiscioglu (2002).

Remark 13.6 The moment conditions, Equation 13.31 assumes $\underset{\sim}{x}_{it}$ are strictly exogenous. However, there could be feedback relations between y_{it} and $\underset{\sim}{x}_{i,t+j}$ as in Cheng and Kwan (2000). If $\underset{\sim}{x}_{it}$ is only weakly exogenous, it does not affect the likelihood approach. But for GMM, instead of defining $\underset{\sim}{q}_{it}$ as $(y_{it}, \ldots, y_{i,t-2}, \underset{\sim}{x}_i)$, we have to redefine $\underset{\sim}{q}_{it}$ as $(y_{it}, \ldots, y_{i,t-2}, \underset{\sim}{x}_{it}', \ldots, \underset{\sim}{x}_{i1}')'$, then Equation 13.29 still holds and GMM can be applied with the redefined $\underset{\sim}{q}_{it}$.

13.6 Likelihood Approach for the Dynamic Fixed Individual-Specific Effects Model

When α_i is treated as fixed constants, we can estimate ρ and $\underset{\sim}{\beta}$ by the GMM method discussed in Section 13.5. A similar likelihood approach can also be implemented on the system (Equation 13.28), which no longer contains the fixed α_i. However, just like the RE case, there is the problem of initial values if T is finite. If the data generating process of Δy_{i1} is no different from the data generating process of Δy_{it}, then

$$\Delta y_{i1} = \theta_{i1} + v_{i1}, \tag{13.37}$$

where $\theta_{i1} = \underset{\sim}{\beta}' \sum_{j=0}^{m-1} \Delta \underset{\sim}{x}_{i,1-j} \rho^j$, $v_{i1} = \sum_{j=0}^{m-1} \Delta \epsilon_{i,1-j} \rho^j$, if the process started at period $-m$. Since $\Delta \underset{\sim}{x}_{i,1-j}$ are unknown, so are θ_{i1}. Treating θ_{i1} as an unknown parameter again introduces incidental parameters. To get around this issue, the expected value of θ_{i1} conditional on $\Delta \underset{\sim}{x}_i$ has to be a function of constant parameters,

$$E(\theta_{i1} \mid \Delta \underset{\sim}{x}_i) = \underset{\sim}{\pi}' \Delta \underset{\sim}{x}_i, \quad i = 1, \ldots, N, \tag{13.38}$$

where $\Delta \underset{\sim}{x}_i' = (\Delta \underset{\sim}{x}_{i2}', \ldots, \Delta \underset{\sim}{x}_{iT}')$. Hsiao, Pesaran, and Tahmiscioglu (2002) have shown that if $\underset{\sim}{x}_{it}$ is generated by

$$\underset{\sim}{x}_{it} = \underset{\sim}{\mu}_i + \underset{\sim}{g} + \sum_{j=0}^{\infty} B_j \underset{\sim}{\xi}_{i,t-j}, \quad \sum |B_j| < \infty, \tag{13.39}$$

then Equation 13.38 holds.

Given Equation 13.38, we may write the system of T equations in the form,

$$\begin{aligned} \Delta y_{i1} &= \underset{\sim}{\pi}' \Delta \underset{\sim}{x}_i + v_{i1}^*, \\ \Delta \underset{\sim}{y}_i &= \Delta \underset{\sim}{y}_{i,-1} \rho + \Delta X_i \underset{\sim}{\beta} + \Delta \underset{\sim}{\epsilon}_i, \qquad i = 1, \ldots, N, \end{aligned} \tag{13.40}$$

where $v_{i1}^* = v_{i1} + (\theta_{i1} - E\theta_{i1})$. By construction, $E(v_{i1}^* \mid \Delta \underset{\sim}{x}_i) = 0$, $E v_{i1}^{*2} = \sigma_{v^*}^2$, $E(v_{i1}^* \Delta \epsilon_{i2}) = -\sigma_\epsilon^2$, and $E(v_{i1}^* \Delta \epsilon_{it}) = 0$, for $t = 3, \ldots, T$. Let the $(T \times 1)$ vector $\Delta \underset{\sim}{\epsilon}_i^{*'} = (v_{i1}^*, \Delta \underset{\sim}{\epsilon}_i')$. The covariance matrix of $\Delta \underset{\sim}{\epsilon}_i^*$ is

$$E \Delta \underset{\sim}{\epsilon}_i^* \Delta \underset{\sim}{\epsilon}_i^{*'} = \sigma_\epsilon^2 \begin{bmatrix} h & -1 & 0 & \cdots & \cdots & \cdot \\ -1 & 2 & -1 & \cdots & \cdot & \cdot \\ 0 & -1 & 2 & \cdots & \cdot & \cdot \\ \cdot & \cdot & \cdot & \cdots & \cdot & \cdot \\ \cdot & \cdot & \cdot & \cdots & \cdot & \cdot \\ 0 & \cdot & \cdot & \cdots & -1 & 2 \end{bmatrix} = \Omega^*, \tag{13.41}$$

where $h = \frac{\sigma_{v^*}^2}{\sigma_\epsilon^2}$. Assuming $\Delta \underset{\sim}{\epsilon}_i^*$ is independently normally distributed with covariance matrix Ω^*, then the likelihood function of $\Delta \underset{\sim}{y}_i^* = (\Delta y_{i1}, \Delta \underset{\sim}{y}_i')'$, $i = 1, \ldots, N$, is in the form

$$(2\pi)^{-\frac{NT}{2}} |\Omega^*|^{-\frac{N}{2}} \exp\left\{ -\frac{1}{2} \sum_{i=1}^{N} \Delta \underset{\sim}{\epsilon}_i^{*'} \Omega^{*-1} \Delta \underset{\sim}{\epsilon}_i^* \right\}, \tag{13.42}$$

where $\Delta \underset{\sim}{\epsilon}_i^{*'} = [\Delta y_{i1} - \underset{\sim}{\pi}' \Delta \underset{\sim}{x}_i, \Delta y_{i2} - \Delta y_{i1} \rho - \Delta \underset{\sim}{x}_{i2}' \underset{\sim}{\beta}, \ldots, \Delta y_{iT} - \Delta y_{i,T-1} \rho - \Delta \underset{\sim}{x}_{iT}' \underset{\sim}{\beta}]$. The likelihood function (Equation 13.42) is a function of fixed number of parameters, hence, the MLE is consistent and asymptotically normally distributed either N or T or both tend to infinity. Therefore, standard t or F tests can be applied.[4]

[4] For further discussion of hypothese testing involving dynamic panel data models, see Harris, Matyas, and Sevestre (2008, Section 8.6).

Conditional on h, the MLE of $\underset{\sim}{\delta} = (\underset{\sim}{\pi}', \rho, \underset{\sim}{\beta}')'$ is identical to the GLS,

$$\hat{\underset{\sim}{\delta}}_{\text{GLS}} = \left(\sum_{i=1}^{N} \Delta\tilde{X}_i' \Omega^{*-1} \Delta\tilde{X}_i\right)^{-1} \left(\sum_{i=1}^{N} \Delta\tilde{X}_i' \Omega^{*-1} \Delta\underset{\sim}{y}_i\right), \tag{13.43}$$

where

$$\Delta\tilde{X}_i = \begin{pmatrix} \Delta\underset{\sim}{x}_i' & 0 & \underset{\sim}{0}' \\ \underset{\sim}{0} & \Delta\underset{\sim}{y}_{i-1} & \Delta X_i \end{pmatrix}. \tag{13.44}$$

When h is unknown, one can use a two-step procedure. In the first step, we regress Δy_{i1} on $\Delta\underset{\sim}{x}_i$ to obtain $\hat{\sigma}^2_{v^*}$ and apply GMM to obtain $\hat{\sigma}^2_{\epsilon}$. In the second step, we substitute estimated $\hat{h}$ for h in Equation 13.43. However, the feasible GLS is not as efficient as GLS (for detail, see Hsiao, Pesaran, and Tahmiscoglu 2002).

13.7 Models with Both Individual- and Time-Specific Additive Effects

When time-specific effects also appear in v_{it} as in Equation 13.2, the estimators ignoring the presence of λ_t like those discussed in Sections 13.13 to 13.6 are no longer consistent when T is finite. For notational ease and without loss of generality, we illustrate the fundamental issues of dynamic model with both individual- and time-specific additive effects model by restricting $\underset{\sim}{\beta} = \underset{\sim}{0}$ in Equation 13.1, thus the model becomes

$$y_{it} = \rho y_{i,t-1} + v_{it}, \tag{13.45}$$

$$v_{it} = \alpha_i + \lambda_t + \epsilon_{it}, i = 1, \ldots, N, t = 1, \ldots, T, y_{i0} \text{ observable.} \tag{13.46}$$

The panel data estimators discussed in Sections 13.5 and 13.6 assume no presence of λ_t (i.e., $\lambda_t = 0 \forall t$). When λ_t are indeed present, those estimators are not consistent if T is finite when $N \to \infty$. For instance, the consistency of GMM (Equation 13.33) is based on the assumption that $\frac{1}{N}\sum_{i=1}^{N} y_{i,t-j}\Delta v_{it}$ converges to the population moments (Equation 13.32). However, if λ_t are also present as in Equation 13.46, this condition is likely to be violated. To see this, taking first difference of Equation 13.45 yields

$$\begin{aligned} \Delta y_{it} &= \rho\Delta y_{i,t-1} + \Delta v_{it} \\ &= \rho\Delta y_{i,t-1} + \Delta\lambda_t + \Delta\epsilon_{it}, \\ i &= 1, \ldots, N, \\ t &= 2, \ldots, T. \end{aligned} \tag{13.47}$$

Although

$$E(y_{i,t-j}\Delta v_{it}) = 0 \text{ for } j = 2, \ldots, t, \tag{13.48}$$

the sample moment, as $N \longrightarrow \infty$,

$$\frac{1}{N}\sum_{i=1}^{N} y_{i,t-j}\Delta v_{it} = \frac{1}{N}\sum_{i=1}^{N} y_{i,t-j}\Delta\lambda_t + \frac{1}{N}\sum_{i=1}^{N} y_{i,t-j}\Delta\epsilon_{it} \tag{13.49}$$

converges to $\bar{y}_{t-j}\Delta\lambda_t$, which in general is not equal to zero, in particular, if y_{it} has mean different from zero,[5] where $\bar{y}_t = \frac{1}{N}\sum_{i=1}^{N} y_{it}$.

To obtain consistent estimators of ρ, we need to take explicit account of the presence of λ_t in addition to α_i. If α_i and λ_t are random and satisfy Equation 13.4, because $Ey_{i0}v_{it} \neq 0$, we either have to write Equation 13.45 conditional on y_{i0} or to complete the system (Equation 13.45) by deriving the marginal distribution of y_{i0}. By continuous substitutions, we have

$$\begin{aligned} y_{i0} &= \frac{1-\rho^m}{1-\rho}\alpha_i + \sum_{j=0}^{m-1}\lambda_{-j}\rho^j + \sum_{j=0}^{m-1}\epsilon_{i,-j}\rho^j \\ &= v_{i0}, \end{aligned} \tag{13.50}$$

assuming the process started at period $-m$.

Under Equation 13.4, $Ey_{i0} = Ev_{i0} = 0$, $\text{Var}\,(y_{i0}) = \sigma_0^2$, $E(v_{i0}v_{it}) = \frac{1-\rho^m}{1-\rho}\sigma_\alpha^2 = c$, $Ev_{it}v_{jt} = d$. Stacking the $T+1$ time series observations for the ith individual into a vector, $\underset{\sim}{y}_i = (y_{i0}, \ldots, y_{iT})'$ and $\underset{\sim}{y}_{i,-1} = (0, y_{i0}, \ldots, y_{i,T-1})'$, $\underset{\sim}{v}_i = (v_{i0}, \ldots, v_{iT})'$. Let $\underset{\sim}{y} = (\underset{\sim}{y}'_1, \ldots, \underset{\sim}{y}'_N)'$, $\underset{\sim}{y}_{-1} = (\underset{\sim}{y}'_{1,-1}, \ldots, \underset{\sim}{y}'_{N,-1})$, $\underset{\sim}{v} = (\underset{\sim}{v}'_1, \ldots, \underset{\sim}{v}'_N)'$, then

$$\underset{\sim}{y} = \underset{\sim}{y}_{-1}\rho + \underset{\sim}{v}, \tag{13.51}$$

$$E\underset{\sim}{v} = \underset{\sim}{0},$$

$$\begin{aligned} E\underset{\sim}{v}\underset{\sim}{v}' &= \sigma_\epsilon^2 I_N \otimes \begin{pmatrix} \omega & \underset{\sim}{0}' \\ \underset{\sim}{0} & I_T \end{pmatrix} + \sigma_\alpha^2 I_N \otimes \begin{pmatrix} 0 & c^*\underset{\sim}{e}'_T \\ c^*\underset{\sim}{e}_T & \underset{\sim}{e}_T\underset{\sim}{e}'_T \end{pmatrix} \\ &\quad + \sigma_\lambda^2 \underset{\sim}{e}_N\underset{\sim}{e}_N \otimes \begin{pmatrix} d^* & \underset{\sim}{0}' \\ \underset{\sim}{0} & I_T \end{pmatrix}, \end{aligned} \tag{13.52}$$

$$\omega = \frac{\sigma_0^2 - d}{\sigma_\epsilon^2}, \; d^* = \frac{d}{\sigma_\lambda^2}, \; c^* = \frac{c}{\sigma_\alpha^2}, \tag{13.53}$$

where $\otimes$ denotes the kronecker product. The system (Equation 13.51) has a fixed number of unknowns $(\rho, \sigma_\epsilon^2, \sigma_\alpha^2, \sigma_\lambda^2, \sigma_0^2, c, d)$ as N and T increase. Therefore, the MLE (or quasi-MLE or GLS) of Equation 13.51 is consistent and asymptotically normally distributed.

[5] For instance, if y_{it} is also a function of exogenous variables as Equation 13.1.

When α_i and λ_t are fixed constants, we note that first differencing only eliminates α_i from the specification. The time-specific effects, $\Delta\lambda_t$, remain at Equation 13.47. To further eliminate $\Delta\lambda_t$, we note that the cross-sectional mean $\Delta y_t = \frac{1}{N}\sum_{i=1}^{N}\Delta y_{it}$ is equal to

$$\Delta y_t = \rho\Delta y_{t-1} + \Delta\lambda_t + \Delta\epsilon_t, \tag{13.54}$$

where $\Delta\epsilon_t = \frac{1}{N}\sum_{i=1}^{N}\Delta\epsilon_{it}$. Taking deviation of Equation 13.47 from Equation 13.54 yields

$$\begin{aligned} \Delta y_{it}^* &= \rho\Delta y_{i,t-1}^* + \Delta\epsilon_{it}^*, \\ i &= 1, \ldots, N, \\ t &= 2, \ldots, T, \end{aligned} \tag{13.55}$$

where $\Delta y_{it}^* = (\Delta y_{it} - \Delta y_t)$ and $\Delta\epsilon_{it}^* = (\Delta\epsilon_{it} - \Delta\epsilon_t)$. The system (Equation 13.55) no longer involves α_i and λ_t.

Since

$$E[y_{i,t-j}\Delta\epsilon_{it}^*] = 0 \text{ for } \begin{array}{l} j = 2, \ldots, t, \\ t = 2, \ldots, T, \end{array} \tag{13.56}$$

the $\frac{1}{2}T(T-1)$ orthogonality conditions can be represented as

$$E(W_i\Delta\underset{\sim}{\xi}_i^*) = \underset{\sim}{0}, \tag{13.57}$$

where $\Delta\underset{\sim}{\xi}_i^* = (\Delta\epsilon_{i2}^*, \ldots, \Delta\epsilon_{iT}^*)'$,

$$W_i = \begin{pmatrix} \underset{\sim}{q}_{i2} & \underset{\sim}{0} & \cdots & \underset{\sim}{0} \\ \underset{\sim}{0} & \underset{\sim}{q}_{i3} & & \\ \cdot & \cdot & \ddots & \\ \vdots & \vdots & & \\ \underset{\sim}{0} & \underset{\sim}{0} & & \underset{\sim}{q}_{iT} \end{pmatrix}, \quad i = 1, \ldots, N,$$

and $\underset{\sim}{q}_{it} = (y_{i0}, y_{i1}, \ldots, y_{i,t-2})', t = 2, 3, \ldots, T$. Following Arellano and Bond (1991), we can propose a generalized method of moments (GMM) estimator,[6]

$$\begin{aligned} \tilde{\rho}_{\text{GMM}} = &\left\{\left[\frac{1}{N}\sum_{i=1}^{N}\Delta\underset{\sim}{\tilde{y}}_{i,-1}^{*\prime}W_i'\right]\hat{\Psi}^{-1}\left[\frac{1}{N}\sum_{i=1}^{N}W_i\Delta\underset{\sim}{\tilde{y}}_{i,-1}^{*}\right]\right\}^{-1} \\ &\left\{\left[\frac{1}{N}\sum_{i=1}^{N}\Delta\underset{\sim}{\tilde{y}}_{i,-1}^{*\prime}W_i'\right]\hat{\Psi}^{-1}\left[\frac{1}{N}\sum_{i=1}^{N}W_i\Delta\underset{\sim}{\tilde{y}}_{i}^{*}\right]\right\}, \end{aligned} \tag{13.58}$$

[6] For ease of exposition, we have only considered the GMM that makes use of orthogonality conditions. For additional moments conditions such as homoscedasticity or initial observations see, e.g., Ahn and Schmidt (1995), Blundell and Bond (1998).

where $\Delta\tilde{\underset{\sim}{y}}_i^* = (\Delta y_{i2}^*, \ldots, \Delta y_{iT}^*)'$, $\Delta\tilde{\underset{\sim}{y}}_{i-1}^* = (\Delta y_{i1}^*, \ldots, \Delta y_{i,T-1}^*)'$, and

$$\hat{\Psi} = \frac{1}{N^2}\left[\sum_{i=1}^{N} W_i \hat{\underset{\sim}{\epsilon}}_i^*\right]\left[\sum_{i=1}^{N} W_i \hat{\underset{\sim}{\epsilon}}_i^*\right]' \tag{13.59}$$

and $\Delta\hat{\underset{\sim}{\epsilon}}_i^* = \Delta\tilde{\underset{\sim}{y}}_i^* - \Delta\tilde{\underset{\sim}{y}}_{i,-1}^*\tilde{\rho}$, and $\tilde{\rho}$ denotes some initial consistent estimator of ρ, say a simple instrumental variable estimator.

The asymptotic covariance matrix of $\tilde{\rho}_{\mathrm{GMM}}$ can be approximated by

$$\text{asy. cov}\,(\tilde{\rho}_{\mathrm{GMM}}) = \left\{\left[\sum_{i=1}^{N} \Delta\tilde{\underset{\sim}{y}}_{i,-1}^{*'} W_i\right]\hat{\Psi}^{-1}\left[\sum_{i=1}^{N} W_i \Delta\tilde{\underset{\sim}{y}}_{i,-1}^*\right]\right\}^{-1}. \tag{13.60}$$

To implement the likelihood approach, we need to complete the system (Equation 13.55) by deriving the marginal distribution of Δy_{i1}^* through continuous substitution,

$$\begin{aligned}\Delta y_{i1}^* &= \sum_{j=0}^{m-1} \Delta\epsilon_{i,1-j}^* \rho^j \\ &= \Delta\tilde{\epsilon}_{i1}^*, \qquad i = 1, \ldots, N.\end{aligned} \tag{13.61}$$

Let $\Delta\underset{\sim}{y}_i^* = (\Delta y_{i1}^*, \ldots, \Delta y_{iT}^*)$, $\Delta\underset{\sim}{y}_i^* = (0, \ldots, \Delta y_{i,T-1}^*)$, $\Delta\tilde{\underset{\sim}{\epsilon}}_i^{*'} = (\Delta\tilde{\epsilon}_{i1}^*, \ldots, \Delta\epsilon_{iT}^*)$, the system

$$\Delta\underset{\sim}{y}_i^* = \Delta\underset{\sim}{y}_{i,-1}^*\rho + \Delta\tilde{\underset{\sim}{\epsilon}}_i^*, \tag{13.62}$$

does not involve α_i and λ_t. The MLE conditional on $\omega = \frac{\mathrm{Var}\,(\Delta y_{i1}^*)}{\sigma_\epsilon^2}$ is identical to the GLS

$$\hat{\rho}_{\mathrm{GLS}} = \left[\sum_{i=1}^{N} \Delta\underset{\sim}{y}_{i,-1}^{*'}\tilde{A}^{-1}\Delta\underset{\sim}{y}_{i,-1}^*\right]^{-1}\left[\sum_{i=1}^{N} \Delta\underset{\sim}{y}_{i,-1}^{*'}\tilde{A}^{-1}\Delta\underset{\sim}{y}_i^*\right]. \tag{13.63}$$

where

$$\tilde{A} = \begin{bmatrix} \omega & -1 & 0 & 0 & \cdots & 0 & 0 \\ -1 & 2 & -1 & 0 & \cdots & \cdot & \cdot \\ 0 & -1 & 2 & -1 & \cdots & \cdot & \cdot \\ \cdot & \cdot & \cdot & \cdot & \cdot & 2 & -1 \\ 0 & \cdot & \cdot & \cdot & & -1 & 2 \end{bmatrix}. \tag{13.64}$$

The GLS is consistent and asymptotically normally distributed with covariance matrix equal to

$$\mathrm{Var}(\hat{\rho}_{\mathrm{GLS}}) = \sigma_\epsilon^2\left[\sum_{i=1}^{N} \Delta\underset{\sim}{y}_{i,-1}^{*'}\tilde{A}^{-1}\Delta\underset{\sim}{y}_{i,-1}^*\right]^{-1}. \tag{13.65}$$

Remark 13.7 The GLS with $\Delta\underset{\sim}{\lambda}$ present is basically of the same form as the GLS without the time-specific effects (i.e., $\Delta\underset{\sim}{\lambda} = \underset{\sim}{0}$) (Hsiao, Pesaran, and Tahmiscioglu 2002), (Equation 13.25). However, there is an important difference between the two. The estimator (Equation 13.63) uses $\Delta y^*_{i,t-1}$ as the regressor for the equation Δy^*_{it} (Equation 13.62), not uses $\Delta y_{i,t-1}$ as the regressor for the equation Δy_{it} (Equation 13.47). If there are indeed common shocks that affect all the cross-sectional units, then the estimator Equation 13.25 is inconsistent while Equation 13.63 is consistent (for detail, see Hsiao and Tahmiscioglu 2008). Note also that even though when there are no time-specific effects, Equation 13.63 remains consistent, although it will not be as efficient as Equation 13.25.

Remark 13.8 The estimator (Equation 13.63) and the estimator Equation 13.58 remain consistent and asymptotically normally distributed when the effects are random because the transformation (Equation 13.54) effectively removes the individual- and time-specific effects from the specification. However, if the effects are indeed random,then the MLE or GLS of Equation 13.51 is more efficient.

Remark 13.9 The GLS (Equation 13.63) assumes known ω. If ω is unknown, one may substitute it by a consistent estimator $\hat{\omega}$, then apply the feasible GLS. However, there is an important difference between the GLS and the feasible GLS in a dynamic setting. The feasible GLS is not asymptotically equivalent to the GLS when T is finite. However, if both N and $T \to \infty$ and $\lim\left(\frac{N}{T}\right) = c > 0$, then the FGLS will be asymptotically equivalent to the GLS. (Hsiao and Tahmiscioglu 2008).

Remark 13.10 The MLE or GLS of Equation 13.63 can also be derived by treating $\Delta\lambda_t$ as fixed parameters in the system (Equation 13.47). Through continuous substitution, we have

$$\Delta y_{i1} = \lambda_1^* + \Delta\tilde{\epsilon}_{i1}, \tag{13.66}$$

where $\lambda_1^* = \sum_{j=0}^{m} \rho^j \Delta\lambda_{1-j}$ and $\Delta\tilde{\epsilon}_{i1} = \sum_{j=0}^{m} \rho^j \Delta\epsilon_{i,1-j}$. Let $\Delta\underset{\sim}{y}_i' = (\Delta y_{i1}, \ldots, \Delta y_{iT})$, $\Delta\underset{\sim}{y}_{i,-1}' = (0, \Delta y_{i1}, \ldots, \Delta y_{i,T-1})$, $\Delta\underset{\sim}{\epsilon}_i' = (\Delta\tilde{\epsilon}_{i1}, \ldots, \Delta\epsilon_{iT})$, and $\Delta\underset{\sim}{\lambda}' = (\lambda_1^*, \Delta\lambda_2, \ldots, \Delta\lambda_T)$, we may write

$$\begin{aligned} \underset{NT\times 1}{\Delta\underset{\sim}{y}} = \begin{pmatrix} \Delta\underset{\sim}{y}_1 \\ \vdots \\ \Delta\underset{\sim}{y}_N \end{pmatrix} &= \begin{pmatrix} \Delta\underset{\sim}{y}_{1,-1} \\ \vdots \\ \Delta\underset{\sim}{y}_{N,-1} \end{pmatrix} \rho + (\underset{\sim}{e}_N \otimes I_T)\Delta\underset{\sim}{\lambda} + \begin{pmatrix} \Delta\underset{\sim}{\epsilon}_1 \\ \vdots \\ \Delta\underset{\sim}{\epsilon}_N \end{pmatrix} \\ &= \Delta\underset{\sim}{y}_{-1}\rho + (\underset{\sim}{e}_N \otimes I_T)\Delta\underset{\sim}{\lambda} + \Delta\underset{\sim}{\epsilon}, \end{aligned} \tag{13.67}$$

If ϵ_{it} is i.i.d. normal with mean 0 and variance σ_ϵ^2, then $\Delta\underset{\sim}{\epsilon}_i'$ is independently normally distributed across i with mean $\underset{\sim}{0}$ and covariance matrix $\sigma_\epsilon^2\tilde{A}$, and $\omega = \frac{\text{Var}\,(\Delta\tilde{\epsilon}_{i1})}{\sigma_\epsilon^2}$.

The log-likelihood function of $\Delta \underset{\sim}{y}$ takes the form

$$\log L = -\frac{NT}{2}\log\sigma_\epsilon^2 - \frac{N}{2}\log|\tilde{A}| - \frac{1}{2\sigma_\epsilon^2}[\Delta\underset{\sim}{y} - \Delta\underset{\sim}{y}_{-1}\rho - (e_N \otimes I_T)\Delta\underset{\sim}{\lambda}]' \\ (I_N \otimes \tilde{A}^{-1})[\Delta\underset{\sim}{y} - \Delta\underset{\sim}{y}_{-1}\rho - (\underset{\sim}{e}_N \otimes I_T)\Delta\underset{\sim}{\lambda}]. \quad (13.68)$$

Taking partial derivative of Equation 13.68 with respect to $\Delta\underset{\sim}{\lambda}$ and solving for $\Delta\underset{\sim}{\lambda}$ yields

$$\Delta\hat{\underset{\sim}{\lambda}} = (N^{-1}\underset{\sim}{e}_N' \otimes I_T)(\Delta\underset{\sim}{y} - \Delta\underset{\sim}{y}_{-1}\rho). \quad (13.69)$$

Substituting Equation 13.69 into Equation 13.68 yields the concentrated log-likelihood function.

$$\log L_c = -\frac{NT}{2}\log\sigma_\epsilon^2 - \frac{N}{2}\log|\tilde{A}| \\ - \frac{1}{2\sigma_\epsilon^2}(\Delta\underset{\sim}{y}^* - \Delta\underset{\sim}{y}_{-1}^*\rho)'(I_N \otimes \tilde{A}^{-1})(\Delta\underset{\sim}{y}^* - \Delta\underset{\sim}{y}_{-1}^*\rho). \quad (13.70)$$

Maximizing Equation 13.69 conditional on ω yields Equation 13.63.

Remark 13.11 When ρ approaches to 1 and σ_α^2 is large relative to σ_ϵ^2, the GMM estimator of the form (Equation 13.68) suffers from the weak instrumental variables issues and performs poorly (e.g., Binder, Hsiao, and Pesaran 2005). On the other hand, the performance of the likelihood or GLS estimator (Equation 13.63) is not affected by these problems.

Remark 13.12 Hahn and Moon (2006) propose a bias corrected estimator as

$$\tilde{\rho}_b = \tilde{\rho}_{cv}^* + \frac{1}{T}(1 + \tilde{\rho}_{cv}^*). \quad (13.71)$$

They show that when $N/T \to c$, as both N and T tend to infinity where $0 < c < \infty$,

$$\sqrt{NT}(\tilde{\rho}_b - \rho) \Longrightarrow N(0, 1 - \rho^2). \quad (13.72)$$

The limited Monte Carlo studies conducted by Hsiao and Tahmiscioglu (2008) to investigate the finite sample properties of the feasible GLS (FGLS), GMM, bias corrected (BC) estimator of Hahn and Moon (2006) have shown that in terms of bias and root mean square errors, FGLS dominates. However, the BC rapidly improves as T increase. In terms of the closeness of actual size to the nominal size, again FGLS dominates and rapidly approaches the nominal size when N or T increases. The GMM also has actual sizes close to nominal sizes except for the cases when ρ is close to unity (here $\rho = 0.8$). The BC has significant size distortion, presumably because of the correction of bias being based on $\hat{\rho}_{cv}^*$ and the use of asymptotic covariance matrix which is significantly downward biased in finite sample.

Remark 13.13 Hsiao and Tahmiscioglu (2008) also compared the FGLS and GMM with and without the correction of time-specific effects in the presence of both individual- and time-specific effects or in the presence of individual-specific effects only. It is interesting to note that when both individual- and time-specific effects are present, the biases and root mean squares errors are large for estimators assuming no time-specific effects. On the other hand, even in the case of no time-specific effects in the true data generating process, there is hardly any efficiency loss for the FGLS or GMM that makes the correction of presumed presence of time-specific effects. Therefore, if an investigator is not sure if the assumption of cross-sectional independence is valid or not, it might be advisable to use estimators that take account both individual- and time-specific effects.

13.8 Estimation of Multiplicative Models

In this section we consider the estimation of Equation 13.1, where v_{it} is assumed to be of the form

$$v_{it} = \alpha_i \lambda_t + \epsilon_{it}. \tag{13.73}$$

When α_i is independently distributed across i with mean 0 and variance σ^2_α and λ_t is independently distributed over t with mean 0 and variance σ^2_λ, $Ev_{it} = 0$, $Ev^2_{it} = \sigma^2_\epsilon + \sigma^2_\alpha \sigma^2_\lambda = \sigma^2_v$, and $Ev_{it}v_{is} = 0$ for $t \neq s$, $Ev_{it}v_{js} = 0$ for $i \neq j$. In other words, Equation 13.1 has error terms that are uncorrelated over time and across individuals, with constant variance σ^2_v. Hence the least squares estimator is consistent and asymptotically normally distributed either N or T or both tend to infinity.

When α_i and λ_t are treated as fixed constants, the MLE are inconsistent if T is finite for the same basic reason as the additive model (Equation 13.2). Ahn, Lee, and Schmidt (2001), Bai (2007), Kiefer (1980), etc., have proposed a nonlinear GMM and iterative LS estimators for the static model with multiplicative effects. Their nonlinear GMM approach can be similarly generalized to obtain a consistent estimator of ρ (e.g., Hsiao 2008).

Let $\theta_t = \lambda_t / \lambda_{t-1}$, then

$$(y_{it} - \theta_t y_{i,t-1}) = \rho(y_{i,t-1} - \theta_t y_{i,t-2}) + (\epsilon_{it} - \theta_t \epsilon_{i,t-1}), \quad t = 2, \dots, T. \tag{13.74}$$

It follows that

$$E[y_{i,t-j}(\epsilon_{it} - \theta_t \epsilon_{i,t-1})] = 0, \quad \text{for } j = 2, \dots, t. \tag{13.75}$$

Let

$$\underset{\frac{T(T-1)}{2}\times(T-1)}{W_i} = \begin{bmatrix} \underset{\sim}{q}_{i2} & \underset{\sim}{0} & \cdot & \cdot & \underset{\sim}{0} \\ \underset{\sim}{0} & \underset{\sim}{q}_{i3} & \cdot & \cdot & \underset{\sim}{0} \\ \underset{\sim}{0} & \underset{\sim}{0} & \cdot & \cdot & \cdot \\ \cdot & \cdot & \cdot & \cdot & \cdot \\ \underset{\sim}{0} & \cdot & \cdot & \cdot & \underset{\sim}{q}_{iT} \end{bmatrix},$$

$$\underset{(T-1)\times(T-1)}{\Theta} = \begin{bmatrix} \theta_2 & 0 & \cdot & \cdot & 0 \\ 0 & \theta_3 & \cdot & \cdot & \cdot \\ \cdot & \cdot & \cdot & \cdot & \cdot \\ \cdot & \cdot & \cdot & \cdot & \theta_T \end{bmatrix},$$

$$\underset{\sim}{q}'_{it} = (y_{i0}, \ldots, y_{i,t-2}), \quad t = 2, \ldots, T,$$

$$\underset{\sim}{\epsilon}_i = (\epsilon_{i2}, \ldots, \epsilon_{iT})', \underset{\sim}{\epsilon}_{i,-1} = (\epsilon_{i1}, \ldots, \epsilon_{i,T-1})'.$$

Then a GMM estimator of ρ and Θ can be obtained from the moment conditions

$$E[W_i(\underset{\sim}{\epsilon}_i - \Theta\underset{\sim}{\epsilon}_{i,-1})] = \underset{\sim}{0}. \tag{13.76}$$

The nonlinear GMM estimators of ρ and Θ amount to applying nonlinear three-stage least squares to the system

$$\underset{\sim}{y}_i = [\rho I_{T-1} + \Theta]\underset{\sim}{y}_{i,-1} - \rho\Theta\underset{\sim}{y}_{i,-2} + \underset{\sim}{\epsilon}_i - \Theta\underset{\sim}{\epsilon}_{i,-1}, \; i = 1, \ldots, N, \tag{13.77}$$

using W_i as instruments, where $\underset{\sim}{y}_i = (y_{i2}, \ldots, y_{iT})'$, $\underset{\sim}{y}_{i,-1} = (y_{i1}, \ldots, y_{i,T-1})'$, and $\underset{\sim}{y}_{i,-2} = (y_{i0}, \ldots, y_{i,T-2})'$.

The nonlinear GMM estimators of ρ and θ_t are consistent and asymptotically normally distributed as $N \to \infty$. From the θ_t, we can solve for λ_t through the normalization rule $\lambda_1 = 1$ or $\sum_{t=1}^{T} \lambda_t^2 = 1$. From ρ and λ_t, we obtain

$$\hat{\alpha}_i = \frac{1}{\sum_{t=1}^{T} \hat{\lambda}_t^2}\left[\sum_{t=1}^{T} \hat{\lambda}_t y_{it} - \hat{\rho}\sum_{t=1}^{T} \hat{\lambda}_t y_{i,t-1}\right], i = 1, \ldots, N. \tag{13.78}$$

The estimator (Equation 13.78) is consistent if $T \to \infty$.

The implementation of nonlinear GMM is quite complicated, Pesaran (2006, 2007) notes that

$$\bar{y}_t = \rho\bar{y}_{t-1} + \bar{\alpha}\lambda_t + \bar{\epsilon}_t, \tag{13.79}$$

where

$$\bar{y}_t = \frac{1}{N}\sum_{i=1}^{N} y_{it}, \bar{\alpha} = \frac{1}{N}\sum_{i=1}^{N} \alpha_i, \bar{\epsilon}_t = \frac{1}{N}\sum_{i=1}^{N} \epsilon_{it}.$$

When $N \to \infty$, $\bar{\epsilon}_t \longrightarrow 0$. Assuming $\bar{\alpha} \neq 0$, substituting $\lambda_t = \bar{\alpha}^{-1}(\bar{y}_t - \rho\bar{y}_{t-1})$ into Equation 13.45 yields,

$$y_{it} = \rho y_{i,t-1} + \gamma_{1i}\bar{y}_t + \gamma_{2i}\bar{y}_{t-1} + \epsilon_{it} \tag{13.80}$$

Therefore, Pesaran (2006, 2007) suggests estimating the cross-sectional mean augment regression (Equation 13.80) and shows that as both N and $T \to \infty$, the least squares estimator of Equation 13.80 yields consistent and asymptotically normally distributed $\hat{\rho}$.

13.9 Test of Additive versus Multiplicative Model

Multiplicative model implies departure from additivity in their effects on outcomes. It is shown by Bai (2007) that the additive model is embedded into the model of multiple common factors with heterogeneous response by letting

$$\underset{\sim}{\alpha}_i = \begin{bmatrix} \alpha_i \\ 1 \end{bmatrix}, \underset{\sim}{\lambda}_t = \begin{bmatrix} 1 \\ \lambda_t \end{bmatrix},$$

then Equation 13.2 becomes

$$v_{it} = \underset{\sim}{\alpha}_i'\underset{\sim}{\lambda}_t + \epsilon_{it}. \tag{13.81}$$

When $N \longrightarrow \infty$, one may solve $\underset{\sim}{\lambda}_t$ from Equation 13.79 that yields

$$\hat{\underset{\sim}{\lambda}}_t = (\bar{\underset{\sim}{\alpha}}\bar{\underset{\sim}{\alpha}}')^{-}\bar{\underset{\sim}{\alpha}}(v_t - \rho\bar{y}_{t-1}), \tag{13.82}$$

where $(\bar{\underset{\sim}{\alpha}}\bar{\underset{\sim}{\alpha}}')^{-}$ denotes the generalized inverse of $(\bar{\underset{\sim}{\alpha}}\bar{\underset{\sim}{\alpha}}')$. Substituting Equation 13.82 into Equation 13.45 again yields Equation 13.80. Therefore, the Pesaran cross-sectional mean augmented regression of Equation 13.80 is consistent whether the unobserved heterogeneity is additive or multiplicative, but Equation 13.80 is inefficient if the unobserved heterogeneities are additive compared to Equation 13.58 or Equation 13.63. However, if the underlying model is multiplicative, Equation 13.80 is consistent, but not Equation 13.58 or Equation 13.63. Therefore, a Hausman type specification test can be proposed to test the null:

H_0: Equation 13.2 holds

versus

H_1: Equation 13.2 does not hold

by considering the test statistic

$$\frac{\hat{\rho}_A - \hat{\rho}_m}{\sqrt{\text{Var}\,(\hat{\rho}_m) - \text{Var}(\hat{\rho}_A)}} \sim N(0, 1), \tag{13.83}$$

where $\hat{\rho}_A$ denotes the efficient estimator of Equation 13.1 under the additive assumption (Equation 13.2) and $\hat{\rho}_m$ is the estimator (Equation 13.1) under the multiplicative assumption (Equation 13.73).

13.10 Concluding Remarks

In this chapter we review three fundamental issues of modeling dynamic panel data in the presence of unobserved heterogeneity across individuals and over time—the fixed effects of modeling unobserved individual- and time-specific heterogeneity versus random effects; additive versus multiplicative effects and the likelihood versus methods of moments approach.

We have not discussed issues of modeling multivariate dynamic panel models (e.g., Binder, Hsiao, and Pesaran (2005), panel unit root tests (e.g., Breitung and Pesaran 2008; Moon and Perron 2004; Phillips and Sul 2003); parameter heterogeneity (e.g., Hsiao and Pesaran 2008), etc. However, in principle, those issues can also be put in these perspectives.

The advantage of the fixed effects specification is that there is no need to specify the relations between the unobserved effects and observed conditional (or explanatory) variables. The disadvantages are that (1) unless both cross-sectional dimension and time dimension of panels are large, the fixed effects specification introduces incidental parameters issues on the individual-specific effects, α_i, if the time dimension is fixed and on the time-specific effects, λ_t if the cross-sectional dimension is small; (2) the impact of time-invariant but individual-specific variables such as gender or socio-demographic background variables with the presence of additive individual-specific effects and the impact of time-specific but individual invariant such as price and some macro-variables with the presence of additive time-specific effects are unidentified; and (3) the fixed effects inference only makes use of within-group variation. The between group information is ignored.

The advantages of random-effects specification are (1) there are no incidental parameter issues; (2) the impacts of observed individual-specific but time-invariant and individual-invariant but time-varying variables can be identified; (3) both the within-group and between group information are used for inference. Since the between group variation in general is much larger than the within group variation, the RE specification can lead to much more efficient use of sample information. The disadvantage is that the relationship between the unobserved effects and observed conditional variables need to be specified. In short, the advantages of random effects specification are the disadvantage of fixed effects specification and the advantages of fixed effects specification are the disadvantages of random effects specification.

Statistical inference procedures for additive effects models are simpler than the multiplicative effects models. However, if the data generating process calls for a multiplicative effects specification, statistical inference procedures based on additive effects specification will be misleading. On the other hand, if the effects are additive, statistical procedures based on multiplicative effects will also be misleading. In this chapter, we have proposed a testing procedure for additive versus multiplicative effects.

Inference procedures based on the likelihood and moments approaches are reviewed. The likelihood approach uses a fixed number of moment conditions. The moment conditions used in the moments approach increase at the order of square of time series dimension of the panel. In finite sample the moments approach is likely to generate larger bias than the likelihood approach as shown in the Monte Carlo by Binder, Hsiao and Pesaran (2005), Hsiao and Tahmiscioglu (2008), Hsiao, Pesaran, and Tahmiscioglu (2002), Ziliak (1997), etc. Moreover, if the observed outcomes in the time dimension is persistent (when the coefficient of lagged variables, ρ, is close to one) or if the variance of individual-specific effects is large relative to overall variance, the moments approach either breaks down or suffers from the weak instrumental variables issue, but the performance of the likelihood approach is not affected.

13.11 Acknowledgment

I would like to thank a referee for helpful comments.

References

Ahn, S. C., and P. Schmidt. 1995. Efficient Estimation of Models for Dynamic Panel Data. *Journal of Econometrics* 68:5–27.

Ahn, S. G., Y. H. Lee, and P. Schmidt. 2001. GMM Estimation of Linear Panel Data Models with Time-Varying Individual Effects. *Journal of Econometrics* 101:219–255.

Amemiya, T., and W. A. Fuller. 1967. A Comparative Study of Alternative Estimators in a Distributed-Lag Model. *Econometrica* 35:509–529.

Amemiya, T., and T. E. MaCurdy. 1986. Instrumental-Variable Estimation of an Error-Components Model. *Econometrica* 54, 869–880.

Anderson, T. W. and C. Hsiao. 1981. Estimation of Dynamic Models with Error Components. *Journal of American Statistical Association* 76:598–606.

Anderson, T. W., and C. Hsiao. 1982. Formulation and Estimation of Dynamic Models Using Panel Data. *Journal of Econometrics* 18:47–82.

Arellano, M., and S. R. Bond. 1991. Some Tests of Specification for Panel Data: Monte Carlo Evidence and an Application to Employment Equations. *Review of Economic Studies* 58:277–297.

Arellano, M., and O. Bover. 1995. Another Look at the Instrumental Variable Estimation of Error-Components Models. *Journal of Econometrics* 68:29–51.

Bai, J. 2009. Panel Data Models with Interactive Fixed Effects. *Econometrica*, 77, 1229–1279.

Bhargava, A., and D. Sargan. 1983. Estimating Dynamic Random Effects Models from Panel Data Covering Short Time Periods. *Econometrica* 51:1635–1659.
Binder, M., C. Hsiao, and M. H. Pesaran. 2005. Estimation and Inference in Short Panel Vector Autoregressions with Unit Roots and Cointegration. *Econometric Theory* 21:795–837.
Blundell, R., and S. Bond. 1998. Initial Conditions and Moment Restrictions in Dynamic Panel Data Models. *Journal of Econometrics* 87:115–143.
Breitung, J., and M. H. Pesaran. 2008. Unit Roots and Cointegration in Panels. In *The Econometrics of Panel Data*. Berlin: Springer. pp. 279–322.
Breusch, T., G. E. Mizon, and P. Schmidt. 1989. Efficient Estimation Using Panel Data. *Econometrica* 57:695–700.
Cheng, L. K., and Y. K. Kwan. 2000. The Location of Foreign Direct Investment in Chinese Regions – Further Analysis of Labor Quality. In *The Role of Foreign Direct Investment in East Asian Economic Development*. T. Ito and A.O. Krueger (eds). Chicago: Chicago University Press. pp. 213–238.
Hahn, J., and G. Kuersteiner. 2002. Asymptotically Unbiased Inference for a Dynamic Panel Model with Fixed Effects When Both n and T are Large. *Econometrica* 70:1639–1659.
Hahn, J., and H. R. Moon. 2006. Reducing Bias of MLE in a Dynamic Panel Model. *Econometric Theory* 22:499–512.
Harris, M. N., L. Matyas, and P. Sevestre. 2008. Dynamic Models for Short Panels. In *The Econometrics of Panel Data*. 3rd ed. L. Matyas and P. Sevestre (eds). Berlin: Springer. pp. 249–278.
Hayakawa, K. 2009. On the Effect of Mean-Nonstationary Initial Conditions in Dynamic Panel Data Models. *Journal of Econometrics* (forthcoming).
Hsiao, C. 2003. *Analysis of Panel Data*. 2nd ed. Econometric Society Monograph 36, New York: Cambridge University Press.
Hsiao, C. 2007. Panel Data Analysis – Advantages and Challenges. *Test* 16:1–22.
Hsiao, C. 2008. Dynamic Panel Data Models with Interactive Effects. *mimeo.*
Hsiao, C., and M. H. Pesaran. 2008. Random Coefficients Models. *The Econometrics of Panel Data*, 3rd ed. L. Matayas and P. Sevestre (eds). Berlin: Springer. pp. 187–216.
Hsiao, C., M. H. Pesaran, and A. K. Tahmiscioglu. 2002. Maximum Likelihood Estimation of Fixed Effects Dynamic Panel Data Models Covering Short Time Periods. *Journal of Econometrics* 109:107–150.
Hsiao, C., and A. K. Tahmiscioglu. 2008. Estimation of Dynamic Panel Data Models with Both Individual- and Time-Specific Effects. *Statistical Planning and Statistics Inference* 138:2698–2721.
Kiefer, N. 1980. Estimation of Fixed Effect Models for Time Series of Cross-Sections with Arbitrary Intertemporal Covariance. *Journal of Econometrics* 14:195–202.
Moon, H. R. and B. Perron. 2004. Testing for a Unit Root in Panels with Dynamic Factors. *Journal of Econometrics* 122:81–126.
Nerlove, M. 2002. *Essays in Panel Data Econometrics*. Cambridge: Cambridge University Press.
Neyman, J., and E. Scott. 1948. Consistent Estimates Based on Partially Consistent Observations. *Econometrica* 16:1–32.
Pesaran, M. H. 2006. Estimation and Inference in Large Heterogeneous Panels with Cross-Section Dependence. *Econometrica* 74:967–1012.
Pesaran, M. H. 2007. A Simple Panel Unit Root Test in the Presence of Cross-Section Dependence. *Journal of Applied Econometrics* 22:265–312.

Phillips, P.C.B., and D. Sul. 2003. Dynamic Panel Estimation and Homogeneity Testing Under Cross-Section Dependence. *Econometrics Journal* 6: 217–259.

Ziliak, J. P. 1997. Efficient Estimation with Panel Data When Instruments Are Predetermined: An Empirical Comparison of Moment-Condition Estimators, *Journal of Business and Economic Statistics* 15:419–431.

14

A Unified Estimation Approach for Spatial Dynamic Panel Data Models: Stability, Spatial Co-integration, and Explosive Roots

Lung-fei Lee and Jihai Yu

CONTENTS

14.1 Introduction

In recent decades, there is growing literature on the estimation of dynamic panel data models (see Phillips and Moon 1999; Hahn and Kuersteiner 2002; Alvarez and Arellano 2003; Hahn and Newey 2004, etc.). For the panel data with spatial interactions, Kapoor, Kelejian, and Prucha (2007) extend the asymptotic analysis of the method of moments estimators to a spatial panel model with error components, where T is finite. Baltagi, Song, Jung, and Koh (2007) consider the testing of spatial and serial dependence in an extended model, where serial correlation on each spatial unit over time and spatial dependence across spatial units are allowed in the disturbances. Su and Yang (2007) study the dynamic panel data with spatial error and random effects. These panel models specify the spatial correlation by including spatially correlated disturbances but do not incorporate a spatial autoregressive term in the regression equation. With large n and moderate or large T, Korniotis (2005) studies a time-space recursive model where only an individual time lag and a spatial time lag are present but not a contemporaneous spatial lag. A general model could be the spatial dynamic panel data (SDPD) where a contemporaneous spatial lag is also included. Yu, de Jong, and Lee (2007, 2008) and Yu and Lee (2010) study, respectively, the spatial cointegration, stable, and unit root SDPD models, where the individual time lag, spatial time lag and contemporaneous spatial lag are all included.

When the SDPD model has time dummy effects, we might need to transform the data to reduce the possible bias caused by the estimation of time effects (see Lee and Yu, 2010a), especially, when n is proportional to T, or n is small relative to T. Yu, de Jong, and Lee (2007) have a different bias correction procedures from that of the stable case in Yu, de Jong, and Lee (2008). In this chapter, we propose a data transformation approach based on a spatial difference operator, which can eliminate the time dummy effects as well as possible unstable and/or explosive components. After the data transformation, we can estimate the model by the method of maximum likelihood (ML) or quasi-maximum likelihood (QML) similar to Yu, de Jong, and Lee (2008), where there are neither time dummy effects, nor unstable and explosive components. We derive the asymptotics for the ML estimator (MLE) and QML estimator (QMLE). We propose a bias correction procedure that can be applied to different types of DGPs.

This chapter is organized as follows. In Section 14.2, the model is presented. We show that the stochastic process can be decomposed into stable, unstable or explosive, and time components. A spatial difference operator motivated by the spatial co-integration can provide a unified data transformation to eliminate the time component and the possible unstable or explosive components. We explain our method of estimation, which is a concentrated QML. Section 14.3 establishes the consistency and asymptotic distribution of the QMLE of the unified transformation approach. A bias correction procedure is also proposed. A Monte Carlo study is conducted in Section 14.4 to investigate finite sample performance of the estimators under different DGPs, and also the power of hypothesis testing of spatial co-integration using this unified approach. Section 14.5 concludes the chapter. Some useful lemmas and proofs are collected in the appendices.

14.2 The Model

14.2.1 The DGP

Consider the general SDPD model:

$$Y_{nt} = \lambda_0 W_n Y_{nt} + \gamma_0 Y_{n,t-1} + \rho_0 W_n Y_{n,t-1} + X_{nt}\beta_0 + \mathbf{c}_{n0} + \alpha_{t0} l_n + V_{nt}, \quad t = 1, 2, \ldots, T, \tag{14.1}$$

where $Y_{nt} = (y_{1t}, y_{2t}, ..., y_{nt})'$ and $V_{nt} = (v_{1t}, v_{2t}, ..., v_{nt})'$ are $n \times 1$ column vectors, and v_{it} is i.i.d. across i and t with zero mean and variance σ_0^2. W_n is an $n \times n$ nonstochastic spatial weights matrix, X_{nt} is an $n \times k$ matrix of nonstochastic regressors, $\mathbf{c}_{n0}$ is an $n \times 1$ column vector of individual fixed effects, α_{t0} is a scalar of time effect, and l_n is an $n \times 1$ column vector of ones.[1] Therefore, the total number of parameters in this model is equal to the sum of the number of individuals n and the number of time periods T, plus the dimension of the common parameters $(\gamma, \rho, \beta', \lambda, \sigma^2)'$ which is $k+4$. In practice, W_n is usually row-normalized with zero diagonals. A row-normalized W_n has the property $W_n l_n = l_n$. The row-normalization of W_n ensures that all the weights are between 0 and 1 and weighting operations can be interpreted as an average of the neighboring values. In this chapter, the row-normalization feature is imposed for our estimation approach.

Define $S_n(\lambda) = I_n - \lambda W_n$ and $S_n \equiv S_n(\lambda_0) = I_n - \lambda_0 W_n$. Then, presuming that S_n is invertible and denoting $A_n = S_n^{-1}(\gamma_0 I_n + \rho_0 W_n)$, Equation 14.1 can be rewritten as

$$Y_{nt} = A_n Y_{n,t-1} + S_n^{-1} X_{nt}\beta_0 + S_n^{-1}\mathbf{c}_{n0} + \alpha_{t0} S_n^{-1} l_n + S_n^{-1} V_{nt}. \tag{14.2}$$

In the SDPD model, when all the eigenvalues of A_n are smaller than 1, we have the stable case. When some eigenvalues of A_n are equal to 1 but not all being 1, we have the spatial co-integration case. When some of them are greater than 1, we have the explosive case. Let $\varpi_n = \text{diag}\{\varpi_{n1}, \varpi_{n2}, ..., \varpi_{nn}\}$ be the $n \times n$ diagonal eigenvalues matrix of W_n such that $W_n = R_n \varpi_n R_n^{-1}$, where R_n is the corresponding eigenvector matrix. As $A_n = S_n^{-1}(\gamma_0 I_n + \rho_0 W_n)$, the eigenvalues matrix of A_n is $D_n = (I_n - \lambda_0 \varpi_n)^{-1}(\gamma_0 I_n + \rho_0 \varpi_n)$ such that $A_n = R_n D_n R_n^{-1}$. When W_n is row-normalized, all the eigenvalues are less than or equal to 1 in the absolute value, where it has definitely some eigenvalues being 1. Let m_n be the number of unit eigenvalues of W_n and let the first m_n eigenvalues of W_n be the unity. Hence, D_n can be decomposed into two parts, one corresponding to the unit eigenvalues of W_n, and the other corresponding to the eigenvalues of W_n which are smaller than 1. Define $J_n = \text{diag}\{\mathbf{1}'_{m_n}, 0, \cdots, 0\}$ with $\mathbf{1}_{m_n}$ being an $m_n \times 1$ vector of ones and $\tilde{D}_n = \text{diag}\{0, \cdots, 0, d_{n,m_n+1}, \cdots, d_{nn}\}$, where $|d_{ni}| < 1$, for $i = m_n + 1, \cdots, n$, are assumed.[2] As $J_n \cdot \tilde{D}_n = \mathbf{0}$, we have $A_n^h = (\frac{\gamma_0+\rho_0}{1-\lambda_0})^h R_n J_n R_n^{-1} + B_n^h$ where $B_n^h = R_n \tilde{D}_n^h R_n^{-1}$ for any $h = 1, 2, \cdots$. Hence, depending on the value of $\frac{\gamma_0+\rho_0}{1-\lambda_0}$, we have three cases. As $|\lambda_0| < 1$, which will be maintained under the Assumption 1 and 3 (see Section 14.3), we have the stable case when $\gamma_0 + \rho_0 + \lambda_0 < 1$; the spatial co-integration case when $\gamma_0 + \rho_0 + \lambda_0 = 1$ but $\gamma_0 \neq 1$; and the explosive case when $\gamma_0 + \rho_0 + \lambda_0 > 1$.

For the stable case, the rates of convergence of QMLEs are $\sqrt{nT}$, as shown in Yu, de Jong, and Lee (2008). For the spatial co-integration case where Y_{nt} and

---

[1] Due to the presence of fixed individual and time effects, the X_{nt} will not include time invariant or individual invariant regressors.

[2] We note that $d_{ni} = (\gamma_0 + \rho_0 \varpi_{ni})/(1 - \lambda_0 \varpi_{ni})$. Hence, if $\gamma_0 + \lambda_0 + \rho_0 < 1$, we have $d_{ni} < 1$ as $|\varpi_{ni}| \leq 1$. Some additional conditions are needed to ensure that $d_{ni} > -1$. See Appendix A.1.

$W_n Y_{nt}$ are spatially co-integrated, Yu, de Jong, and Lee (2007) show that the QMLEs for such a model are $\sqrt{nT}$ consistent and asymptotically normal, but, the presence of the unstable components will make the estimators' asymptotic variance matrix singular. Consequently, a linear combination of the spatial and dynamic effects estimates can converge at a higher rate.[3] In addition to the above stable case and the spatial co-integration case, we may also have an explosive case in the event that some eigenvalues of A_n are greater than unity in the absolute value.[4] In this chapter, we propose a unified transformation approach that can be used to estimate all three cases, namely, stable, spatial co-integrated, and explosive cases.

In earlier studies of the SDPD model, Yu, de Jong, and Lee (2007, 2008) consider the QMLE of the model with only the individual fixed effects. Subsequently, Lee and Yu (2010a) study the SDPD model with additional time effect when the process is stable. They propose a data transformation based on the deviation from cross-sectional mean, $I_n - \frac{1}{n} l_n l_n'$, to eliminate the time effects. That approach may be applied to study the unstable SDPD models with time effects but might not be able to eliminate unstable or explosive components. In this chapter, we report the use of a spatial difference operator, $I_n - W_n$, which may not only eliminate the time dummy effects, but also the possible unstable or explosive components, generated from the spatial co-integration or explosive roots. This implies that the spatial difference transformation can be applied to DGPs with stability, spatial co-integration, or explosive roots. The asymptotics of the resulting estimates can then be easily established for these DGPs. Thus, the transformation $I_n - W_n$ provides a unified estimation procedure for SDPD models.[5]

Denote $W_n^u = R_n J_n R_n^{-1}$. Then, for $t \geq 0$, Y_{nt} can be decomposed into a sum of a possible stable part, a possible unstable or explosive part, and a time effect part (see Appendix A.2 for proof)

$$Y_{nt} = Y_{nt}^u + Y_{nt}^s + Y_{nt}^\alpha, \tag{14.3}$$

[3] When $\gamma_0 + \lambda_0 + \rho_0 = 1$ and $\gamma_0 = 1$, the asymptotic properties of estimators are considered in Yu and Lee (2010). The QML estimate of the dynamic coefficient is $\sqrt{nT^3}$ consistent and the estimates of other parameters are $\sqrt{nT}$ consistent, and they are all asymptotically normal. Also, the sum of the contemporaneous and dynamic spatial effects will converge at $\sqrt{nT^3}$ rate.

[4] For the autoregressive AR(1) process in time series, asymptotic properties of the ordinary least square estimator have been investigated in White (1958, 1959), Anderson (1959), Nielsen (2001, 2005) and Phillips and Magdalinos (2007). For the SDPD due to its complexity, properties of a possible QMLE have not been investigated.

[5] We note that the spatial difference operator can be applied to cross-sectional units. However, its function is different from the time difference operator for a time series. The spatial difference operator does not eliminate pure time series unit root or explosive roots. Thus, the unified approach cannot be applied to the pure unit root SDPD models in Yu and Lee (2010).

where

$$Y_{nt}^s = \sum_{h=0}^{\infty} B_n^h S_n^{-1}(\mathbf{c}_{n0} + X_{n,t-h}\beta_0 + V_{n,t-h}),$$

$$Y_{nt}^u = W_n^u \left\{ \left(\frac{\gamma_0 + \rho_0}{1 - \lambda_0} \right)^{t+1} Y_{n,-1} + \frac{1}{(1-\lambda_0)} \left[\sum_{h=0}^{t} \left(\frac{\gamma_0 + \rho_0}{1 - \lambda_0} \right)^h (\mathbf{c}_{n0} + X_{n,t-h}\beta_0 + V_{n,t-h}) \right] \right\},$$

$$Y_{nt}^\alpha = \frac{1}{(1-\lambda_0)} l_n \sum_{h=0}^{t} \alpha_{t-h,0} (\frac{\gamma_0 + \rho_0}{1 - \lambda_0})^h.$$

The Y_{nt}^u can be an unstable component when $\frac{\gamma_0+\rho_0}{1-\lambda_0} = 1$, which occurs when $\gamma_0 + \rho_0 + \lambda_0 = 1$ and $\lambda_0 \neq 1$. When $\gamma_0 + \rho_0 + \lambda_0 > 1$, it implies $\frac{\gamma_0+\rho_0}{1-\lambda_0} > 1$ and, hence, Y_{nt}^u can be explosive. The Y_{nt}^α can be rather complicated as it depends on what exactly the time dummies represent. The Y_{nt} can be explosive when α_{t0} represents some explosive functions of t, even when $\frac{\gamma_0+\rho_0}{1-\lambda_0}$ were smaller than 1. Without a specific time structure for α_{t0}, it is desirable to eliminate this component for the estimation. The Y_{nt}^s can be a stable component unless $\gamma_0 + \rho_0 + \lambda_0$ is much larger than 1. If the sum $\gamma_0 + \rho_0 + \lambda_0$ were too big, some of the eigenvalues d_{ni} in Y_{nt}^s might become larger than 1.

14.2.2 Data Transformation

Both the deviation from the cross-sectional mean $I_n - \frac{1}{n} l_n l_n'$ and the spatial difference operator $I_n - W_n$ can eliminate the Y_{nt}^α component in Y_{nt}. The transformation $I_n - W_n$ can be motivated via a feature of spatial co-integration below. Because $(I_n - W_n)l_n = 0$, $(I_n - W_n)Y_{nt}^\alpha = 0$. The $(I_n - W_n)Y_{nt}$ does not involve time dummies. In addition, because $W_n^u = R_n J_n R_n^{-1}$, it follows that $(I_n - W_n)W_n^u = R_n(I_n - D_n)J_n R_n^{-1} = 0$, and $(I_n - W_n)Y_{nt}^u = 0$. Therefore, $(I_n - W_n)Y_{nt} = (I_n - W_n)Y_{nt}^s$. That is, the transformation $I_n - W_n$ can eliminate not only time dummies but also the unstable component. Therefore, after the $(I_n - W_n)$ transformation, we will end up with the following equation:

$$(I_n - W_n)Y_{nt} = \lambda_0 W_n(I_n - W_n)Y_{nt} + \gamma_0(I_n - W_n)Y_{n,t-1} + \rho_0 W_n(I_n - W_n)Y_{n,t-1} + (I_n - W_n)X_{nt}\beta_0 + (I_n - W_n)\mathbf{c}_{n0} + (I_n - W_n)V_{nt}. \tag{14.4}$$

The variance of $(I_n - W_n)V_{nt}$ is $\sigma_0^2\Sigma_n$, where $\Sigma_n = (I_n - W_n)(I_n - W_n)'$. This transformed equation has less degrees of freedom than n. Denote the degree of freedom of Equation 14.4 as n^*. Then, n^* is the rank of the variance matrix of $(I_n - W_n)V_{nt}$, which is the number of nonzero eigenvalues of Σ_n. Hence,

$n^* = n - m_n$ is also the number of non-unit eigenvalues[6] of W_n. Thus, the transformed variables do not have time effects and are all stable even when $\lambda_0 + \gamma_0 + \rho_0$ is equal to or greater than 1.

Let $[F_n, H_n]$ be the orthonormal matrix of eigenvectors and Λ_n be the diagonal matrix of nonzero eigenvalues of Σ_n such that $\Sigma_n F_n = F_n \Lambda_n$ and $\Sigma_n H_n = 0$. That is, the columns of F_n consist of eigenvectors of nonzero eigenvalues and those of H_n are for zero-eigenvalues of Σ_n. The F_n is an $n \times n^*$ matrix and Λ_n is an $n^* \times n^*$ diagonal matrix. Denote $W_n^* = \Lambda_n^{-1/2} F_n' W_n F_n \Lambda_n^{1/2}$ which is an $n^* \times n^*$ matrix. As is derived in Appendix A.3, we have

$$Y_{nt}^* = \lambda_0 W_n^* Y_{nt}^* + \gamma_0 Y_{n,t-1}^* + \rho_0 W_n^* Y_{n,t-1}^* + X_{nt}^* \beta_0 + \mathbf{c}_{n0}^* + V_{nt}^*, \tag{14.5}$$

where $Y_{nt}^* = \Lambda_n^{-1/2} F_n' (I_n - W_n) Y_{nt}$ and other variables are defined accordingly. Note that this transformed Y_{nt}^* is an n^* dimensional vector. Thus, at each t, after the removal of the time dummy variables as well as the unstable or explosive components in Y_{nt}, the remaining observations at period t have only n^* degrees of freedom. While the sum of the coefficients $\lambda_0 + \gamma_0 + \rho_0$ of this transformed equation can be equal to or greater than 1, the eigenvalues of W_n^* are exactly those eigenvalues of W_n not equal to the unity (see Appendix A.4) but less than 1 in the absolute value. It follows that the eigenvalues of $A_n^* = (I_{n^*} - \lambda_0 W_n^*)^{-1}(\gamma_0 I_{n^*} + \rho_0 W_n^*)$ are all less than 1 in the absolute values even when $\lambda_0 + \gamma_0 + \rho_0 = 1$ with $|\lambda_0| < 1$ and $|\gamma_0| < 1$. For the explosive case with $\lambda_0 + \gamma_0 + \rho_0 > 1$, the eigenvalue of A_n^* can be less than 1 only if $\frac{\rho_0 + \lambda_0}{1 - \gamma_0} < \frac{1}{\varpi_{\max 1}}$, where $\varpi_{\max 1}$ is the maximum positive eigenvalue of W_n less than the unity (see Appendix A.1). Hence, the transformed model (Equation 14.5) is a stable one as long as $\lambda_0 + \gamma_0 + \rho_0$ is not much bigger than 1.[7]

The transformation $I_n - W_n$ for the case with $\gamma_0 + \rho_0 + \lambda_0 = 1$ but $\gamma_0 \neq 1$ has an interpretation as a spatial co-integrating matrix for elements of Y_{nt}. Denote time difference as $\Delta Y_{nt} = Y_{nt} - Y_{n,t-1}$. The reduced form Equation 14.2) implies that $\Delta Y_{nt} = (A_n - I_n) Y_{n,t-1} + S_n^{-1}(X_{nt}\beta_0 + \mathbf{c}_{n0} + V_{nt} + \alpha_{t0} l_n)$. For the case $\lambda_0 + \gamma_0 + \rho_0 = 1$ with $\gamma_0 \neq 1$, $A_n - I_n = (I_n - \lambda_0 W_n)^{-1}(\gamma_0 I_n + \rho_0 W_n) - I_n = (1 - \gamma_0)(I_n - \lambda_0 W_n)^{-1}(W_n - I_n)$. Hence, we have a vector error correction model (VECM) representation of Equation 14.2 as

$$\Delta Y_{nt} = (1 - \gamma_0)(I_n - \lambda_0 W_n)^{-1}(W_n - I_n) Y_{n,t-1} + S_n^{-1}(X_{nt}\beta_0 + \mathbf{c}_{n0} + V_{nt} + \alpha_{t0} l_n).$$

The matrix $I_n - W_n = R_n(I_n - \varpi_n) R_n^{-1}$ has its rank equal to the number of eigenvalues of W_n different from 1. With the VECM representation, one may

[6] This is so, because (1) the set K_n of eigenvectors corresponding to the zero eigenvalues of $(I_n - W_n)(I_n - W_n)'$ is the same as that of $(I_n - W_n)'$; (2) the dimension of K_n is the number of unit eigenvalues of W_n'; (3) $W_n = R_n \varpi_n R_n^{-1}$ if and only if $W_n' = R_n^{-1\prime} \varpi_n R_n'$, i.e., the eigenvalues of W_n and W_n' are the same.

[7] Similar to Yu, de Jong, and Lee (2007) for the spatial co-integration case, we assume that the eigenvalues of W_n with their absolute values less than 1 are bounded away from 1 for all n. Appendix A.1 provides sufficient conditions on the parameters of the model, which can imply this regularity condition.

regard $I_n - W_n$ as a co-integrating matrix with the co-integration rank as the number of non-unit eigenvalues of W_n. Hence, this transformation method has exploited the spatial co-integration of Y_{nt}'s for the estimation.

14.2.3 The Log-Likelihood Function

Suppose that V_{nt} is normally distributed as $N(0, \sigma_0^2 I_n)$, the transformed V_{nt}^* in Equation 14.5 will be $N(0, \sigma_0^2 I_{n^*})$. Denote $\delta = (\gamma, \rho, \beta')'$, $\theta = (\delta', \lambda)'$ and $S_n^*(\lambda) = I_{n^*} - \lambda W_n^*$. The log-likelihood function for Y_{nt}^* in Equation 14.5 is

$$\ln L_{n,T}(\theta, \mathbf{c}_n^*) = -\frac{n^*T}{2}\ln 2\pi - \frac{n^*T}{2}\ln\sigma^2 + T\ln|S_n^*(\lambda)| \\ -\frac{1}{2\sigma^2}\sum_{t=1}^{T} V_{nt}^{*\prime}(\theta, \mathbf{c}_n^*)V_{nt}^*(\theta, \mathbf{c}_n^*), \tag{14.6}$$

where $V_{nt}^*(\theta, \mathbf{c}_n^*) = S_n^*(\lambda)Y_{nt}^* - Z_{nt}^*\delta - \mathbf{c}_n^*$, $Z_{nt}^* = (Y_{n,t-1}^*, W_n^*Y_{n,t-1}^*, X_{nt}^*)$. In order to use Equation 14.6 for an effective estimation, the determinant and inverse of $S_n^*(\lambda)$ are needed. As is derived in Appendix A.4, using $S_n^*(\lambda) = \Lambda_n^{-1/2}F_n'S_n(\lambda)F_n\Lambda_n^{1/2}$, we have

$$|S_n^*(\lambda)| = \frac{1}{(1-\lambda)^{n-n^*}}|S_n(\lambda)|, \text{ and } S_n^{*-1}(\lambda) = \Lambda_n^{-1/2}F_n'S_n^{-1}(\lambda)F_n\Lambda_n^{1/2}. \tag{14.7}$$

Hence, the computation of the determinant of $S_n^*(\lambda)$ is not more complicated than $S_n(\lambda)$. Also,

$$\begin{aligned} V_{nt}^*(\theta, \mathbf{c}_n^*) &= S_n^*(\lambda)Y_{nt}^* - Z_{nt}^*\delta - \mathbf{c}_n^* \\ &= \Lambda_n^{-1/2}F_n'S_n(\lambda)F_nF_n'(I_n - W_n)Y_{nt} - \Lambda_n^{-1/2}F_n'(I_n - W_n)Z_{nt}\delta \\ &\quad - \Lambda_n^{-1/2}F_n'(I_n - W_n)\mathbf{c}_n \\ &= \Lambda_n^{-1/2}F_n'(I_n - W_n)[S_n(\lambda)Y_{nt} - Z_{nt}\delta - \mathbf{c}_n] \\ &= \Lambda_n^{-1/2}F_n'(I_n - W_n)V_{nt}(\theta, \mathbf{c}_n), \end{aligned}$$

by using $F_nF_n' + H_nH_n' = I_n$ and $H_n'(I_n - W_n) = \mathbf{0}$, where $Z_{nt} = (Y_{n,t-1}, W_nY_{n,t-1}, X_{nt})$ and $V_{nt}(\theta, \mathbf{c}_n) = S_n(\lambda)Y_{nt} - Z_{nt}\delta - \mathbf{c}_n$. Hence,

$$V_{nt}^{*\prime}(\theta, \mathbf{c}_n^*)V_{nt}^*(\theta, \mathbf{c}_n^*) = V_{nt}'(\theta, \mathbf{c}_n)(I_n - W_n)'\Sigma_n^+(I_n - W_n)V_{nt}(\theta, \mathbf{c}_n), \tag{14.8}$$

where $\Sigma_n^+ = F_n\Lambda_n^{-1}F_n'$ is the generalized inverse of $\Sigma_n = (I_n - W_n)(I_n - W_n)'$. By using Equation 14.7 and 14.8, the log-likelihood function (Equation 14.6) for Y_{nt}^* can be expressed in terms of Y_{nt} as

$$\ln L_{n,T}(\theta, \mathbf{c}_n) = -\frac{n^*T}{2}\ln(2\pi\sigma^2) - (n-n^*)T\ln(1-\lambda) + T\ln|S_n(\lambda)| - \frac{1}{2\sigma^2}\sum_{t=1}^{T}(S_n(\lambda)Y_{nt} - Z_{nt}\delta - \mathbf{c}_n)'(I_n - W_n)'\Sigma_n^+(I_n - W_n) \times (S_n(\lambda)Y_{nt} - Z_{nt}\delta - \mathbf{c}_n). \quad (14.9)$$

Hence, after the transformation, the QML method is to estimate the SDPD model with only individual effects with n^* cross-section units and T time periods, where Equation 14.6 is the objective function. Alternatively, one may maximize Equation 14.9 expressed in terms of the original variables. However, although the components of V_{nt} are i.i.d. in the model, the elements of V_{nt}^* might not be independent (they are uncorrelated). The asymptotic analysis in Yu, de Jong, and Lee (2008) may not be directly carried over to the transformed model with the disturbances V_{nt}^*.[8] As Equation 14.6 is equivalent to Equation 14.9, we can analyze the asymptotic distribution of the estimator via Equation 14.9.

Using first order conditions, we concentrate out $\mathbf{c}_n$ in Equation 14.9 to obtain the concentrated likelihood function in terms of θ. For an $n \times 1$ vector at period t, Υ_{nt}, we define the deviation from time means as $\tilde{\Upsilon}_{nt} = \Upsilon_{nt} - \bar{\Upsilon}_{nT}$ and $\tilde{\tilde{\Upsilon}}_{n,t-1} = \Upsilon_{n,t-1} - \bar{\Upsilon}_{nT,-1}$, where $\bar{\Upsilon}_{nT} = \frac{1}{T}\sum_{t=1}^{T}\Upsilon_{nt}$ and $\bar{\Upsilon}_{nT,-1} = \frac{1}{T}\sum_{t=1}^{T}\Upsilon_{n,t-1}$. The concentrated log-likelihood is

$$\ln L_{n,T}(\theta) = -\frac{n^*T}{2}\ln 2\pi - \frac{n^*T}{2}\ln\sigma^2 - (n-n^*)T\ln(1-\lambda) + T\ln|I_n - \lambda W_n| - \frac{1}{2\sigma^2}\sum_{t=1}^{T}\tilde{V}'_{nt}(\theta)(I_n - W_n)'\Sigma_n^+(I_n - W_n)\tilde{V}_{nt}(\theta), \quad (14.10)$$

where $\tilde{V}_{nt}(\theta) = S_n(\lambda)\tilde{Y}_{nt} - \tilde{Z}_{nt}\delta$ and $(I_n - W_n)\tilde{V}_{nt}(\theta) = (I_n - W_n)[S_n(\lambda)\tilde{Y}_{nt} - \tilde{Z}_{nt}\delta - \tilde{\alpha}_t l_n]$ because $(I_n - W_n)l_n = \mathbf{0}$. At θ_0, $\tilde{V}_{nt} = S_n\tilde{Y}_{nt} - \tilde{Z}_{nt}\delta_0$. For Equation 14.10, its first- and second-order derivatives are Equation A.16 and A.17 in Appendix C.2.

14.3 Asymptotic Properties of QMLE

For our analysis of the asymptotic properties of estimators, we make the following assumptions. Denote $J_n^* = (I_n - W_n)'\Sigma_n^+(I_n - W_n)$. We note that J_n^* is an orthonormal projector with rank n^* (see Appendix A.5).

[8] One could not treat the components of V_{nt}^* as if they were independent when the disturbances are not normally distributed. Furthermore, it is not clear whether W_n^* and $A_n^* = (I_{n^*} - \lambda_0 W_n^*)^{-1}(\gamma_0 I_{n^*} + \rho_0 W_n^*)$ would be uniformly bounded in both row and column sums even though W_n and A_n are.

Assumption 1 W_n is a row-normalized nonstochastic spatial weights matrix with zero diagonals.

Assumption 2 The disturbances $\{v_{it}\}$, $i = 1, 2, ..., n$ and $t = 1, 2, ..., T$, are i.i.d. across i and t with zero mean, variance σ_0^2 and $E|v_{it}|^{4+\eta} < \infty$ for some $\eta > 0$.

Assumption 3 $S_n(\lambda)$ is invertible for all $\lambda \in \Lambda$. Furthermore, Λ is compact and the true parameter λ_0 is in the interior of Λ.

Assumption 4 The elements of X_{nt} are nonstochastic and bounded, uniformly in n and t, and the limit of $\frac{1}{nT}\sum_{t=1}^{T}\tilde{X}_{nt}'J_n^*\tilde{X}_{nt}$ exists and is nonsingular.

Assumption 5 W_n is uniformly bounded in row and column sums in the absolute value (for short, UB).[9] Also $S_n^{-1}(\lambda)$ is UB, uniformly in $\lambda \in \Lambda$.

Assumption 6 $\sum_{h=1}^{\infty} abs(B_n^h)$ is UB, where $[abs(B_n)]_{ij} = |B_{n,ij}|$.

Assumption 7 n^* is a nondecreasing function of T and T goes to infinity.

Assumption 1 is a standard normalization assumption in spatial econometrics. In many empirical applications, the rows of W_n sum to 1, which ensures that all the weights are between 0 and 1. Assumption 2 provides regularity assumptions for v_{it}. Assumption 3 guarantees that Equation 14.2 is valid. When exogenous variables X_{nt} are included in the model, it is convenient to assume that their elements are uniformly bounded[10] as in Assumption 4. Assumption 5 is originated by Kelejian and Prucha (1998, 2001) and is also used in Lee (2004, 2007). The uniform boundedness of W_n and $S_n^{-1}(\lambda)$ is a condition that limits the spatial correlation to a manageable degree. Assumption 6 is the absolute summability condition and row/column sum boundedness condition, which will play an important role for asymptotic properties of QML estimator. In order to justify the absolute summability of B_n, a sufficient condition is $\|B_n\| < 1$ for any matrix norm (see Horn and Johnson (1985), Corollary 5.6.16) that satisfies $\|B_n\| = \|abs(B_n)\|$. When $\|B_n\| < 1$, $\sum_{h=0}^{\infty} B_n^h$ exists and can be defined as $(I_n - B_n)^{-1}$. Assumption 7 allows two cases: (1) $n^* \to \infty$ as $T \to \infty$; (2) n^* can remain finite as $T \to \infty$. Because (2) is similar to a vector autoregressive (VAR) model, our main interest is in (1). If Assumption 7 holds, then we say that $n^*, T \to \infty$ simultaneously. These assumptions are similar to those in Yu, de Jong, and Lee (2008).

14.3.1 Consistency

For the log-likelihood function Equation 14.10 divided by the effective sample size n^*T, we have the corresponding $Q_{n,T}(\theta) = \mathrm{E}\max_{\mathbf{c}_n} \frac{1}{n^*T}\ln L_{n,T}(\theta, \mathbf{c}_n)$.

---

[9] We say a (sequence of $n \times n$) matrix P_n is uniformly bounded in row and column sums if $\sup_{n\geq 1}\|P_n\|_\infty < \infty$ and $\sup_{n\geq 1}\|P_n\|_1 < \infty$, where $\|P_n\|_\infty \equiv \sup_{1\leq i\leq n}\sum_{j=1}^{n}|p_{ij,n}|$ is the row sum norm and $\|P_n\|_1 = \sup_{1\leq j\leq n}\sum_{i=1}^{n}|p_{ij,n}|$ is the column sum norm.

[10] If X_{nt} is allowed to be stochastic, appropriate moment conditions can be imposed instead.

Hence,[11]

$$\begin{aligned} Q_{n,T}(\theta) &= \frac{1}{n^*T}\mathrm{E}\ln L_{n,T}(\theta) \\ &= -\frac{1}{2}\ln 2\pi - \frac{1}{2}\ln\sigma^2 - \frac{n-n^*}{n^*}\ln(1-\lambda) + \frac{1}{n^*}\ln|S_n(\lambda)| \qquad (14.11) \\ &\quad - \frac{1}{2\sigma^2}\frac{1}{n^*T}\mathrm{E}\left(\sum_{t=1}^{T}\tilde{V}'_{nt}(\theta)J_n^*\tilde{V}_{nt}(\theta)\right). \end{aligned}$$

It is shown in Appendix D.2 that, under Assumptions 1–7, $\frac{1}{n^*T}\ln L_{n,T}(\theta) - Q_{n,T}(\theta) \xrightarrow{p} 0$ uniformly in $\theta \in \Theta$ and $Q_{n,T}(\theta)$ is uniformly equicontinuous for $\theta \in \Theta$. For the identification, denote the information matrix $\Sigma_{\theta_0,nT} = -\mathrm{E}(\frac{1}{n^*T}\frac{\partial^2 \ln L_{n,T}(\theta_0)}{\partial\theta\partial\theta'})$. If $\Sigma_{\theta_0,nT}$ is nonsingular and $-\mathrm{E}(\frac{1}{n^*T}\frac{\partial^2 \ln L_{n,T}(\theta)}{\partial\theta\partial\theta'})$ has full rank for θ in some neighborhood $N(\theta_0)$ of θ_0, the parameters are locally identified (see Rothenberg 1971). Denote $\mathcal{H}_{nT} = \frac{1}{n^*T}\sum_{t=1}^{T}(\tilde{Z}_{nt}, G_n\tilde{Z}_{nt}\delta_0)'J_n^*(\tilde{Z}_{nt}, G_n\tilde{Z}_{nt}\delta_0)$ and $G_n^* = W_n^*S_n^{*-1}$. Using Lemma 15 in Yu, de Jong, and Lee (2008),

$$\begin{aligned} \Sigma_{\theta_0,nT} &= \frac{1}{\sigma_0^2}\begin{pmatrix} \mathrm{E}\mathcal{H}_{nT} & \mathbf{0}_{(k+3)\times 1} \\ \mathbf{0}_{1\times(k+3)} & 0 \end{pmatrix} \\ &\quad + \begin{pmatrix} \mathbf{0}_{(k+2)\times(k+2)} & \mathbf{0}_{(k+2)\times 1} & \mathbf{0}_{(k+2)\times 1} \\ \mathbf{0}_{1\times(k+2)} & \frac{1}{n^*}\left[tr(G_n^{*\prime}G_n^*) + tr(G_n^{*2})\right] & \frac{1}{\sigma_0^2 n^*}tr(G_n^*) \\ \mathbf{0}_{1\times(k+2)} & \frac{1}{\sigma_0^2 n^*}tr(G_n^*) & \frac{1}{2\sigma_0^4} \end{pmatrix} \qquad (14.12) \\ &\quad + O\left(\frac{1}{T}\right), \end{aligned}$$

which is nonsingular if $\mathrm{E}\mathcal{H}_{nT}$ is nonsingular or $\frac{1}{n^*}[tr(G_n^{*\prime}G_n^*) + tr(G_n^{*2}) - \frac{2tr^2(G_n^*)}{n^*}]$ is positive (see Appendix D.1). Also, its rank does not change in a small neighborhood of θ_0 (see Equation 14.49).

When $\lim_{T\to\infty}\mathrm{E}\mathcal{H}_{nT}$ is nonsingular, the parameters are identified.

Theorem 14.1 *Under Assumptions 1–7, if* $\lim_{T\to\infty}\mathrm{E}\mathcal{H}_{nT}$ *is nonsingular,* θ_0 *is identified and* $\hat{\theta}_{nT} \xrightarrow{p} \theta_0$.

Proof *See Appendix D.2.* ■

When $\lim_{T\to\infty}\mathrm{E}\mathcal{H}_{nT}$ is singular, identification can still be obtained from the following theorem. Denote $\sigma_n^2(\lambda) = \frac{\sigma_0^2}{n^*}tr(S_n'^{-1}S_n'(\lambda)J_n^*S_n(\lambda)S_n^{-1})$.

[11] Because $W_n = R_n\varpi_nR_n^{-1}$, $|S_n(\lambda)| = |I_n - \lambda\varpi_n| = (1-\lambda)^{m_n}\prod_{j=m_n+1}^{n}(1-\lambda\varpi_{nj})$. Therefore, $\frac{1}{n^*}\ln|S_n(\lambda)| - \frac{n-n^*}{n^*}\ln(1-\lambda) = \frac{1}{n^*}\sum_{j=m_n+1}^{n}(1-\lambda\varpi_{nj})$ shows that the division by n^* is proper.

Theorem 14.2 *Under Assumptions 1–7, if* $\lim_{n^* \to \infty}(\frac{1}{n^*} \ln |\sigma_0^2 S_n^{*-1\prime} S_n^{*-1}| - \frac{1}{n^*} \ln |\times \sigma_n^2(\lambda) S_n^{*\prime -1}(\lambda) S_n^{*-1}(\lambda)|) \neq 0$ *for* $\lambda \neq \lambda_0$, *then* θ_0 *is identified*[12] *and* $\hat{\theta}_{nT} \xrightarrow{p} \theta_0$.

Proof See Appendix D.3. ■

14.3.2 Asymptotic Distribution

As $Z_{nt} = (Y_{n,t-1}, W_n Y_{n,t-1}, X_{nt})$, we can decompose $(I_n - W_n)\tilde{Z}_{nt}$ such that

$$(I_n - W_n)\tilde{Z}_{nt} = (I_n - W_n)\tilde{Z}_{nt}^{(c)} - ((I_n - W_n)\bar{U}_{nT,-1}, \ (I_n - W_n)W_n\bar{U}_{nT,-1}, \ \mathbf{0}_{n\times k}), \tag{14.13}$$

where $\tilde{Z}_{nt}^{(c)} = ((\overset{\approx}{\mathcal{X}}_{n,t-1} + \mathcal{U}_{n,t-1}), \ (W_n\overset{\approx}{\mathcal{X}}_{n,t-1} + W_n\mathcal{U}_{n,t-1}), \ \tilde{X}_{nt})$ with $\overset{\approx}{\mathcal{X}}_{n,t-1} = \mathcal{X}_{n,t-1} - \bar{\mathcal{X}}_{nT,-1}, \mathcal{X}_{nt} \equiv \sum_{h=0}^{\infty} B_n^h S_n^{-1} X_{n,t-h}$ and $\mathcal{U}_{nt} \equiv \sum_{h=0}^{\infty} B_n^h S_n^{-1} V_{n,t-h}$. Hence, $(I_n - W_n)\tilde{Z}_{nt}$ has two components: one is $(I_n - W_n)\tilde{Z}_{nt}^{(c)}$, which is uncorrelated with V_{nt}; the remaining one can be correlated with V_{nt} when $t \leq T-1$. Here, after the data transformation by $I_n - W_n$, the unstable or explosive components and time component in $\tilde{Z}_{nt}$ are all eliminated. Therefore, from Equation 14.45, the score can be decomposed into two parts such that

$$\frac{1}{\sqrt{n^*T}} \frac{\partial \ln L_{n,T}(\theta_0)}{\partial \theta} = \frac{1}{\sqrt{n^*T}} \frac{\partial \ln L_{n,T}^{(c)}(\theta_0)}{\partial \theta} - \Delta_{nT}, \tag{14.14}$$

where

$$\frac{1}{\sqrt{n^*T}} \frac{\partial \ln L_{n,T}^{(c)}(\theta_0)}{\partial \theta} = \begin{pmatrix} \frac{1}{\sigma_0^2}\frac{1}{\sqrt{n^*T}} \sum_{t=1}^{T} \tilde{Z}_{nt}^{(c)\prime} J_n^* V_{nt} \\ \frac{1}{\sigma_0^2}\frac{1}{\sqrt{n^*T}} \sum_{t=1}^{T} (G_n \tilde{Z}_{nt}^{(c)}\delta_0)' J_n^* V_{nt} + \frac{1}{\sigma_0^2}\frac{1}{\sqrt{n^*T}} \sum_{t=1}^{T} (V_{nt}' G_n' J_n^* V_{nt} - \sigma_0^2 tr G_n^*) \\ \frac{1}{2\sigma_0^4}\frac{1}{\sqrt{n^*T}} \sum_{t=1}^{T} (V_{nt}' J_n^* V_{nt} - n^*\sigma_0^2) \end{pmatrix}, \tag{14.15}$$

[12] For our asymptotic analysis, finite n^* is allowed as long as T is tending to infinity, even though that is not an interesting case for SAR models. When n^* is finite, the condition is $\frac{1}{n^*} \ln |\sigma_0^2 S_n^{*-1\prime} S_n^{*-1}| - \frac{1}{n^*} \ln |\sigma_n^2(\lambda) S_n^{*\prime -1}(\lambda) S_n^{*-1}(\lambda)| \neq 0$ for $\lambda \neq \lambda_0$.

and

$$\Delta_{nT} = \sqrt{\frac{n^*}{T}} \begin{pmatrix} \frac{1}{\sigma_0^2}\frac{T}{n^*}(J_n^*\bar{U}_{nT,-1},\ J_n^* W_n\bar{U}_{nT,-1},\ \mathbf{0})'\bar{V}_{nT} \\ \frac{1}{\sigma_0^2}\frac{T}{n^*}(J_n^* G_n(\bar{U}_{nT,-1},\ W_n\bar{U}_{nT,-1},\ \mathbf{0})\delta_0)'\bar{V}_{nT} + \frac{1}{\sigma_0^2}\frac{T}{n^*}\bar{V}'_{nT}G'_nJ_n^*\bar{V}_{nT} \\ \frac{1}{2\sigma_0^4}\frac{T}{n^*}\bar{V}'_{nT}J_n^*\bar{V}_{nT} \end{pmatrix}. \tag{14.16}$$

Similarly to Yu, de Jong, and Lee (2008), the variance matrix of $\frac{1}{\sqrt{n^*T}}\frac{\partial \ln L_{n,T}^{(c)}(\theta_0)}{\partial\theta}$ is equal to

$$\mathrm{E}\left(\frac{1}{\sqrt{n^*T}}\frac{\partial \ln L_{n,T}^{(c)}(\theta_0)}{\partial\theta}\cdot\frac{1}{\sqrt{n^*T}}\frac{\partial \ln L_{n,T}^{(c)}(\theta_0)}{\partial\theta'}\right) = \Sigma_{\theta_0,nT} + \Omega_{\theta_0,n} + O(T^{-1}), \tag{14.17}$$

where $\Sigma_{\theta_0,nT}$ is in Equation 14.12 and

$$\Omega_{\theta_0,n} = \frac{\mu_4 - 3\sigma_0^4}{\sigma_0^4}\begin{pmatrix} \mathbf{0}_{(k+2)\times(k+2)} & \mathbf{0}_{(k+2)\times 1} & \mathbf{0}_{(k+2)\times 1} \\ \mathbf{0}_{1\times(k+2)} & \frac{1}{n^*}\sum_{i=1}^{n}(G_n^{*2})_{ii} & \frac{1}{2\sigma_0^2 n^*}tr(G_n^*) \\ \mathbf{0}_{1\times(k+2)} & \frac{1}{2\sigma_0^2 n^*}tr(G_n^*) & \frac{1}{4\sigma_0^4} \end{pmatrix}$$

is a symmetric matrix with μ_4 being the fourth moment of v_{it}. When V_{nt} is normally distributed, $\Omega_{\theta_0,n} = 0$ because $\mu_4 - 3\sigma_0^4 = 0$. Denote $\Sigma_{\theta_0} = \lim_{T\to\infty}\Sigma_{\theta_0,nT}$ and $\Omega_{\theta_0} = \lim_{T\to\infty}\Omega_{\theta_0,n}$. The asymptotic distribution of $\frac{1}{\sqrt{n^*T}}\frac{\partial \ln L_{n,T}^{(c)}(\theta_0)}{\partial\theta}$ can be derived from the central limit theorem for martingale difference arrays (Lemma 14.3). For the term Δ_{nT}, from Equation 14.36 in Lemma 14.1 and Equation 14.38 in Lemma 14.2, $\Delta_{nT} = \sqrt{\frac{n^*}{T}}a_{\theta_0,n} + O(\sqrt{\frac{n^*}{T^3}}) + O_p(\frac{1}{\sqrt{T}})$ where

$$a_{\theta_0,n} = \begin{pmatrix} \frac{1}{n^*}tr\left(\left(J_n^*\sum_{h=0}^{\infty}B_n^h\right)S_n^{-1}\right) \\ \frac{1}{n^*}tr\left(W_n\left(J_n^*\sum_{h=0}^{\infty}B_n^h\right)S_n^{-1}\right) \\ \mathbf{0}_{k\times 1} \\ \frac{1}{n^*}\gamma_0 tr(G_n\left(J_n^*\sum_{h=0}^{\infty}B_n^h\right)S_n^{-1}) + \frac{1}{n^*}\rho_0 tr(G_nW_n\left(J_n^*\sum_{h=0}^{\infty}B_n^h\right)S_n^{-1}) + \frac{1}{n^*}trG_n^* \\ \frac{1}{2\sigma_0^2} \end{pmatrix} \tag{14.18}$$

is $O(1)$. It is shown in Appendix D.4 that, under Assumptions 1–7, $\frac{1}{\sqrt{n^*T}} \frac{\partial \ln L_{n,T}(\theta_0)}{\partial \theta} + \Delta_{nT} \xrightarrow{d} N(0, \Sigma_{\theta_0} + \Omega_{\theta_0})$.

To get the asymptotic distribution of the estimates, we need the following additional assumption.

Assumption 8. $\lim_{T\to\infty} \mathrm{E}\mathcal{H}_{nT}$ is nonsingular or $\lim_{n\to\infty} \frac{1}{n^*}[tr(G_n^{*\prime}G_n^*)+tr(G_n^{*2}) - \frac{2tr^2(G_n^*)}{n^*}] > 0$.

Assumption 8 is a condition for the nonsingularity of the limit of the information matrix Σ_{θ_0}. When $\lim_{T\to\infty} \mathrm{E}\mathcal{H}_{nT}$ is singular,[13] as long as we have $\lim_{n\to\infty} \frac{1}{n^*}[tr(G_n^{*\prime}G_n^*) + tr((G_n^*)^2) - \frac{2tr^2(G_n^*)}{n^*}] > 0$, the information matrix Σ_{θ_0} is still nonsingular (see Appendix D.1). Hence, for the second order derivatives of the log-likelihood function, under Assumption 1–8, we have $\frac{1}{n^*T}\frac{\partial^2 \ln L_{n,T}(\theta)}{\partial\theta\partial\theta'} - \frac{1}{n^*T}\frac{\partial^2 \ln L_{n,T}(\theta_0)}{\partial\theta\partial\theta'} = \|\theta-\theta_0\| \cdot O_p(1)$, and $\frac{1}{n^*T}\frac{\partial^2 \ln L_{n,T}(\theta_0)}{\partial\theta\partial\theta'} - \frac{\partial^2 Q_{n,T}(\theta_0)}{\partial\theta\partial\theta'} = O_p(\frac{1}{\sqrt{n^*T}})$ from Appendix C.3. Thus, we have the following theorem for the asymptotic distribution of $\hat{\theta}_{nT}$.

Theorem 14.3 *Under Assumptions 1–8,*

$$\sqrt{n^*T}(\hat{\theta}_{nT} - \theta_0) + \sqrt{\frac{n^*}{T}} b_{\theta_0,nT} + O_p\left(\max\left(\sqrt{\frac{n^*}{T^3}}, \sqrt{\frac{1}{T}}\right)\right)$$

$$\xrightarrow{d} N(0, \Sigma_{\theta_0}^{-1}(\Sigma_{\theta_0} + \Omega_{\theta_0})\Sigma_{\theta_0}^{-1}), \tag{14.19}$$

where $b_{\theta_0,nT} = \Sigma_{\theta_0,nT}^{-1} a_{\theta_0,n}$ *is* $O(1)$.

When $\frac{n^*}{T} \to 0$, $\sqrt{n^*T}(\hat{\theta}_{nT} - \theta_0) \xrightarrow{d} N(0, \Sigma_{\theta_0}^{-1}(\Sigma_{\theta_0} + \Omega_{\theta_0})\Sigma_{\theta_0}^{-1})$.

When $\frac{n^*}{T} \to c < \infty$, $\sqrt{n^*T}(\hat{\theta}_{nT} - \theta_0) + \sqrt{c} b_{\theta_0,nT} \xrightarrow{d} N(0, \Sigma_{\theta_0}^{-1}(\Sigma_{\theta_0} + \Omega_{\theta_0})\Sigma_{\theta_0}^{-1})$.

When $\frac{n^*}{T} \to \infty$, $T(\hat{\theta}_{nT} - \theta_0) + b_{\theta_0,nT} \xrightarrow{p} 0$.

Proof *See Appendix D.4.* ■

14.3.3 Bias Correction

From Equation (14.19), the QML estimator has the leading bias $-\frac{1}{T} b_{\theta_0,nT}$ where $b_{\theta_0,nT} = \Sigma_{\theta_0,nT}^{-1} \cdot a_{\theta_0,n}$ and the confidence interval will not be centered when $\frac{n^*}{T} \to c < \infty$. Furthermore, when T is relatively smaller than n^*, the presence of $b_{\theta_0,nT}$ causes $\hat{\theta}_{nT}$ to have a degenerate distribution. An analytical bias reduction procedure can be used to correct this bias of the estimate. Define the

[13] The $\lim_{T\to\infty} \mathrm{E}\mathcal{H}_{nT}$ can be singular if, for example, $\delta_0 = \mathbf{0}$.

bias corrected estimator as

$$\hat{\theta}^1_{nT} = \hat{\theta}_{nT} - \frac{\hat{B}_{nT}}{T}, \tag{14.20}$$

where, from Theorem 14.3, $\hat{B}_{nT} = [-\Sigma^{-1}_{\theta,nT} \cdot a_{\theta,n}]|_{\theta=\hat{\theta}_{nT}}$. We show that when $n^*/T^3 \to 0$, $\hat{\theta}^1_{nT}$ is $\sqrt{n^*T}$ consistent and asymptotically centered normal even when $n^*/T \to \infty$.

For the asymptotic properties of the bias corrected estimator, we need the following additional assumption.

Assumption 9. $\sum_{h=0}^{\infty} B_n^h(\theta)$ and $\sum_{h=1}^{\infty} h B_n^{h-1}(\theta)$ are uniformly bounded in either row sum or column sums, uniformly in a neighborhood of θ_0.

Assumption 9 can be justified through Lemma 14.5 in Appendix B. Our result for the bias corrected estimator is as follows.

Theorem 14.4 *If* $\frac{n^*}{T^3} \to 0$, *under Assumptions 1–9,* $\sqrt{n^*T}(\hat{\theta}^1_{nT} - \theta_0) \xrightarrow{d} N(0, \Sigma^{-1}_{\theta_0} + \Sigma^{-1}_{\theta_0}\Omega_{\theta_0}\Sigma^{-1}_{\theta_0})$.

Proof *See Appendix D.5.* ■

Hence, if T grows faster than $n^{*1/3}$, the analytical bias adjusted estimator is asymptotically normal and centered properly around θ_0. For the case $\frac{n}{T} \to c$, $\hat{\theta}^1_{nT}$ has removed the asymptotic bias $b_{\theta_0,nT}$. Note that $\frac{n}{T} \to c$ implies $T/n^{*1/3} \to \infty$. For the case $\frac{n^*}{T} \to \infty$, as long as $T/n^{*1/3} \to \infty$, $\hat{\theta}^1_{nT}$ is $\sqrt{n^*T}$ consistent, which is also an improvement upon the T consistency of $\hat{\theta}_{nT}$. Thus, $\hat{\theta}^1_{nT}$ might have better performance, especially when n^* is much larger than T.

14.3.4 Testing

For the unified transformation approach with a bias correction, we have $\sqrt{n^*T}(\hat{\theta}^1_{nT} - \theta_0) \xrightarrow{d} N(0, \Sigma^{-1}_{\theta_0} + \Sigma^{-1}_{\theta_0}\Omega_{\theta_0}\Sigma^{-1}_{\theta_0})$. Hence, we can use the bias corrected estimate $\hat{\theta}^1_{nT}$ for the statistical inference of $\gamma_0 + \rho_0 + \lambda_0$. Let $\Sigma_{\hat{\theta}^1_{nT},nT}$ and $\Omega_{\hat{\theta}^1_{nT},n}$ be consistent estimates for $\Sigma_{\theta_0,nT}$ and $\Omega_{\theta_0,n}$. We can construct t-statistic to test the null of spatial co-integration, i.e., $\gamma_0 + \rho_0 + \lambda_0 = 1$. Denote $r = (1, 1, 0_{1\times k_x}, 1, 0)'$. With $\theta = (\gamma, \rho, \beta', \lambda, \sigma^2)$, we are testing $r'\hat{\theta}^1_{nT} = 1$. The test statistic is

$$t = \frac{\sqrt{n^*T} \cdot (r'\hat{\theta}^1_{nT} - 1)}{\sqrt{r'(\Sigma^{-1}_{\hat{\theta}^1_{nT},nT} + \Sigma^{-1}_{\hat{\theta}^1_{nT},nT}\Omega_{\hat{\theta}^1_{nT},n}\Sigma^{-1}_{\hat{\theta}^1_{nT},nT})r}} \xrightarrow{d} N(0,1), \tag{14.21}$$

because $\sqrt{n^*T} \cdot (r'\hat{\theta}^1_{nT} - 1) \xrightarrow{d} N(0, r'(\Sigma^{-1}_{\theta_0} + \Sigma^{-1}_{\theta_0}\Omega_{\theta_0}\Sigma^{-1}_{\theta_0})r)$ and $\Sigma^{-1}_{\hat{\theta}^1_{nT},nT} + \Sigma^{-1}_{\hat{\theta}^1_{nT},nT}\Omega_{\hat{\theta}^1_{nT},n}\Sigma^{-1}_{\hat{\theta}^1_{nT},nT} \xrightarrow{p} \Sigma^{-1}_{\theta_0} + \Sigma^{-1}_{\theta_0}\Omega_{\theta_0}\Sigma^{-1}_{\theta_0}$. We present a simulation in the next section to investigate the finite sample performance of the test statistic in terms of its significance level and power, under one-sided or two-sided tests.

14.4 Monte Carlo Results

We conduct a Monte Carlo experiment to evaluate the performance of the bias corrected MLE of this unified approach and compare them with other estimation methods under different DGPs:

$$\begin{aligned} \text{A}: \quad Y_{nt} &= \lambda_0 W_n Y_{nt} + \gamma_0 Y_{n,t-1} + \rho_0 W_n Y_{n,t-1} + X_{nt}\beta_0 + \mathbf{c}_{n0} \\ &\quad + V_{nt}, \ \lambda_0 + \gamma_0 + \rho_0 < 1, \end{aligned} \tag{14.22}$$

$$\begin{aligned} \text{B}: \quad Y_{nt} &= \lambda_0 W_n Y_{nt} + \gamma_0 Y_{n,t-1} + \rho_0 W_n Y_{n,t-1} + X_{nt}\beta_0 + \mathbf{c}_{n0} + \alpha_{t0} l_n \\ &\quad + V_{nt}, \ \lambda_0 + \gamma_0 + \rho_0 < 1, \end{aligned} \tag{14.23}$$

$$\begin{aligned} \text{C}: \quad Y_{nt} &= \lambda_0 W_n Y_{nt} + \gamma_0 Y_{n,t-1} + \rho_0 W_n Y_{n,t-1} + X_{nt}\beta_0 + \mathbf{c}_{n0} \\ &\quad + V_{nt}, \ \lambda_0 + \gamma_0 + \rho_0 = 1, \ \gamma_0 \neq 1, \end{aligned} \tag{14.24}$$

$$\begin{aligned} \text{D}: \quad Y_{nt} &= \lambda_0 W_n Y_{nt} + \gamma_0 Y_{n,t-1} + \rho_0 W_n Y_{n,t-1} + X_{nt}\beta_0 + \mathbf{c}_{n0} + \alpha_{t0} l_n \\ &\quad + V_{nt}, \ \lambda_0 + \gamma_0 + \rho_0 = 1, \ \gamma_0 \neq 1. \end{aligned} \tag{14.25}$$

The DGPs A and B are stable SDPD models with or without time dummy effects. The C and D are spatial co-integrated SDPD models with or without time dummy effects. We will also consider subsequently DGPs with explosive roots in E and F. We generate samples using $\theta_0 = (0.2, 0.2, 1, 0.2, 1)'$ for the stable cases and $\theta_0 = (0.4, 0.2, 1, 0.4, 1)'$ for the spatial co-integration cases where $\theta_0 = (\gamma_0, \rho_0, \beta_0', \lambda_0, \sigma_0^2)'$, and X_{nt}, $\mathbf{c}_{n0}$, $\alpha_{T0} = (\alpha_1, \alpha_2, \cdots, \alpha_T)'$ and V_{nt} are generated from independent normal distributions.[14] The spatial weights matrix we use is a block diagonal matrix formed by a row-normalized queen matrix.[15] We use $T = 10, 50$, and $n = 18, 54$.

For each set of generated sample observations, we use two methods: one is the corresponding estimation method without any transformation when the model does not have time dummies, or using the deviation from group mean transformation when the model includes time dummies, and the other is the unified transformation method. We obtain the MLE $\hat{\theta}_{nT}$, construct the bias corrected estimator $\hat{\theta}^1_{nT}$ and evaluate the bias $\hat{\theta}^1_{nT} - \theta_0$. We do this 1000

[14] We generated the data with $20 + T$ periods and then take the last T periods as our sample. And the initial value is generated as $N(0, I_n)$ in the simulation.

[15] We choose the spatial weights matrix such that it contains unit eigenvalues. We use the block diagonal matrix where each block uses the same weights matrix. By increasing the number of blocks, the number of unit eigenvalues of the block diagonal matrix will also increase, but the percentage remains a constant. In our simulation, when $n = 18$, $n^* = 16$; when $n = 54$, $n^* = 48$.

times. We also compare the empirical standard deviation (SD) and the empirical mean square error (RMSE) of these 1000 estimators. Also, a coverage probability (CP) is reported.[16] With different values of n and T, finite sample properties of the bias corrected estimators[17] are summarized in Tables 14.1 and 14.2. Table 14.1 presents the results for the stable SDPD models; and Table 14.2 is for the spatial co-integration cases.

Because the unified transformation method will lose more degrees of freedom than the other methods, we expect less precision for the estimates from the unified transformation approach than the others. For our MC design with blocks of the queen matrix, the use of the unified transformation will result in more loss of degrees of freedom than that of the deviation from the group mean transformation for the models with time effects. It is of interest to see that the estimators by the unified transformation method perform well, and they are a little bit worse than the corresponding estimators in the loss of precision. All the estimates have small biases. The CPs are adequate except for some cases with small $T = 10$.

The unified transformation method would be of more interest for the explosive roots case. We conduct a simulation to check the performance of the unified estimator when the DGP is explosive:

$$\text{E}: \quad Y_{nt} = \lambda_0 W_n Y_{nt} + \gamma_0 Y_{n,t-1} + \rho_0 W_n Y_{n,t-1} + X_{nt}\beta_0 + \mathbf{c}_{n0} + V_{nt}, \ \lambda_0 + \gamma_0 + \rho_0 > 1, \tag{14.26}$$

$$\text{F}: \quad Y_{nt} = \lambda_0 W_n Y_{nt} + \gamma_0 Y_{n,t-1} + \rho_0 W_n Y_{n,t-1} + X_{nt}\beta_0 + \mathbf{c}_{n0} + \alpha_{t0} l_n + V_{nt}, \ \lambda_0 + \gamma_0 + \rho_0 > 1, \tag{14.27}$$

where $\theta_0 = (0.4, 0.4, 1, 0.4, 1)'$. Finite sample properties of both estimators are summarized in Table 14.3 for the bias corrected estimators. We can see that even though we have explosive roots in the DGP, the unified approach can still yield estimators with good finite sample performances, i.e., the biases are small and the CPs are adequate. However, if we use the QMLE without any transformation when the model does not have time dummies, or use the deviation from group mean transformation when the model includes time dummies, the estimates' Biases, SD, and RMSE become very large and the CPs are nearly zero when T is large.

Finally, we present the simulation result of the size and power of the hypothesis testing of spatial co-integration, i.e., $H_0 : \lambda_0 + \gamma_0 + \rho_0 = 1$. We run 1000 repetitions to calculate the power for $n = 54$ and $T = 10$ or 50, where the power is obtained with a 1% or 5% significance levels. We first use the

[16] The coverage probability is obtained by using the estimated analytical standard errors of the estimators in each repitition.

[17] For the estimators before bias correction, they have a larger bias than the corresponding bias corrected estimators. As the comparison between the unified transformation estimators and the corresponding estimators are similar to the counterpart of bias corrected estimators, we do not report the tables of results to save space.

TABLE 14.1

Performance of Estimators When the DGP Is Stable

	T	n	Estimator		γ	ρ	β	λ	σ^2
No Time Dummy in the DGP (Equation 14.22):									
(1)	10	54	A	Bias	−0.0010	−0.0015	0.0016	−0.0086	−0.0288
				SD	0.0320	0.0659	0.0451	0.0517	0.0592
				RMSE	0.0439	0.0926	0.0619	0.0721	0.0860
				CP	0.9400	0.9040	0.9290	0.9410	0.8500
	10	54	Unified	Bias	−0.0010	−0.0002	0.0001	−0.0108	−0.0305
				SD	0.0375	0.1409	0.0494	0.1139	0.0674
				RMSE	0.0515	0.1908	0.0676	0.1535	0.0981
				CP	0.9250	0.9360	0.9300	0.9780	0.8470
(2)	50	18	A	Bias	−0.0007	−0.0011	−0.0009	−0.0025	−0.0043
				SD	0.0235	0.0476	0.0337	0.0393	0.0470
				RMSE	0.0317	0.0652	0.0458	0.0536	0.0647
				CP	0.9570	0.9410	0.9480	0.9510	0.9230
	50	18	Unified	Bias	−0.0006	0.0004	−0.0025	−0.0077	−0.0062
				SD	0.0274	0.1034	0.0370	0.0873	0.0534
				RMSE	0.0371	0.1414	0.0505	0.1176	0.0736
				CP	0.9450	0.9400	0.9440	0.9470	0.9250
(3)	50	54	A	Bias	−0.0002	−0.0009	0.0000	−0.0007	−0.0015
				SD	0.0136	0.0275	0.0195	0.0227	0.0272
				RMSE	0.0182	0.0370	0.0259	0.0311	0.0377
				CP	0.9570	0.9500	0.9620	0.9320	0.9320
	50	54	Unified	Bias	0.0002	0.0018	0.0001	−0.0015	−0.0017
				SD	0.0158	0.0596	0.0214	0.0504	0.0309
				RMSE	0.0211	0.0800	0.0284	0.0684	0.0424
				CP	0.9480	0.9520	0.9600	0.9420	0.9320
Time Dummy in the DGP (Equation 14.23):									
(1)	10	54	B	Bias	−0.0036	−0.0000	0.0016	−0.0066	−0.0283
				SD	0.0323	0.0700	0.0455	0.0550	0.0597
				RMSE	0.0452	0.0987	0.0632	0.0769	0.0871
				CP	0.9190	0.9160	0.9260	0.9280	0.8620
	10	54	Unified	Bias	−0.0043	−0.0010	0.0008	−0.0086	−0.0312
				SD	0.0375	0.1405	0.0495	0.1140	0.0673
				RMSE	0.0519	0.1950	0.0690	0.1532	0.0968
				CP	0.9270	0.9170	0.9200	0.9760	0.8600
(2)	50	18	B	Bias	−0.0009	−0.0025	−0.0024	−0.0013	−0.0046
				SD	0.0242	0.0595	0.0347	0.0498	0.0484
				RMSE	0.0330	0.0804	0.0477	0.0683	0.0662
				CP	0.9510	0.9460	0.9430	0.9310	0.9380
	50	18	Unified	Bias	0.0005	0.0074	−0.0030	−0.0023	−0.0253
				SD	0.0273	0.1033	0.0370	0.0873	0.0523
				RMSE	0.0375	0.1411	0.0507	0.1178	0.0732
				CP	0.9440	0.9420	0.9410	0.9480	0.9330
(3)	50	54	B	Bias	0.0002	−0.0008	0.0002	0.0007	−0.0015
				SD	0.0136	0.0290	0.0196	0.0241	0.0269
				RMSE	0.0185	0.0395	0.0268	0.0334	0.0373
				CP	0.9470	0.9410	0.9360	0.9270	0.9400
	50	54	Unified	Bias	0.0004	0.0003	−0.0001	−0.0033	−0.0031
				SD	0.0158	0.0596	0.0214	0.0504	0.0309
				RMSE	0.0215	0.0811	0.0291	0.0693	0.0421
				CP	0.9500	0.9410	0.9430	0.9430	0.9400

Note: $\theta_0 = (0.2, 0.2, 1, 0.2, 1)'$

TABLE 14.2

Performance of Estimators When the DGP Is Spatial Co-integrated

	T	n	Estimator		γ	ρ	β	λ	σ^2
No Time Dummy in the DGP (Equation 14.24):									
(1)	10	54	C	Bias	0.0065	0.0518	0.0073	0.0007	−0.0330
				SD	0.0314	0.0531	0.0452	0.0405	0.0594
				RMSE	0.0447	0.0899	0.0626	0.0589	0.0876
				CP	0.9090	0.7640	0.9190	0.9160	0.8330
	10	54	Unified	Bias	−0.0023	0.0023	−0.0005	−0.0173	−0.0354
				SD	0.0354	0.1353	0.0490	0.1113	0.0659
				RMSE	0.0494	0.1843	0.0674	0.1560	0.0982
				CP	0.9180	0.9300	0.9240	0.9230	0.8350
(2)	50	18	C	Bias	0.0001	0.0046	−0.0006	−0.0046	−0.0039
				SD	0.0224	0.0365	0.0338	0.0314	0.0473
				RMSE	0.0303	0.0495	0.0460	0.0425	0.0651
				CP	0.9550	0.9450	0.9470	0.9530	0.9280
	50	18	Unified	Bias	−0.0013	−0.0005	−0.0024	−0.0078	−0.0062
				SD	0.0249	0.0969	0.0367	0.0851	0.0525
				RMSE	0.0337	0.1306	0.0501	0.1171	0.0725
				CP	0.9460	0.9470	0.9420	0.9520	0.9280
(3)	50	54	C	Bias	0.0003	0.0033	0.0003	−0.0035	−0.0010
				SD	0.0129	0.0210	0.0195	0.0181	0.0274
				RMSE	0.0175	0.0287	0.0260	0.0250	0.0380
				CP	0.9510	0.9380	0.9620	0.9360	0.9340
	50	54	Unified	Bias	−0.0002	0.0002	0.0002	−0.0002	−0.0016
				SD	0.0143	0.0558	0.0212	0.0491	0.0304
				RMSE	0.0193	0.0757	0.0282	0.0675	0.0418
				CP	0.9470	0.9490	0.9600	0.9430	0.9310
Time Dummy in the DGP (Equation 14.25):									
(1)	10	54	D	Bias	0.0030	0.0483	0.0059	0.0006	−0.0323
				SD	0.0316	0.0557	0.0456	0.0435	0.0597
				RMSE	0.0450	0.0917	0.0638	0.0631	0.0882
				CP	0.8550	0.6790	0.8880	0.8600	0.8500
	10	54	Unified	Bias	−0.0054	0.0020	−0.0000	−0.0160	−0.0364
				SD	0.0354	0.1349	0.0490	0.1114	0.0658
				RMSE	0.0501	0.1871	0.0686	0.1559	0.0969
				CP	0.9160	0.9110	0.9240	0.9220	0.8510
(2)	50	18	D	Bias	−0.0004	0.0017	−0.0024	−0.0039	−0.0050
				SD	0.0226	0.0441	0.0347	0.0404	0.0483
				RMSE	0.0308	0.0594	0.0477	0.0560	0.0661
				CP	0.9560	0.9480	0.9400	0.9310	0.9340
	50	18	Unified	Bias	−0.0007	0.0032	−0.0029	−0.0024	−0.0059
				SD	0.0248	0.0967	0.0367	0.0851	0.0525
				RMSE	0.0339	0.1323	0.0503	0.1171	0.0721
				CP	0.9530	0.9450	0.9410	0.9500	0.9290
(3)	50	54	D	Bias	0.0004	0.0015	0.0003	−0.0017	−0.0014
				SD	0.0130	0.0219	0.0197	0.0193	0.0276
				RMSE	0.0175	0.0297	0.0269	0.0266	0.0375
				CP	0.9360	0.8810	0.9240	0.9180	0.9430
	50	54	Unified	Bias	−0.0001	−0.0004	−0.0000	−0.0027	−0.0030
				SD	0.0143	0.0557	0.0212	0.0491	0.0304
				RMSE	0.0195	0.0758	0.0288	0.0685	0.0415
				CP	0.9420	0.9430	0.9370	0.9350	0.9400

Note: $\theta_0 = (0.4, 0.2, 1, 0.4, 1)'$

TABLE 14.3

Performance of Estimators When the DGP Is Explosive

	T	n	Estimator		γ	ρ	β	λ	σ^2
No Time Dummy in the DGP (Equation 14.26):									
(1)	10	54	A	Bias	0.0053	0.0395	0.0049	−0.0336	−0.0241
				SD	0.0336	0.0584	0.0465	0.0422	0.0626
				RMSE	0.0340	0.0705	0.0467	0.0540	0.0670
				CP	0.9200	0.8890	0.9270	0.8630	0.9230
	10	54	Unified	Bias	−0.0018	0.0031	−0.0007	−0.0196	−0.0360
				SD	0.0379	0.1382	0.0504	0.1201	0.0716
				RMSE	0.0380	0.1382	0.0504	0.1217	0.0801
				CP	0.9170	0.9310	0.9270	0.9100	0.8070
(2)	50	18	A	Bias	******	******	2.4973	−0.0624	******
				SD	******	******	264.78	0.2958	******
				RMSE	******	******	264.79	0.3023	******
				CP	0.0150	0.0090	0.0140	0.0130	0.0110
	50	18	Unified	Bias	−0.0013	−0.0013	−0.0025	−0.0088	−0.0065
				SD	0.0246	0.0931	0.0373	0.0878	0.0543
				RMSE	0.0246	0.0931	0.0374	0.0882	0.0547
				CP	0.9480	0.9440	0.9420	0.9260	0.9050
(3)	50	54	A	Bias	******	******	−4.1263	−0.0668	******
				SD	******	******	724.64	0.3096	******
				RMSE	******	******	724.66	0.3167	******
				CP	0.0010	0.0000	0.0000	0.0010	0.0000
	50	54	Unified	Bias	−0.0004	−0.0006	0.0002	−0.0005	−0.0016
				SD	0.0139	0.0557	0.0203	0.0510	0.0315
				RMSE	0.0139	0.0557	0.0203	0.0510	0.0315
				CP	0.9450	0.9380	0.9600	0.9250	0.9130
Time dummy in the DGP (Equation 14.27):									
(1)	10	54	B	Bias	0.0021	0.0386	0.0037	−0.0305	−0.0257
				SD	0.0346	0.0635	0.0482	0.0462	0.0639
				RMSE	0.0347	0.0743	0.0483	0.0554	0.0689
				CP	0.9190	0.8870	0.9240	0.8880	0.9100
	10	54	Unified	Bias	−0.0049	0.0029	−0.0003	−0.0191	−0.0371
				SD	0.0390	0.1435	0.0529	0.1200	0.0688
				RMSE	0.0394	0.1435	0.0529	0.1216	0.0782
				CP	0.9120	0.9060	0.9230	0.9090	0.8090
(2)	50	18	B	Bias	******	******	−4.0205	−0.0478	******
				SD	******	******	105.34	0.2891	******
				RMSE	******	******	105.41	0.2931	******
				CP	0.1030	0.0640	0.0960	0.0790	0.0660
	50	18	Unified	Bias	−0.0011	0.0014	−0.0030	−0.0033	−0.0061
				SD	0.0248	0.0972	0.0378	0.0885	0.0536
				RMSE	0.0248	0.0972	0.0379	0.0885	0.0540
				CP	0.9520	0.9390	0.9430	0.9260	0.9110
(3)	50	54	B	Bias	******	******	−35.49	−0.0596	******
				SD	******	******	835.56	0.3128	******
				RMSE	******	******	836.31	0.3184	******
				CP	0.0020	0.0000	0.0010	0.0040	0.0000
	50	54	Unified	Bias	−0.0001	−0.0009	−0.0001	−0.0030	−0.0031
				SD	0.0143	0.0553	0.0215	0.0521	0.0308
				RMSE	0.0143	0.0553	0.0215	0.0522	0.0310
				CP	0.9410	0.9370	0.9380	0.9220	0.9270

Note: 1. $\theta_0 = (0.4, 0.4, 1, 0.4, 1)'$.

2. ****** denotes an explosive number, which is of the order 10^{11} for the column of σ^2, and 10^5 for other columns.

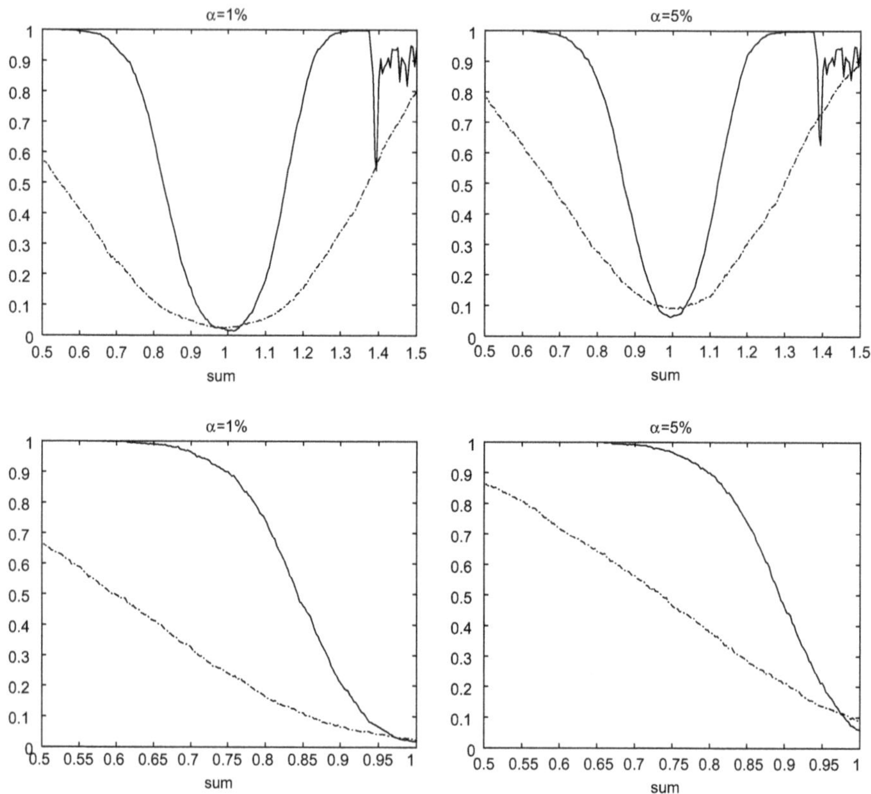

Note: 1. -.-.-. denotes the power curve for T = 10, and —— denotes the power curve for T = 50.
2. The first row is for the two-sided tests and the second row is for the one-sided tests.

FIGURE 14.1
Power curves under the unified approach for $H_0 : \gamma_0 + \rho_0 + \lambda_0 = 1$.

unified approach to get the power curves. The results are in Figure 14.1. For the two-sided tests, the sum $\lambda_0 + \gamma_0 + \rho_0$ under the alternative hypothesis ranges from 0.65 to 1.35 with a $\frac{0.7}{200}$ increment; for the one-sided test with $H_1 : \lambda_0 + \gamma_0 + \rho_0 < 1$, the sum $\lambda_0 + \gamma_0 + \rho_0$ ranges from 0.65 to 1.0 with a $\frac{0.35}{200}$ increment. From Figure 14.2, we can see that the empirical sizes[18] are close to the theoretical ones and the tests are more powerful when $T = 50$ than those for the small $T = 10$. The power seems reasonable for the large $T = 50$. We run additional simulations where we use the corresponding estimation method without any transformation. Figure 14.2 is the counterparts[19] of Table 14.1.

[18] For the empirical size, the $T = 10$ case has 2.4%, 2.2%, 9.1%, and 8.8% from the first row to the second row, and the $T = 50$ case has 1.6%, 1.7%, 6.5%, and 5.8%. As the significance level are 1%, 1%, 5%, and 5% correspondingly, a larger T will yield empirical sizes closer to the theoretical values.

[19] For the first row in Table 14.2, when the sum $\lambda_0 + \gamma_0 + \rho_0$ is much larger than 1 (i.e., the process is explosive), the estimates might not be available due to overflow without the unified transformation. Hence, for the two-sided power curves, we allow the sum only up to 1.3.

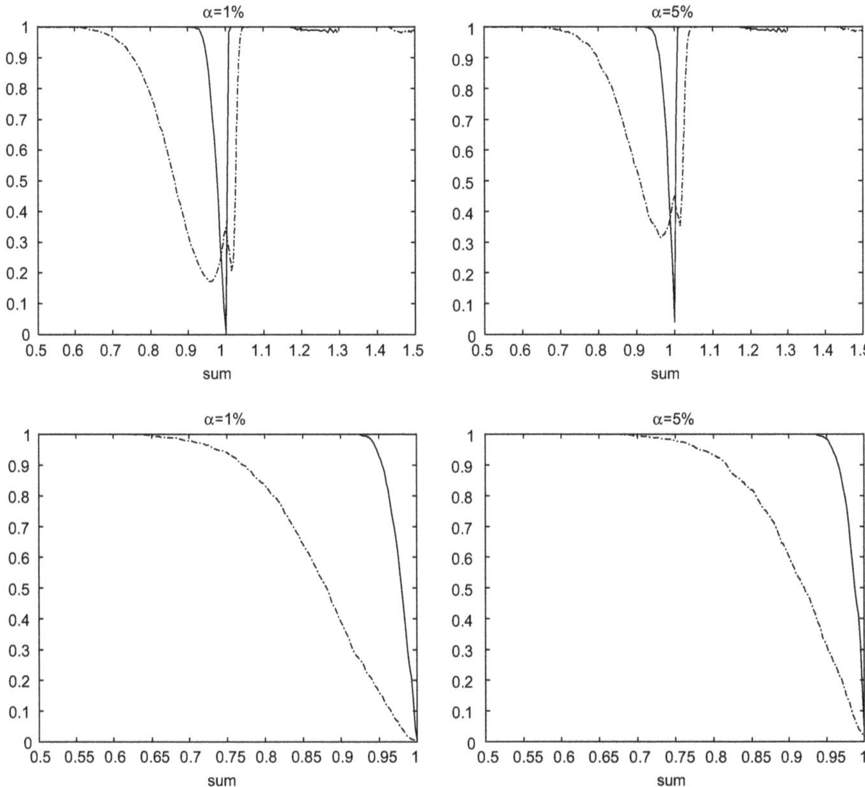
α=1%
α=5%
α=1%
α=5%
sum
sum
sum
sum
Note: 1. -.-.-. denotes the power curve for T = 10, and —— denotes the power curve for T = 50.
2. The first row is for the two-sided tests and the second row is for the one-sided tests.

FIGURE 14.2
Power curves under Yu, de Jong, and Lee (2007) for $H_0 : \gamma_0 + \rho_0 + \lambda_0 = 1$.

We can see that, when $\lambda_0 + \gamma_0 + \rho_0 < 1$, the test is more powerful by using the corresponding method without any transformation; when $\lambda_0 + \gamma_0 + \rho_0 > 1$, the power curves are irregular and we need to rely on the unified approach for the inferences.[20]

14.5 Conclusion

This chapter establishes asymptotic properties of QMLEs for SDPD models with both time and individual fixed effects when both the number of individuals n and the number of time periods T can be large. Instead of using different

[20] For the empirical size, the $T = 10$ case has 34.8%, 0.3%, 44.9%, and 1.5% from the first row to the second row in Table 14.2, and the $T = 50$ case has 1.1%, 0.8%, 4%, and 4%. Hence, when T is small, the empirical sizes could be far away from the theoretical values.

estimation methods depending on whether the DGP has time effects or not and whether the DGP is stable or not, we propose a data transformation approach to eliminate both the time effects and the possible unstable or explosive effects. The transformation is motivated by the possible co-integration relationship in the SDPD model, which is implied by the unit eigenvalues in the spatial weights matrix W_n. Unlike the co-integration in the multi-variate time series, the co-integrating vector is known and does not need to be estimated. With the proposed data transformation, the possible unstable or explosive components and time effects can be eliminated.

The transformation uses the co-integrating matrix. The effective sample size n^* after transformation corresponds to the co-integration rank, which is the number of eigenvalues not equal to the unity. This transformation is of particular value when the process may contain explosive roots, as usual estimation methods can be poorly performed under such a situation. For the unified approach, when T is relatively larger than n^*, the estimators are $\sqrt{n^*T}$ consistent and asymptotically centered normal; when n^* is asymptotically proportional to T, the estimators are $\sqrt{n^*T}$ consistent and asymptotically normal, but the limit distribution is not centered around 0; when T is relatively smaller than n^*, the estimators are consistent with rate T and have a degenerate limit distribution. We also propose a bias correction for our estimators. We show that when T grows faster than $n^{*1/3}$, the correction will asymptotically eliminate the bias and yield a centered confidence interval. Monte Carlo experiments have demonstrated a desirable finite sample performance of the estimator. A test statistic for testing possible spatial co-integration is also considered. In Lee and Yu (2010b), this unified estimation approach is applied to study the market integration in Keller and Shiue (2007) with the SDPD model and test for the spatial co-integration.

Appendices

A Some Notes

A.1 The Eigenvalues of A_n: Three Cases of the DGP

From Subsection 14.2.1, the eigenvalues matrix of A_n can be decomposed as $D_n = \frac{\gamma_0+\rho_0}{1-\lambda_0}J_n + \tilde{D}_n$, where $J_n = \text{diag}\{1_{m_n}, 0, \cdots, 0\}$ and $\tilde{D}_n = \text{diag}\{0, \cdots, 0, d_{n,m_n+1}, \cdots, d_{nn}\}$ with $|d_{ni}| < 1$. Hence, $A_n^h = (\frac{\gamma_0+\rho_0}{1-\lambda_0})^h R_n J_n R_n^{-1} + B_n^h$ with $B_n^h = R_n \tilde{D}_n^h R_n^{-1}$. As $d_{ni} = \frac{\gamma_0+\rho_0\omega_{ni}}{1-\lambda_0\omega_{ni}}$, the derivative of $d_{ni} = \frac{\gamma_0+\rho_0\omega_{ni}}{1-\lambda_0\omega_{ni}}$ as a function of ω_{ni} is $\frac{\partial(\frac{\gamma_0+\rho_0\omega_{ni}}{1-\lambda_0\omega_{ni}})}{\partial\omega_{ni}} = \frac{\rho_0+\gamma_0\lambda_0}{(1-\lambda_0\omega_{ni})^2}$. Thus, d_{ni} is a monotonic function of ω_{ni}. Our setting assumes that $|d_{ni}| < 1$ whenever $d_{ni} \neq 1$. This requirement can be satisfied with appropriate restriction on the parameter space of ρ_0, γ_0 and λ_0 as shown below.

The case with $\rho_0 + \gamma_0\lambda_0 = 0$ implies that d_{ni} is a constant function of ω_{ni}. As $|\lambda_0| < 1$ (implied by Assumptions 1 and 3), the derivative is zero if and only if $\rho_0 + \gamma_0\lambda_0 = 0$, i.e., $\rho_0 = -\lambda_0\gamma_0$. In this situation, $d_{ni} = \frac{\gamma_0+\rho_0\omega_{ni}}{1-\lambda_0\omega_{ni}} = \gamma_0$, and all $|d_{ni}| < 1$ if $|\gamma_0| < 1$.[21] The d_{ni} is a strictly increasing function of ω_{ni} if and only if $\rho_0 + \lambda_0\gamma_0 > 0$; otherwise it is a strictly decreasing function of ω_{ni} when $\rho_0 + \lambda_0\gamma_0 < 0$. Let $\gamma_0 + \rho_0 + \lambda_0 = 1 + a$, where a is a constant. We have the stable case when $\gamma_0 + \rho_0 + \lambda_0 < 1$; the spatial cointegration case when $\gamma_0 + \rho_0 + \lambda_0 = 1$ but $\gamma_0 \neq 1$; and the explosive case when $\gamma_0 + \rho_0 + \lambda_0 > 1$. The condition $\rho_0 + \gamma_0\lambda_0 > 0$ (< 0) is equivalent to $(1 - \gamma_0)(1 - \lambda_0) > -a$ $(< -a)$ because $(1 - \gamma_0)(1 - \lambda_0) = \rho_0 + \gamma_0\lambda_0 - a$.

Assume that d_{ni} is an increasing function of ω_{ni}. As W_n is row-normalized, $-1 \leq \omega_{ni} \leq 1$ for all i. With the relation $d_{ni} = \frac{\gamma_0+\rho_0\omega_{ni}}{1-\lambda_0\omega_{ni}}$ on $[-1, 1]$, $d_{ni} = \frac{\gamma_0-\rho_0}{1+\lambda_0}$ at $\omega_{ni} = -1$, and $d_{ni} = \frac{\gamma_0+\rho_0}{1-\lambda_0}$ at $\omega_{ni} = 1$. Hence, the smallest eigenvalue of A_n will be greater than or equal to $\frac{\gamma_0-\rho_0}{1+\lambda_0}$, and the largest eigenvalue will occur at $\omega_{ni} = 1$. Hence, the possible range of d_{ni} with ω_{ni} in $[-1, 1]$ is $[\frac{\gamma_0-\rho_0}{1+\lambda_0}, \frac{\gamma_0+\rho_0}{1-\lambda_0}]$. The smallest eigenvalue of A_n will be greater than -1 if

$$\frac{\gamma_0 - \rho_0}{1 + \lambda_0} > -1 \Leftrightarrow 1 + \gamma_0 + \lambda_0 > \rho_0 \Leftrightarrow 1 - \rho_0 > -\frac{a}{2}.$$

Also, whenever $\omega_{ni} < \frac{1-\gamma_0}{\rho_0+\lambda_0}$, the corresponding $d_{ni} < 1$. This is so, because the critical value ω^* such that $\frac{\gamma_0+\rho_0\omega^*}{1-\lambda_0\omega^*} = 1$ is at $\omega^* = \frac{1-\gamma_0}{\rho_0+\lambda_0} = 1 - \frac{a}{(\rho_0+\lambda_0)}$.

In summary, for any eigenvalue ω_{ni} of W_n (with $|\omega_{ni}| \leq 1$), the corresponding eigenvalue of A_n is $d_{ni} = \frac{\gamma_0+\rho_0\omega_{ni}}{1-\lambda_0\omega_{ni}}$. Under the situation $(1-\gamma_0)(1-\lambda_0) > -a$, we have $d_{ni} < 1$ if $\omega_{ni} < 1 - \frac{a}{\rho_0+\lambda_0}$; and $d_{ni} > -1$ if $1 - \rho_0 > -\frac{a}{2}$.

Hence, we have the following sufficient conditions for three cases in our studies. Assume that $|\lambda_0| < 1$ and $(1 - \gamma_0)(1 - \lambda_0) > -a$.

1. Stable case: $a < 0$. If $\rho_0 + \lambda_0 > 0$, all $d_{ni} \leq 1$ (because $\omega_{ni} < 1 - \frac{a}{\rho_0+\lambda_0}$); if $1 - \rho_0 > -\frac{a}{2}$, $-1 < d_{ni}$.
2. Spatial co-integration case: $a = 0$. When $\omega_{ni} = 1$, $d_{ni} = 1$; when $\omega_{ni} < 1$ and $1 - \rho_0 > 0$, then $|d_{ni}| < 1$.
3. Explosive case: $a > 0$. When $\omega_{ni} = 1$, $d_{ni} > 1$; when $\omega_{ni} < 1 - \frac{a}{\rho_0+\lambda_0} = \frac{1-\gamma_0}{\rho_0+\lambda_0}$, $|d_{ni}| < 1$; furthermore, with $1 - \rho_0 > -\frac{a}{2}$, $|d_{ni}| < 1$.

A.2 Decomposition

From Equation 14.2, by iterative substitution, we have

$$Y_{nt} = A_n^{t+1}Y_{n,-1} + \sum_{h=0}^{t} A_n^h S_n^{-1}(\mathbf{c}_{n0} + X_{n,t-h}\beta_0 + V_{n,t-h} + \alpha_{t-h,0}l_n).$$

[21] For this special case, the model becomes $Y_{nt} = \gamma_0 Y_{n,t-1} + S_n^{-1}(X_{nt}\beta_0 + \mathbf{c}_{n0} + \alpha_{t0}l_n + V_{nt})$. Hence, this case is $Y_{nt} = \gamma_0 Y_{n,t-1} + S_n^{-1}X_{nt}\beta_0 + \frac{\alpha_{t0}}{1-\lambda_0}l_n + \epsilon_{nt}$, where $\epsilon_{nt} = \lambda_0 W_n \epsilon_{nt} + \mathbf{c}_{n0} + V_{nt}$ has the panel disturbance structure in Kapoor, Kelejian, and Prucha (2007). This model is close to the one considered in Su and Yang (2007) except for the resulting regressor term.

As $S_n^{-1}l_n = \frac{1}{1-\lambda_0}l_n$ and $A_n = S_n^{-1}(\gamma_0 I_n + \rho_0 W_n) = (\gamma_0 I_n + \rho_0 W_n)S_n^{-1}$, using $W_n l_n = l_n$, we have $A_n^h S_n^{-1} l_n = \frac{1}{1-\lambda_0}(\frac{\gamma_0+\rho_0}{1-\lambda_0})^h l_n$. By $A_n^h = (\frac{\gamma_0+\rho_0}{1-\lambda_0})^h R_n J_n R_n^{-1} + B_n^h$ and $R_n J_n R_n^{-1} S_n^{-1} = S_n^{-1} R_n J_n R_n^{-1} = \frac{1}{1-\lambda_0} R_n J_n R_n^{-1}$ (see Proposition B.4 in Yu, de Jong, and Lee 2007), the above equation can be written as

$$Y_{nt} = A_n^{t+1} Y_{n,-1} + \sum_{h=0}^{t} B_n^h S_n^{-1}(\mathbf{c}_{n0} + X_{n,t-h}\beta_0 + V_{n,t-h}) + \frac{1}{1-\lambda_0}\sum_{h=0}^{t}\left(\frac{\gamma_0+\rho_0}{1-\lambda_0}\right)^h$$

$$\times\, \alpha_{t-h,0} l_n + \frac{1}{1-\lambda_0}\sum_{h=0}^{t}\left(\frac{\gamma_0+\rho_0}{1-\lambda_0}\right)^h R_n J_n R_n^{-1}(\mathbf{c}_{n0} + X_{n,t-h}\beta_0 + V_{n,t-h}).$$

For $A_n^{t+1} Y_{n,-1}$, we have $A_n^{t+1} Y_{n,-1} = (\frac{\gamma_0+\rho_0}{1-\lambda_0})^{t+1} R_n J_n R_n^{-1} Y_{n,-1} + B_n^{t+1} Y_{n,-1}$, where

$$B_n^{t+1} Y_{n,-1} = \sum_{h=t+1}^{\infty} B_n^h S_n^{-1}(\mathbf{c}_{n0} + X_{n,t-h}\beta_0 + V_{n,t-h}) + \frac{1}{1-\lambda_0}\sum_{h=t+1}^{\infty} \alpha_{t-h,0} B_n^h l_n,$$

using $B_n A_n = B_n^2$ and $B_n S_n^{-1} = S_n^{-1} B_n$. The item with $B_n^h l_n$ is zero. Because R_n is the eigenvectors matrix of W_n and its first column is l_n, we have $R_n^{-1} l_n = e_{n1}$ which is the first unit vector. As $\tilde{D}_n e_{n1} = 0$, it follows that $B_n l_n = 0$. Hence, we can decompose Y_{nt} as $Y_{nt} = Y_{nt}^u + Y_{nt}^s + Y_{nt}^\alpha$, which is Equation 14.3.

The Y_{nt}^s represents a stable component as the eigenvalues of B_n can be less than unity in absolute value for many parameter values (see Appendix A.1). The Y_{nt}^α captures the component due to time dummies. As $|\frac{\gamma_0+\rho_0}{1-\lambda_0}| < 1$ if and only if $-1 < \gamma_0 + \rho_0 + \lambda_0 < 1$ because $\lambda_0 < 1$, Y_{nt}^u is also stable when $\gamma_0 + \rho_0 + \lambda_0 < 1$. But when $\gamma_0 + \rho_0 + \lambda_0 = 1$ (> 1), then $\frac{\gamma_0+\rho_0}{1-\lambda_0} = 1$ (> 1) and Y_{nt}^u may represent the unstable or explosive components.

A.3 Data Transformation

We can transform Equation 14.1 by $I_n - W_n$ into Equation 14.4, where the remaining $(I_n - W_n)\mathbf{c}_{n0}$ can be regarded as the individual effects. A special feature of the transformed Equation 14.4 is that the variance matrix of $(I_n - W_n)V_{nt}$ is equal to $\sigma_0^2 \Sigma_n \equiv \sigma_0^2 (I_n - W_n)(I_n - W_n)'$, which is singular. Hence, there is a linear dependence among the elements of $(I_n - W_n)V_{nt}$. An effective estimation method shall eliminate the linear dependence. This can be done with the eigenvalues and eigenvectors decomposition (see, e.g., Theil 1971, Chapter 6).

Let $[F_n, H_n]$ be the orthonormal matrix of eigenvectors and Λ_n be the diagonal matrix of nonzero eigenvalues of Σ_n such that $\Sigma_n F_n = F_n \Lambda_n$ and $\Sigma_n H_n = 0$. That is, the columns of F_n consist of eigenvectors of nonzero eigenvalues and those of H_n are for zero-eigenvalues of Σ_n. Let n^* be the number of nonzero eigenvalues. The F_n is an $n \times n^*$ matrix and Λ_n is an $n^* \times n^*$ diagonal matrix. Thus,

$$\begin{array}{llll} \Sigma_n F_n = F_n \Lambda_n, & F_n' F_n = I_{n^*}, & \Sigma_n H_n = \mathbf{0}, & H_n' H_n = I_{n-n^*}, \\ F_n' H_n = \mathbf{0}, & F_n F_n' + H_n H_n' = I_n, & F_n \Lambda_n F_n' = \Sigma_n. & \end{array} \tag{14.28}$$

Because $\Sigma_n H_n = 0$, it implies that $(I_n - W_n)' H_n = 0$. In turn, $W_n(I_n - W_n) = W_n(F_n F_n' + H_n H_n')(I_n - W_n) = W_n F_n F_n'(I_n - W_n)$. Denote $W_n^* = \Lambda_n^{-1/2} F_n' W_n F_n \Lambda_n^{1/2}$ which is a $n^* \times n^*$ matrix. This matrix can be regarded as a spatial weights matrix for the following transformed equation:

$$Y_{nt}^* = \lambda_0 W_n^* Y_{nt}^* + \gamma_0 Y_{n,t-1}^* + \rho_0 W_n^* Y_{n,t-1}^* + X_{nt}^* \beta_0 + \mathbf{c}_{n0}^* + V_{nt}^*, \tag{14.29}$$

where $Y_{nt}^* = \Lambda_n^{-1/2} F_n'(I_n - W_n) Y_{nt}$ and other variables are defined correspondingly. Note that this transformed Y_{nt}^* is an n^* dimensional vector. Hence, after the transformation, the observations at time period t have only n^* degrees of freedom. Equation 14.29 shall provide the structural parameters for estimation. This equation is in the format of a typical SAR model in panel data, where the number of observations is $n^* T$.

A.4 Determinant and Inverse of $S_n^*(\lambda) \equiv I_{n^*} - \lambda W_n^*$

We note that $S_n^* = \Lambda_n^{-1/2} F_n' S_n F_n \Lambda_n^{1/2}$. Let μ be a scalar. Because $(I_n - W_n) \cdot H_n = 0$,

$$\begin{aligned} &[F_n, H_n]'(\mu I_n - W_n)[F_n, H_n] \\ &= \begin{pmatrix} \mu I_{n^*} - F_n' W_n F_n & -F_n' W_n H_n \\ -H_n' W_n F_n & \mu I_{n-n^*} - H_n' W_n H_n \end{pmatrix} = \begin{pmatrix} \mu I_{n^*} - F_n' W_n F_n & -F_n' W_n H_n \\ 0 & (\mu - 1) I_{n-n^*} \end{pmatrix}. \end{aligned}$$

Hence, $|\mu I_n - W_n| = (\mu - 1)^{n-n^*} |\mu I_{n^*} - F_n' W_n F_n|$. Because $|\mu I_{n^*} - W_n^*| = |\mu I_{n^*} - \Lambda_n^{-1/2} F_n' W_n F_n \Lambda_n^{1/2}| = |\mu I_{n^*} - F_n' W_n F_n|$, $|\mu I_n - W_n| = (\mu - 1)^{n-n^*} |\mu I_{n^*} - W_n^*|$. As W_n has $(n - n^*)$ unit eigenvalues, the eigenvalues of W_n^* are exactly the remaining eigenvalues of W_n, which are less than unity in the absolute value. Furthermore,

$$|S_n^*(\lambda)| = \frac{1}{(1-\lambda)^{n-n^*}} |S_n(\lambda)|. \tag{14.30}$$

Thus, the tractability in computing the determinant of $S_n^*(\lambda)$ is exactly that of $S_n(\lambda)$. When W_n is constructed as a weights matrix that is row-normalized from an original symmetric matrix, Ord (1975) has suggested a computationally tractable method for the evaluation of $|S_n(\lambda)|$ at various λ for the ML method. This is useful for evaluating the determinant of $S_n^*(\lambda)$ even though the row sums of W_n^* may not even be unity.

Furthermore, a SAR model is an equilibrium model in the sense that the observed outcomes are determined by the equation. That is, the matrix $S_n^*(\lambda)$ shall be invertible. For the transformed equation (Equation 14.29), $S_n^*(\lambda)$ is invertible as long as the original matrices $S_n(\lambda)$ in Equation 14.1 is invertible. We can see that

$$S_n^{*-1}(\lambda) = \Lambda_n^{-1/2} F_n' S_n^{-1}(\lambda) F_n \Lambda_n^{1/2}, \tag{14.31}$$

because

$$S_n^*(\lambda) \cdot \Lambda_n^{-1/2} F_n' S_n^{-1}(\lambda) F_n \Lambda_n^{1/2} = \Lambda_n^{-1/2} F_n' S_n(\lambda) F_n F_n' S_n^{-1}(\lambda) F_n \Lambda_n^{1/2}$$
$$= \Lambda_n^{-1/2} F_n' S_n(\lambda)(I_n - H_n H_n') S_n^{-1}(\lambda) F_n \Lambda_n^{1/2}$$
$$= I_{n^*} - \Lambda_n^{-1/2} F_n' S_n(\lambda) H_n H_n' S_n^{-1}(\lambda) F_n \Lambda_n^{1/2} = I_{n^*},$$

as $H_n' W_n = H_n'$, $H_n' S_n^{-1}(\lambda) = \frac{1}{1-\lambda} H_n'$ and $H_n' F_n = \mathbf{0}$.

A.5 About $tr(G_n^*(\lambda))$

We have $G_n^*(\lambda) = \Lambda_n^{-1/2} F_n' G_n(\lambda) F_n \Lambda_n^{1/2}$. This is so because, from Equation 14.31,

$$G_n^*(\lambda) = W_n^* S_n^{-1*}(\lambda) = \Lambda_n^{-1/2} F_n' W_n F_n F_n' S_n^{-1}(\lambda) F_n \Lambda_n^{1/2}$$
$$= \Lambda_n^{-1/2} F_n' W_n (I_n - H_n H_n') S_n^{-1}(\lambda) F_n \Lambda_n^{1/2}$$
$$= \Lambda_n^{-1/2} F_n' W_n S_n^{-1}(\lambda) F_n \Lambda_n^{1/2} - \Lambda_n^{-1/2} F_n' W_n H_n H_n' S_n^{-1}(\lambda) F_n \Lambda_n^{1/2}$$
$$= \Lambda_n^{-1/2} F_n' W_n S_n^{-1}(\lambda) F_n \Lambda_n^{1/2} = \Lambda_n^{-1/2} F_n' G_n(\lambda) F_n \Lambda_n^{1/2},$$

because $H_n' S_n^{-1}(\lambda) F_n = \frac{1}{1-\lambda} H_n' F_n = \mathbf{0}$. Hence,

$$tr(G_n^*(\lambda)) = tr(F_n' G_n(\lambda) F_n) = tr[G_n(\lambda)(I_n - H_n H_n')] = tr(G_n(\lambda)) - \frac{n - n^*}{1-\lambda}, \tag{14.32}$$

where the last equality holds because $H_n' W_n = H_n'$ and $H_n' S_n^{-1}(\lambda) = \frac{1}{1-\lambda} H_n'$ implies that

$$tr(G_n(\lambda) H_n H_n') = tr(H_n' G_n(\lambda) H_n) = tr(H_n' W_n S_n^{-1}(\lambda) H_n) = \frac{1}{1-\lambda} tr(H_n' H_n)$$
$$= \frac{n - n^*}{1-\lambda}.$$

As $G_n^{*2}(\lambda) = \Lambda_n^{-1/2} F_n' G_n(\lambda) F_n F_n' G_n(\lambda) F_n \Lambda_n^{1/2}$, we have

$$tr(G_n^{*2}(\lambda)) = tr(F_n' G_n(\lambda) F_n F_n' G_n(\lambda) F_n) = tr(G_n(\lambda) F_n F_n' G_n(\lambda) F_n F_n')$$
$$= tr(G_n(\lambda)(I_n - H_n H_n') G_n(\lambda)(I_n - H_n H_n')).$$

Using $H_n' G_n(\lambda) = \frac{1}{(1-\lambda)} H_n'$ and $H_n' H_n = I_{n-n^*}$, we have $[G_n(\lambda)(I_n - H_n H_n')]^2 = [G_n(\lambda)]^2 [I_n - H_n H_n']$ and

$$tr(G_n^{*2}(\lambda)) = tr(G_n^2(\lambda)) - \frac{n - n^*}{(1-\lambda)^2}, \tag{14.33}$$

because $H_n' G_n^2(\lambda) H_n = \frac{1}{(1-\lambda)^2} H_n' H_n = \frac{1}{(1-\lambda)^2} I_{n-n^*}$. In terms of the eigenvalues of W_n, as $W_n = R_n \varpi R_n^{-1}$, $tr(G_n^*(\lambda)) = \sum_{j=m_n+1}^{n} \frac{\varpi_{nj}}{1-\lambda\varpi_{nj}}$ and $tr(G_n^{*2}(\lambda)) = \sum_{j=m_n+1}^{n} \frac{\varpi_{nj}^2}{(1-\lambda\varpi_{nj})^2}$.

Also, as $J_n^* = (I_n - W_n)'\Sigma_n^+(I_n - W_n)$ and $(I_n - W_n)G_n(\lambda) = G_n(\lambda)(I_n - W_n)$, Equation 14.32 implies that

$$\begin{aligned} tr(J_n^* G_n(\lambda)) &= tr(G_n(\lambda)(I_n - W_n)(I_n - W_n)'F_n\Lambda_n^{-1}F_n') \\ &= tr(G_n(\lambda)F_nF_n') = tr(G_n(\lambda)(I_n - H_nH_n')) \\ &= tr(G_n^*(\lambda)). \end{aligned} \tag{14.34}$$

For J_n^*, we have $tr(J_n^*) = tr((I_n - W_n)'F_n\Lambda_n^{-1}F_n'(I_n - W_n)) = tr(\Lambda_n^{-1}\Lambda_n) = n^*$ by using Equation 14.28. The J_n^* is an orthogonal projector. This is so, because J_n^* is symmetric and $J_n^*J_n^* = (I_n - W_n)'\Sigma_n^+(I_n - W_n)\cdot(I_n - W_n)'\Sigma_n^+(I_n - W_n) = (I_n - W_n)'\Sigma_n^+\Sigma_n\Sigma_n^+(I_n - W_n) = (I_n - W_n)'\Sigma_n^+(I_n - W_n) = J_n^*$.

B Lemmas for Some Statistics in the Model

The following lemmas can be found in Yu, de Jong, and Lee (2008). These lemmas provide orders for relevant terms in the score and the Hessian matrix of the log-likelihood function. They include also a CLT for linear and quadratic forms of disturbances. Denote $\mathbb{U}_{nt} = \sum_{h=1}^{\infty} P_{nh}V_{n,t+1-h}$, where $\{P_{nh}\}_{h=1}^{\infty}$ is a sequence of $n \times n$ nonstochastic square matrices.

Assumption A1 The disturbances $\{v_{it}\}$, $i = 1, 2, ..., n$ and $t = 1, 2, ..., T$, are i.i.d. across i and t with zero mean, variance σ_0^2 and $\mathrm{E}|v_{it}|^{4+\eta} < \infty$ for some $\eta > 0$.
Assumption A2 $\sum_{h=1}^{\infty} abs(P_{nh})$ is UB.
Assumption A3 The elements of $n \times 1$ vector D_{nt} are nonstochastic and bounded, uniformly in n and t.
Assumption A4 n is a nondecreasing function of T and T goes to infinity.

Lemma 14.1 *Under Assumptions A1 and A4, for an $n \times n$ nonstochastic matrix $\mathcal{B}_n$, uniformly bounded in row and column sums,*

$$\frac{1}{nT}\sum_{t=1}^{T} V_{nt}'\mathcal{B}_nV_{nt} - \mathrm{E}(\frac{1}{nT}\sum_{t=1}^{T} V_{nt}'\mathcal{B}_nV_{nt}) = O_p\left(\frac{1}{\sqrt{nT}}\right), \tag{14.35}$$

$$\frac{1}{n}\bar{V}_{nT}'\mathcal{B}_n\bar{V}_{nT} - \mathrm{E}(\frac{1}{n}\bar{V}_{nT}'\mathcal{B}_n\bar{V}_{nT}) = O_p\left(\frac{1}{\sqrt{nT^2}}\right), \tag{14.36}$$

and

$$\frac{1}{nT}\sum_{t=1}^{T} \tilde{V}_{nt}'\mathcal{B}_n\tilde{V}_{nt} - \mathrm{E}(\frac{1}{nT}\sum_{t=1}^{T} \tilde{V}_{nt}'\mathcal{B}_n\tilde{V}_{nt}) = O_p\left(\frac{1}{\sqrt{nT}}\right), \tag{14.37}$$

where $\mathrm{E}(\frac{1}{nT}\sum_{t=1}^{T} V_{nt}'\mathcal{B}_nV_{nt}) = O(1)$, $\mathrm{E}(\frac{1}{n}\bar{V}_{nT}'\mathcal{B}_n\bar{V}_{nT}) = O(T^{-1})$ *and* $\mathrm{E}(\frac{1}{nT}\sum_{t=1}^{T} \tilde{V}_{nt}'\mathcal{B}_n\tilde{V}_{nt}) = O(1)$.

Lemma 14.2 *Under Assumptions A1, A2, and A4,*

$$\sqrt{\frac{T}{n}}(\bar{\mathbb{U}}'_{nT,-1}\bar{V}_{nT} - \mathrm{E}(\bar{\mathbb{U}}'_{nT,-1}\bar{V}_{nT})) = O_p\left(\frac{1}{\sqrt{T}}\right), \tag{14.38}$$

where $\sqrt{\frac{T}{n}}\mathrm{E}(\bar{\mathbb{U}}'_{nT,-1}\bar{V}_{nT}) = \sqrt{\frac{n}{T}}\frac{1}{n}\sigma_0^2 tr\left(\sum_{h=1}^{\infty} P_{nh}\right) + O\left(\sqrt{\frac{n}{T^3}}\right)$.

For the lemma that follows, we will consider the following form:

$$Q_{nT} = \sum_{t=1}^{T}(\mathbb{U}'_{n,t-1}V_{nt} + D'_{nt}V_{nt} + V'_{nt}\mathcal{B}_n V_{nt} - \sigma_0^2 tr(\mathcal{B}_n)) = \sum_{t=1}^{T}\sum_{i=1}^{n} z_{nt,i},$$

where $\mathcal{B}_n$ is a $n \times n$ nonstochastic symmetric matrix which is UB, and $z_{nt,i} = (u_{i,t-1} + d_{nti})v_{it} + b_{n,ii}(v_{it}^2 - \sigma_0^2) + 2(\sum_{j=1}^{i-1} b_{n,ij}v_{jt})v_{it}$, where $b_{n,ij}$ is the (i, j) element of $\mathcal{B}_n$ and d_{nti} is the ith element of D_{nt}. Then, for the mean and variance of Q_{nT}, $\mu_{Q_{nT}} = 0$ and

$$\sigma_{Q_{nT}}^2 = T\sigma_0^4 tr\left(\sum_{h=1}^{\infty} P'_{nh}P_{nh}\right) + \sigma_0^2\sum_{t=1}^{T} D'_{nt}D_{nt}$$
$$+T\left((\mu_4 - 3\sigma_0^4)\sum_{i=1}^{n} b_{n,ii}^2 + 2\sigma_0^4 tr(\mathcal{B}_n^2)\right) + 2\mu_3\sum_{t=1}^{T}\sum_{i=1}^{n} d_{nti}b_{n,ii},$$

where $\mu_s = Ev_{it}^s$ for $s = 3, 4$.

Lemma 14.3 *Under Assumptions A1, A2, A3, A4, and that $\mathcal{B}_n$ is UB, if the sequence $\frac{1}{nT}\sigma_{Q_{nT}}^2$ is bounded away from zero, then, $\frac{Q_{nT}}{\sigma_{Q_{nT}}} \xrightarrow{d} N(0, 1)$.*

Denote $Z_{nt} = (Y_{n,t-1}, W_nY_{n,t-1}, X_{nt})$, we are going to provide some lemmas related to $(I_n - W_n)\tilde{Z}_{nt}$, $(I_n - W_n)\bar{Z}_{nT}$ and $\tilde{V}_{nt}$, $\bar{V}_{nT}$ of the model Equation 14.1.

Lemma 14.4 *Under Assumptions 1–7, for an $n \times n$ nonstochastic UB matrix $\mathcal{B}_n$,*

$$\frac{1}{nT}\sum_{t=1}^{T}\tilde{Z}'_{nt}(I_n - W_n)'\mathcal{B}_n(I_n - W_n)\tilde{Z}_{nt} - \mathrm{E}\frac{1}{nT}\sum_{t=1}^{T}\tilde{Z}'_{nt}(I_n - W_n)'\mathcal{B}_n(I_n - W_n)\tilde{Z}_{nt}$$
$$= O_p\left(\frac{1}{\sqrt{nT}}\right), \tag{14.39}$$

and

$$\frac{1}{nT}\sum_{t=1}^{T}\tilde{Z}'_{nt}(I_n - W_n)'\mathcal{B}_n(I_n - W_n)\tilde{V}_{nt} - \mathrm{E}\frac{1}{nT}\sum_{t=1}^{T}\tilde{Z}'_{nt}(I_n - W_n)'\mathcal{B}_n(I_n - W_n)\tilde{V}_{nt}$$
$$= O_p\left(\frac{1}{\sqrt{nT}}\right), \tag{14.40}$$

where $\mathrm{E}\frac{1}{nT}\sum_{t=1}^{T}\tilde{Z}'_{nt}(I_n - W_n)'\mathcal{B}_n(I_n - W_n)\tilde{Z}_{nt}$ *is* $O(1)$ *and* $\mathrm{E}\frac{1}{nT}\sum_{t=1}^{T}\tilde{Z}'_{nt}(I_n - W_n)'\mathcal{B}_n(I_n - W_n)\tilde{V}_{nt}$ *is* $O\left(\frac{1}{T}\right)$.

Lemma 14.5 *If* $\|B_n(\theta_0))\|_\infty < 1$ *(resp:* $\|B_n(\theta_0))\|_1 < 1$*), then the row sum (resp: column sum) of* $\sum_{h=0}^{\infty}B_n^h(\theta)$ *and* $\sum_{h=1}^{\infty}hB_n^{h-1}(\theta)$ *are bounded uniformly in n and in a neighborhood of* θ_0.

C Concentrated QML of the Transformation Approach

C.1 Reduced Form of Equation 14.1

From Equation 14.1, we have $Y_{nt} = S_n^{-1}(Z_{nt}\delta_0 + \mathbf{c}_{n0} + \alpha_t l_n + V_{nt})$ and $W_nY_{nt} = G_nZ_{nt}\delta_0 + G_n\mathbf{c}_{n0} + \alpha_tG_nl_n + G_nV_{nt}$. By using $S_n^{-1} = I_n + \lambda_0G_n$, $Y_{nt} = Z_{nt}\delta_0 + \lambda_0G_nZ_{nt}\delta_0 + S_n^{-1}\mathbf{c}_{n0} + \alpha_tS_n^{-1}l_n + S_n^{-1}V_{nt}$. With $S_n^{-1}l_n = \frac{1}{1-\lambda_0}l_n$ and $(I_n - W_n)l_n = \mathbf{0}$,

$$\tilde{Y}_{nt} = \tilde{Z}_{nt}\delta_0 + \lambda_0G_n\tilde{Z}_{nt}\delta_0 + \frac{\tilde{\alpha}_t}{1-\lambda_0}l_n + S_n^{-1}\tilde{V}_{nt},$$

and

$$(I_n - W_n)\tilde{Y}_{nt} = (I_n - W_n)\tilde{Z}_{nt}\delta_0 + \lambda_0(I_n - W_n)G_n\tilde{Z}_{nt}\delta_0 + (I_n - W_n)S_n^{-1}\tilde{V}_{nt}. \tag{14.41}$$

Similarly, as $W_n\tilde{Y}_{nt} = G_n\tilde{Z}_{nt}\delta_0 + \tilde{\alpha}_tG_nl_n + G_n\tilde{V}_{nt}$,

$$(I_n - W_n)W_n\tilde{Y}_{nt} = (I_n - W_n)G_n\tilde{Z}_{nt}\delta_0 + (I_n - W_n)G_n\tilde{V}_{nt}, \tag{14.42}$$

because $(I_n - W_n)G_nl_n = \frac{1}{1-\lambda_0}(I_n - W_n)l_n = \mathbf{0}$.

C.2 FOC and SOC of the Concentrated Log-Likelihood

Denote $J_n^* = (I_n - W_n)'\Sigma_n^+(I_n - W_n)$ and $G_n^* = W_n^* S_n^{*-1}$. By using $tr G_n(\lambda) - tr(G_n^*(\lambda)) = \frac{n-n^*}{1-\lambda}$ and $tr(G_n^2(\lambda)) - tr(G_n^{*2}(\lambda)) = \frac{n-n^*}{(1-\lambda)^2}$ (see Appendix A.5), the first-order derivatives of Equation 14.10 are

$$\frac{\partial \ln L_{n,T}(\theta)}{\partial \theta} = \begin{pmatrix} \frac{1}{\sigma^2}\sum_{t=1}^{T}(J_n^*\tilde{Z}_{nt})'\tilde{V}_{nt}(\theta) \\ \frac{1}{\sigma^2}\sum_{t=1}^{T}((J_n^* W_n\tilde{Y}_{nt})'\tilde{V}_{nt}(\theta)) - T tr G_n^*(\lambda) \\ \frac{1}{2\sigma^4}\sum_{t=1}^{T}(\tilde{V}_{nt}'(\theta)J_n\tilde{V}_{nt}(\theta) - n^*\sigma^2) \end{pmatrix}, \tag{14.43}$$

and the second order derivatives are

$$\frac{\partial^2 \ln L_{n,T}(\theta)}{\partial\theta\partial\theta'}$$

$$= -\begin{pmatrix} \frac{1}{\sigma^2}\sum_{t=1}^{T}\tilde{Z}_{nt}'J_n^*\tilde{Z}_{nt} & \frac{1}{\sigma^2}\sum_{t=1}^{T}\tilde{Z}_{nt}'J_n^*W_n\tilde{Y}_{nt} & \frac{1}{\sigma^4}\sum_{t=1}^{T}\tilde{Z}_{nt}'J_n^*\tilde{V}_{nt}(\theta) \\ * & \frac{1}{\sigma^2}\sum_{t=1}^{T}((W_n\tilde{Y}_{nt})'J_n^*W_n\tilde{Y}_{nt}) + T tr((G_n^*(\lambda))^2) & \frac{1}{\sigma^4}\sum_{t=1}^{T}(W_n\tilde{Y}_{nt})'J_n^*\tilde{V}_{nt}(\theta) \\ * & * & -\frac{n^*T}{2\sigma^4} + \frac{1}{\sigma^6}\sum_{t=1}^{T}\tilde{V}_{nt}'(\theta)J_n^*\tilde{V}_{nt}(\theta) \end{pmatrix}. \tag{14.44}$$

At θ_0,

$$\frac{1}{\sqrt{n^*T}}\frac{\partial \ln L_{n,T}(\theta_0)}{\partial\theta}$$

$$= \begin{pmatrix} \frac{1}{\sigma_0^2}\frac{1}{\sqrt{n^*T}}\sum_{t=1}^{T}\tilde{Z}_{nt}'J_n^*\tilde{V}_{nt} \\ \frac{1}{\sigma_0^2}\frac{1}{\sqrt{n^*T}}\sum_{t=1}^{T}(G_n\tilde{Z}_{nt}\delta_0)'J_n^*\tilde{V}_{nt} + \frac{1}{\sigma_0^2}\frac{1}{\sqrt{n^*T}}\sum_{t=1}^{T}(\tilde{V}_{nt}'G_n'J_n^*\tilde{V}_{nt} - \sigma_0^2 tr G_n^*) \\ \frac{1}{2\sigma_0^4}\frac{1}{\sqrt{n^*T}}\sum_{t=1}^{T}(\tilde{V}_{nt}'J_n^*\tilde{V}_{nt} - n^*\sigma_0^2) \end{pmatrix}, \tag{14.45}$$

which is a linear and quadratic form of $\tilde{V}_{nt}$. For the information matrix,

$$\Sigma_{\theta_0,nT} = \frac{1}{\sigma_0^2}\begin{pmatrix} \mathrm{E}\mathcal{H}_{nT} & \mathbf{0}_{(k+3)\times 1} \\ \mathbf{0}_{1\times(k+3)} & 0 \end{pmatrix}$$

$$+ \begin{pmatrix} \mathbf{0}_{(k+2)\times(k+2)} & \mathbf{0}_{(k+2)\times 1} & \mathbf{0}_{(k+2)\times 1} \\ \mathbf{0}_{1\times(k+2)} & \frac{1}{n^*}\left[tr(G_n' J_n^* G_n) + tr((G_n^*)^2)\right] & \frac{1}{\sigma_0^2 n^*} tr(J_n^* G_n) \\ \mathbf{0}_{1\times(k+2)} & \frac{1}{\sigma_0^2 n^*} tr(J_n^* G_n) & \frac{1}{2\sigma_0^4} \end{pmatrix}$$

$$- \begin{pmatrix} \mathbf{0}_{(k+2)\times(k+2)} & * & * \\ \frac{1}{\sigma_0^2 n^*}\mathrm{E}(G_n \bar{V}_{nT})' J_n^* \bar{Z}_{nT} & \frac{2}{\sigma_0^2 n^*}\mathrm{E}[(G_n \bar{Z}_{nT}\delta_0)' J_n^* G_n \bar{V}_{nT}] + \frac{1}{n^* T} tr(G_n' J_n^* G_n) & * \\ \frac{1}{\sigma_0^4 n^*}\mathrm{E}(\bar{Z}_{nT}' J_n^* \bar{V}_{nT})' & \frac{1}{\sigma_0^4 n^*}\mathrm{E}[(G_n \bar{Z}_{nT}\delta_0)' J_n^* \bar{V}_{nT}]' + \frac{1}{\sigma_0^2 n^* T} tr(J_n^* G_n) & \frac{1}{T}\frac{1}{\sigma_0^4} \end{pmatrix}.$$

C.3 About $-\frac{1}{n^*T}\frac{\partial^2 \ln L_{nT}(\theta)}{\partial\theta\partial\theta'}$

Denote $\|\theta - \theta_0\|$ as the Euclidean norm of $\theta - \theta_0$, and Θ_1 as a neighborhood of θ_0, then, we have

$$\frac{1}{n^*T}\frac{\partial^2 \ln L_{nT}(\theta)}{\partial\theta\partial\theta'} - \frac{1}{n^*T}\frac{\partial^2 \ln L_{nT}(\theta_0)}{\partial\theta\partial\theta'} = \|\theta - \theta_0\| \cdot O_p(1), \tag{14.46}$$

$$\frac{1}{n^*T}\frac{\partial^2 \ln L_{nT}(\theta_0)}{\partial\theta\partial\theta'} + \Sigma_{\theta_0,nT} = O_p\left(\frac{1}{\sqrt{n^*T}}\right), \tag{14.47}$$

$$\sup_{\theta\in\Theta}\left|\frac{1}{n^*T}\frac{\partial^2 \ln L_{nT}(\theta)}{\partial\theta\partial\theta'} - \frac{1}{n^*T}\mathrm{E}\frac{\partial^2 \ln L_{nT}(\theta)}{\partial\theta\partial\theta'}\right|_{ij} = O_p\left(\frac{1}{\sqrt{n^*T}}\right), \tag{14.48}$$

and

$$\sup_{\theta\in\Theta_1}\left|\frac{1}{n^*T}\mathrm{E}\frac{\partial^2 \ln L_{nT}(\theta)}{\partial\theta\partial\theta'} + \Sigma_{\theta_0,nT}\right|_{ij} = \sup_{\theta\in\Theta_1}\|\theta - \theta_0\| \cdot O(1) \tag{14.49}$$

for all $i, j = 1, 2, \cdots, k+4$. These are Equation A.11 to Equation A.14 in Yu, de Jong, and Lee (2008).

D Proofs for Claims and Theorems

D.1 Proof of nonsingularity of the information matrix

The result can be proved by using an argument by contradiction. For $\Sigma_{\theta_0} \equiv \lim_{T\to\infty}\Sigma_{\theta_0,nT}$, where $\Sigma_{\theta_0,nT}$ is Equation 14.12, we shall prove that $\Sigma_{\theta_0}\alpha = 0$

implies $\alpha = 0$, where $\alpha = (\alpha_1', \alpha_2, \alpha_3)'$, α_2, α_3 are scalars and α_1 is $(k+2) \times 1$ vector. If this is true, then, columns of Σ_{θ_0} would be linear independent so that Σ_{θ_0} would be nonsingular. Denote $\mathcal{H}_\delta = \text{plim}_{T\to\infty}\frac{1}{n^*T}\sum_{t=1}^T \tilde{Z}_{nt}' J_n^* \tilde{Z}_{nt}$, $\mathcal{H}_{\delta\lambda} = \text{plim}_{T\to\infty}\frac{1}{n^*T}\sum_{t=1}^T \tilde{Z}_{nt}' J_n^* G_n \tilde{Z}_{nt}\delta_0$, $\mathcal{H}_{\lambda\delta} = \mathcal{H}_{\delta\lambda}'$ and $\mathcal{H}_\lambda = \text{plim}_{T\to\infty}\frac{1}{n^*T}\sum_{t=1}^T (G_n\tilde{Z}_{nt}\delta_0)' J_n^* G_n \tilde{Z}_{nt}\delta_0$. Then

$$\Sigma_{\theta_0} = \frac{1}{\sigma_0^2}\begin{pmatrix} \mathcal{H}_\delta & \mathcal{H}_{\delta\lambda} & \mathbf{0}_{(k+2)\times 1} \\ \mathcal{H}_{\lambda\delta} & E\mathcal{H}_\lambda + \lim_{n\to\infty}\frac{\sigma_0^2}{n^*}[tr(G_n' J_n^* G_n) + tr((G_n^*)^2)] & \lim_{n\to\infty}\frac{1}{n^*}tr(J_n^* G_n) \\ \mathbf{0}_{1\times(k+2)} & \lim_{n\to\infty}\frac{1}{n^*}tr(J_n^* G_n) & \frac{1}{2\sigma_0^2} \end{pmatrix}.$$

Hence, $\Sigma_{\theta_0}\alpha = 0$ implies

$$\mathcal{H}_\delta \times \alpha_1 + \mathcal{H}_{\delta\lambda} \times \alpha_2 = 0,$$

$$\frac{1}{\sigma_0^2}\mathcal{H}_{\lambda\delta} \times \alpha_1 + \left(\frac{1}{\sigma_0^2}\mathcal{H}_\lambda + \lim_{n\to\infty}\frac{1}{n^*}\left[tr(G_n' J_n^* G_n) + tr((G_n^*)^2)\right]\right) \times \alpha_2 + \lim_{n\to\infty}\frac{1}{\sigma_0^2 n^*}tr(J_n^* G_n) \times \alpha_3 = 0,$$

$$\lim_{n\to\infty}\frac{1}{n^*}tr(J_n^* G_n) \times \alpha_2 + \frac{1}{2\sigma_0^2} \times \alpha_3 = 0.$$

From the first equation, $\alpha_1 = -(\mathcal{H}_\delta)^{-1}\mathcal{H}_{\delta\lambda} \times \alpha_2$; from the third equation, $\alpha_3 = -2\lim_{n\to\infty}\frac{\sigma_0^2}{n^*}tr(J_n^* G_n) \times \alpha_2$. By eliminating α_1 and α_3, the remaining equation becomes

$$\left\{\left(\frac{1}{\sigma_0^2}(\mathcal{H}_\lambda - \mathcal{H}_{\lambda\delta}\mathcal{H}_\delta^{-1}\mathcal{H}_{\delta\lambda})\right) + \lim_{n\to\infty}\frac{1}{n^*}\left[tr(G_n' J_n^* G_n) + tr((G_n^*)^2) - 2\frac{tr^2(J_n^* G_n)}{n^*}\right]\right\} \times \alpha_2 = 0.$$

Using Equation 14.34 and that J_n^* is idempotent, denote $C_n = G_n^* - \frac{tr(G_n^*)}{n^*}$, we have

$$tr(G_n' J_n^* G_n) + tr((G_n^*)^2) - 2\frac{tr^2(J_n^* G_n)}{n^*} = tr(G_n^{*\prime} G_n^*) + tr((G_n^*)^2) - 2\frac{tr^2(G_n^*)}{n^*}$$
$$= \frac{1}{2}tr(C_n' + C_n)(C_n' + C_n)',$$

which is nonnegative. Hence, if the limit of $E\mathcal{H}_{nT}$ is nonsingular or the limit of $\frac{1}{n^*}(tr(G_n^{*\prime}G_n^*) + tr((G_n^*)^2) - 2\frac{tr^2(G_n^*)}{n^*})$ is nonzero, we have $\alpha_2 = 0$ and hence $\alpha = 0$. This proves the nonsingularity of Σ_{θ_0}. ■

D.2 Proof of Theorem 14.1

To prove $\frac{1}{n^*T}\ln L_{n,T}(\theta) - Q_{n,T}(\theta) \xrightarrow{p} 0$ *uniformly in* θ *in any compact parameter space* Θ:

From $\tilde{V}_{nt}(\theta) \equiv S_n(\lambda)\tilde{Y}_{nt} - \tilde{Z}_{nt}\delta - \tilde{\alpha}_t l_n$ and $\tilde{V}_{nt} = S_n\tilde{Y}_{nt} - \tilde{Z}_{nt}\delta_0 - \tilde{\alpha}_{t0}l_n$, using $J_n^* l_n = \mathbf{0}$, we have $J_n^*\tilde{V}_{nt}(\theta) = J_n^*\tilde{V}_{nt} - (\lambda - \lambda_0)J_n^* W_n\tilde{Y}_{nt} - J_n^*\tilde{Z}_{nt}(\delta - \delta_0)$. As Θ is compact and σ^2 is bounded away from zero in Θ, by Lemma 14.1 and 14.4,

$$\frac{1}{n^*T}\ln L_{n,T}(\theta) - Q_{n,T}(\theta)$$

$$= -\frac{1}{2\sigma^2}\left(\frac{1}{n^*T}\sum_{t=1}^{T}\tilde{V}'_{nt}(\theta)J_n^*\tilde{V}_{nt}(\theta) - \frac{1}{n^*T}\mathrm{E}\sum_{t=1}^{T}\tilde{V}'_{nt}(\theta)J_n^*\tilde{V}_{nt}(\theta)\right) \xrightarrow{p} 0$$

uniformly in θ in Θ.

To prove $Q_{n,T}(\theta)$ *is uniformly equicontinuous in* θ *in any compact parameter space* Θ:

For $Q_{n,T}(\theta)$ in Equation 14.11, as $J_n^*\tilde{V}_{nt}(\theta) = J_n^*[S_n(\lambda)\tilde{Y}_{nt} - \tilde{Z}_{nt}\delta]$ and $\tilde{Y}_{nt} = S_n^{-1}\tilde{Z}_{nt}\delta_0 + S_n^{-1}\tilde{V}_{nt} + \frac{\tilde{\alpha}_{t0}}{1-\lambda_0}l_n$,

$$J_n^*\tilde{V}_{nt}(\theta) = J_n^*[S_n(\lambda)S_n^{-1}\tilde{Z}_{nt}\delta_0 - \tilde{Z}_{nt}\delta + S_n(\lambda)S_n^{-1}\tilde{V}_{nt}]$$

because $J_n^* l_n = \mathbf{0}$. Hence,

$$\begin{aligned}\mathrm{E}\frac{1}{n^*T}\sum_{t=1}^{T}\tilde{V}'_{nt}(\theta)J_n^*\tilde{V}_{nt}(\theta) &= \frac{1}{n^*T}\mathrm{E}\sum_{t=1}^{T}(S_n(\lambda)S_n^{-1}\tilde{Z}_{nt}\delta_0 - \tilde{Z}_{nt}\delta)'J_n^*(S_n(\lambda)\\ &\quad\times S_n^{-1}\tilde{Z}_{nt}\delta_0 - \tilde{Z}_{nt}\delta) + \frac{1}{n^*}\frac{T-1}{T}\sigma_0^2 tr\\ &\quad\times (S_n^{-1'}S_n'(\lambda)J_n^*S_n(\lambda)S_n^{-1}) + \frac{2}{n^*T}\mathrm{E}\sum_{t=1}^{T}\\ &\quad\times (S_n(\lambda)S_n^{-1}\tilde{Z}_{nt}\delta_0 - \tilde{Z}_{nt}\delta)'J_n^*S_n(\lambda)S_n^{-1}\tilde{V}_{nt}.\end{aligned} \tag{14.50}$$

With these terms, similar to Lee and Yu (2010a), it can be shown that $Q_{n,T}(\theta)$ is uniformly equicontinuous in θ in any compact parameter space Θ.

To prove the identification:

As $tr J_n^* = n^*$, $\mathrm{E}\sum_{t=1}^{T}\tilde{V}'_{nt}J_n^*\tilde{V}_{nt} = n^*(T-1)\sigma_0^2$ from Lemma 14.1. Hence, $\frac{1}{n^*T}\mathrm{E}\ln L_{n,T}(\theta) - \frac{1}{n^*T}\mathrm{E}\ln L_{n,T}(\theta_0) = -\frac{1}{2}(\ln\sigma^2 - \ln\sigma_0^2) + \frac{1}{n^*}\ln|S_n(\lambda)| - \frac{1}{n^*}\ln|S_n| - \frac{n-n^*}{n^*}(\ln(1-\lambda) - \ln(1-\lambda_0)) - (\frac{1}{2\sigma^2}\frac{1}{n^*T}\sum_{t=1}^{T}E\,\tilde{V}'_{nt}(\theta)J_n^*\tilde{V}_{nt}(\theta) - \frac{T-1}{2T})$. By using $S_n(\lambda)S_n^{-1} = I_n + (\lambda_0 - \lambda)G_n$, from Equation 14.50, $\frac{1}{n^*T}\mathrm{E}\ln L_{n,T}(\theta) - \frac{1}{n^*T}\mathrm{E}\ln L_{n,T}(\theta_0) = T_{1,n}(\lambda,\sigma^2) - \frac{1}{2\sigma^2}T_{2,n,T}(\delta,\lambda) + O(T^{-1})$, where

$$\begin{aligned}T_{1,n}(\lambda,\sigma^2) &= -\frac{1}{2}(\ln\sigma^2 - \ln\sigma_0^2) + \frac{1}{n^*}\ln|S_n(\lambda)| - \frac{1}{n^*}\ln|S_n| - \frac{n-n^*}{n^*}\\ &\quad\times(\ln(1-\lambda) - \ln(1-\lambda_0)) - \frac{1}{2\sigma^2}(\sigma_n^2(\lambda) - \sigma^2),\end{aligned}$$

and

$$T_{2,n,T}(\delta,\lambda) = \frac{1}{n^*T}\sum_{t=1}^{T} \mathrm{E}\{[\tilde{Z}_{nt}(\delta_0-\delta) + (\lambda_0-\lambda)G_n\tilde{Z}_{nt}\delta_0]' J_n^* \times [\tilde{Z}_{nt}(\delta_0-\delta) + (\lambda_0-\lambda)G_n\tilde{Z}_{nt}\delta_0]\},$$

where $\sigma_n^2(\lambda) = \frac{\sigma_0^2}{n^*} tr(S_n^{-1\prime} S_n'(\lambda) J_n^* S_n(\lambda) S_n^{-1})$. Consider the pure spatial process $Y_{nt} = \lambda_0 W_n Y_{nt} + \alpha_t l_n + V_{nt}$ for a single period t. With similar data transformation as in Equation 14.5, the log-likelihood function of this process is

$$\ln L_{p,n}(\lambda,\sigma^2) = -\frac{n^*}{2}\ln 2\pi - \frac{n^*}{2}\ln\sigma^2 - (n-n^*)\ln(1-\lambda) + \ln|S_n(\lambda)| - \frac{1}{2\sigma^2} V_{nt}'(\lambda) J_n^* V_{nt}'(\lambda), \tag{14.51}$$

where $V_{nt}(\lambda) = S_n(\lambda)Y_{nt}$. Let $\mathrm{E}_p(\cdot)$ be the expectation operator for Y_{nt} based on this pure spatial autoregressive process. It follows that

$$\mathrm{E}_p(\frac{1}{n^*}\ln L_{p,n}(\lambda,\sigma^2)) - \mathrm{E}_p(\frac{1}{n^*}\ln L_{p,n}(\lambda_0,\sigma_0^2)) = -\frac{1}{2}(\ln\sigma^2 - \ln\sigma_0^2) + \frac{1}{n^*}\ln|S_n(\lambda)| - \frac{1}{n^*}\ln|S_n| - \frac{n-n^*}{n^*}(\ln(1-\lambda) - \ln(1-\lambda_0)) - \frac{1}{2\sigma^2}(\sigma_n^2(\lambda) - \sigma^2),$$

which equals to $T_{1,n}(\lambda,\sigma^2)$. By the information inequality, $\ln L_{p,n}(\lambda,\sigma^2) - \ln L_{p,n}(\lambda_0,\sigma_0^2) \leq 0$. Thus, $T_{1,n}(\lambda,\sigma^2) \leq 0$ for any (λ,σ^2).

For $T_{2,n,T}(\delta,\lambda)$, it is a quadratic function of δ and λ. Under the assumed condition that $\lim_{T\to\infty}\mathrm{E}\mathcal{H}_{nT}$ is nonsingular, $\lim_{T\to\infty} T_{2,n,T}(\delta,\lambda) > 0$ whenever $(\delta,\lambda) \neq (\delta_0,\lambda_0)$. So, (δ,λ) is globally identified. Given λ_0, σ_0^2 is also the unique maximizer of $T_{1,n}(\lambda_0,\sigma^2)$ for any given n^*. In the event that $n^*\to\infty$, σ_0^2 is the unique maximizer of $\lim_{T\to\infty} T_{1,n}(\lambda_0,\sigma^2)$. Hence, $(\delta,\lambda,\sigma^2)$ is globally identified.

By combining the results above together, the consistency follows. ■

D.3 Proof of Theorem 14.2

When the limit of $\mathrm{E}\mathcal{H}_{nT}$ is singular, δ_0 and λ_0 cannot be identified from $T_{2,n,T}(\delta,\lambda)$ in Appendix D.2. Identification requires that the limit of $T_{1,n}(\lambda,\sigma^2)$ is strictly less than zero whenever $(\lambda,\sigma^2) \neq (\lambda_0,\sigma_0^2)$. Thus, the identification will just be from the likelihood function Equation 14.51. By concentrating out

σ^2 in Equation 14.51, we have the concentrated log-likelihood function

$$\begin{aligned}\ln L_{p,n}(\lambda) &= -\frac{n^*}{2}(\ln(2\pi)+1) - \frac{n^*}{2}\ln\hat{\sigma}^2_{nt}(\lambda) - (n-n^*)\ln(1-\lambda) + \ln|S_n(\lambda)| \\ &= -\frac{n^*}{2}(\ln(2\pi)+1) - \frac{n^*}{2}\ln\hat{\sigma}^2_{nt}(\lambda) + \ln|S_n^*(\lambda)|\end{aligned}$$

from Equation 14.30, where $\hat{\sigma}^2_{nt}(\lambda) = \frac{1}{n^*}V'_{nt}(\lambda)J_n^*V_{nt}(\lambda)$. Also, we have the corresponding $Q_n(\lambda) = \max_{\sigma^2} \mathrm{E}(\ln L_{p,n}(\lambda,\sigma^2)) = -\frac{n^*}{2}(\ln(2\pi)+1) - \frac{n^*}{2}\ln\sigma^2_n(\lambda) + \ln|S_n^*(\lambda)|$. Identification of λ_0 requires that $\lim_{n\to\infty}\frac{1}{n^*}[Q_n(\lambda) - Q_n(\lambda_0)] \neq 0$ whenever $\lambda \neq \lambda_0$, which is equivalent to

$$\frac{1}{n^*}\ln\left|\sigma_0^2 S_n^{*\prime-1}S_n^{*-1}\right| - \frac{1}{n^*}\ln\left|\sigma_n^2(\lambda)S_n^{*\prime-1}(\lambda)S_n^{*-1}(\lambda)\right| \neq 0 \text{ for } \lambda \neq \lambda_0.$$

After λ_0 is identified, σ_0^2 is then identified. Also, given λ_0, δ_0 can then be identified from $\lim_{T\to\infty} T_{2,n,T}(\delta,\lambda)$. Combined with uniform convergence and equicontinuity, the consistency follows. ■

D.4 Proof of Theorem 14.3

From Equation 14.13,

$$J_n^*\tilde{Z}_{nt} = J_n^*\tilde{Z}_{nt}^{(c)} - (J_n^*\bar{U}_{nT,-1},\ J_n^*W_n\bar{U}_{nT,-1},\ \mathbf{0}_{n\times k}), \tag{14.52}$$

where $J_n^*\tilde{Z}_{nt}^{(c)}$ is uncorrelated with V_{nt} and the remaining term is correlated with V_{nt} when $t \leq T-1$. For the score decomposition $\frac{1}{\sqrt{n^*T}}\frac{\partial \ln L_{n,T}(\theta_0)}{\partial\theta} = \frac{1}{\sqrt{n^*T}}\frac{\partial \ln L_{n,T}^{(c)}(\theta_0)}{\partial\theta} - \Delta_{nT}$ in Equation 14.14, the first term is a linear and quadratic form of V_{nt}, and the asymptotic distribution can be derived from the CLT for martingale difference arrays (Lemma 14.3). Hence,

$$\frac{1}{\sqrt{n^*T}}\frac{\partial \ln L_{n,T}(\theta_0)}{\partial\theta} + \Delta_{nT} \xrightarrow{d} N(0, \Sigma_{\theta_0} + \Omega_{\theta_0}).$$

For Δ_{nT}, from Equation 14.36 in Lemma 14.1 and Equation 14.38 in Lemma 14.2, we have $\Delta_{nT} = \sqrt{\frac{n^*}{T}}a_{\theta_0,n} + O(\sqrt{\frac{n^*}{T^3}}) + O_p(\frac{1}{\sqrt{T}})$ where $a_{\theta_0,n}$ specified in Equation 14.18 is $O(1)$.

The Taylor expansion gives $\sqrt{n^*T}(\hat{\theta}_{nT} - \theta_0) = (-\frac{1}{n^*T}\frac{\partial^2 \ln L_{n,T}(\bar{\theta}_{nT})}{\partial\theta\partial\theta'})^{-1}\frac{1}{\sqrt{n^*T}} \times \frac{\partial \ln L_{n,T}(\theta_0)}{\partial\theta}$, where $\bar{\theta}_{nT}$ lies between θ_0 and $\hat{\theta}_{nT}$. Similar to Lee and Yu (2010a), we have $\hat{\theta}_{nT} - \theta_0 = O_p(\max(\frac{1}{\sqrt{n^*T}}, \frac{1}{T}))$. Using the fact that $(-\frac{1}{n^*T}\frac{\partial^2 \ln L_{n,T}(\bar{\theta}_{nT})}{\partial\theta\partial\theta'})^{-1} = \Sigma_{\theta_0,nT}^{-1} + O_p(\max(\frac{1}{\sqrt{n^*T}}, \frac{1}{T}))$, given that $\Sigma_{\theta_0,nT}$ is nonsingular and its inverse is of order $O(1)$, we have

$$\sqrt{n^*T}(\hat{\theta}_{nT}-\theta_0) = \left(-\frac{1}{n^*T}\frac{\partial^2 \ln L_{n,T}(\bar{\theta}_{nT})}{\partial\theta\partial\theta'}\right)\cdot\left(\frac{1}{\sqrt{n^*T}}\frac{\partial \ln L^{(c)}_{n,T}(\theta_0)}{\partial\theta} - \Delta_{nT}\right)$$

$$= \Sigma^{-1}_{\theta_0,nT}\cdot\frac{1}{\sqrt{n^*T}}\frac{\partial \ln L^{(c)}_{n,T}(\theta_0)}{\partial\theta} + O_p\left(\max\left(\frac{1}{\sqrt{n^*T}},\frac{1}{T}\right)\right)$$

$$\cdot\frac{1}{\sqrt{n^*T}}\frac{\partial \ln L^{(c)}_{n,T}(\theta_0)}{\partial\theta} - \Sigma^{-1}_{\theta_0,nT}\cdot\Delta_{nT}$$

$$-O_p\left(\max\left(\frac{1}{\sqrt{n^*T}},\frac{1}{T}\right)\right)\cdot\Delta_{nT},$$

which implies that

$$\sqrt{n^*T}(\hat{\theta}_{nT}-\theta_0) + \Sigma^{-1}_{\theta_0,nT}\cdot\Delta_{nT} + O_p\left(\max\left(\frac{1}{\sqrt{n^*T}},\frac{1}{T}\right)\right)\cdot\Delta_{nT}$$

$$= (\Sigma^{-1}_{\theta_0,nT} + o_p(1))\cdot\frac{1}{\sqrt{n^*T}}\frac{\partial \ln L^{(c)}_{n,T}(\theta_0)}{\partial\theta}. \tag{14.53}$$

As $\Sigma_{\theta_0} = \lim_{T\to\infty}\Sigma_{\theta_0,nT}$ exists, then using $\Delta_{nT} = \sqrt{\frac{n^*}{T}}a_{\theta_0,n} + O(\sqrt{\frac{n^*}{T^3}}) + O_p(\frac{1}{\sqrt{T}})$ with $a_{\theta_0,n} = O(1)$ and $\frac{1}{\sqrt{n^*T}}\frac{\partial \ln L^{(c)}_{n,T}(\theta_0)}{\partial\theta} \xrightarrow{d} N(0,\Sigma_{\theta_0}+\Omega_{\theta_0})$, the result in the theorem follows. ■

D.5 Proof for Theorem 14.4

Theorem 14.3 states that $\sqrt{n^*T}(\theta_{nT}-\theta_0) + \sqrt{\frac{n^*}{T}}b_{\theta_0,nT} + O_p(\max(\sqrt{\frac{n^*}{T^3}},\frac{1}{\sqrt{T}})) \xrightarrow{d} N(0,\Sigma^{-1}_{\theta_0}(\Sigma_{\theta_0}+\Omega_{\theta_0})\Sigma^{-1}_{\theta_0})$. As the bias corrected estimator is $\hat{\theta}^1_{nT} = \hat{\theta}_{nT} + \frac{1}{T}(-\frac{1}{n^*T}\mathrm{E}\frac{\partial^2 \ln L_{nT}(\hat{\theta}_{nT})}{\partial\theta\partial\theta'})^{-1}\cdot a_n(\hat{\theta}_{nT})$ where $a_n(\theta) = a_{\theta,n}$, we have $\sqrt{n^*T}(\hat{\theta}^1_{nT}-\theta_0) \xrightarrow{d} N(0,\Sigma^{-1}_{\theta_0}(\Sigma_{\theta_0}+\Omega_{\theta_0})\Sigma^{-1}_{\theta_0})$ if

$$\sqrt{\frac{n^*}{T}}\left(\left(-\frac{1}{n^*T}\mathrm{E}\frac{\partial^2 \ln L_{nT}(\hat{\theta}_{nT})}{\partial\theta\partial\theta'}\right)^{-1} a_n(\hat{\theta}_{nT}) - \Sigma^{-1}_{\theta_0,nT}a_n(\theta_0)\right) \xrightarrow{p} 0 \tag{14.54}$$

and $\frac{n^*}{T^3} \to 0$. Similar to Lee and Yu (2010a), Equation 14.54 can be proved under the assumed regularity conditions. ■

References

Alvarez, J., and M. Arellano. 2003. The time series and cross-section asymptotics of dynamic panel data estimators. *Econometrica* 71:1121–1159.

Amemiya, T. 1971. The estimation of the variances in a variance-components model. *International Economic Review* 12:1–13.

Anderson, T. W. 1959. On asymptotic distributions of estimates of parameters of stochastic difference equations. *Annals of Mathematical Statistics* 30:676–687.

Baltagi, B., S. H. Song, and W. Koh. 2003. Testing panel data regression models with spatial error correlation. *Journal of Econometrics* 117:123–150.

Baltagi, B., S. H. Song, B. C. Jung, and W. Koh. 2007. Testing for serial correlation, spatial autocorrelation and random effects using panel data. *Journal of Econometrics* 140:5–51.

Hahn, J., and G. Kuersteiner. 2002. Asymptotically unbiased inference for a dynamic panel model with fixed effects when both n and T are Large. *Econometrica* 70:1639–1657.

Hahn, J., and W. Newey. 2004. Jackknife and analytical bias reduction for nonlinear panel models. *Econometrica* 72:1295–1319.

Hamilton, J. 1994. *Times Series Analysis.* Princeton, NJ: Princeton University Press.

Horn, R., and C. Johnson. 1985. *Matrix Algebra.* New York: Cambridge University Press.

Kapoor, M., H. H. Kelejian, and I. R. Prucha. 2007. Panel data models with spatially correlated error components. *Journal of Econometrics* 140:97–130.

Kelejian, H. H., and I. R. Prucha. 1998. A generalized spatial two-stage least squares procedure for estimating a spatial autoregressive model with autoregressive disturbance. *Journal of Real Estate Finance and Economics* 17(1):99–121.

Kelejian, H. H., and I. R. Prucha. 2001. On the asymptotic distribution of the Moran I test statistic with applications. *Journal of Econometrics* 104:219–257.

Keller, W., and C. H. Shiue. 2007. The origin of spatial interaction. *Journal of Econometrics* 140:304–332.

Korniotis, G. M. 2005. A dynamic panel estimator with both fixed and spatial effects. Manuscript, Department of Finance, University of Notre Dame, South Bend, IN.

Lee, L. F. 2004. Asymptotic distributions of quasi-maximum likelihood estimators for spatial econometric models. *Econometrica* 72:1899–1925.

Lee, L. F. 2007. GMM and 2SLS estimation of mixed regressive, spatial autoregressive models. *Journal of Econometrics* 137:489–514.

Lee, L. F., and J. Yu. 2010a. A spatial dynamic panel data model with both time and individual fixed effects. *Econometric Theory* 26:564–597.

Lee, L. F., and J. Yu. 2010b. Some recent developments in spatial panel data models. *Regional Science and Urban Economics* 40:255–271.

Nielsen, B. 2001. The asymptotic distribution of unit root tests of unstable autoregressive processes. *Econometrica* 69:211–219.

Nielsen, B. 2005. Strong consistency results for least squares estimators in general vector autoregressions with deterministic terms. *Econometric Theory* 21:534–561.

Ord, J. K. 1975. Estimation methods for models of spatial interaction. *Journal of the American Statistical Association* 70:120–297.

Phillips, P. C. B., and T. Magdalinos. 2007. Limit theory for moderate deviations from a unit root. *Journal of Econometrics* 136:115–130.

Phillips, P. C. B., and H. R. Moon. 1999. Linear regression limit theory for nonstationary panel data. *Econometrica* 67:1057–1111.

Rothenberg, T. J. 1971. Identification in parametric models. *Econometrica* 39:577–591.

Sims, C. A., J. H. Stock, and M. W. Watson. 1990. Inference in linear time series models with some unit roots. *Econometrica* 58:113–144.

Su, L., and Z. Yang. 2007. QML estimation of dynamic panel data models with spatial errors. Manuscript, Singapore Management University.

Theil, H. 1971. *Principles of Econometrics.* New York: John Wiley & Sons.

Wallace, T. D., and A. Hussain. 1969. The use of error components models in combining cross-section and time-series data. *Econometrica* 37:55–72.

White, J. S. 1958. The limiting distribution of the serial correlation coefficient in the explosive case I. *Annals of Mathematical Statistics* 29:1188–1197.

White, J. S. 1959. The limiting distribution of the serial correlation coefficient in the explosive case II. *Annals of Mathematical Statistics* 30:831–834.

White, H. 1994. *Estimation, Inference and Specification Analysis.* New York: Cambridge University Press.

Yu, J., R. de Jong, and L. F. Lee. 2007. Quasi-maximum likelihood estimators for spatial dynamic panel data with fixed effects when both n and T are large: a nonstationary case. Manuscript, The Ohio State University, Columbus, OH.

Yu, J., R. de Jong, and L. F. Lee. 2008. Quasi-maximum likelihood estimators for spatial dynamic panel data with fixed effects when both n and T are large. *Journal of Econometrics* 146:118–134.

Yu, J., and L. F. Lee. 2010. Estimation of unit root spatial dynamic panel data models. *Econometric Theory,* forthcoming. doi:10.1017/s0266466609990600

15

Spatial Panels

Badi H. Baltagi

CONTENTS

15.1 Introduction

Economists are interested in spill-over effects and externalities. Spatial models allow simple econometric methods for modeling these spill-over effects. For example, you spend more money on police in one neighborhood, you may increase the crime in an adjacent neighborhood. This externality is dependent on contiguity of the neighborhoods, their common borders, or the distance between these neighborhoods. The same idea can be applied for the analysis of welfare or trade. If California is generous in providing welfare to its residents, this may attract welfare recipients from adjacent states. Gravity models of trade use distance, common border, common language, culture and history, common colonizer, common currency, to see if these things enhance trade. These may be interpreted as distances that are economic, historic, or cultural in nature. In sum, these metrics can be used in a spatial economic model to explain crime or trade or dependency on welfare.

Spatial models deal with correlation across spatial units usually in a cross-section setting; see Anselin (1988, 2001) and Anselin and Bera (1998) for a nice introduction to this literature. Panel data models allow the researcher to control for heterogeneity across these units; see Baltagi (2008a). Spatial panel models can control for *both* heterogeneity and spatial correlation; see for example Baltagi, Song, and Koh (2003) for a joint test of spatial correlation and heterogeneity using panel data. Recent spatial panel data applications in economics include household level survey data from villages observed over time to study nutrition (see Case 1991); per capita expenditures on police to study their effect on reducing crime across counties (see Kelejian and Robinson 1992); the productivity of public capital like roads and highways in the private sector across U.S. states (see Holtz-Eakin 1994); hedonic housing equations using residential sales (see Bell and Bockstael 2000); unemployment clustering with respect to different social and economic metrics (see Conley and Topa 2002); spatial price competition in the wholesale gasoline markets (see Pinkse, Slade, and Brett 2002); and foreign direct investment (see Baltagi, Egger and Pfaffermayr 2007).

Usually one does not worry about cross-section correlation in randomly drawn samples at the individual level. However, when one starts looking at a cross-section of countries, regions, states, counties, etc., these aggregate units are likely to exhibit cross-sectional correlation that have to be dealt with. There is an extensive literature using spatial statistics that deals with this type of correlation. Spatial dependence models may use a metric of economic distance which provides cross-sectional data with a structure similar to that provided by the time index in time series. With the increasing availability of micro as well as macro level panel data, spatial panel data models are becoming increasingly attractive in empirical economic research. The recent literature on spatial panel data models with error components adopts two alternative spatial autoregressive error processes. One specification assumes that only the remainder error term is spatially correlated but the individual effects are not (Anselin 1988; Baltagi, Song, and Koh 2003; Anselin, Le Gallo, and Jayet 2008; we refer to this as the Anselin model). The other specification assumes that both the individual and remainder error components follow the same spatial error process (see Kapoor, Kelejian, and Prucha 2007; we refer to this as the KKP model). Maximum likelihood (ML) estimation, even in its simplest form entails substantial computational problems when the number of cross-sectional units N is large. Kelejian and Prucha (1999) suggested a generalized moments (GM) estimation method which is computationally feasible even when N is large. Kapoor, Kelejian, and Prucha (2007) generalized this GM procedure from cross-section to panel data and derived its large sample properties when T is fixed and $N \rightarrow \infty$. Baltagi, Egger, and Pfaffermayr (2008a) introduced a generalized spatial panel data model which nests these two alternative processes in a more general model. They derive LM tests of the generalized model against its restricted alternatives and study their size and power performance against LR tests. In a companion paper, Baltagi, Egger, and Pfaffermayr (2008b) compare the performance of ML estimates of these models under misspecification and suggest a pretest estimator based on the LM tests derived by Baltagi, Egger, and Pfaffermayr (2008a). They show that misspecified MLE can cause substantial loss in MSE where as the pretest estimator performs well, ranking a close second to the true MLE. Monte Carlo experiments are performed to shed some light on the performance of say the Anselin MLE when the true specification is that of KKP, and vice versa. Also, to see how robust is the MLE of the general spatial panel model to *overspecification*, i.e., if the true model is KKP or Anselin. Conversely, how the Anselin and KKP maximum likelihood estimates are affected by *underspecification* of the general model. Since the researcher does not know the true model, the Monte Carlo experiments show that the pretest estimator is a viable second best alternative to the true MLE in practice.

The outline of this chapter is as follows: Section 15.2 introduces the spatial error component regression model and the associated methods of estimation in these models including maximum likelihood and generalized method of moments. Section 15.3 introduces an encompassing spatial error component model and the associated tests for the restricted models. Section 15.4 discusses prediction in the context of spatial panel models, while Section 15.5 studies the performance of various panel unit root tests when spatial correlation across the panel is present. Section 15.6 gives some recent developments in this area and further thoughts for future research.

15.2 Spatial Error Component Regression Model

One can model the spatial correlation as well as the heterogeneity across countries using a spatial error component regression model:

$$y_{ti} = X'_{ti}\beta + u_{ti}, \qquad i = 1, \ldots, N; t = 1, \ldots, T, \tag{15.1}$$

where y_{ti} is the observation on the ith country for the tth time period, X_{ti} denotes the $(k \times 1)$ vector of observations on the nonstochastic regressors and u_{ti} is the regression disturbance. In vector form, the disturbance vector is assumed to have random country effects as well as spatially autocorrelated remainder disturbances, see Anselin (1988):

$$u_t = \mu + \epsilon_t \tag{15.2}$$

with

$$\epsilon_t = \rho W \epsilon_t + \nu_t \tag{15.3}$$

where $\mu' = (\mu_1, \ldots, \mu_N)$ denote the vector of random country effects which are assumed to be $\text{IIN}(0, \sigma^2_\mu)$. ρ is the scalar spatial autoregressive coefficient with $|\rho| < 1$. W is a known $(N \times N)$ spatial weight matrix whose diagonal elements are zero.[1] W also satisfies the condition that $(I_N - \rho W)$ is nonsingular.

[1] In the simplest case, the weights matrix is binary, with $w_{ij} = 1$ when i and j are neighbors and $w_{ij} = 0$ when they are not. By convention, diagonal elements are null: $w_{ii} = 0$ and the weights are usually standardized such that the elements of each row sum to 1. Alternatively, W could be based on physical distances such as port to port or capital to capital, or commuting distances; see Anselin (1988) for more details on the properties of this W matrix.

$\nu_t' = (\nu_{t1}, \ldots, \nu_{tN})$, where ν_{ti} is assumed to be IIN$(0, \sigma_\nu^2)$ and also independent of μ_i. One can rewrite ϵ_t as

$$\epsilon_t = (I_N - \rho W)^{-1}\nu_t = B^{-1}\nu_t \tag{15.4}$$

where $B = I_N - \rho W$ and I_N is an identity matrix of dimension N. The model can be rewritten in matrix notation as

$$y = X\beta + u \tag{15.5}$$

where y is now of dimension $(NT \times 1)$, X is $(NT \times k)$, β is $(k \times 1)$ and u is $(NT \times 1)$. X is assumed to be of full column rank and its elements are assumed to be bounded in absolute value. The error can be written in vector form as

$$u = (\iota_T \otimes I_N)\mu + (I_T \otimes B^{-1})\nu \tag{15.6}$$

where $\nu' = (\nu_1', \ldots, \nu_T')$. Under these assumptions, the variance–covariance matrix for u is given by

$$\Omega = \sigma_\mu^2(J_T \otimes I_N) + \sigma_\nu^2(I_T \otimes (B'B)^{-1}), \text{ and } J_T \text{ is } \alpha (T \times T) \text{ matrix of ones.} \tag{15.7}$$

This matrix can be rewritten as

$$\Omega = \sigma_\nu^2\left[\bar{J}_T \otimes (T\phi I_N + (B'B)^{-1}) + E_T \otimes (B'B)^{-1}\right] = \sigma_\nu^2\Sigma \tag{15.8}$$

where $\phi = \sigma_\mu^2/\sigma_\nu^2$, $\bar{J}_T = J_T/T$ and $E_T = I_T - \bar{J}_T$. Using results in Wansbeek and Kapteyn (1983), Σ^{-1} is given by

$$\Sigma^{-1} = \bar{J}_T \otimes (T\phi I_N + (B'B)^{-1})^{-1} + E_T \otimes B'B. \tag{15.9}$$

Also, $|\Sigma| = |T\phi I_N + (B'B)^{-1}| \cdot |(B'B)^{-1}|^{T-1}$. Under the assumption of normality, the log-likelihood function for this model was derived by Anselin (1988, p. 154) as

$$\begin{aligned} L &= -\frac{NT}{2}\ln 2\pi\sigma_\nu^2 - \frac{1}{2}\ln|\Sigma| - \frac{1}{2\sigma_\nu^2}u'\Sigma^{-1}u \\ &= -\frac{NT}{2}\ln 2\pi\sigma_\nu^2 - \frac{1}{2}\ln[|T\phi I_N + (B'B)^{-1}|] + \frac{(T-1)}{2}\ln|B'B| \\ &\quad -\frac{1}{2\sigma_\nu^2}u'\Sigma^{-1}u \end{aligned} \tag{15.10}$$

with $u = y - X\beta$. For a derivation of the first-order conditions of MLE as well as the LM test for $\rho = 0$ for this model; see Anselin (1988). As an extension to this work, Baltagi, Song, and Koh (2003) derived the joint LM test for spatial error correlation as well as random country effects. Additionally, they derived conditional LM tests, which test for random country effects given the presence of spatial error correlation. Also, spatial error correlation given the presence of random country effects. These conditional LM tests are an alternative to the one directional LM tests that test for random country effects ignoring the presence of spatial error correlation or the one directional LM tests for spatial error correlation ignoring the presence of random country effects. Extensive Monte Carlo experiments are conducted to study the performance of these LM tests as well as the corresponding Likelihood Ratio tests. Baltagi, Song, Jung and Koh (2007) generalize the Baltagi, Song, and Koh (2003) paper by allowing for serial correlation over time for each spatial unit and spatial dependence across these units at a particular point in time. In addition, the model allows for heterogeneity across the spatial units through random effects. Testing for any one of these symptoms ignoring the other two is shown to lead to misleading results. Baltagi, Song, and Kwon (2009) extend these LM statistics to a panel data regression model with heteroskedastic as well as spatially correlated disturbances. A joint LM test for homoskedasticity and no spatial correlation is derived. In addition, a conditional LM test for no spatial correlation given heteroskedasticity, as well as a conditional LM test for homoskedasticity given spatial correlation, are also derived. These LM tests are compared with marginal LM tests that ignore heteroskedasticity in testing for spatial correlation, or spatial correlation in testing for homoskedasticity. Monte Carlo results show that these LM tests as well as their LR counterparts, perform well even for small N and T. However, misleading inference can occur when using marginal rather than joint or conditional LM tests when spatial correlation or heteroskedasticity is present.

Baltagi and Liu (2008) derive a joint LM test which simultaneously tests for the absence of spatial lag dependence and random individual effects in a panel data regression model. This is an extension of the above model to allow for spatial lag dependence in the dependent variable, i.e.,

$$y_t = \lambda W y_t + X_t\beta + u_t, \qquad i = 1, \dots, N; \quad t = 1, \dots, T$$

where $y_t' = (y_{t1}, \dots, y_{tN})$ is a vector of observations on the dependent variables for N regions or households at time $t = 1, \dots, T$. λ is a scalar spatial autoregressive coefficient and W is a known $N \times N$ spatial weight matrix whose diagonal elements are zero. W also satisfies the condition that $(I_N - \lambda W)$ is nonsingular for all $|\lambda| < 1$. X_t is an $N \times k$ matrix of observations on k explanatory variables at time t. $u_t' = (u_{t1}, \dots, u_{tN})$ is a vector of disturbances following an error component model as described in Equation 15.2. It turns out that this LM statistic is the sum of two standard LM statistics. The first one tests for the absence of spatial lag dependence ignoring the random individual effects, and the second one tests for the absence of random individual effects ignoring the spatial lag dependence. Baltagi and Liu (2008) derive two conditional LM tests. The first one tests for the absence of random individual effects allowing for the possible presence of spatial lag dependence. The second one tests for the absence of spatial lag dependence allowing for the possible presence of random individual effects.

As an alternative to the MLE, generalized method of moments have been proposed for spatial cross-section models by Conley (1999) and Kelejian and Prucha (1999) and an application of the latter method to housing data is given in Bell and Bockstael (2000). Frees (1995) derives a distribution-free test for spatial correlation in panels. This is based on Spearman-rank correlation across pairs of cross-section disturbances. Driscoll and Kraay (1998) show through Monte Carlo simulations that the presence of even modest spatial dependence can impart large bias to OLS standard errors when N is large. They present conditions under which a simple modification of the standard nonparametric time series covariance matrix estimator yields estimates of the standard errors that are robust to general forms of spatial and temporal dependence as $T \to \infty$. However, if T is small, they conclude that the problem of consistent nonparametric covariance matrix estimation is much less tractable. Parametric corrections for spatial correlation are possible only if one places strong restrictions on their form, i.e., knowing W. For typical micropanels with N much larger than T, estimating this correlation is impossible without imposing restrictions, since the number of spatial correlations increases at the rate N^2, while the number of observations grow at rate N. Even for macropanels where $N = 100$ countries observed over $T = 20$ to 30 years, N is still larger than T and prior restrictions on the form of spatial correlation are still needed.

ML estimation, even in its simplest form entails substantial computational problems when the number of cross-sectional units N is large. Kelejian and Prucha (1999) suggested a generalized moments (GM) estimation method which is computationally feasible even when N is large. Kapoor, Kelejian, and Prucha (2007) generalized this GM procedure from cross-section to panel data and derived its large sample properties when T is fixed and $N \to \infty$.

The basic regression model is the same as above; however, the disturbance term u follows the first order spatial autoregressive process

$$u = \rho(I_T \otimes W)u + \epsilon \tag{15.11}$$

with

$$\epsilon = (\iota_T \otimes I_N)\mu + \nu \tag{15.12}$$

where μ, ν and W were defined earlier. This is different from the Anselin (1988) specification described above since it also allows the individual country effects μ to be spatially correlated.

Defining $\bar{u} = (I_T \otimes W)u$, $\bar{\bar{u}} = (I_T \otimes W)\bar{u}$ and $\bar{\epsilon} = (I_T \otimes W)\epsilon$, Kapoor, Kelejian, and Prucha (2007) suggest a GM estimator based on the following six moment conditions

$$
\begin{aligned}
&E[\epsilon' Q\epsilon/N(T-1)] = \sigma_\nu^2\\
&E[\bar{\epsilon}' Q\bar{\epsilon}/N(T-1)] = \sigma_\nu^2 \operatorname{tr}(W'W)/N\\
&E[\bar{\epsilon}' Q\epsilon/N(T-1)] = 0\\
&E(\epsilon' P\epsilon/N) = T\sigma_\mu^2 + \sigma_\nu^2 = \sigma_1^2\\
&E(\bar{\epsilon}' P\bar{\epsilon}/N) = \sigma_1^2 \operatorname{tr}(W'W)/N\\
&E(\bar{\epsilon}' P\epsilon/N) = 0
\end{aligned}
\tag{15.13}
$$

where, $\epsilon = u - \rho\bar{u}$ and $\bar{\epsilon} = \bar{u} - \rho\bar{\bar{u}}$, substituting these expressions in the six moment conditions we obtain a system of six equations involving the second moments of $u, \bar{u}$ and $\bar{\bar{u}}$. Under the random effects specification considered, the OLS estimator of β is consistent. Using $\widehat{\beta}_{OLS}$ one gets a consistent estimator of the disturbances $\widehat{u} = y - X\widehat{\beta}_{OLS}$. The GM estimator of σ_1^2, σ_ν^2 and ρ is the solution of the sample counterpart of these six equations.

Kapoor, Kelejian, and Prucha (2007) suggest three GM estimators. The first involves only the first three moments which do not involve σ_1^2 and yield estimates of ρ and σ_ν^2. The fourth moment condition is then used to solve for σ_1^2 given estimates of ρ and σ_ν^2. Kapoor, Kelejian, and Prucha (2007) give the conditions needed for the consistency of this estimator as $N \to \infty$. The second GM estimator is based upon weighing the moment equations by the inverse of a properly normalized variance–covariance matrix of the sample moments evaluated at the true parameter values. A simple version of this weighting matrix is derived under normality of the disturbances. The third GM estimator is motivated by computational considerations and replaces a component of the weighting matrix for the second GM estimator by an identity matrix. They perform Monte Carlo experiments comparing MLE and these three GM estimation methods. They find that, on average, the RMSE of ML and their weighted GM estimators are quite similar. However, the first unweighted GM estimator has a RMSE that is 17%–14% larger than that of the weighted GM estimators. For an application of this GM estimator to Foreign Direct Investment (FDI), see Baltagi, Egger, and Pfaffermayr (2007) and to the spatial competition in excise taxation among U.S. states, see Egger, Pfaffermayr, and Winner (2005). Fingleton (2008) extends the GM estimator of Kapoor, Kelejian, and Prucha (2007) to the spatial moving average panel data model. The generalized moments estimator has the advantage that is computationally less demanding than MLE, especially as N gets large.

15.3 A Generalized Spatial Error Component Model

More recently, Baltagi, Egger, and Pfaffermayr (2008a) suggest a generalized spatial panel model which encompasses the Anselin (1988) and Kapoor, Kelejian, and Prucha (2007) models and allows for spatial correlation in the individual and remainder error components that may have different spatial autoregressive parameters. They derive the maximum likelihood estimator (MLE) for this more general spatial panel model when the individual effects are assumed to be random. This in turn allows the researcher to test whether this generalized model reduces to (1) the Anselin model, (2) the Kapoor, Kelejian, and Prucha model, or (3) a simple random effects model that ignores the spatial correlation in the residuals. Baltagi, Egger, and Pfaffermayr (2008a) derive the corresponding LM and LR tests for these three hypotheses and compare their size and power performance using Monte Carlo experiments.

In fact, Baltagi, Egger, and Pfaffermayr (2008a) consider the following generalized spatial error components model:

$$\begin{aligned} \mathbf{y} &= \mathbf{X}\beta + \mathbf{u} \\ \mathbf{u} &= \mathbf{Z}_{\mu}\mathbf{u}_1 + \mathbf{u}_2 \\ \mathbf{u}_1 &= \rho_1 \mathbf{W}_N \mathbf{u}_1 + \mu \\ \mathbf{u}_2 &= \rho_2 \mathbf{W} \mathbf{u}_2 + \nu. \end{aligned} \tag{15.14}$$

This is a balanced panel, which consists of $n = NT$ observations, where N is the number of unique cross-sectional units, while T is the number of time periods. The $(n \times 1)$ vector $\mathbf{y}$ denotes the dependent variable, $\mathbf{X}$ is an $(n \times K)$ matrix of nonstochastic exogenous variables. β is the corresponding $(K \times 1)$ parameter vector. $\mathbf{Z}_{\mu} = \iota_T \otimes \mathbf{I}_N$ denotes the design matrix for the $(N \times 1)$ vector of random individual effects $\mathbf{u}_1$. ι_T is a $(T \times 1)$ vector of ones and $\mathbf{I}_N$ is an identity matrix of dimension N. The vector of individual effects μ is assumed to be i.i.d. $N(0, \sigma_{\mu}^2 \mathbf{I}_N)$, while the $(n \times 1)$ vector of remainder disturbances ν is assumed to be i.i.d. $N(0, \sigma_{\nu}^2 \mathbf{I}_n)$. Furthermore, the elements of μ and ν are assumed to be independent of each other. Both $\mathbf{u}_1$ and $\mathbf{u}_2$ are spatially correlated involving the same spatial weight matrix $\mathbf{W}_N$ for each time period, but with different spatial autocorrelation parameters ρ_1 and ρ_2, respectively. $\mathbf{W}_N$ exhibits zero diagonal elements, the remaining entries are usually assumed to decline with distance. The eigenvalues of $\mathbf{W}_N$ are bounded and smaller than 1 in absolute value. The latter assumption holds for the row normalized $\mathbf{W}_N$. It also holds for the maximum-row normalized spatial weights matrices. This assumption also implies that all row and column sums of $\mathbf{W}_N$ are uniformly bounded in absolute value. In addition, we assume that $|\rho_r| < 1$ for $r = 1, 2$. The data are ordered such that $i = 1, \ldots, N$ is the fast index and $t = 1, \ldots, T$ is the slow one. The spatial weights matrix for the panel is then given by $\mathbf{W} = \mathbf{I}_T \otimes \mathbf{W}_N$, which is block diagonal and of dimension $(n \times n)$.

This model encompasses both the KKP model, which assumes that $\rho_1 = \rho_2$, and the Anselin model, which maintains that $\rho_1 = 0$. The familiar random effects (RE) panel data model without any spatial correlation is represented by $\rho_1 = \rho_2 = 0$ (see Baltagi 2008a).

In order to derive the $(n \times n)$ variance–covariance of the generalized model, we define $\mathbf{A} = (\mathbf{I}_N - \rho_1 \mathbf{W}_N)$ and $\mathbf{B} = (\mathbf{I}_N - \rho_2 \mathbf{W}_N)$. This allows us to write

$$\mathbf{u}_1 = \mathbf{A}^{-1}\mu \sim N\big(0, \sigma_\mu^2 (\mathbf{A}'\mathbf{A})^{-1}\big) \tag{15.15}$$

$$\mathbf{u}_2 = (\mathbf{I}_T \otimes \mathbf{B}^{-1})\nu \sim N(0, \sigma_\nu^2 \big(\mathbf{I}_T \otimes (\mathbf{B}'\mathbf{B})^{-1}\big)). \tag{15.16}$$

and

$$\begin{aligned}\Omega_u &= E(\mathbf{u}\mathbf{u}') = E[(\mathbf{Z}_\mu \mathbf{u}_1 + \mathbf{u}_2)(\mathbf{Z}_\mu \mathbf{u}_1 + \mathbf{u}_2)'] \\ &= \bar{\mathbf{J}}_T \otimes [T\sigma_\mu^2 (\mathbf{A}'\mathbf{A})^{-1} + \sigma_\nu^2 (\mathbf{B}'\mathbf{B})^{-1}] + \sigma_\nu^2 (\mathbf{E}_T \otimes (\mathbf{B}'\mathbf{B})^{-1}) = \sigma_\nu^2 \Sigma_u.\end{aligned} \tag{15.17}$$

This uses the fact that $E[\mathbf{u}_1 \mathbf{u}_2'] = \mathbf{0}$ since μ and ν are independent by assumption. Note that $\mathbf{Z}_\mu \mathbf{Z}_\mu' = \mathbf{J}_T \otimes \mathbf{I}_N$, where $\mathbf{J}_T$ again denotes a $(T \times T)$ matrix of ones. We define $\mathbf{E}_T = \mathbf{I}_T - \bar{\mathbf{J}}_T$, where $\bar{\mathbf{J}}_T = \mathbf{J}_T / T$ is the averaging matrix over T. The inverse of Ω_u can then be obtained from the inverse of smaller dimension $(N \times N)$ matrices as follows:

$$\begin{aligned}\Omega_u^{-1} &= \bar{\mathbf{J}}_T \otimes [T\sigma_\mu^2 (\mathbf{A}'\mathbf{A})^{-1} + \sigma_\nu^2 (\mathbf{B}'\mathbf{B})^{-1}]^{-1} + \frac{1}{\sigma_\nu^2}(\mathbf{E}_T \otimes (\mathbf{B}'\mathbf{B})) \\ &= \frac{1}{\sigma_\nu^2}\left[\bar{\mathbf{J}}_T \otimes [\frac{T\sigma_\mu^2}{\sigma_\nu^2}(\mathbf{A}'\mathbf{A})^{-1} + (\mathbf{B}'\mathbf{B})^{-1}]^{-1} + (\mathbf{E}_T \otimes (\mathbf{B}'\mathbf{B}))\right] = \frac{1}{\sigma_\nu^2}\Sigma_u^{-1}\end{aligned} \tag{15.18}$$

Furthermore, $\det[\Omega_u] = \det[T\sigma_\mu^2 (\mathbf{A}'\mathbf{A})^{-1} + \sigma_\nu^2 (\mathbf{B}'\mathbf{B})^{-1}] \det[\sigma_\nu^2 (\mathbf{B}'\mathbf{B})^{-1}]^{T-1}$. Assuming normality of the disturbances the log-likelihood function of the unrestricted model is given by

$$\begin{aligned}L\big(\beta, \sigma_\nu^2, \sigma_\mu^2, \rho_1, \rho_2\big) = &-\frac{NT}{2}\ln 2\pi - \frac{1}{2}\ln\det\big[T\sigma_\mu^2 (\mathbf{A}'\mathbf{A})^{-1} + \sigma_\nu^2 (\mathbf{B}'\mathbf{B})^{-1}\big] \\ &-\frac{T-1}{2}\ln\det\big(\sigma_\nu^2 (\mathbf{B}'\mathbf{B})^{-1}\big) - \frac{1}{2}\mathbf{u}'\Omega_u^{-1}\mathbf{u},\end{aligned} \tag{15.19}$$

where $\mathbf{u} = \mathbf{y} - \mathbf{X}\beta$. For the special case of $\rho_1 = 0$, this implies that $\mathbf{A} = \mathbf{I}_N$ and the restricted log-likelihood function reduces to the one considered by Anselin (1988, p. 154):

$$\begin{aligned}L_{\mathrm{A}}\big(\beta, \sigma_\nu^2, \sigma_\mu^2, \rho_2\big) = &-\frac{NT}{2}\ln 2\pi\sigma_\nu^2 - \frac{1}{2}\ln\det[T\sigma_\mu^2 \mathbf{I}_N + \sigma_\nu^2 (\mathbf{B}'\mathbf{B})^{-1}]^{-1} \\ &+\frac{T-1}{2}\ln\det(\mathbf{B}'\mathbf{B}) - \frac{1}{2}\mathbf{u}'\Omega_{u,\mathrm{A}}^{-1}\mathbf{u}\end{aligned} \tag{15.20}$$

$$\Omega_{u,\mathrm{A}}^{-1} = \frac{1}{\sigma_\nu^2}\left[\bar{\mathbf{J}}_T \otimes \left(\frac{T\sigma_\mu^2}{\sigma_\nu^2}\mathbf{I}_N + (\mathbf{B}'\mathbf{B})^{-1}\right)^{-1}\right] + \frac{1}{\sigma_\nu^2}\big[\mathbf{E}_T \otimes (\mathbf{B}'\mathbf{B})\big].$$

For the alternative case with $\rho_1 = \rho_2 = \rho \neq 0$, $\mathbf{A} = \mathbf{B}$ and we obtain the log-likelihood representation of the KKP estimator:

$$
\begin{aligned}
L_{\text{KKP}}\left(\beta, \sigma_\nu^2, \sigma_\mu^2, \rho\right) &= -\frac{NT}{2}\ln 2\pi\sigma_\nu^2 - \frac{N}{2}\ln\left(\frac{\sigma_1^2}{\sigma_\nu^2}\right) \\
&\quad + \frac{T}{2}\ln\det(\mathbf{B}'\mathbf{B}) - \frac{1}{2}\mathbf{u}'\Omega_{u,\text{KKP}}^{-1}\mathbf{u} \\
\Omega_{u,\text{KKP}}^{-1} &= \frac{1}{T\sigma_\mu^2 + \sigma_\nu^2}[\bar{\mathbf{J}}_T \otimes (\mathbf{B}'\mathbf{B})] + \frac{1}{\sigma_\nu^2}[\mathbf{E}_T \otimes (\mathbf{B}'\mathbf{B})]. \qquad (15.21)
\end{aligned}
$$

Finally, with $\rho_1 = \rho_2 = 0$, the log-likelihood reduces to the one representing the familiar RE model without any spatial autocorrelation:

$$
\begin{aligned}
L_{\text{RE}}\left(\beta, \sigma_\nu^2, \sigma_\mu^2\right) &= -\frac{NT}{2}\ln 2\pi\sigma_\nu^2 - \frac{N}{2}\ln\frac{\sigma_1^2}{\sigma_\nu^2} - \frac{1}{2}\mathbf{u}'\Omega_{u,\text{RE}}^{-1}\mathbf{u} \\
\Omega_{u,\text{RE}}^{-1} &= \frac{1}{T\sigma_\mu^2 + \sigma_\nu^2}(\bar{\mathbf{J}}_T \otimes \mathbf{I}_N) + \frac{1}{\sigma_\nu^2}(\mathbf{E}_T \otimes \mathbf{I}_N). \qquad (15.22)
\end{aligned}
$$

The pretest estimator is based on a sequence of LM tests derived by Baltagi, Egger, and Pfaffermayr (2008b). Specifically, the following hypotheses were considered:

$$
\begin{aligned}
&H_0^A : \rho_1 = \rho_2 = 0 \text{ vs. } H_1^A : \text{at least one of the } \rho_1 \text{ or } \rho_2 \neq 0 \\
&H_0^B : \rho_1 = \rho_2 \text{ vs. } H_1^B : \rho_1 \neq \rho_2 \\
&H_0^C : \rho_1 = 0 \text{ vs. } H_1^C : \rho_1 \neq 0 \qquad (15.23)
\end{aligned}
$$

First, we test $H_0^A; \rho_1 = \rho_2 = 0$, to see whether there is no spatial correlation in the error term. If H_0^A is not rejected, the pretest estimator reverts to the random effects MLE. In case H_0^A is rejected, we test $H_0^B; \rho_1 = \rho_2$. If H_0^B is not rejected, the pretest estimator reverts to the KKP MLE. Otherwise, $\rho_1 \neq 0$ or $\rho_2 \neq 0$ and $\rho_1 \neq \rho_2$. Next, we test $H_0^C; \rho_1 = 0$. In case H_0^C is not rejected, the pretest estimator reverts to the Anselin MLE. If H_0^C is rejected, the pretest estimator reverts to the MLE of the general model considered by Baltagi, Egger, and Pfaffermayr (2008b). In other words,

$$
\begin{aligned}
\widehat{\beta}_{\text{pretest}} &= \widehat{\beta}_{\text{RE,MLE}} \text{ if } H_0^A \text{ is not rejected} \\
&= \widehat{\beta}_{\text{KKP,MLE}} \text{ if } H_0^A \text{ is rejected, and } H_0^B \text{ is not rejected} \\
&= \widehat{\beta}_{\text{Anselin,MLE}} \text{ if } H_0^A \text{ and } H_0^B \text{ are rejected, and } H_0^C \text{ is not rejected} \\
&= \widehat{\beta}_{\text{General,MLE}} \text{ if } H_0^A \text{ and } H_0^B \text{ and } H_0^C \text{ are rejected.} \qquad (15.24)
\end{aligned}
$$

It has to be emphasized that the pretest estimator becomes the MLE of the general model when all three hypotheses are rejected. Also, it is the MLE of the RE model when H_0^A is not rejected. Hence changing the sequence of tests for H_0^B and H_0^C will not affect the number of times the pretest estimator reverts to the MLE of the RE or General model. This affects only the number of times the pretest estimator reverts to the Anselin or KKP ML estimators. In using the same data set to select the estimator to use based on a series of tests makes the statistical properties of the resulting pretest estimator difficult to derive.[2]

LM tests for these hypotheses were derived by Baltagi, Egger, and Pfaffermayr (2008a) under the assumption of normality. For H_0^A the LM-test statistic is given by

$$LM_A = \frac{1}{2b_A\widetilde{\sigma}_1^4}G_A^2 + \frac{1}{2b_A(T-1)\widetilde{\sigma}_\nu^4}M_A^2, \qquad (15.25)$$

where $\widetilde{\sigma}_1^2 = T\widetilde{\sigma}_\mu^2 + \widetilde{\sigma}_\nu^2$, $b_A = tr[(\mathbf{W}_N' + \mathbf{W}_N)^2]$, $G_A = \widetilde{\mathbf{u}}'\{\bar{\mathbf{J}}_T \otimes (\mathbf{W}_N' + \mathbf{W}_N)\}\widetilde{\mathbf{u}}$, and $M_A = \widetilde{\mathbf{u}}'\{\mathbf{E}_T \otimes (\mathbf{W}_N' + \mathbf{W}_N)\}\widetilde{\mathbf{u}}$. Here, $\widetilde{\mathbf{u}} = \mathbf{y} - \mathbf{X}\widetilde{\beta}_{\text{mle,re}}$ denotes the vector of restricted ML residuals under H_0^A. Baltagi, Egger, and Pfaffermayr (2008a) show that under H_0^A, the LM_A statistic is asymptotically distributed as χ_2^2.

For H_0^B, the LM-test statistic is given by

$$LM_B = \frac{1}{2b_B\overline{\sigma}_1^4}G_B^2 + \frac{1}{2b_B\overline{\sigma}_\nu^4(T-1)}M_B^2, \qquad (15.26)$$

with $G_B = \overline{\mathbf{u}}'(\bar{\mathbf{J}}_T \otimes \mathbf{F})\overline{\mathbf{u}} - \overline{\sigma}_1^2 tr[\mathbf{D}]$, $M_B = \overline{\mathbf{u}}'(\mathbf{E}_T \otimes \mathbf{F})\overline{\mathbf{u}} - \overline{\sigma}_\nu^2(T-1)tr[\mathbf{D}]$, $\mathbf{D} = (\mathbf{W}_N'\overline{\mathbf{A}} + \overline{\mathbf{A}}'\mathbf{W}_N)(\overline{\mathbf{A}}'\overline{\mathbf{A}})^{-1}$ and $\mathbf{F} = \mathbf{W}_N'\overline{\mathbf{A}} + \overline{\mathbf{A}}'\mathbf{W}_N$. Also, $b_B = tr[\mathbf{D}^2] - (tr[\mathbf{D}])^2/N$, $\overline{\sigma}_1^2 = \frac{\overline{\mathbf{u}}'\{\bar{\mathbf{J}}_T \otimes (\overline{\mathbf{A}}'\overline{\mathbf{A}})\}\overline{\mathbf{u}}}{N}$ and $\overline{\sigma}_\nu^2 = \frac{\overline{\mathbf{u}}'\{\mathbf{E}_T \otimes (\overline{\mathbf{A}}'\overline{\mathbf{A}})\}\overline{\mathbf{u}}}{N(T-1)}$. Here, $\overline{\mathbf{u}} = \mathbf{y} - \mathbf{X}\widetilde{\beta}_{\text{mle,KKP}}$ denotes the vector of restricted ML residuals under H_0^B. The LM_B statistic is asymptotically distributed as χ_1^2 under H_0^B.

Finally, to test H_0^C, we let $\mathbf{C}_1 = [T\widehat{\sigma}_\mu^2\mathbf{I}_N + \widehat{\sigma}_\nu^2(\widehat{\mathbf{B}}'\widehat{\mathbf{B}})^{-1}]^{-1}$, and $\mathbf{C}_2 = (\mathbf{W}_N' + \mathbf{W}_N)$. The corresponding LM test for H_0^C, which has no simple closed form representation is given by

$$LM_C = \widehat{d}_{\rho_1}^2 J_{33}^{-1}, \qquad (15.27)$$

where

$$\widehat{d}_{\rho_1} = \left.\frac{\partial L}{\partial \rho_1}\right|_{H_0^B} = -\frac{1}{2}T\widehat{\sigma}_\mu^2 tr[\mathbf{C}_1\mathbf{C}_2] + \frac{1}{2}\widehat{\sigma}_\mu^2\widehat{\mathbf{u}}'\{\mathbf{J}_T \otimes \mathbf{C}_1\mathbf{C}_2\mathbf{C}_1\}\widehat{\mathbf{u}},$$

$\widehat{\mathbf{u}} = \mathbf{y} - \mathbf{X}\widehat{\beta}_{\text{mle,Anselin}}$ denotes the vector of restricted ML residuals under H_0^C, i.e., the Anselin model, and J_{33}^{-1} is the (3,3) element of the inverse of the information matrix described in Baltagi, Egger, and Pfaffermayr (2008a).

Given that the researcher does not know the true model, Baltagi, Egger, and Pfaffermayr (2008b) recommend the pretest estimator which performed well in Monte Carlo experiments no matter what the true underlying model. In fact this pretest estimator was a close second in MSE performance to the true MLE. Additionally, the Monte Carlo experiments shed some light on the performance of the Anselin MLE when the true model is KKP, and vice versa. Ignoring spatial correlation in panel data and performing RE MLE leads to considerable loss in MSE efficiency. When the true model is a general spatial panel model with $\rho_1 \neq \rho_2 \neq 0$, both KKP and Anselin MLE impose

2 Pretest estimators in econometrics are surveyed in Giles and Giles (1993).

wrong restrictions on the ρ parameters, which in turn, introduce bias and lead to bad MSE performance of the resulting MLEs. Fortunately, this does not translate fully into bad MSE performance for the regression coefficients. The pretest estimator of the regression coefficients always performs better than the misspecified MLE and is recommended in practice.

15.4 Forecasts Using Panel Data with Spatial Error Correlation

The literature on forecasting is rich with time series applications, but this is not the case for spatial panel data applications. Exceptions are Baltagi and Li (2004, 2006) with applications to forecasting sales of cigarette and liquor per capita for U.S. states over time. In order to explain how spatial autocorrelation may arise in the demand for cigarettes, we note that cigarette prices vary among states primarily due to variation in state taxes on cigarettes. Border effect purchases not included in the cigarette demand equation can cause spatial autocorrelation among the disturbances. In forecasting sales of cigarettes, the spatial autocorrelation due to neighboring states and the individual heterogeneity across states is taken explicitly into account. Baltagi and Li (2004) derive the best linear unbiased predictor for the random error component model with spatial correlation using a simple demand equation for cigarettes based on a panel of 46 states over the period 1963–1992. They compare the performance of several predictors of the states demand for cigarettes for 1 year and 5 years ahead. The estimators whose predictions are compared include OLS, fixed effects ignoring spatial correlation, fixed effects with spatial correlation, random effects GLS estimator ignoring spatial correlation and random effects estimator accounting for the spatial correlation. Based on the RMSE criteria, the fixed effects and the random effects spatial estimators gave the best out of sample forecast performance.

Best linear unbiased prediction (BLUP) in panel data using an error component model have been surveyed in Baltagi (2008b). However, these panel forecasting applications do not deal with spatial dependence across the panel units. Following Baltagi and Li (2004), Baltagi, Bresson, and Pirotte (2010) compare various forecasts using panel data with spatial error correlation. This is done using a Monte Carlo setup rather than empirical applications. The true data generating process is assumed to be a simple error component regression model with spatial remainder disturbances of the autoregressive or moving average type. The best linear unbiased predictor is compared with other forecasts ignoring spatial correlation, or ignoring heterogeneity due to the individual effects. The paper checks the performance of these forecasts under misspecification of the spatial error process, different spatial weight matrices, and various sample sizes.

Goldberger (1962) has shown that, for a given Ω, the best linear unbiased predictor (BLUP) for the ith individual at a future period $T+\tau$ is given by:

$$\widehat{y}_{i,T+\tau} = X_{i,T+\tau}\widehat{\beta}_{GLS} + \omega'\Omega^{-1}\widehat{u}_{\mathrm{GLS}} \tag{15.28}$$

where $\omega = E[u_{i,T+\tau}u]$ is the covariance between the future disturbance $u_{i,T+\tau}$ and the sample disturbances u. $\widehat{\beta}_{\text{GLS}}$ is the GLS estimator of β based on Ω and $\widehat{u}_{\text{GLS}}$ denotes the corresponding GLS residual vector. For the error component without spatial autocorrelation ($\rho = 0$), this BLUP reduces to

$$\widehat{y}_{i,T+\tau} = X_{i,T+\tau}\widehat{\beta}_{\text{GLS}} + \frac{\sigma_\mu^2}{\sigma_1^2}\left(\iota_T' \otimes l_i'\right)\widehat{u}_{\text{GLS}} \tag{15.29}$$

where $\sigma_1^2 = T\sigma_\mu^2 + \sigma_v^2$ and l_i is the ith column of I_N. The typical element of the last term of Equation 15.29 is $(T\theta)\overline{u}_{i.,\text{GLS}}$, where $\overline{u}_{i.,\text{GLS}} = \sum_{t=1}^{T}\widehat{u}_{ti,\text{GLS}}/T$ and $\theta = \sigma_\mu^2/\sigma_v^2$; see Baltagi (2008b). Therefore, the BLUP of $y_{i,T+\tau}$ for the RE model modifies the usual GLS forecasts by adding a fraction of the mean of the GLS residuals corresponding to the ith individual. In order to make this forecast operational, $\widehat{\beta}_{\text{GLS}}$ is replaced by its feasible GLS estimate and the variance components are replaced by their feasible estimates.

Baltagi and Li (2004, 2006) derived the BLUP correction term when both error components and spatial autocorrelation are present and ϵ_t follows a SAR process. So, the predictor for the SAR is given by:

$$\begin{aligned}\widehat{y}_{i,T+\tau} &= X_{i,T+\tau}\widehat{\beta}_{\text{MLE}} + \theta\left(\iota_T' \otimes l_i'C_1^{-1}\right)\widehat{u}_{\text{MLE}}\\ &= X_{i,T+\tau}\widehat{\beta}_{\text{MLE}} + T\theta\sum_{j=1}^{N}c_{1,j}\overline{u}_{j.,\text{MLE}}\end{aligned} \tag{15.30}$$

where c_{1j} is the jth element of the ith row of C_1^{-1} with $C_1 = [T\theta I_N + (B'B)^{-1}]$ and $\overline{u}_{j.,\text{MLE}} = \sum_{t=1}^{T}\widehat{u}_{tj,\text{MLE}}/T$. In other words, the BLUP of $y_{i,T+\tau}$ adds to $X_{i,T+\tau}\widehat{\beta}_{\text{MLE}}$ a weighted average of the MLE residuals for the N individuals averaged over time. The weights depend upon the spatial matrix W_N and the spatial autoregressive coefficient ρ. To make these predictors operational, we replace θ and ρ by their estimates from the RE-spatial MLE with SAR. When there are no random individual effects, so that $\sigma_\mu^2 = 0$, then $\theta = 0$ and the BLUP prediction terms drop out completely from Equation 15.30. In these cases, Ω reduces to $\sigma_v^2[I_T \otimes (B'B)^{-1}]$ for SAR, and the corresponding MLE for these models yield the pooled spatial MLE with SAR remainder disturbances. This result can be extended to the spatial moving average model (SMA); see Baltagi, Bresson, and Pirotte (2010).

For the Kapoor, Kelejian, and Prucha (2007) model, the BLUP of $y_{i,T+\tau}$ for the SAR-RE also modifies the usual GLS forecasts by adding a fraction of the mean of the GLS residuals corresponding to the ith individual. More specifically, the predictor is given by

$$\begin{aligned}\widehat{y}_{i,T+\tau} &= X_{i,T+\tau}\widehat{\beta}_{\text{FGLS}} + \left(\frac{\sigma_\mu^2}{\sigma_1^2}\right)b_i\left(\iota_T' \otimes B_N\right)\widehat{u}_{\text{FGLS}}\\ &= X_{i,T+\tau}\widehat{\beta}_{\text{FGLS}} + \left(\frac{\sigma_\mu^2}{\sigma_1^2})(\iota_T' \otimes l_i'\right)\widehat{u}_{\text{FGLS}}\end{aligned} \tag{15.31}$$

where b_i is the ith row of the matrix B_N^{-1}. This holds because $b_i(\iota'_T \otimes B_N) = (1 \otimes b_i)(\iota'_T \otimes B_N) = (\iota'_T \otimes l'_i)$, where l'_i is the ith row of I_N as defined above. $B_N^{-1} B_N = I_N$ and therefore $b_i B_N = l'_i$. This proof applies to both the Kapoor, Kelejian, and Prucha (2007) SAR-RE specification and the Fingleton (2008) SMA-RE specification. Therefore, the BLUP of $y_{i,T+\tau}$ for the SAR-RE and the SMA-RE, like the usual RE model with no spatial effects, modifies the usual GLS forecasts by adding a fraction of the mean of the GLS residuals corresponding to the ith individual. While the predictor formula is the same, the MLEs for these specifications yield different estimates which in turn yield different residuals and hence different forecasts.

The results of the Monte Carlo study by Baltagi, Bresson, and Pirotte (2010) find that when the true DGP is RE with a SAR or SMA remainder disturbances, estimators that ignore heterogeneity/spatial correlation perform badly in RMSE forecasts. Accounting for heterogeneity improves the forecast performance by a big margin and accounting for spatial correlation improves the forecast but by a smaller margin. Ignoring both leads to the worst forecasting performance. Heterogeneous estimators based on averaging perform worse than homogeneous estimators in forecasting performance. This performance improves with a larger sample size and seems robust to the type of spatial error structure imposed on the remainder disturbances. These Monte Carlo experiments confirm earlier empirical studies that report similar findings.

15.5 Panel Unit Root Tests and Spatial Dependence

Baltagi, Bresson, and Pirotte (2007) studied the performance of panel unit root tests when spatial effects are present that account for cross-section correlation. Monte Carlo simulations show that there can be considerable size distortions in panel unit root tests when the true specification exhibits spatial error correlation.

Panel data unit root tests have been proposed as alternative more powerful tests than those based on individual time series unit roots tests; see Baltagi (2008a) and Breitung and Pesaran (2008) for some recent reviews of this literature. One of the advantages of panel unit root tests is that their asymptotic distribution is standard normal. This is in contrast to individual time series unit roots which have nonstandard asymptotic distributions. But these tests are not without their critics. The first generation panel unit root tests assumed cross-section independence. These tests include the one proposed by Levin, Lin, and Chu (2002), hereafter denoted by LLC, where the null hypothesis is that each individual time series contains a unit root against the alternative that each time series is stationary. As Maddala (1999) pointed out, the null may be fine for testing convergence in growth among countries, but the alternative restricts every country to converge at the same rate. Im, Pesaran, and Shin (2003), hereafter denoted by IPS, allow for heterogeneous panels and propose panel unit root tests which are based on the average of the individual ADF unit root tests computed from each time series. The null hypothesis is that each individual time series contains a unit root while the alternative allows for some but not all of the individual series to have unit roots. One major criticism of both the LLC and IPS tests is that they require cross-sectional independence. This is a restrictive assumption given the cross-section correlation and spillovers across countries, states, and regions.

Maddala and Wu (1999) and Choi (2001) proposed combining the p-values from the individual unit root ADF tests applied to each time series. Once again, these tests follow a standard normal limiting distribution. They have the advantage that N, the number of cross sections, can be finite or infinite; the time series can be of different length; and the alternative allows some groups to have unit roots while others may not.

Recent studies that try to account for cross-sectional dependence in panel unit root testing include the following: Chang (2002) who explored the nonlinear IV methodology to solve the inferential difficulties in the panel unit root testing which arise from the intrinsic heterogeneities and dependencies of panel models. Chang (2002) suggests an average of individual nonlinear IV t-ratio statistics of the autoregressive coefficient obtained from using an integrable transformation of the lagged level as instrument. These methods assume cross-sectional correlation in the innovation terms driving the autoregressive processes. Choi (2002), on the other hand, generalizes the three unit root tests (inverse chi-square, inverse normal and logit) to the case where the cross-sectional correlation is modeled by error component models. The tests are formulated by combining p-values from the ADF test applied to each individual time series whose stochastic trend components and cross-sectional correlations are eliminated using GLS-demeaning and GLS-detrending. Choi (2002) shows that the combination tests have a standard normal limiting distributions under the sequential asymptotics $T \to \infty$ and $N \to \infty$.

To avoid the restrictive nature of cross-section demeaning procedure, Bai and Ng (2004), and Phillips and Sul (2003), among others, propose dynamic factor models by allowing the common factors to have differential effects on cross-section units. Phillips and Sul's model is a one-factor model where the factor is independently distributed across time. They propose a moment-based method to eliminate the common factor which is different from principal components. More specifically, in the context of a residual one-factor model, Phillips and Sul (2003) provide an orthogonalization procedure which in effect asymptotically eliminates the common factors before preceding to the application of standard unit root tests. Pesaran (2007) suggests a simple way of getting rid of cross-sectional dependence that does not require the estimation of factor loading. His method is based on augmenting the usual ADF regression with the lagged cross-sectional mean and its first-difference to capture the cross-sectional dependence that arises through a single factor model.

Baltagi, Bresson, and Pirotte (2007) run Monte Carlo simulations to compare the empirical size of panel unit root tests with and without spatial error dependence. The structure of the dependence is based on some commonly used spatial error processes: the spatial autoregressive (SAR) and the spatial moving average (SMA) error process and the spatial error components model (SEC). For each experiment, they perform nine panel unit root test statistics: the Levin, Lin, and Chu test (2002), the Breitung (2000) test, the Im, Pesaran, and Shin test (2003), the Maddala and Wu test (1999), the Choi tests (2001, 2002) with and without cross-sectional correlation, the Chang IV test (2002), the Phillips and Sul test (2003), and the Pesaran test (2007). The experiments include a case of no spatial correlation as well as four types of spatial correlation (SAR, SMA, SEC1, and SEC3), with two values of the parameters indicating weak versus strong spatial dependence. They also consider 10 weight matrices, differing in their degree of sparseness, four pairs of (N, T) and two models including individual effects and individual deterministic trends. Even with this modest design, the total number of experiments considered is 1600. They find that ignoring spatial dependence when present can seriously bias the size of panel unit root tests.

15.6 Extensions

Elhorst (2003) considers the ML estimation of a fixed and random effects panel data model extended either to include spatial error autocorrelation or a spatially lagged dependent variable. This is also extended to the case of random coefficients model. In another paper, Elhorst (2005) considers the estimation of a fixed effects dynamic panel data model extended either to include spatial error autocorrelation or a spatially lagged dependent variable. The latter models are first differenced to eliminate the fixed effects and then the unconditional likelihood function is derived taking into account the density function of the first-differenced observations on each spatial unit. Lee and Yu (2010) consider the estimation of a SAR panel model with fixed effects and SAR disturbances. If T is finite but N is large, they show that direct ML estimation of all the parameters including the fixed effects will yield consistent estimators except for the variance of disturbances. Using a transformation that eliminates the individual fixed effects, they provide consistent estimates for all the parameters including the variance of disturbances. The transformation approach is shown to be a conditional likelihood approach if the disturbances are normally distributed. Next, they extend their results to the SAR model with both individual and time-fixed effects. In this case, the transformation approach yields consistent estimators of all the parameters when either N or T are large. For the direct approach, consistency of the variance parameter requires both N and T to be large and consistency of other parameters requires N to be large. Monte Carlo results are provided illustrating the finite sample properties of the various estimators with N and/or T being small or moderately large.

Yu, de Jong, and Lee (2007, 2008) study the asymptotic properties of quasi-maximum likelihood estimators for spatial dynamic panel data with fixed effects when both the number of individuals N and the number of time periods T are large. They cover both the stationary and nonstationary cases. When the roots in the DGP are not all unitary, the estimators' rates of convergence will be the same as the stationary case, and the estimators can be asymptotically normal. In fact, for the distribution of the common parameters, when T is asymptotically large relative to N, the estimators are $\sqrt{NT}$ consistent and asymptotically normal, with the limiting distribution centered around 0. When N is asymptotically proportional to T, the estimators are $\sqrt{NT}$ consistent and asymptotically normal, but the limiting distribution is not centered around 0. When N is large relative to T, the estimators are consistent with rate T, and have a degenerate limiting distribution. Compared to the stationary case, the estimators' rate of convergence will be the same, but the asymptotic variance matrix will be driven by the nonstationary component and it is singular. Consequently, a linear combination of the spatial and dynamic effects can converge at a higher rate. They also propose a bias correction which performs well when T grows faster than $N^{1/3}$.

Pesaran and Tosetti (2008) study large panel data sets where even after conditioning on common observed effects the cross-section units might remain dependently distributed. This could be due to unobserved common factors and/or spatial effects. They introduce the concepts of time-specific weak and strong cross-section dependence and show that the commonly used spatial models are examples of weak cross-section dependence. Pesaran's (2006) common correlated effects (CCE) estimator of panel data model with a multifactor error structure continues to provide consistent estimates of the slope coefficient, even in the presence of spatial error processes.

This chapter highlights some of the recent research in spatial panels. Due to space limitations, several applications and related extensions have not been discussed. Hopefully, this will entice the reader to read more papers on this subject and spur some needed research in this area.

15.7 Acknowledgment

A preliminary version of this chapter was presented as a keynote speech at the 13th African Econometric Society meeting held at the University of Pretoria, South Africa, July 9–11, 2008. Also as the keynote address for the 10th Econometrics and Statistics Symposium held at Ataturk University, Turkey, May 27–29, 2009, and in a session in honor of Cheng Hsiao at the 15th International Conference on Panel Data at the University of Bonn, Germany, July 3–5, 2009. I would like to thank my coauthors Georges Bresson, Alain Pirotte, Dong Li, Seuck Heun Song, Peter Egger, Michael Pfaffermayer, Byoung Cheol Jung, Jae Hyeok Kwon, and Won Koh for allowing me to draw freely on our work.

References

Anselin, L. 1988. *Spatial Econometrics: Methods and Models*. Dordrecht: Kluwer Academic Publishers.

Anselin, L. 2001. Spatial econometrics. In B. Baltagi, (ed.). *A Companion to Theoretical Econometrics*. pp. 310–330. Oxford, U.K.: Blackwell.

Anselin, L., and A. K. Bera. 1998. Spatial dependence in linear regression models with an introduction to spatial econometrics. In: A. Ullah, D.E.A. Giles (eds). *Handbook of Applied Economic Statistics*. New York: Marcel Dekker.

Anselin, L., J. Le Gallo, and H. Jayet. 2008. Spatial panel econometrics. In L. Mátyás and P. Sevestre (eds.). *The Econometrics of Panel Data: Fundamentals and Recent Developments in Theory and Practice*, Chap. 19. Berlin: Springer, pp. 625–660.

Bai, J., and S. Ng. 2004. A PANIC attack on unit roots and cointegration. *Econometrica* 72:1127–1177.

Baltagi, B. H. 2008a. *Econometric Analysis of Panel Data*. Chichester, U.K.: Wiley.

Baltagi, B. H. 2008b. Forecasting with panel data. *Journal of Forecasting* 27:153–173.

Baltagi, B. H., G. Bresson, and A. Pirotte. 2007. Panel unit root tests and spatial dependence. *Journal of Applied Econometrics* 22:339–360.

Baltagi, B. H., G. Bresson, and A. Pirotte. 2010. Forecasting with spatial panel data. *Computational Statistics and Data Analysis* (forthcoming).

Baltagi, B. H., and D. Li. 2004. Prediction in the panel data model with spatial correlation. In L. Anselin, R. J. G. M. Florax, and S. J. Rey (eds.). *Advances in Spatial Econometrics: Methodology, Tools and Applications* Chap. 13. Berlin: Springer, pp. 283–295.

Baltagi, B. H., and D. Li. 2006. Prediction in the panel data model with spatial correlation: The case of liquor. *Spatial Economic Analysis* 1:175–185.

Baltagi, B. H., and L. Liu. 2008. Testing for random effects and spatial lag dependence in panel data models. *Statistics and Probability Letters* 17:3304–3306.

Baltagi, B. H., S. H. Song, and W. Koh. 2003. Testing panel data models with spatial error correlation. *Journal of Econometrics* 117:123–150.

Baltagi, B. H., P. Egger, and M. Pfaffermayr. 2007. Estimating models of complex FDI: Are there third country effects? *Journal of Econometrics* 140:260–281.

Baltagi, B. H., P. Egger, and M. Pfaffermayr. 2008a. A generalized spatial panel data model with random effects. Working paper, Syracuse University, Department of Economics and Center for Policy Research, Syracuse, NY.

Baltagi, B. H., P. Egger, and M. Pfaffermayr. 2008b. A Monte Carlo study for pure and pretest estimators of a panel data model with spatially autocorrelated disturbances, *Annales d'Économie et de Statistique* 87/88:11–38.

Baltagi, B. H., S. H. Song, and J. H. Kwon. 2009. Testing for heteroskedasticity and spatial correlation in a random effects panel data model. *Computational Statistics and Data Analysis* 53:2897–2922.

Baltagi, B. H., S. H. Song, B. C. Jung, and W. Koh. 2007. Testing for serial correlation, spatial autocorrelation and random effects using panel data. *Journal of Econometrics* 140:5–51.

Bell, K. P., and N. R. Bockstael. 2000. Applying the generalized-moments estimation approach to spatial problems involving microlevel data. *Review of Economics and Statistics* 82:72–82.

Breitung, J. 2000. The local power of some unit root tests for panel data. *Advances in Econometrics* 15:161–177.

Breitung, J., and M. H. Pesaran. 2008. Unit roots and cointegration in panels, In L. Mátyás and P. Sevestre (eds.). *The Econometrics of Panel Data: Fundamentals and Recent Developments in Theory and Practice*, Chap. 9, Berlin: Springer, pp. 279–322.

Case, A. C. 1991. Spatial patterns in household demand. *Econometrica* 59:953–965.

Chang, Y. 2002. Nonlinear IV unit root tests in panels with cross sectional dependency. *Journal of Econometrics* 110:261–292.

Choi, I. 2001. Unit root tests for panel data. *Journal of International Money and Finance* 20:249–272.

Choi, I. 2002. Instrumental variables estimation of a nearly nonstationary, heterogeneous error component model. *Journal of Econometrics* 109:1–32.

Conley, T. G. 1999. GMM estimation with cross sectional dependence. *Journal of Econometrics* 92:1–45.

Conley, T. G., and G. Topa. 2002. Socio-economic distance and spatial patterns in unemployment. *Journal of Applied Econometrics* 17:303–327.

Driscoll, J. C., and A. C. Kraay. 1998. Consistent covariance matrix estimation with spatially dependent panel data. *Review of Economics and Statistics* 80:549–560.

Egger, P., M. Pfaffermayr, and H. Winner. 2005. An unbalanced spatial panel data approach to US state tax competition. *Economics Letters* 88:329–335.

Elhorst, J. P. 2003. Specification and estimation of spatial panel data models. *International Regional Science Review* 26:244–268.

Elhorst, J. P. 2005. Unconditional maximum likelihood estimation of linear and log-linear dynamic models for spatial panels. *Geographical Analysis* 37:85–106.

Fingleton, B. 2008. A generalized method of moments estimator for a spatial panel model with an endogeneous spatial lag and spatial moving average errors. *Spatial Economic Analysis* 3(1):27–44.

Frees, E. W. 1995. Assessing cross-sectional correlation in panel data. *Journal of Econometrics* 69:393–414.

Giles, J. A., and D. E. A. Giles. 1993. Pre-test estimation and testing in econometrics: Recent developments. *Journal of Economic Surveys* 7:145–197.

Goldberger, A. S. 1962. Best linear unbiased prediction in the generalized linear regression model. *Journal of the American Statistical Association* 57:369–375.

Holtz-Eakin, D. 1994. Public-sector capital and the productivity puzzle. *Review of Economics and Statistics* 76:12–21.

Im, K. S., M. H. Pesaran, and Y. Shin. 2003. Testing for unit roots in heterogeneous panels. *Journal of Econometrics* 115:53–74.

Kapoor, M., H. H. Kelejian, and I. R. Prucha. 2007. Panel data models with spatially correlated error components. *Journal of Econometrics* 140:97–130.

Kelejian, H. H., and I. R. Prucha. 1999. A generalized moments estimator for the autoregressive parameter in a spatial model. *International Economic Review* 40:509–533.

Kelejian, H. H., and D. P. Robinson. 1992. Spatial autocorrelation: A new computationally simple test with an application to per capita county police expenditures. *Regional Science and Urban Economics* 22:317–331.

Lee, L. F., and J. Yu. 2010. Estimation of spatial autoregressive panel data models with fixed effects. *Journal of Econometrics* 154:165–185.

Levin, A., C. F. Lin, and C. Chu. 2002. Unit root test in panel data: Asymptotic and finite sample properties. *Journal of Econometrics* 108:1–24.

Maddala, G. S. 1999. On the use of panel data methods with cross country data. *Annales D'Économie et de Statistique* 55–56:429–448.

Maddala, G. S., and S. Wu. 1999. A comparative study of unit root tests with panel data and a new simple test. *Oxford Bulletin of Economics and Statistics* 61:631–652.

Pesaran, M. H. 2006. Estimation and inference in large heterogenous panels with multifactor error structure. *Econometrica* 74:967–1012.

Pesaran, M. H. 2007. A simple panel unit root test in the presence of cross section dependence. *Journal of Applied Econometrics* 27:265–312.

Pesaran, M. H., and E. Tosetti. 2008. Large panels with common factors and spatial correlations. Working paper, Faculty of Economics, Cambridge University.

Phillips, P. C. B., and D. Sul. 2003. Dynamic Panel Estimation and homogeneity testing under cross section dependence. *Econometrics Journal* 6:217–259.

Pinkse, J., M. E. Slade, and C. Brett. 2002. Spatial price competition: A semiparametric approach. *Econometrica* 70:1111–1153.

Wansbeek, T. J., and A. Kapteyn. 1983. A note on spectral decomposition and maximum likelihood estimation of ANOVA models with balanced data. *Statistics and Probability Letters* 1:213–215.

Yu, J., R. de Jong, and L. F. Lee. 2007. Quasi-maximum likelihood estimators for spatial dynamic panel data with fixed effects when both n and T are large: A nonstationary case. Working paper, Ohio State University, Department of Economics.

Yu, J., R. de Jong, and L. F. Lee. 2008. Quasi-maximum likelihood estimators for spatial dynamic panel data with fixed effects when both n and T are large. *Journal of Econometrics* 146:118–134.

16

Nonparametric and Semiparametric Panel Econometric Models: Estimation and Testing

Liangjun Su and Aman Ullah

CONTENTS

16.1 Introduction

There exists enormous literature on the development of panel data models in the last five decades or so. The readers are referred to Arellano (2003), Hsiao (2003), and Baltagi (2008) for an overview of this literature. Nevertheless, these books only focus on the study of parametric panel data models which can be misspecified. Estimators from misspecified models are often inconsistent, invalidating the subsequent statistical inference. For this reason, we also observe a rapid growth of the literature on nonparametric (NP) and semiparametric (SP) panel data models in the last 15 years. For an early review on this latter literature, the readers are referred to Ullah and Roy (1998). See also Ai and Li (2008) whose survey focuses on partially linear and limited dependent NP and SP panel data models.

In this chapter, we review the recent literature on nonparametric and semiparametric panel data models. Given the space limitation, it is impossible to survey all the important developments in this literature. We choose to focus on the following areas:

- Nonparametric panel data models with random effects
- Nonparametric panel data models with fixed effects
- Partially linear panel data models
- Varying coefficient panel data models
- Nonparametric panel data models with cross-section dependence
- Nonseparable nonparametric panel data models

The first two areas are limited to the conventional nonparametric panel data models with one-way error component structure:

$$y_{it} = m(x_{it}) + \varepsilon_{it}, i = 1, \ldots, n, t = 1, \ldots T, \tag{16.1}$$

where x_{it} is a $p \times 1$ random vector, $m(\cdot)$ is unknown smooth function, ε_{it} is the disturbance term that exists the one-way error component structure:

$$\varepsilon_{it} = \alpha_i + u_{it}. \tag{16.2}$$

Here, α_i represents the cross-sectional heterogeneity parameters, and u_{it} is the idiosyncratic error term. As in the parametric framework, α_i can be treated as either random or fixed so that we will have random effects or fixed effects nonparametric panel data models.

Given the notorious "curse of dimensionality" problem in the nonparametric literature, applications of Equation 16.1 may be limited in practice. This motivates the fast developments of two classes of semiparametric panel data models, namely, partially linear panel data models and varying coefficient panel data models. In Section 16.4, we study the estimation of the following partially linear panel data models

$$y_{it} = x_{it}'\beta_0 + m(z_{it}) + \alpha_i + u_{it}, i = 1, \ldots, n, t = 1, \ldots, T, \tag{16.3}$$

where x_{it} and z_{it} are of dimensions $p \times 1$ and $q \times 1$, respectively, β_0 is a $p \times 1$ vector of unknown parameters, $m(\cdot)$ is an unknown smooth function, α_i and u_{it} are as defined above. In Section 16.5, we study the estimation of the following varying coefficient panel data models

$$y_{it} = x_{it}'m(z_{it}) + \alpha_i + u_{it} = \sum_{d=1}^{p} x_{it,d}m_d(z_{it}) + \alpha_i + u_{it} \tag{16.4}$$

where the covariate z_{it} is a $q \times 1$ vector, $x_{it} = (x_{it,1}, \ldots, x_{it,p})'$, and $m(\cdot) = (m_1(\cdot), \ldots, m_p(\cdot))'$ has p unknown smooth functions.

The literature on the estimation of parametric panel data models with cross-section dependence has been growing rapidly in the last decade. See Pesaran (2006) and Bai (2009) and the references therein. In Section 16.6 we consider the estimation of m_i in

$$y_{it} = m_i(x_{it}) + \gamma_{1i}' f_{1t} + \gamma_{2i}' f_{2t} + \varepsilon_{it}, i = 1, \ldots, n, t = 1, \ldots, T, \tag{16.5}$$

where $m_i(\cdot)$ is an unknown smooth function from, f_{1t} is a $q_1 \times 1$ vector of observed common factors, f_{2t} is a $q_2 \times 1$ vector of unobserved common factors, γ_{1i} and γ_{2i} are factor loadings, ε_{it} is the usual idiosyncratic disturbance. Since $\gamma_{2i}' f_{2t} + \varepsilon_{it}$ is treated as the error term, we say it exhibits multifactor error structure. Specification tests can be conducted to test the homogeneous relationship (m_i does not depend on i) and the existence of cross-section dependence.

All previous works assume that the unobserved heterogeneity and idiosyncratic error term enter the nonparametric panel data model additively. In Section 16.7, we focus on the estimation of the following two models

$$y_{it} = m(x_{it}, \alpha_i) + u_{it} \tag{16.6}$$

and

$$y_{it} = m(x_{it}, \alpha_i, u_{it}) \tag{16.7}$$

where both $m(\cdot, \cdot)$ and $m(\cdot, \cdot, \cdot)$ are unknown functions, and α_i and u_{it} are as defined above. Clearly, Equation 16.6 is a partially separable model because the idiosyncratic disturbance enters the model additively; Equation 16.7 is fully nonseparable. We also remark that specification testing can be developed to test the monotonicity of the response variable in the individual heterogeneity parameter.

Throughout the chapter, we restrict our attention to the balanced panel. We use $i = 1, \ldots, n$ to denote an individual and $t = 1, \ldots, T$ to denote time, but keep in mind that in some applications, the index t may not really mean time. For example, i may denote a family and t a specific child in the family. Unless otherwise stated, all asymptotic theories are established by passing n to infinity. T may also pass to infinity in some scenarios, say, in some dynamic panel data models or the panel data models with cross-section dependence. For a natural number a, we use I_a to denote an $a \times a$ identity matrix and l_a an $a \times 1$ vector of ones. $\otimes$ and $\odot$ denote the Kronecker and Hadarmard products, respectively.

16.2 Nonparametric Panel Data Models with Random Effects

In this section, we consider nonparametric panel data models with random effects:

$$y_{it} = m(x_{it}) + \alpha_i + u_{it}, i = 1, \ldots, n, t = 1, \ldots, T, \tag{16.8}$$

where x_{it} is $p \times 1$ vector of exogenous variables, α_i is independently and identically distributed (i.i.d.) $(0, \sigma_\alpha^2)$, u_{jt} is i.i.d. $(0, \sigma_u^2)$, and α_i and u_{jt} are uncorrelated for all $i,\ j = 1, \ldots, n$ and $t = 1, \ldots, T$. We remark that some of these assumptions can be relaxed and specification testing is also possible.

Let $\varepsilon_{it} = \alpha_i + u_{it}$, $\varepsilon_i = (\varepsilon_{i1}, \ldots, \varepsilon_{iT})'$ and $\varepsilon_i = (\varepsilon_1, \ldots, \varepsilon_n)'$. Then $\Sigma \equiv E(\varepsilon_i\varepsilon_i') = \sigma_u^2 I_T + \sigma_\alpha^2 l_T l_T'$ and $\Omega \equiv E(\varepsilon\varepsilon') = I_n \otimes \Sigma$. We first discuss local linear least squares (LLLS) estimator of m and its first-order derivatives by ignoring the information contained in the variance–covariance matrix Ω and then proceed to the more efficient estimation of m and its derivatives by exploring the information on Ω.

16.2.1 Local Linear Least Squares Estimator

A local linear approximation of the model (Equation 16.8) can be written as

$$\begin{aligned} y_{it} &\approx m(x) + (x_{it} - x)'\beta(x) + \alpha_i + u_{it} \\ &= x_{it}(x)\delta(x) + \alpha_i + u_{it} \end{aligned}$$

where x_{it} is "close" to x, $x_{it}(x) = (1 (x_{it} - x)')'$, $\beta(x) = \partial m(x)/\partial x$, and $\delta(x) = (m(x)\ \beta(x)')'$. In a vector form, we can write

$$Y \approx \mathbf{X}(x)\delta(x) + \varepsilon \tag{16.9}$$

where $Y = (y_{11}, \ldots, y_{1T}, \ldots, y_{n1}, \ldots, y_{nT})'$, and $\mathbf{X}(x) = ((x_{11}(x), \ldots, x_{1T}(x), \ldots, x_{n1}(x), \ldots, x_{nT}(x))'$.

Let $K_h(x) = h^{-p}K(x/h)$, where K is a kernel function and $h \equiv h(n)$ is a bandwidth parameter. Then the LLLS estimator of $\delta(x)$ is obtained by choosing δ to minimize

$$(Y - \mathbf{X}(x)\delta)'\mathbf{K}(x)(Y - \mathbf{X}(x)\delta), \tag{16.10}$$

where $\mathbf{K}(x) = \text{diag}(K_h(x_{11} - x), \ldots, K_h(x_{1T} - x), \ldots, K_h(x_{n1} - x), \ldots, K_h(x_{nT} - x))$ is an $nT \times nT$ diagonal matrix. The solution to this minimization problem is given by

$$\hat{\delta}(x) = [\mathbf{X}(x)'\mathbf{K}(x)\mathbf{X}(x)]^{-1}\mathbf{X}(x)'\mathbf{K}(x)Y. \tag{16.11}$$

Denote the first component of $\hat{\delta}(x)$ as $\hat{m}(x)$ which estimates $m(x)$. It is straightforward to study the asymptotic properties of $\hat{\delta}(x)$ and $\hat{m}(x)$; see, e.g., see Li and Racine (2007).

16.2.2 More Efficient Estimation

Clearly, the estimator in Equation 16.11 ignores the information on Ω. To incorporate this, we can define a weighted LLLS estimator of $\delta(x)$ by choosing δ to minimize

$$[Y - \mathbf{X}(x)\delta)]'\mathbf{W}(x)[Y - \mathbf{X}(x)\delta)]$$

which gives

$$\hat{\delta}_{\mathbf{W}}(x) = [\mathbf{X}(x)'\mathbf{W}(x)\mathbf{X}(x)]^{-1}\mathbf{X}(x)'\mathbf{W}(x)Y \tag{16.12}$$

where $\mathbf{W}(x)$ is a kernel-based weight matrix; see Henderson and Ullah (2005). Lin and Carroll (2000) have considered $\mathbf{W}(x) = \mathbf{K}(x)^{1/2}\Omega^{-1}\mathbf{K}(x)^{1/2}$ and $\mathbf{W}(x) = \Omega^{-1}\mathbf{K}(x)$, and Ullah and Roy (1998) have suggested $\mathbf{W}(x) = \Omega^{-\frac{1}{2}}\mathbf{K}(x)\Omega^{-\frac{1}{2}}$. When Ω is a diagonal matrix, these choices of $\mathbf{W}(x)$ are the same.

For an operational estimate, we need to estimate Ω. For this purpose, define

$$\hat{\sigma}_1^2 = \frac{T}{n}\sum_{i=1}^{n}\bar{\hat{\varepsilon}}_i^2, \hat{\sigma}_u^2 = \frac{1}{n(T-1)}\sum_{i=1}^{n}\sum_{t=1}^{T}(\hat{\varepsilon}_{it} - \bar{\hat{\varepsilon}}_i)^2 \tag{16.13}$$

where $\bar{\hat{\varepsilon}}_i = T^{-1}\sum_{t=1}^{T}\hat{\varepsilon}_{it}$ and $\hat{\varepsilon}_{it} = y_{it} - \hat{m}(x_{it})$ is the LLLS residual. Noting that $\hat{\sigma}_1^2$ and $\hat{\sigma}_u^2$ estimate $\sigma_1^2 = T\sigma_\alpha^2 + \sigma_u^2$ and σ_u^2, respectively, we can estimate σ_α^2 by $\hat{\sigma}_\alpha^2 = \frac{1}{T}(\hat{\sigma}_1^2 - \hat{\sigma}_u^2)$. With these estimates, one can obtain an estimate $\hat{\Omega}$ of Ω with σ_α^2 and σ_u^2 replaced by $\hat{\sigma}_\alpha^2$ and $\hat{\sigma}_u^2$, respectively. The operational estimator of $\delta(x)$ is given by

$$\hat{\delta}_{\hat{\mathbf{W}}}(x) = [\mathbf{X}(x)'\hat{\mathbf{W}}(x)\mathbf{X}(x)]^{-1}\mathbf{X}(x)'\hat{\mathbf{W}}(x)Y \tag{16.14}$$

where $\hat{\mathbf{W}}(x)$ is $\mathbf{W}(x)$ with Ω replaced by $\hat{\Omega}$. However, Lin and Carroll (2000) demonstrate that one cannot achieve asymptotic improvement over the LLLS estimator by such weighted LLLS estimation. Henderson and Ullah (2008) also find similar observations in their Monte Carlo study by comparing these weighted estimators. They also show that the following two-step estimator of Reckstuhl, Welsh, and Carroll (2000) is more efficient than the above weighted estimators as well as the conventional LLLS estimator.

This two-step estimator of Ruckstuhl, Welsh, and Carroll (2000) is developed as follows. Let us write Equation 16.8 in vector form:

$$Y = \mathbf{m}(X) + \varepsilon, \tag{16.15}$$

where $X = (x_{11}, \ldots, x_{1T}, \ldots, x_{n1}, \ldots, x_{nT})'$, $\mathbf{m}(X) = (m(x_{11}), \ldots, m(x_{1T}), \ldots, m(x_{n1}), \ldots, m(x_{nT}))'$, $\varepsilon = \alpha \otimes l_T + U$, $U = (u_{11}, \ldots, u_{1T}, \ldots, u_{n1}, \ldots, u_{nT})'$. Multiplying both sides of Equation 16.15 by $\Omega^{-\frac{1}{2}}$ yields

$$\begin{aligned} \Omega^{-\frac{1}{2}}Y &= \Omega^{-\frac{1}{2}}\mathbf{m}(X) + \Omega^{-\frac{1}{2}}\varepsilon \\ &= \Omega^{-\frac{1}{2}}\mathbf{m}(X) - \mathbf{m}(X) + \mathbf{m}(X) + \Omega^{-\frac{1}{2}}\varepsilon \end{aligned}$$

or

$$Y^* = \mathbf{m}(X) + \Omega^{-\frac{1}{2}}\varepsilon \tag{16.16}$$

where $Y^* = \Omega^{-\frac{1}{2}}Y + (I - \Omega^{-\frac{1}{2}})\mathbf{m}(X)$ is the transformed variable and $\Omega^{-\frac{1}{2}}\varepsilon$ has an identity variance–covariance matrix. However, Y^* is not observed. So, a feasible estimator based on this transformed model can be obtained via a two-step procedure. In the first step we can run the LLLS regression Y on X to obtain the estimate $\hat{m}(x)$ of $m(x)$ at each data point and the residuals, based on which we can obtain consistent estimate $\hat{\Omega}$ of Ω as discussed above. This gives $\hat{Y}^* = \hat{\Omega}^{-\frac{1}{2}}Y + (I - \hat{\Omega}^{-\frac{1}{2}})\hat{\mathbf{m}}(X)$, where $\hat{\mathbf{m}}(X) = (\hat{m}(x_{11}), \ldots, \hat{m}(x_{1T}), \ldots, \hat{m}(x_{n1}), \ldots, \hat{m}(x_{nT}))'$. In the second step, we run the LLLS regression of $\hat{Y}^*$ on X. Such two-step estimation performs better than the weighted LLLS estimator (Henderson and Ullah 2008). The asymptotic property of this type of two-step estimators is established in Su and Ullah (2007). See also Martins-Filho and Yao (2009) and Su, Ullah, and Wang (2010) for related research along this line.

16.3 Nonparametric Panel Data Model with Fixed Effects

In this section, we consider the following nonparametric panel data model with fixed effects

$$y_{it} = m(x_{it}) + \alpha_i + u_{it}, i = 1, \ldots, n, t = 1, \ldots, T, \tag{16.17}$$

where the covariate (regressor) x_{it} is of dimension $p \times 1$, $m(\cdot)$ is an unknown smooth function, α_i's are fixed effects heterogeneity parameters, and u_{it} is i.i.d. with zero mean, finite variance σ_u^2 and independent of x_{jt} for all i, j, and t. We assume $\sum_{i=1}^{n} \alpha_i = 0$ (so that $\alpha_1 = -\sum_{i=2}^{n} \alpha_i$) for the purpose of identification. Also, for the sake of simplicity, x_{it} is strictly exogenous. We are interested in consistent estimation of $m(\cdot)$ and its first-order derivative.

Following the notation in the previous section, we can approximate the model in Equation 16.17 as follows

$$Y \approx \mathbf{X}(x)\delta(x) + D\alpha + U \tag{16.18}$$

where $\alpha = (\alpha_2, \ldots, \alpha_n)'$, $D = (I_n \otimes l_T)d_n$, $d_n = [-l_{n-1}I_{n-1}]'$, and other notations are as defined above. Note that α contains heterogeneity parameters that may be correlated with the idiosyncratic error term u_{it} and the regressor x_{it} as well. So the LLLS estimator is generally inconsistent in this case.

16.3.1 Profile Least Squares Estimators

We argue that $\delta(x)$ in Equation 16.18 can be estimated by using the idea of profile least squares. There are two alternative approaches here. In the first approach, one can profile out the individual effects parameter α and consider the concentrated least squares for $\delta(x)$. In the second approach, one profiles out the nonparametric component $\delta(x)$ and consider the concentrated least squares for α. We discuss the first approach, followed by the second approach.

For the moment, we pretend α is known and then we can estimate $\delta(x)$ in Equation 16.18 by choosing δ to minimize the following criterion function

$$[Y - \mathbf{X}(x)\delta - D\alpha]'\mathbf{K}(x)[Y - \mathbf{X}(x)\delta - D\alpha]. \tag{16.19}$$

We denote the solution to the above minimization problem as $\delta_\alpha(x)$, which is the LLLS estimator of $\delta(x)$ by regressing $y_{it} - \alpha_i$ on x_{it}. It is easy to verify that

$$\delta_\alpha(x) = S(x)(Y - D\alpha) \tag{16.20}$$

where

$$S(x) = [\mathbf{X}(x)'\mathbf{K}(x)\mathbf{X}(x)]^{-1}\mathbf{X}(x)'\mathbf{K}(x) \tag{16.21}$$

is a $(p+1) \times nT$ matrix. In particular, the LLLS estimator of $m(x)$ is given by

$$m_\alpha(x) = e_1'\delta_\alpha(x) = e_1'S(x)(Y - D\alpha) = s(x)'(Y - D\alpha) \tag{16.22}$$

where $e_1 = (1, 0, \ldots, 0)'$ is a $(p+1) \times 1$ vector, and $s(x)' = e_1'S(x)$.

However, $\delta_\alpha(x)$ is not operational since it depends on the unknown parameter α. This motivates us to profile out the nonparametric component $m(x)$ in Equation 16.17. Note that Equation 16.17 can be written as

$$Y = \mathbf{m}(X) + D\alpha + U \tag{16.23}$$

To profile out $\mathbf{m}(X)$ in the above regression, we consider choosing α to minimize the following criterion function

$$[Y - D\alpha - \mathbf{m}_\alpha(X)]'[Y - D\alpha - \mathbf{m}_\alpha(X)] = (Y^* - D^*\alpha)'(Y^* - D^*\alpha), \tag{16.24}$$

where

$$\mathbf{m}_\alpha(X) = [m_\alpha(x_{11}) \cdots m_\alpha(x_{1T}), \ldots, m_\alpha(x_{n1}) \cdots m_\alpha(x_{nT})] = \mathbf{S}(Y - D\alpha),$$
$$Y^* = (I_{nT} - \mathbf{S})Y,$$
$$D^* = (I_{nT} - \mathbf{S})D,$$

$\mathbf{S} = (s_{11}, \ldots, s_{1T}, \ldots, s_{n1}, \ldots, s_{nT})'$ is an $nT \times nT$ matrix, and $s_{it} = s(x_{it})$. Then the solution to the above minimization problem is given by

$$\hat{\alpha} = (D^* D^*)^{-1} D^* Y^* = (D'QD)^{-1} D'QY, \tag{16.25}$$

where $Q = (I_{nT} - \mathbf{S})'(I_{nT} - \mathbf{S})$. The estimator for α_1 is $\hat{\alpha}_1 = -\sum_{i=2}^{n} \hat{\alpha}_i$, where $\hat{\alpha} = (\hat{\alpha}_2, \ldots, \hat{\alpha}_n)'$.

The profile least squares estimator for $\delta(x)$ and $m(x)$ are given respectively by

$$\hat{\delta}(x) = \delta_{\hat{\alpha}}(x) = S(x)(Y - D\hat{\alpha}) = S(x)MY \tag{16.26}$$

and

$$\hat{m}(x) = m_{\hat{\alpha}}(x) = s(x)(Y - D\hat{\alpha}) = s(x)MY \tag{16.27}$$

where $M = I_{NT} - D(D'QD)^{-1}D'Q$ is an $nT \times nT$ matrix such that $MD = 0$. The asymptotic properties of $\hat{\delta}(x)$ have been studied in Su and Ullah (2006) in the framework of partially linear panel data models.

An alternative way to obtain the estimates of α and $\delta(x)$ is to profile out α first by choosing α to minimize the following criterion function:

$$[Y - \mathbf{X}(x)\delta(x) - D\alpha]'\mathbf{K}(x)[Y - \mathbf{X}(x)\delta(x) - D\alpha]. \tag{16.28}$$

The solution to this minimization problem is given by

$$\tilde{\alpha}(x) = [D'\mathbf{K}(x)D]^{-1}D'\mathbf{K}(x)[Y - \mathbf{X}(x)\delta(x)]. \tag{16.29}$$

In the second stage, we substitute $\tilde{\alpha}(x)$ in Equation 16.28 to obtain the following concentrated weighted least squares objective function

$$[Y - \mathbf{X}(x)\delta(x)]'\mathbf{K}^*(x)[Y - \mathbf{X}(x)\delta(x)] \tag{16.30}$$

where $\mathbf{K}^*(x) = M(x)\mathbf{K}(x)M(x)$ and $M(x) = I_{nT} - D(D'\mathbf{K}(x)D)^{-1}D'K(x)$ is such that $M(x)D = 0$. Choosing $\delta(x)$ to minimize Equation 16.30 yields the solution

$$\tilde{\delta}(x) = [\mathbf{X}(x)'\mathbf{K}^*(x)\mathbf{X}(x)]^{-1}\mathbf{X}(x)'\mathbf{K}^*(x)Y.$$

See Sun, Carroll, and Li (2009) for this estimator in a more general framework and its asymptomatic properties. An operational estimator of $\alpha(x)$ is obtained by substituting $\delta(x)$ with $\tilde{\delta}(x)$ in Equation 16.29. This approach, however, does not provide an estimator of α.

16.3.2 Measure of Goodness-of-Fit

Now we present the measure of goodness-of-fit in the fixed effects model which can be similarly defined in other types of models. Let $\hat{\mathbf{m}}(X) = (\hat{m}(x_{11}), \ldots, \hat{m}(x_{1T}), \ldots, \hat{m}(x_{n1}), \ldots, \hat{m}(x_{nT}))'$, and $\hat{U} = Y - \hat{\mathbf{m}}(X) - D\hat{\alpha}$. Noting that $\hat{\mathbf{m}}(X) = \mathbf{S}MY$ and $\hat{\alpha} = (D'QD)^{-1}D'QY$, we have

$$\begin{aligned} Y &= \hat{\mathbf{m}}(X) + D\hat{\alpha} + \hat{U} \\ &= \mathbf{S}MY + D(D'QD)^{-1}D'QY + \hat{U} \\ &= \mathbf{S}MY + (I_{nT} - M)Y + \hat{U} = \hat{Y} + \hat{U}, \end{aligned}$$

where $\hat{Y} = [I_{nT} + (\mathbf{S} - I_{nT})M]Y$ is the stack of the fitted values, and thus $\hat{U} = (I_{nT} - \mathbf{S})MY$. Under the assumption that u_{it} is i.i.d. across both i and t, we can estimate its variance σ_u^2 by

$$\hat{\sigma}_u^2 = \frac{\hat{U}'\hat{U}}{tr(N)} = \frac{Y'NY}{tr(N)}$$

where $N = M'QM$. Conditional on X, we have

$$E\left(\hat{\sigma}_u^2 | X\right) = \sigma_u^2 + \frac{1}{tr(N)}\mathbf{m}(X)'N\mathbf{m}(X). \tag{16.31}$$

Thus, $\hat{\sigma}_u^2$ is unbiased only if $N\mathbf{m}(X) = 0$. In general, we can establish only the consistency of $\hat{\sigma}_u^2$ for σ_u^2.

A global goodness-of-fit measure can be defined as

$$R^2 = \frac{\hat{Y}'\hat{Y}}{Y'Y}, \tag{16.32}$$

or obtained by calculating the square of correlation between Y and $\hat{Y}$. However, this may not have the same interpretation as in the case of linear regression model because $Y'Y = \hat{Y}'\hat{Y} + \hat{U}'\hat{U} + 2\hat{U}'\hat{Y}$ but $\hat{U}'\hat{Y}$ is not guaranteed to be zero.

In view of the above problem, we propose an alternative way to construct a goodness-of-fit measure as follows. First, we define a local R^2 and then the global R^2. We write from Equation 16.18

$$\begin{aligned} Y &= \mathbf{X}(x)\hat{\delta}(x) + D\hat{\alpha} + \hat{U}_x \\ &= \mathbf{Z}(x)\hat{\gamma}(x) + \hat{U}_x, \end{aligned} \tag{16.33}$$

where $\mathbf{Z}(x) = [\mathbf{X}(x)D]$, $\hat{\gamma}(x) = [\hat{\delta}'(x)\hat{\alpha}']'$, and $\hat{U}_x \equiv Y - \mathbf{Z}(x)\hat{\gamma}(x)$. Then

$$\begin{aligned} (Y - LY)'\mathbf{K}(x)(Y - LY) &= [Z(x)\hat{\gamma}(x) - LY]'\mathbf{K}(x)[Z(x)\hat{\gamma}(x) - LY] \\ &\quad + \hat{U}_x'\mathbf{K}(x)\hat{U}_x \end{aligned} \tag{16.34}$$

where $L = l_{nT}l_{nT}'/(nT)$, $\mathbf{K}(x)$ is a diagonal matrix with typical elements $K_h(x_{it} - x)/(nT\hat{f}(x))$ for $i = 1, \ldots, n$, and $t = 1, \ldots, T$, $\hat{f}(x) = (nT)^{-1}$

$\sum_{i=1}^{n}\sum_{t=1}^{T} K_h(x_{it} - x)$. Observe that $\hat{\gamma}(x)$ can be written as

$$\hat{\gamma}(x) = A(x)Y, \ A(x) = \begin{pmatrix} S(x)M \\ (DQD)^{-1}D'Q \end{pmatrix}. \tag{16.35}$$

Thus we can write Equation 16.34 as

$$Y'N_1(x)Y = Y'N_2(x)Y + Y'N_3(x)Y \tag{16.36}$$

where $N_1(x) = (I_{nT} - L)'K(x)(I_{nT} - L)$, $N_2(x) = [I_{nT} - \mathbf{Z}(x)A(x) - L]'\mathbf{K}(x)[I_{nT} - \mathbf{Z}(x)A(x) - L]$, $N_3 = [I_{nT} - \mathbf{Z}(x)A(x)]'\ \mathbf{K}(x)[I_{nT} - \mathbf{Z}(x)A(x)]$, and $N_2(x)N_3(x) = 0$. It follows that

$$TSS(x) = SSR(x) + RSS(x) \tag{16.37}$$

where $TSS(x) = Y'N_1(x)Y$, $SSR(x) = Y'N_2(x)Y$, and $RSS(x) = Y'N_3(x)Y$.

Thus Equation 16.37 represents a local analysis of variance (ANOVA) so that we can define a local R^2 as

$$R^2(x) = \frac{SSR(x)}{TSS(x)} = 1 - \frac{RSS(x)}{TSS(x)} \tag{16.38}$$

where $0 \leq R^2 \leq 1$ by construction. Further, a global R^2 can be defined as

$$R^2 = \frac{SSR}{TSS} = 1 - \frac{RSS}{TSS} \tag{16.39}$$

where $SSR = \int_x SSR(x)\hat{f}(x)dx$, $TSS = \int_x TSS(x)\hat{f}(x)dx$ and $RSS = \int_x RSS(x)\hat{f}(x)dx$. It is worth pointing out that $TSS = \sum_{i=1}^{n}\sum_{t=1}^{T}(y_{it} - \bar{y})^2$ where $\bar{y} = (nT)^{-1}\sum_{i=1}^{n}\sum_{t=1}^{T} y_{it}$.

16.3.3 Differencing Method

Let $\Delta y_{it} = y_{it} - y_{i,t-1}$. Δu_{it} is similarly defined. As in the usual differencing method, we can consider subtracting the model in Equation 16.17 for time t from that for time $t-1$ so that

$$\Delta y_{it} = m(x_{it}) - m(x_{i,t-1}) + \Delta u_{it} \tag{16.40}$$

or subtracting the equation for time t from that for time 1 so that

$$y_{it} - y_{i1} = m(x_{it}) - m(x_{i1}) + u_{it} - u_{i1}. \tag{16.41}$$

Another method, which is conventional, removes the fixed effects by deducting each equation from the cross-time average. This gives

$$y_{it} - \frac{1}{T}\sum_{t=1}^{T} y_{it} = m(x_{it}) - \frac{1}{T}\sum_{s=1}^{T} m(x_{is}) + u_{it} - \frac{1}{T}\sum_{s=1}^{T} u_{is} \tag{16.42}$$

or

$$y_{it}^{*} = \sum_{s=1}^{T} d_{ts} m(x_{is}) + u_{it}^{*} \tag{16.43}$$

where $d_{ts} = -\frac{1}{T}$ if $s \neq t$ and $1 - \frac{1}{T}$ otherwise, and $\sum_{s=1}^{T} d_{ts} = 0$ for all t, $y_{it}^{*} = y_{it} - T^{-1}\sum_{t=1}^{T} y_{it}$, and $u_{it}^{*} = u_{it} - T^{-1}\sum_{t=1}^{T} u_{it}$.

For each i, the right-hand sides of Equations 16.40 to 16.42 contain linear combination of $m(x_{is})$, $s = 1, \ldots, T$. We discuss the estimation corresponding to each of these differencing methods. To proceed, it is worth mentioning that some components of the function $m(\cdot)$ may not be fully identified via differencing methods. For example, if $m(x_{it}) = a + m_1(x_{it})$, then the difference will wipe out a and hence we can only estimate $m(x_{it})$ under some identification restriction. Similar issues arise when we consider the case of varying functional coefficient models later on if differencing methods are called upon.

For the first differencing (FD) model in Equation 16.40, Li and Stengos (1996) suggest estimation of $m(x_{it}, x_{i,t-1}) = m(x_{it}) - m(x_{i,t-1})$ by doing a local linear regression of Δy_{it} on x_{it} and $x_{i,t-1}$. Then we can obtain estimates of $m(x)$ by the method of estimating nonparametric additive models, e.g., by the marginal integration method of Linton and Nielson (1995) or by the backfitting method. For example, after we obtain estimates $\hat{m}(x, x_{i,t-1})$ of $m(x, x_{i,t-1})$ for $i = 1, \ldots, n$, and $t = 2, \ldots, T$, we can estimate $m(x)$ by $\hat{m}(x) = (n(T-1))^{-1}\sum_{i=1}^{n}\sum_{t=2}^{T}\hat{m}(x, x_{i,t-1})$, apart from the concerns discussed above for the differencing method. (See Hu, Wang and Carroll (2004) for a comparison of the two methods.) We also note that this method suffers from the "curse of dimensionality" problem in calculating $\hat{m}(x, x_{i,t-1})$ because it involves estimating a $2p$-dimensional nonparametric object. In view of this, Baltagi and Li (2002) obtain consistent estimators of $m(x)$ by considering the first differencing method and using series approximation for the nonparametric component.

Based on the differencing model in Equation 16.41, Henderson, Carroll, and Li (2008) propose an iterative kernel estimator of $m(x)$ and establish the asymptotic normality for their estimator. But this estimator is also subject to the comments on differencing given above. Since this method is elaborated in detail in Li and Racine (2007), we skip it for brevity.

Now we consider eliminating the fixed effects via the sample average over time. Following Equation 16.42, we write

$$y_{it} - \bar{y}_i = m(x_{it}) - \bar{m}_i + u_{it} - \bar{u}_i$$

where $\bar{y}_i = T^{-1}\sum_{t=1}^{T} y_{it}$, $\bar{u}_i = T^{-1}\sum_{t=1}^{T} u_{it}$, and $\bar{m}_i = T^{-1}\sum_{t=1}^{T} m(x_{it})$. Then writing $m(x_{it}) \approx m(x) + (x_{it} - x)'\beta(x)$ with $\beta(x) = \partial m(x)/\partial x$, we get

$$y_{it} - \bar{y}_i \approx (x_{it} - \bar{x}_i)'\beta(x) + u_{it} - \bar{u}_i,$$

where $\overline{x}_i = T^{-1}\sum_{t=1}^{T} x_{it}$. The local linear within-group estimator of $\beta(x)$ then follows as

$$\hat{\beta}_W(x) = \left(\sum_{i=1}^{n}\sum_{t=1}^{T}(x_{it}-\overline{x}_i)(x_{it}-\overline{x}_i)'K_h(x_{it}-x)\right)^{-1} \sum_{i=1}^{n}\sum_{t=1}^{T}(x_{it}-\overline{x}_i)(y_{it}-\overline{y}_i)K_h(x_{it}-x).$$

Similarly, if we use the first differencing method, then the local linear estimator of $\beta(x)$ for some fixed element x in $\{x_{it}, i = 1, \ldots, n; t = 1, \ldots, T\}$ is given by

$$\hat{\beta}_D(x) = \left(\sum_{i=1}^{n}\sum_{t=1}^{T}\Delta x_{it}\Delta x_{it}'K_h(x_{it}-x)\right)^{-1}\sum_{i=1}^{n}\sum_{t=1}^{T}\Delta x_{it}\Delta y_{it}K_h(x_{it}-x).$$

Lee and Mukherjee (2008) study the asymptotic properties of the above two estimators. For the case where x_{it} is a scalar random variable (i.e., $p = 1$), they show that under some standard assumptions,

$$E[\hat{\beta}_W(x) - \beta(x)|X] = \frac{m^{(2)}(x)[\mu_1(x)\mu_2(x)+\mu_3(x)]}{2\left[\mu_1^2(x)+\mu_2(x)\right]} + O_p(h^2)$$

and

$$E[\hat{\beta}_D(x) - \beta(x)|X] = \frac{m^{(2)}(x)\mu_3(x)}{2\mu_2(x)} + O_p(h^2),$$

where $\mu_j(x) = E(x_{it}-x)^j < \infty$ for $j = 1, 2, 3$, and $m^{(2)}(x) = \partial^2 m(x)/\partial x^2$.

It is clear from the above expressions that both the conventional within-group estimator and first-difference estimator are inconsistent because as $n \longrightarrow \infty$ and $h \longrightarrow 0$ we have a nondegenerating bias. This bias, however, is zero when the true regression function $m(x)$ is linear in x or x_{it} is symmetric around the point of evaluation x such that $\mu_j(x) = 0$ for $j = 1$ and 3. As Lee and Mukherjee (2008) observed, the nonvanishing biases arise because the difference equations are not locally weighted by the differenced variables whereas the original model is a local approximation around the point x of the original variable x_{it}. In other words, the differenced equations are initially localized around a value of x_{it} without considering the rest of values x_{is}, $s \neq t$. But $|x_{is} - x|$ cannot be small enough uniformly over all i and $s \neq t$ such that $\max_{i,s}|x_{is} - x| < Ch$ for some $C < \infty$, so that the differenced remainder terms cannot be tending to zero. Here the remainder term is $\bar{R}_{it} = (T-1)^{-1}\sum_{s=1,s\neq t}^{T} R_{is}(x)$ when $x = x_{it}$, where $R_{is}(x) = \frac{1}{2}m^{(2)}(x_{is}^*)(x_{is}-x)^2$ and x_{is}^* lies between x_{is} and x. Obviously, the biases do not vanish even when $T \longrightarrow \infty$. Again, this is due to the local approximation of $m(x)$ at given x_{it} as indicated in the kernel weight function $K_h(x_{it}-x)$, but the local estimator involves the average of $(x_{is}-x)$ for all i and $s \neq t$.

We notice that the estimator $\hat{\beta}_W(x)$, based on conventional within average differencing, was introduced in Ullah and Roy (1998), whereas the estimator $\hat{\beta}_D(x)$ is based on the first differencing method in Li and Stengos (1996) and Mundra (2005). In views of this, Mukherjee (2002) and Mukherjee and Ullah (2003) (also Henderson and Ullah 2005, p. 406) proposed elimination of the fixed effects by taking the within differencing in using local weighted average at x.

Define the locally weighted averages as

$$\bar{x}_i(x) = \frac{\sum_{s=1,s\neq t}^{T} x_{is} K_h(x_{is} - x)}{\sum_{s=1,s\neq t}^{T} K_h(x_{is} - x)}, \quad \text{and } \bar{y}_i(x) = \frac{\sum_{s=1,s\neq t}^{T} y_{is} K_h(x_{is} - x)}{\sum_{s=1,s\neq t}^{T} K_h(x_{is} - x)}.$$

The local-within leave-one-out estimator of $\beta(x)$ for $x = x_{it}$ is given by

$$\tilde{\beta}(x) = \left[\sum_{i=1}^{n} \sum_{s=1,s\neq t}^{T} x_{is}^*(x) x_{is}^*(x)' K_h(x_{is} - x)\right]^{-1} \sum_{i=1}^{n} \sum_{s=1,s\neq t}^{T} x_{is}^*(x) y_{is}^*(x) K_h(x_{is} - x),$$

where $x_{is}^*(x) = x_{is} - \bar{x}_i(x)$ and $y_{is}^*(x) = y_{is} - \bar{y}_i(x)$. Clearly, this estimator is the solution to the problem

$$\min_{\beta} \sum_{i=1}^{n} \sum_{s=1,s\neq t}^{T} [y_{is}^*(x) - x_{it}^*(x)'\beta]^2 K_h(x_{is} - x)$$

For $p = 1$, Lee and Mukherjee (2008) provide the following results under the standard regularity assumptions: (1) u_{it} is i.i.d. with mean 0 and variance σ^2 and it is independent of α_i and x_{it} for all i and t, (2) α_i is i.i.d., (3) x_{it} is i.i.d. with probability density function (p.d.f.) $f(x)$ whose support is bounded, and for the interior point x, it is twice differentiable with bounded second-order derivative, (4) $m(x)$ is twice differentiable with bounded second-order derivative, (5) K is compactly supported, bounded, and symmetric second-order kernel, and (6) $h \longrightarrow 0$ as $nh \longrightarrow 0$, $Th \longrightarrow 0$ and $nTh^3 \longrightarrow 0$ as $n, T \longrightarrow \infty$. Under these assumptions,

$$E[\tilde{\beta}(x) - \beta(x)|X] = \frac{h^2}{2}\left(\frac{m^{(2)}(x) f^{(1)}(x)}{f(x)}\right)\left(\frac{\kappa_4 - \kappa_2^2}{\kappa_2}\right) + O(h^2)$$

$$\text{Var}(\tilde{\beta}(x)|X) = \frac{1}{nTh^3}\left(\frac{\sigma^2}{f(x)}\right)\frac{\omega_2}{\kappa_2^2} + O_p\left(\frac{1}{nTh^3}\right)$$

where $f^{(1)}(x) = \partial f(x)/\partial x$, $\omega_2 = \int x^2 K(x)^2 dx$, and $\kappa_l = \int x^l K(x) dx$ for $l = 2, 4$. Further, using the above results one can show that the optimal bandwidth in minimizing $MSE(\tilde{\beta}(x))$ is proportional to $(nT)^{-1/7}$. If $m(x)$ is three times differentiable then in the bias of $\tilde{\beta}(x)$ we add an additional term $h^2 m^{(3)}(x)\kappa_4/(6\kappa_2)$, where $m^{(3)}(x) = \partial^3 m(x)/\partial x^3$. These results show that for the local weighted average differencing the orders of magnitudes of bias and variance are the same as those of the local linear derivative estimator.

See Pagan and Ullah (1999) and Li and Racine (2007). However, the magnitude of bias differs with $-h^2 m^{(2)}(x) f^{(1)}(x)\kappa_2/(2f(x))$ which arises due to the local weighted average differencing, but the magnitude of variance remains the same.

A similar idea can be applied to the case of time differenced model. Lee and Mukherjee (2008) suggest estimating $\beta(x)$ by

$$\hat{\beta}(x) = \min_{\beta} \sum_{i=1}^{n} \sum_{t=1}^{T} (\Delta y_{it} - \beta \Delta x_{it})^2 K_h(x_{it} - x, x_{i,t-1} - x).$$

But this method does not go through when the model has time-heterogeneity.

Finally, although the estimator of $m(x)$ is not directly obtained from the objective function, an estimator of $m(x)$ could be written as

$$\tilde{m}(x) = \frac{1}{n} \sum_{i=1}^{n} \tilde{m}_i(x)$$

where $\tilde{m}_i(x) = \bar{y}_i(x) - \tilde{\beta}(x)\bar{x}_i(x)$. See Lee and Mukherjee (2008) for an alternative proposal. The properties of $\tilde{m}_i(x)$ are not yet known, also the asymptotic normality of $\tilde{\beta}(x)$.

16.3.4 Series Estimation

The above estimation procedures are invalid if x_{it} contains lagged dependent variables. Lee (2008) considers series estimation of the following nonparametric dynamic panel data model:

$$y_{it} = m(y_{i,t-1}) + \alpha_i + u_{it}, \tag{16.44}$$

where α_i can be eliminated via first differencing or within-group difference. Let $m^*(y_{i,t-1}) = m(y_{i,t-1}) - T^{-1}\sum_{s=1}^{T} m(y_{i,s-1})$ and similarly define y_{it}^* and u_{it}^*. Then we have the within-group transformation of the above model as follows:

$$y_{it}^* = m^*(y_{i,t-1}) + u_{it}^*. \tag{16.45}$$

Lee's (2008) series estimator of m is based on the above within-group transformation. Under the assumption that $\lim_{n,T\to\infty} n/T = \kappa \in (0, \infty)$, he finds that the series estimator is asymptotically biased and proposes a bias-corrected series estimator. Asymptotic normality is also established.

16.3.5 A Nonparametric Hausman Test

To test the random effects against the fixed effects specification in the model $y_{it} = m(x_{it}) + \alpha_i + u_{it}$, we can specify the null and alternative hypotheses as

$$H_0 : E(\alpha_i | x_{i1}, \ldots, x_{iT}) = 0 \text{ a.s. versus } H_1 : \text{ the negation of } H_0,$$

where a.s. is an abbreviation for almost surely. If we maintain the assumption that $E(u_{it}|x_{i1}, \ldots, x_{iT}) = 0$, the null hypothesis can also be written as

$$H_0 : E(\varepsilon_{it}|x_{i1}, \ldots, x_{iT}) = 0 \text{ a.s.}$$

where $\varepsilon_{it} = \alpha_i + u_{it}$. Then one can propose a test based on the sample analogue of

$$J = E\{\varepsilon_{it} E(\varepsilon_{it}|x_{it}) f(x_{it})\}$$

where $f(\cdot)$ is the p.d.f. of x_{it} because $J = 0$ under H_0 and $J = E\{[E(\varepsilon_{it}|x_{it})]^2 f(x_{it})\} > 0$ under H_1. A feasible test statistic is given by

$$J_n = \frac{1}{nT} \sum_{i=1}^{n} \sum_{t=1}^{T} \hat{\varepsilon}_{it} \hat{E}_{-it}(\hat{\varepsilon}_{it}|x_{it}) \hat{f}_{-it}(x_{it})$$

where $\hat{\varepsilon}_{it}$ is the residual from the random effects regression, $\hat{f}_{-it}(x_{it})$ and $\hat{E}_{-it}(\hat{\varepsilon}_{it}|x_{it})$ are leave-one-out kernel estimates of $f(x_{it})$ and $E(\varepsilon_{it}|x_{it})$, respectively, by using observations on $\{x_{it}, \hat{\varepsilon}_{it}\}$. This test statistic is considered in Henderson, Carroll, and Li (2008). But they do not provide a formal asymptotic distributional analysis. Instead, they propose a bootstrap method to obtain the critical values and demonstrate through simulations that J_n works reasonably well in finite samples.

16.4 Partially Linear Panel Data Models

In this section, we review the literature on partially linear panel data models. We focus on the following model

$$y_{it} = x_{it}'\beta_0 + m(z_{it}) + \alpha_i + u_{it}, i = 1, \ldots, n, t = 1, \ldots, T, \qquad (16.46)$$

where x_{it} and z_{it} are of dimensions $p \times 1$ and $q \times 1$, respectively, β_0 is a $p \times 1$ vector of unknown parameters, $m(\cdot)$ is an unknown smooth function, α_i is random or fixed effects, and u_{it} is the idiosyncratic disturbance. We will first discuss the estimation of Equation 16.46 when α_i represents the random effects and then the fixed effects model. We also comment on extensions and specification tests.

16.4.1 Partially Linear Panel Data Models with Random Effects

Let $\varepsilon_{it} = \alpha_i + u_{it}$. We can rewrite Equation 16.46 as

$$y_{it} = x_{it}'\beta_0 + m(z_{it}) + \varepsilon_{it}. \qquad (16.47)$$

In the literature, it is frequently assumed that

$$E(\varepsilon_{it}|z_{it}) = 0. \qquad (16.48)$$

Note that this assumption does not rule out the dependence between x_{it} and ε_{it}. As a matter of fact, some or all the components of x_{it} may be correlated with the error ε_{it}. Li and Stengos (1996) discuss the estimation of Equation 16.46 for the case of random effects model.

Under the assumption in Equation 16.48, we can take conditional expectation of Equation 16.46 given z_{it} on both sides to yield

$$E(y_{it}|z_{it}) = E(x_{it}|z_{it})'\beta_0 + m(z_{it}). \tag{16.49}$$

Subtracting Equation 16.49 from Equation 16.46, we have

$$Y_{it} = X'_{it}\beta_0 + \varepsilon_{it}. \tag{16.50}$$

Let $Y_{it} = y_{it} - E(y_{it}|z_{it})$ and $X_{it} = x_{it} - E(x_{it}|z_{it})$. So Equation 16.50 is a linear panel data model with dependent variable Y_{it} and independent variable X_{it}. If (Y_{it}, X_{it}) were observable, we can estimate β_0 by the parametric methods. For simplicity, we assume that there exists an instrumental variable (IV) $w_{it} \in \mathbb{R}^p$, such that

$$E(\varepsilon_{it}|w_{it}, z_{it}) = 0 \quad \text{and} \quad E(x'_{it}w_{it}) \neq 0. \tag{16.51}$$

We then can estimate β_0 by the IV method[1]:

$$\tilde{\beta} = (W'X)^{-1}W'Y = \beta_0 + (W'X)^{-1}W'\varepsilon, \tag{16.52}$$

where $W_{it} = w_{it} - E(w_{it}|z_{it})$, $Y = (Y_{11}, \ldots, Y_{1T}, \ldots, Y_{n1}, \ldots, Y_{nT})'$, X, W, and ε are similarly defined. Under Equation 16.51, we have $E(\varepsilon_{it}|W_{it}) = 0$, so the IV estimator $\tilde{\beta}$ is consistent. Nevertheless, it is infeasible since the conditional expectations $E(y_{it}|z_{it})$, $E(x_{it}|z_{it})$, and $E(w_{it}|z_{it})$ are unknown to us. As before, these conditional expectations can be consistently estimated using nonparametric methods. To avoid random denominator problem, we choose to use the marginal p.d.f. $f(\cdot)$ of z_{it} as the weighting function as in Li and Stengos (1996).

Multiplying Equation 16.50 by $f_{it} = f(z_{it})$, we have

$$Y_{it}f_{it} = (X_{it}f_{it})'\beta_0 + \varepsilon_{it}f_{it}. \tag{16.53}$$

Now one can estimate the unknown finite dimensional parameter β_0 by regressing $Y_{it}f_{it}$ on $X_{it}f_{it}$ using $W_{it}f_{it}$ as an IV. The infeasible IV estimator is obtain

$$\tilde{\beta}_f = \left(\sum_{i=1}^{n}\sum_{t=1}^{T} W_{it}X'_{it}f_{it}^2\right)^{-1} \sum_{i=1}^{n}\sum_{t=1}^{T} W_{it}Y_{it}f_{it}^2. \tag{16.54}$$

It is easy to show that $\tilde{\beta}_f$ is asymptotically normally distributed, i.e.,

$$\sqrt{n}(\tilde{\beta}_f - \beta_0) \xrightarrow{d} N(0, \Phi_f^{-1}\Psi_f\Phi_f^{-1}), \tag{16.55}$$

[1] If the dimension of w_{it} is $l \geq p$, the IV estimator of β_0 is given by $\tilde{\beta}_1 = (X'W(W'W)W'X)^{-1} X'W(W'W)W'Y$.

where $\Phi_f = T^{-1}\sum_{t=1}^{T} E[W_{it}X_{it}' f_{it}^2]$, and $\Psi_f = T^{-2}\sum_{t=1}^{T}\sum_{s=1}^{T} E(u_{it}u_{is}W_{it}W_{it}' f_{it}^2 f_{is}^2)$.

To proceed, we estimate f_{it} by $\hat{f}(z_{it}) = (nT)^{-1}\sum_{j=1}^{n}\sum_{s=1}^{T} K_{it,js}$ and $E(y_{it}|z_{it})$ by $\hat{y}_{it} = (nT)^{-1}\sum_{j=1}^{n}\sum_{s=1}^{T} y_{js}K_{it,js}/\hat{f}(z_{it})$, where $K_{it,js} = K_h(z_{it} - z_{js})$. The estimators $\hat{x}_{it}$ and $\hat{w}_{it}$ of $E(x_{it}|z_{it})$ and $E(w_{it}|z_{it})$ are similarly defined. A feasible estimator of β_0 can be obtained by replacing Y_{it}, X_{it}, Z_{it}, and f_{it} with $y_{it} - \hat{y}_{it}$, $x_{it} - \hat{x}_{it}$, $w_{it} - \hat{w}_{it}$, and $\hat{f}(z_{it})$. This leads to the following feasible density-weighed estimator of β_0:

$$\hat{\beta}_f = \left(\sum_{i=1}^{n}\sum_{t=1}^{T}(w_{it} - \hat{w}_{it})(x_{it} - \hat{x}_{it})'\hat{f}(z_{it})^2\right)^{-1}\sum_{i=1}^{n}\sum_{t=1}^{T}(w_{it}-\hat{w}_{it})(y_{it}-\hat{y}_{it})\hat{f}(z_{it})^2. \tag{16.56}$$

Under some regularity conditions, Li and Stengos (1996) have established the asymptotic normality of $\widehat{\beta}_f$:

$$\sqrt{n}(\hat{\beta}_f - \beta_0) \xrightarrow{d} N\big(0, \Phi_f^{-1}\Psi_f\Phi_f^{-1}\big).$$

For statistical inference on β_0, we need to estimate the asymptotic variance-covariance of $\hat{\beta}_f$ consistently, which is straightforward.

After obtaining a $\sqrt{n}$-consistent estimator $\hat{\beta}_f$ of β_0, we can estimate $m(z)$ consistently by $\hat{m}(z) = (nT)^{-1}\sum_{j=1}^{n}\sum_{s=1}^{T}(y_{js} - x_{js}'\hat{\beta}_f)K_{\tilde{h}}(z_{js} - z)/\tilde{f}(z)$, where the bandwidth $\tilde{h}$ is typically different from h, and $\tilde{f}(z) = (nT)^{-1}\sum_{j=1}^{n}\sum_{s=1}^{T} K_{\tilde{h}}(z_{js} - z)$. Since the nonparametric kernel estimator has a slower convergence rate than the parametric $\sqrt{n}$-rate, it is easy to show $\hat{m}(z)$ has the same asymptotic distribution as $\tilde{m}(z) = (nT)^{-1}\sum_{j=1}^{n}\sum_{s=1}^{T}(y_{js} - x_{js}'\beta_0)K_h(z_{js} - z)/\tilde{f}(z)$.

It is worth mentioning the above method works in a variety of applications. In particular, it allows x_{it} to contain lagged dependent variable. Nevertheless, the above IV estimator of β_0 is generally inefficient. When the error follows a one-way error component structure in the partially linear panel data model, Li and Ullah (1998) propose a feasible semiparametric generalized least squares (GLS) type estimator for estimating β_0 and show that is asymptotically more efficient than the semiparametric ordinary least squares (OLS) type estimator. They also discuss the case for which the regressor of the parametric component is correlated with the error, and propose an IV GLS-type semiparametric estimator. They show that their estimator for the finite dimensional parameter is efficient. For brevity, we refer the reader directly to their paper. Also, see the paper by Baltagi and Li (2002) which proposed new IV estimators having substantial efficiency gains over the one suggested by Li and Stengos (1996) and Li and Ullah (1998).

16.4.2 Partially Linear Panel Data Models with Fixed Effects

We now discuss the estimation of Equation 16.46 when α_i represents the fixed effect. For the identification purpose, we can impose $\sum_{i=1}^{n}\alpha_i = 0$. For simplicity, we assume that z_{it} is strictly exogenous but allow x_{it} to be correlated with the error term u_{it}. We are interested in consistent estimation of β_0 and $m(\cdot)$. As usual, we focus on the case where n is approaching infinity and T is fixed.

In principle, we can apply Li and Stengos (1996) or the method introduced in the previous section to estimate the fixed effect model. From Equation 16.46, we can take the first difference as in the linear panel data model to obtain

$$y_{it} - y_{i,t-1} = (x_{it} - x_{i,t-1})'\beta_0 + [m(z_{it}) - m(z_{i,t-1})] + (u_{it} - u_{i,t-1}), \quad (16.57)$$

or

$$Y_{it} = X_{it}'\beta_0 + M(z_{it}, z_{i,t-1}) + U_{it}, \quad (16.58)$$

where $Y_{it} = y_{it} - y_{i,t-1}$, $X_{it} = x_{it} - x_{i,t-1}$, $U_{it} = u_{it} - u_{i,t-1}$, and $M(z_{it}, z_{i,t-1}) = m(z_{it}) - m(z_{i,t-1})$. Equation 16.58 is basically the same as Equation 16.47 except that we know that U_{it} has a moving average structure. Nevertheless, this approach has several drawbacks. First, in order to eliminate $M(z_{it}, z_{i,t-1})$, they suggest estimating $E(Y_{it}|z_{it}, z_{i,t-1})$ and $E(X_{it}|z_{it}, z_{i,t-1})$ by the nonparametric kernel method. This suffers from the "curse of dimensionality" because it ignores the additive structure of Equation 16.57 and requires the kernel function to be defined on $\mathbb{R}^{2q}$ instead of $\mathbb{R}^q$. Secondly, although they propose a method to estimate the finite dimensional parameter β_0 and their method can estimate $M(z_{it}, z_{i,t-1})$, they did not suggest how to estimate the original unknown function $m(z_{it})$. For this reason, Baltagi and Li (2002) consider series estimation of the model that imposes the additive structure of $M(z_{it}, z_{i,t-1}) = m(z_{it}) - m(z_{i,t-1})$.

In matrix form, Equation 16.58 can be rewritten as

$$Y = X\beta_0 + M + U \quad (16.59)$$

where Y is an $nT \times 1$ vector with typical element Y_{it}, and X, M, and U are similarly defined. Let Z denote an $nT \times q$ matrix with typical row given by z_{it}.

A function $\xi(z_{it}, z_{i,t-1})$ is said to be an additive class of functions $\mathcal{M}$ if $\xi(z_{it}, z_{i,t-1}) = m(z_{it}) - m(z_{i,t-1})$, $m(\cdot)$ is twice differentiable in the interior of its support $\mathcal{Z}$, which is a compact subset of $\mathbb{R}^q$ and $E[m^2(z_{it})] < \infty$. We will use series $p^L(z)$ of $L \times 1$ dimension to approximate $m(z)$, where $L = L(n)$. The approximation function $p^L(z)$ has the following properties: (a) $p^L(z, \tilde{z}) \equiv p^L(z) - p^L(\tilde{z}) \in \mathcal{M}$; (b) as L grows, there is a linear combination of $p^L(z, \tilde{z})$ that can approximate any function in $\mathcal{M}$ arbitrarily well in the sense of mean squared error. Therefore, $p^L(z)$ approximates $m(z)$ and $p^L(z, \tilde{z}) \equiv p^L(z) - p^L(\tilde{z})$ approximates $M(z, \tilde{z}) = m(z) - m(\tilde{z})$:

$$p^L(z_{it}, z_{i,t-1}) = \begin{pmatrix} p_1(z_{it}) - p_1(z_{i,t-1}) \\ p_2(z_{it}) - p_2(z_{i,t-1}) \\ \vdots \\ p_L(z_{it}) - p_L(z_{i,t-1}) \end{pmatrix}. \quad (16.60)$$

For notational simplicity, define $p^L_{it} = p^L(z_{it}, z_{i,t-1})$ and $P = (p^L_{11}, p^L_{12}, \ldots, p^L_{1T}, \ldots, p^L_{n1}, p^L_{n2}, \ldots, p^L_{nT})'$. Clearly, P is a $nT \times L$ matrix.

For any scalar or vector function $g(z)$, denote $E_{\mathcal{M}}(g(z))$ the projection onto the additive function space $\mathcal{M}$ (under the L_2 norm). That is, $E_{\mathcal{M}}(g(z))$ is an element that belongs to $\mathcal{M}$ and it is the closest function to $g(z)$ in the L_2 norm for all the functions in L_2 in $\mathcal{M}$. Define $\theta(z) = E(X|Z = z)$ and $h(z) = E_{\mathcal{M}}(\theta(z))$.

Let $\bar{P} = P(P'P)^{-}P'$, where $(\cdot)^{-}$ denotes any symmetric generalized inverse. Let $\tilde{A} = \bar{P}A = P\beta_A$, where $\beta_A = (P'P)^{-}P'A$. Premultiplying Equation 16.59 by $\overline{P}$ yields

$$\tilde{Y} = \tilde{X}\beta_0 + \tilde{M} + \tilde{V}. \tag{16.61}$$

Subtracting Equation 16.61 from Equation 16.59 by $\overline{P}$ leads to

$$Y - \tilde{Y} = (X - \tilde{X})\beta_0 + (M - \tilde{M}) + (V - \tilde{V}). \tag{16.62}$$

We estimate β_0 by the least squares regression of $Y - \tilde{Y}$ on $(X - \tilde{X})$:

$$\hat{\beta} = [(X - \tilde{X})'(X - \tilde{X})]^{-1}(X - \tilde{X})'(Y - \tilde{Y}). \tag{16.63}$$

Upon obtaining $\hat{\beta}$, we can estimate $m(z)$ by

$$\hat{m}(z) = p^L(z)'\hat{\gamma} \tag{16.64}$$

where $\hat{\gamma} = (P'P)^{-}P'(Y - X\hat{\beta})$.

Let $\epsilon_{it} = X_{it} - h(z_{it})$, where $h(z_{it}) = E_{\mathcal{M}}(\theta(z_{it}))$. Let $\Phi = T^{-1}\sum_{t=1}^{T} E(\epsilon_{it}\epsilon_{it}')$ and $\Psi = T^{-1}\sum_{t=1}^{T} E(\sigma^2(X_{it}, Z_{it})\epsilon_{it}\epsilon_{it}')$ where $\sigma^2(X_{it}, Z_{it}) = E[V_{it}^2|X_{it}, Z_{it}]$. Baltagi and Li (2002) prove the following asymptotic normality of $\hat{\beta}$:

$$\sqrt{n}(\hat{\beta} - \beta_0) \xrightarrow{d} N(\Phi^{-1}\Psi\Phi^{-1}).$$

Baltagi and Li (2002) also establish the consistency rate of $\hat{m}(z)$ but not the asymptotic normality.

If x_{it} contains the lagged dependent variable, then the above estimation procedure has to be modified. For example, consider the following partially linear dynamic panel data model

$$y_{it} = \beta_{0,1}y_{i,t-1} + \beta_{0,2}'x_{it}^{(2)} + m(z_{it}) + u_{it}, i = 1, \ldots, n, t = 1, \ldots, T, \tag{16.65}$$

where $x_{it}^{(2)}$ is x_{it} excluding its first element $y_{i,t-1}$. Assume the existence of an IV $w_{it} \in \mathbb{R}^l$ with $l \geq p$ such that

$$E(U_{it}|w_{it}, z_{it}) = 0, \quad \text{and} \quad \text{Cov}(w_{it}, X_{it}) \neq 0. \tag{16.66}$$

We can estimate $\beta_0 = (\beta_{0,1}, \beta_{0,2}')'$ by the IV method for the case $l = p$:

$$\hat{\beta}_{IV} = [(W - \tilde{W})'(X - \tilde{X})]^{-}(W - \tilde{W})'(Y - \tilde{Y}), \tag{16.67}$$

and estimate $m(z)$ by

$$\hat{m}_{IV}(z) = p^L(z)'\hat{\gamma}_{IV} \tag{16.68}$$

where $\hat{\gamma}_{IV} = (P'P)^- P'(Y - X\beta_{IV})$. The asymptotic normality of $\hat{\beta}_{IV}$ is established in Baltagi and Li (2002). See also Baltagi and Li (2000). Obviously, in the case where all elements in X_{it} are exogenous, we can simply set $w_{it} = X_{it}$, and the results will be the same as discussed above.

When $z_{it} = y_{i,t-1}$, Equation 16.46 becomes the partially linear dynamic model studied by Lee (2008). He establishes the asymptotic normality of a bias-corrected series estimator of β_0.

16.4.3 Extensions

Traditionally, the dependent variable in a partially linear model is a continuous random variable. This may not be the case in applications. Lin and Carroll (2001a, 2001b) consider a generalized partially linear panel data model using generalized estimating equations. Given the covariates x_{it} and z_{it}, they assume that the mean μ_{it} of the dependent variable y_{it} satisfies

$$g(\mu_{it}) = x'_{it}\beta_0 + m(z_{it}), \tag{16.69}$$

where z_{it} may be time dependent or not, and g is some known link function. They develop kernel estimating equations for the nonparametric component $m(\cdot)$ and profile estimating equations for the parametric component β_0.

If the dimension q of z_{it} is large, the estimation of the parametric and nonparametric components in Equation 16.46 becomes difficult. In this case, we can consider the following additive partially linear panel data models

$$y_{it} = x'_{it}\beta_0 + m_1(z_{it,1}) + \cdots + m_q(z_{it,q}) + \alpha_i + u_{it}, i = 1, \ldots, n, t = 1, \ldots, T, \tag{16.70}$$

where β_0, x_{it}, α_i, and u_{it} are defined as above, $m_l(\cdot)$, $l = 1, \ldots, q$, are unknown smooth functions. Obviously the individual functions $m_l(\cdot)$, $l = 1, \ldots, q$, are not identified without further conditions. In the literature on kernel estimation, one may assume that $E[m_l(z_{it,l})] = 0$ whereas in the literature on series estimation, it seems convenient to assume that $m_l(0) = 0$ for $l = 2, \ldots, q$. Li (2000) considers the series estimation of the above model in the cross-section framework. It seems straightforward to extend his method to the panel framework.

16.4.4 Specification Tests

Various specification tests can be conducted for partially linear models. These include tests for correct specification of functional forms, tests for random effects versus fixed effects, tests for individual effects, tests for serial correlation, and tests for heteroskedasticity in the disturbance terms, etc. Despite the importance of specification testing in panel data models, only few papers consider this.

Henderson, Carroll, and Li (2008) consider testing the functional form by considering the following possible specifications

$$y_{it} = x'_{it}\beta_0 + z'_{it}\gamma_0 + \varepsilon_{it}, \tag{16.71}$$

$$y_{it} = x'_{it}\beta_0 + m(z_{it}) + \varepsilon_{it}, \tag{16.72}$$

$$y_{it} = g(x_{it}, z_{it}) + \varepsilon_{it}, \tag{16.73}$$

where the definitions of parameters and functions are self-evident. The three pairs of null and alternative hypotheses are

$$H_0^a : \text{Equation 16.71 versus } H_1^a : \text{Equation 16.72,}$$
$$H_0^b : \text{Equation 16.71 versus } H_1^b : \text{Equation 16.73,}$$
$$H_0^c : \text{Equation 16.72 versus } H_1^c : \text{Equation 16.73,}$$

where for example, "H_0^a : Equation 16.71" means the model in Equation 16.71 is the true model under the first null hypothesis H_0^a. For each case, they estimate the models under the null and alternative and compared the squared distance between the estimated models. For example, to test H_0^a : Equation 16.71 versus H_1^a : Equation 16.72, they estimate both Equations 16.71 and 16.72 and base their test statistic on

$$J_n^a = \frac{1}{nT}\sum_{i=1}^{n}\sum_{t=1}^{T}[x'_{it}\tilde{\beta} + z'_{it}\tilde{\gamma} - x'_{it}\hat{\beta} - \hat{m}(z_{it})]^2$$

where $\tilde{\beta}$ and $\tilde{\gamma}$ are estimates of β_0 and γ_0 under H_0^a, $\hat{\beta}$ and $\hat{m}(z_{it})$ are estimates of β_0 and $m(z_{it})$ under H_1^a. Without deriving the asymptotic distribution for such a test statistic, they propose a bootstrap method to obtain the critical values and demonstrate through simulations that the proposed tests work fairly well in finite samples.

Li and Hsiao (1998) consider testing serial correlation in a partially linear panel data models that could allow lagged dependent variables as explanatory variables. They consider the following model

$$y_{it} = x'_{it}\beta_0 + m(z_{it}) + u_{it}. \tag{16.74}$$

where variables are defined as above and u_{it} satisfies $E(u_{it}|x_{it}, z_{it}) = 0$ a.s. The null hypothesis is

$$H_0 : u_{it} \text{ is a martingale difference sequence (m.d.s.).}$$

Clearly, under the above null hypothesis, u_{it} cannot contain the individual effects. Based on the residuals from the above partially linear model, they propose three test statistics that test zero first-order serial correlation, higher-order serial correlations, and individual effects, respectively. These test statistics have either asymptotic normal or chi-square distribution under the null hypothesis of an m.d.s. error process.

16.5 Varying Coefficient Panel Data Models

In this section, we review the literature on varying coefficient models. We consider the following model

$$\begin{aligned} y_{it} &= x_{it}'m(z_{it}) + \alpha_i + u_{it} \\ &= \sum_{d=1}^{p} x_{it,d}m_d(z_{it}) + \alpha_i + u_{it} \end{aligned} \tag{16.75}$$

where the covariate z_{it} is a $q \times 1$ vector, $x_{it} = (x_{it,1}, \ldots, x_{it,p})'$, $m(\cdot) = (m_1(\cdot), \ldots, m_p(\cdot))'$ has p unknown smooth functions, and u_{it} is i.i.d. with zero mean and finite variance σ_u^2. We make explicit assumptions on the dependence of α_i and u_{it} on the covariates x_{it} and z_{it} only when needed. The model in Equation 16.75 is useful where the response parameter (slope coefficient) depends on the variable z_{it}. For example, in a wage equation, y_{it} denotes the logarithm of wage, x_{it} denotes the years of schooling (education), and the rate of return to education may depend on the individual characteristic z_{it}. In a special case where $p = 1$, $x_{it} = 1$ for all i and t, and α_i can be correlated with x_{it} and u_{it}, model Equation 16.75 reduces to the conventional fixed effects panel data models considered by Su and Ullah (2006) and Henderson, Carroll, and Li (2008).

Note that the model in Equation 16.75 includes the partially linear model as special cases: $x_{it}'m(z_{it}) = m_1(z_{it}) + \tilde{x}_{it}'\beta_0$, where $x_{it} = (1, \tilde{x}_{it}')'$, and $m(z_{it}) = (m_1(z_{it}), \beta_0')'$ for some real-valued function m_1 and $(p-1) \times 1$ vector β_0. The latter model was considered by Li and Hsiao (1998) and Kniesner and Li (2002) who assumes that $E(u_{it}|z_{it}, x_{it}) = 0$. Li and Stengos (1996) and Baltagi and Li (2002) considered the same model but allowed $E(u_{it}|x_{it}) \neq 0$. See Section 16.4.

16.5.1 Profile Least Squares Method

We first consider the estimation of m in Equation 16.75 when α_i is treated as fixed effects which can be correlated with either x_{it} or u_{it}.

For any given z and $d \in \{1, 2, \ldots, p\}$, it follows from a first order Taylor expansion that

$$\begin{aligned} m_d(z_{it}) &\approx m_d(z) + (z_{it} - z)'\beta_d(z) \\ &= z_{it}(z)'\delta_d(z) \end{aligned}$$

where $z_{it}(z) = (1\ (z_{it} - z)')'$, $\delta_d(z) = (m_d(z)\ \beta_d(z)')'$, and $\beta_d(z) = \partial m_d(z)/\partial z$. Then following the LLLS estimation procedure in Section 16.2, we can write the estimate of $\delta(z) = (m_1(z), \ldots, m_p(z), \beta_1(z)', \ldots, \beta_p(z)')'$ as

$$\delta_\alpha(z) = \min_{\delta}(Y^* - \tilde{X}\delta)'\mathbf{K}(z)(Y^* - \tilde{X}\delta)$$

where $Y^* = Y - D\alpha$, $\tilde{X} = (X_{11}, \ldots, X_{1T}, \ldots, X_{n1}, \ldots, X_{nT})'$ is an $nT \times p(q+1)$ matrix with $X_{it} = X_{it}(z) = (X_{it}', X_{it}' \otimes (z_{it} - z)')'$, and $\mathbf{K}(z) = \text{diag}$

$(K_h(z_{11}-z),\ldots,K_h(z_{1T}-z),\ldots,K_h(z_{n1}-z),\ldots,K_h(z_{nT}-z))$ is an $nT\times nT$ diagonal matrix. So, given α, the LLLS estimator of $\delta(z)$ is simply

$$\hat{\delta}_\alpha(z)=[\tilde{X}'\mathbf{K}(z)\tilde{X}]^{-1}\tilde{X}'\mathbf{K}(z)Y^*.$$

In particular, the estimator of $m(z)=(m_1(z),\ldots,m_p(z))'$ is given by

$$\hat{m}_\alpha(z)=\mathbf{e}\hat{\delta}_\alpha(z)=\mathbf{s}(z)(Y-D\alpha)$$

where $\mathbf{e}=(I_p,\mathbf{0}_{p\times pq})$ is a $p\times p(q+1)$ selection matrix with $\mathbf{0}_{p\times pq}$ denoting a $p\times pq$ matrix of zeros, and $\mathbf{s}(z)=\mathbf{e}S(z)=\mathbf{e}(\tilde{X}'\mathbf{K}(z)\tilde{X})^{-1}\tilde{X}'\mathbf{K}(z)$ is a $p\times nT$ matrix.

Let $Z=(z_{11},\ldots,z_{1T},\ldots,z_{n1},\ldots,z_{nT})'$. We can write $m(x_{it},z_{it})\equiv x_{it}'m(z_{it})$, $i=1,\ldots,n$, $t=1,\ldots,T$, in vector form as

$$\mathbf{m}(X,Z)=\sum_{d=1}^{p}\mathbf{x}_d\odot\mathbf{m}_d(\mathbf{z})$$

where $\mathbf{x}_d=(x_{11,d},\ldots,x_{1T,d},\ldots,x_{n1,d},\ldots,x_{nT,d})'$, $\mathbf{m}_d(Z)=(m_d(z_{11}),\ldots,m_d(z_{1T}),\ldots,m_d(z_{n1}),\ldots,m_d(z_{nT}))'$ for $d=1,\ldots,p$, and $\odot$ is the Hadarmard product. Thus

$$\hat{\mathbf{m}}_\alpha(X,Z)=\sum_{d=1}^{p}\mathbf{x}_d\odot\hat{\mathbf{m}}_{\alpha,d}(Z)=\sum_{d=1}^{p}\mathbf{x}_d\odot(\mathbf{S}_d(z)(Y-D\alpha))$$

where $\hat{\mathbf{m}}_{\alpha,d}(\mathbf{z})=(\hat{m}_{\alpha,d}(z_{11}),\ldots,\hat{m}_{\alpha,d}(z_{1T}),\ldots,\hat{m}_{\alpha,d}(z_{n1}),\ldots,\hat{m}_{\alpha,d}(z_{nT}))'$, $\hat{m}_{\alpha,d}(z)$ is the dth element of $\hat{m}_\alpha(z)$: $\hat{m}_{\alpha,d}(z)=e_d'\hat{m}_\alpha(z)=e_d'\mathbf{s}(z)(Y-D\alpha)$ with e_d being a $p\times 1$ vector with 1 in the dth element and 0 elsewhere, and $\mathbf{S}_d(Z)'=(\mathbf{s}(z_{11})'e_d,\ldots,\mathbf{s}(z_{1T})'e_d,\ldots,\mathbf{s}(z_{n1})',\ldots,\mathbf{s}(z_{nT})'e_d)$. Noting that

$$\hat{\mathbf{m}}_\alpha(X,Z)=\left(\sum_{d=1}^{p}(\mathbf{x}_d\otimes l_{nT}')\odot\mathbf{S}_d(Z)\right)(Y-D\alpha),$$

the estimate of α is given by

$$\hat{\alpha}=(D'Q_1D)^{-1}D'Q_1Y$$

where $Q_1=(I_{nT}-\sum_{d=1}^{p}(\mathbf{x}_d\otimes l_{nT}')\odot\mathbf{S}_d(Z))'(I_{nT}-\sum_{d=1}^{p}(\mathbf{x}_d\otimes l_{nT}')\odot\mathbf{S}_d(Z))$. Further, the estimator for $\delta(z)$ and $m(z)$ follows by $\hat{\delta}_{\hat{\alpha}}(z)$ and $\hat{m}_{\hat{\alpha}}(z)$, respectively.

Sun, Carroll, and Li (2009) suggest an alternative profile least squares estimator for the above model by profiling out the nonparametric component m. They also propose a test for testing a random effects model against a fixed effects alternative model. Notice that if the vector z_{it} contains both the discrete and continuous variables, then Su, Chen, and Ullah (2009) can be extended to this panel framework.

16.5.2 Differencing Method

As in Section 16.3, we can consider subtracting the model (Equation 16.75) for time t from that of time $t-1$ so that

$$\Delta y_{it} = x'_{it} m(z_{it}) - x'_{i,t-1} m(z_{i,t-1}) + \Delta u_{it} \tag{16.76}$$

or subtracting the equation from time t from that for time 1 so that

$$y_{it} - y_{i1} = x'_{it} m(z_{it}) - x'_{i1} m(z_{i1}) + u_{it} - u_{i1}. \tag{16.77}$$

Alternatively, the within-group differencing method yields

$$y_{it} - \frac{1}{T}\sum_{t=1}^{T} y_{it} = x'_{it} m(z_{it}) - \frac{1}{T}\sum_{t=1}^{T} x'_{it} m(z_{it}) + u_{it} - \frac{1}{T}\sum_{t=1}^{T} u_{it} \tag{16.78}$$

or

$$y^*_{it} = \sum_{s=1}^{T} d_{ts} x'_{is} m(x_{is}) + u^*_{it} \tag{16.79}$$

where $d_{ts} = -\frac{1}{T}$ if $s \neq t$ and $1 - \frac{1}{T}$ otherwise, and $\sum_{s=1}^{T} d_{ts} = 0$ for all t, $y^*_{it} = y_{it} - \frac{1}{T}\sum_{t=1}^{T} y_{it}$, and $u^*_{it} = u_{it} - \frac{1}{T}\sum_{t=1}^{T} u_{it}$.

For each i, the right-hand side of Equations 16.76 to 16.78 contains linear combination of $x'_{it} m(z_{it})$ for different t. If there is an intercept term in x_{it} and $m_1(z_{it})$ is the first element of $m(z_{it})$, then the difference of the first element of $x'_{it} m(z_{it}) = \sum_{d=1}^{p} x_{it,d} m_d(z_{it})$ gives $m_1(z_{it}) - m_1(z_{i,t-1})$. This is an additive function with the same functional form (strong assumption) at different times. The kernel estimation requires some backfitting algorithms or marginal integration method to recover the unknown function, which causes computation burden as well as complications in asymptotic analyses. Further, we have to be aware of the issues raised in Section 16.3.3, and use locally weighted averages in Equation 16.78.

For this reason, Sun, Carroll, and Li (2009) focus on the profile estimation of m. But we believe that the asymptotic analyses based on differencing methods in the conventional fixed effects panel data models can be extended to this model.

16.5.3 Nonparametric GMM Estimation

In the above model $E(u_{it}|z_{it}) = 0$ and $E(u_{it}|x_{it}) = 0$. However, in various economic models $E(u_{it}|x_{it}) \neq 0$, for example, when x_{it} is correlated with u_{it} (endogeneity), x_{it} has measurement errors, and x_{it} has lagged dependent variable. The result for the case of $E(u_{it}|z_{it}) \neq 0$ has not been developed yet.

The IV estimation of the general model $m(x_{it}, z_{it}) = x'_{it} m(z_{it})$ has been considered by Das (2005), Cai et al. (2006), and Cai and Xiong (2006) for discrete and continuous variables in the cross-sectional setup. In addition, with no endogeneity, this model is covered in González, Teräsvirta, and van

Dijk (2005) and the threshold nondynamic model in Hansen (1999). Here we present the nonparametric GMM estimation of Cai and Li (2008) for this model.

Cai and Li (2008) consider the model

$$y_{it} = x_{it}'m(z_{it}) + \varepsilon_{it} \tag{16.80}$$

where ε_{it} plays the role of $\alpha_i + u_{it}$ in Equation 16.75, $E(\varepsilon_{it}|z_{it}) = 0$, and $E(\varepsilon_{it}|x_{it}) \neq 0$. Note that $E(y_{it}|x_{it}, z_{it}) \neq x_{it}'m(z_{it})$ because $E(\varepsilon_{it}|x_{it}) \neq 0$. Let w_{it} be the $l \times 1$ instrument variables such that

$$E(\varepsilon_{it}|v_{it}) = 0, \tag{16.81}$$

where $v_{it} = (w_{it}', z_{it}')'$. Multiplying both sides of Equation 16.81 by $\pi(v_{it}) = E(x_{it}|v_{it})$ and taking expectations, conditional on $z_{it} = z$, we have

$$E[\pi(v_{it})y_{it}|z_{it}] = E[\pi(v_{it})x_{it}'|z_{it} = z]m(z) = E[\pi(v_{it})\pi(v_{it})'|z_{it} = z]m(z).$$

This gives $m(z) = \{E[\pi(v_{it})\pi(v_{it})'|z_{it} = z]\}^{-1}E[\pi(v_{it})y_{it}|z_{it} = z]$ under the assumption of positive definiteness of $E[\pi(v_{it})\pi(v_{it})'|z_{it} = z]$. This assumption guarantees that $m(\cdot)$ is identified locally. To obtain the estimator of $m(\cdot)$, one can consider a two stage nonparametric procedure. At the first stage $\hat{\pi}(v_{it})$ is obtained by a nonparametric estimation of x_{it} on v_{it}. Then at the second stage, one estimate $m(\cdot)$ based on the varying coefficient model: $y_{it} \approx \hat{\pi}(v_{it})'m(z_{it}) + \varepsilon_{it}$. The asymptotic property of such a two-stage nonparametric estimator is, however, quite complicated.

In viewing this, Cai and Li (2008) propose a one-step nonparametric GMM (NPGMM) estimation of $m(z)$. According to this, an $m_1 \times 1$ vector function $g(v_{it})$ is chosen such that

$$E[g(v_{it})\varepsilon_{it}|v_{it}] = E[g(v_{it})\{y_{it} - x_{it}'m(z_{it})\}|v_{it}] = 0. \tag{16.82}$$

Let us write the sample GMM orthogonality conditions based on the local linear approximation of $m(z_{it})$ in a neighborhood of z as

$$\sum_{i=1}^{n}\sum_{t=1}^{T} g(v_{it})(y_{it} - V_{it}'\delta)K_h(z_{it} - z) = 0, \tag{16.83}$$

where

$$V_{it} = \begin{pmatrix} x_{it} \\ x_{it} \otimes (z_{it} - z) \end{pmatrix}$$

is an $m_2 \times 1$ vector with $m_2 = p(q+1)$, $\delta = \delta(z)$ is an $m_2 \times 1$ vector of parameters whose true value corresponds to $(m(z)', \partial m_1(z)/\partial z', \ldots, \partial m_p(z)/\partial z')'$. When $m_1 \geq m_2$, the solution to δ is given by

$$\hat{\delta}(z) = (P'P)^{-1}P'Q$$

where

$$P = \frac{1}{nT}\sum_{i=1}^{n}\sum_{t=1}^{T} g(v_{it})V_{it}'K_h(z_{it}-z), \text{ and } Q = \frac{1}{nT}\sum_{i=1}^{n}\sum_{t=1}^{T} g(v_{it})K_h(z_{it}-z)y_{it}.$$

This gives the NPGMM estimate of $m(z)$ and its first-order derivatives $\partial m_j(z)/\partial z$ for $j = 1, \ldots, p$. This is one-stage estimator which is simpler compared to the two-stage NP estimator described above and studied in Cai et al. (2006) in the cross-sectional setup. Note that the two-stage estimation involves an NP regression of higher dimensions, and requires two smoothing parameters compared to one-step NP estimation which only needs one smoothing component. If the dimension of w_{it} is higher than that of z_{it}, one expects that the one-step estimator has much better finite sample performance than that for the two-stage estimator. When there is no endogenous variables ($w_{it} = x_{it}$) then one can choose $g(v_{it}) = V_{it}$. In this case the GMM conditions become

$$\sum_{i=1}^{n}\sum_{t=1}^{T} V_{it}K_h(z_{it}-z)(y_{it}-V_{it}'\delta) = 0,$$

which is the normal equation of the following LLLS problem of the varying coefficient model:

$$\min_{\delta}\sum_{i=1}^{n}\sum_{t=1}^{T}(y_{it}-V_{it}'\delta)^2K_h(z_{it}-z)$$

and it gives the ordinary LLLS estimator studied above.

For the choice of $g(v_{it})$ one solution is to consider the $p(q+1)\times 1$ vector

$$g(v_{it}) = \begin{pmatrix} w_{it} \\ w_{it}\otimes(z_{it}-z)/h \end{pmatrix}$$

In this case $m_1 = l(q+1) \geq m_2$ implies $l \geq p$. Although it is simple, it may not be optimal. The optimality could be developed by using results analogous to those in Newey (1990) and Ai and Chen (2003).

Under the usual assumptions such as that $h \longrightarrow 0$ and $nh^q \longrightarrow \infty$ as $n \longrightarrow \infty$, K is a symmetric, nonnegative, and bounded second-order kernel, and that $E[g(v_{it})g(v_{it})'|z_{it}=z]$ is positive definite, Cai and Li (2008) showed that, for fixed T,

$$\sqrt{nTh^q}\left[H(\hat{\delta}-\delta) - \frac{h^2}{2}\begin{pmatrix} B_m(z) \\ \mathbf{0}_{pq\times 1}\end{pmatrix} + o_p(h^2)\right] \xrightarrow{d} N\left(\mathbf{0}_{p(q+1)\times 1}, \frac{\Delta}{f(z)}\right),$$

where $H = \text{diag}(I_p, hI_{qp})$ is an $m_2 \times m_2$ matrix, $B_m(z) = \int A(u,z)K(u)du$ is a $p\times 1$ vector, $A(u,z) = (u'm_1^{(2)}(z)u, \ldots, u'm_p^{(2)}(z)u)'$, $m_j^{(2)}(z) = d^2m_j(z)/dzdz'$, $\Omega = \Omega(z) = E(w_{it}x_{it}'|z_{it}=z)$, and $\Delta = \text{diag}\{v_0\Omega_m, \Omega_m \otimes [\mu_2^{-1}(K)\ \mu_2(K^2)\ \mu_2^{-1}(K)]\}$ with $\Omega_m = (\Omega'\Omega)^{-1}\Omega'\Omega_1\Omega(\Omega'\Omega)^{-1}$, $\Omega_1 = \Omega_1(z) = \text{Var}(w_{it}\varepsilon_{it}|z_{it}=z)$,

$\mu_2(K) = \int uu'K(u)du$, and $v_0 = \int K^2(u)du$. If $T \longrightarrow \infty$, then

$$\sqrt{nTh^q}\left[\hat{m}(z) - m(z) - \frac{h^2}{2}B_m(z) + o_p(h^2)\right] \xrightarrow{d} N\left(\mathbf{0}_{p\times 1}, \frac{v_0\Omega_m}{f(z)}\right),$$

where $\hat{m}(z)$ as the first p elements $\hat{\delta}(z)$ is the estimator of $m(z)$. For details in proofs and the assumptions, see Cai and Li (2008). It is clear from the above results that $\hat{m}(z)$ has the same leading bias and variance for both finite and large T cases. Therefore the asymptotic MSE (T is fixed or large) is the same and the optimal h is proportional to $(nT)^{-1/(p+4)}$. However, when T is large and n is small, some modification in the results may be needed.

Finally, when $w_{it} = x_{it}$, we have

$$\sqrt{nTh^q}\left[\hat{m}(z) - m(z) - \frac{h^2}{2}B_m(z) + O_p(h^2)\right] \xrightarrow{d} N(\mathbf{0}_{p\times 1}, v_0 f^{-1}(z)\Omega_m^*(z))$$

where $\Omega_m^*(z) = [E(x_{it}x_{it}'|z_{it} = z)]^{-1}E[\sigma^2(v_{it})x_{it}x_{it}'|z_{it} = z][E(x_{it}x_{it}'|z_{it} = z)]^{-1}$ and $\sigma^2(v_{it}) = \mathrm{Var}(\varepsilon_{it}|v_{it})$.

The efficiency property of Cai et al.'s (2006) two-stage estimator compared to the single-stage estimator is not fully known. For special cases of asymptotic efficiency, see Cai and Li (2008, p. 1333).

The above estimation procedure is valid when ε_{it} is serially correlated and/or x_{it} contains lagged dependent variables. But as remarked earlier, it is unclear how to estimate the model if z_{it} contains the lagged dependent variable. We conjecture that it may be easier to establish the asymptotic theory for estimators based on series method rather than the kernel method.

In a recent paper Tran and Tsionas (2010) considered a two-step NP GMM estimation with a general wighting matrix , and where n is large but T is fixed. They claim that their two-step estimation may lead to potential gain in asymptotic efficiency. They also analyze the finite sample efficiency of their estimator and provide an empirical application.

In addition, Cai and Xiong (2006) have considered the following varying coefficient IV model

$$\begin{aligned} Y &= m(x, z_1) + u \\ &= m_1(z_{11})'z_{12} + m_2(z_{11})'x_1 + \beta_1'z_{13} + \beta_2'x_2 + u \end{aligned}$$

where $x = (x_1', x_2')'$ is a vector of endogenous variables, $z_1 = (z_{11}', z_{12}', z_{13}')'$ is a vector of exogenous variables, $z = (z_1', z_2')$ with z_2 being a vector of IVs, and $E(u|z) = 0$. If there is no endogenous variable, this model becomes the partially varying coefficient model studied by Ahmad, Leelahanon, and Li (2005) and that of the model in Cai et al. (2006) if the parametric part is absent. And if x is a discrete endogenous variable, then the model is as studied by Das (2005), as a special case. The estimation of the above model and its asymptotic properties are developed in Cai and Xiong (2006) in the cross-sectional setup, which can be potentially extended to the panel data framework.

16.5.4 Testing Random Effects versus Fixed Effects

Based on their profile least squares estimates, Sun, Carroll, and Li (2009) propose a test of random effects against fixed effects in model Equation 16.75. The null hypothesis is

$$H_0 : E(\alpha_i | x_{i1}, \ldots, x_{iT}, z_{i1}, \ldots, z_{iT}) = 0 \text{ a.s.}$$

and the alternative hypothesis H_1 is the negation of H_0. Their test statistic is based on the weighted squared difference between the random effects and fixed effects estimators, where the weights are used to get around the random denominator issue in the kernel literature. They show that their test statistic is asymptotically normally distributed under the null and diverges to infinity under the fixed alternative.

16.6 Nonparametric Panel Data Models with Cross-Section Dependence

In this section, we consider a semiparametric panel data model with cross-section dependence:

$$y_{it} = m_i(x_{it}) + \gamma_{1i}' f_{1t} + e_{it}, i = 1, \ldots, n, t = 1, \ldots, T \tag{16.84}$$

where x_{it} is a $p \times 1$ vector of observed individual-specific regressors, $m_i(\cdot)$ is an unknown smooth function form, f_{1t} is a $q_1 \times 1$ vector of observed common factors, and γ_{1i}, $i = 1, \ldots, n$, are factor loadings. Here we assume that f_{1t} includes the intercept term and impose the condition $E[m_i(x_{it})] = 0$ in order to identify $m_i(\cdot)$. The error term e_{it} in Equation 16.84 follows the multi-factor structure

$$e_{it} = \gamma_{2i}' f_{2t} + \varepsilon_{it}, \tag{16.85}$$

where f_{2t} is a $q_2 \times 1$ vector of unobserved common factors, ε_{it} is the idiosyncratic error assumed to be independently distributed of (x_{it}, f_{1t}, f_{2t}), and γ_{2i}, $i = 1, \ldots, n$, are factor loadings. We are interested in the estimation of $m_i(\cdot)$ in the presence of multifactor error structure.

Like Pesaran (2006), the unobserved factor f_{2t} could be correlated with (x_{it}, f_{1t}). To allow for such a possibility, we adopt the following fairly general model for the individual-specific regressors,

$$x_{it} = \Gamma_{1i}' f_{1t} + \Gamma_{2i}' f_{2t} + v_{it}, \tag{16.86}$$

where Γ_{1i} and Γ_{2i} are $q_1 \times p$ and $q_2 \times p$ factor loading matrices, and v_{it} is a $p \times 1$ vector of individual-specific components of x_{it}.

The model specified in Equations 16.84 to 16.86 is fairly general and includes a variety of panel data models as special cases. First, Pesaran's (2006) model corresponds to the case where $m_i(x) = \beta_i' x$ for some $d \times 1$ vector β_i so that model (Equation 16.84) becomes $y_{it} = \beta_i' x_{it} + \gamma_{1i}' f_{1t} + e_{it}$. Second, it includes the conventional fixed or random effects models and the models of Bai (2009) in particular. Third, it includes the usual nonparametric panel data model $y_{it} = m(x_{it}) + \alpha_i + \upsilon_t + \varepsilon_{it}$, where the individual effects α_i and the time effects υ_t enter the model additively. Huang (2006) studies the kernel estimation of Equation 16.84 when the unobserved factor f_{2t} in Equation 16.85 is a scalar random variable.

16.6.1 Common Correlated Effect (CCE) Estimator

Let $\overline{x}_t \equiv n^{-1} \sum_{i=1}^n x_{it}$ and $\overline{y}_t \equiv n^{-1} \sum_{i=1}^n y_{it}$. Then Equations 16.84 to 16.86 imply that

$$\begin{pmatrix} \overline{x}_t \\ \overline{y}_t \end{pmatrix} = \begin{pmatrix} \overline{\Gamma}_1' \\ \overline{\gamma}_1' \end{pmatrix} f_{1t} + \begin{pmatrix} \overline{\Gamma}_2' \\ \overline{\gamma}_2' \end{pmatrix} f_{2t} + \begin{pmatrix} \overline{\upsilon}_t \\ \overline{m}_t + \overline{\varepsilon}_t \end{pmatrix}, \tag{16.87}$$

where $\overline{\Gamma}_1$, $\overline{\Gamma}_2$, $\overline{\gamma}_1$, $\overline{\gamma}_2$, $\overline{\upsilon}_t$, and $\overline{\varepsilon}_t$ are sample averages of Γ_{1i}, Γ_{1i}, γ_{1i}, γ_{2i}, υ_{it}, and ε_{it} over i, respectively, and $\overline{m}_t = n^{-1} \sum_{i=1}^n m_i(x_{it})$. Let $\overline{\Gamma}_2^* \equiv (\overline{\Gamma}_2, \overline{\gamma}_2)$. Premultiplying both sides of Equation 16.87 by $\overline{\Gamma}_2^*$ and solving for f_{2t} yields

$$f_{2t} = (\overline{\Gamma}_2^* \overline{\Gamma}_2^{*\prime})^{-1} \overline{\Gamma}_2 \left(\begin{pmatrix} \overline{x}_t \\ \overline{y}_t \end{pmatrix} - \begin{pmatrix} \overline{\Gamma}_1' \\ \overline{\gamma}_1' \end{pmatrix} f_{1t} - \begin{pmatrix} \overline{\upsilon}_t \\ \overline{g}_t + \overline{\varepsilon}_t \end{pmatrix} \right) \tag{16.88}$$

provided that

$$\text{rank}(\overline{\Gamma}_2^*) = q_2 \leq p + 1 \text{ for sufficiently large } n. \tag{16.89}$$

As $n \to \infty$, $\overline{\upsilon}_t \xrightarrow{p} 0$, $\overline{\varepsilon}_t \xrightarrow{p} 0$ and $\overline{m}_t \xrightarrow{p} 0$ for each t under weak conditions. It follows

$$f_{2t} - (\overline{\Gamma}_2^* \overline{\Gamma}_2^{*\prime})^{-1} \overline{\Gamma}_2^* \left(\begin{pmatrix} \overline{x}_t \\ \overline{y}_t \end{pmatrix} - \begin{pmatrix} \overline{\Gamma}_1' \\ \overline{\gamma}_1' \end{pmatrix} f_{1t} \right) \xrightarrow{p} 0 \text{ as } n \to \infty. \tag{16.90}$$

The last line suggests that we can use $h_t \equiv (f_{1t}', \overline{x}_t', \overline{y}_t)'$ as observable proxies for f_{2t} and consider the following semiparametric regression:

$$y_{it} \approx m_i(x_{it}) + \vartheta_i' h_t + e_{it}. \tag{16.91}$$

Clearly, Equation 16.91 is an additive semiparametric model and series method has its advantage over the kernel method. For this reason, Su and Jin (2010) propose to estimate $m_i(\cdot)$ by sieve method.

To proceed, let $\{p_l(x),\ l = 1, 2, \ldots\}$ denote a sequence of known basis functions that can approximate any square-integrable function of x very well. Let $L \equiv L(T)$ be some integer such that $L \to \infty$ as $T \to \infty$. Let $p^L(x) = (p_1(x), p_2(x), \ldots, p_L(x))'$, $p_{it} = p^L(x_{it})$, and $p_i = (p_{i1}, p_{i2}, \ldots, p_{iT})'$. Under fairly weak conditions, we can approximate $m_i(x)$ in Equation 16.91 very well by $\alpha_{m_i}' p^L(x)$ for some $L \times 1$ vector α_{m_i}.

To estimate α_{m_i}, we run the regression of y_{it} on $p^L(x_{it})$ and $h_t \equiv (f'_{1t}, \overline{x}'_t, \overline{y}_t)'$

$$y_{it} = \alpha'_{m_i} p^L(x_{it}) + \vartheta'_i h_t + u_{it} \tag{16.92}$$

where u_{it} is the new error term. Let $y_i = (y_{i1}, y_{i2}, \ldots, y_{iT})'$, $h = (h_1, h_2, \ldots, h_T)'$, and $u_i = (u_{i1}, u_{i2}, \ldots, u_{iT})'$. We can rewrite Equation 16.92 in vector form

$$y_i = p_i \alpha_{m_i} + h\vartheta_i + u_i. \tag{16.93}$$

By the formula for partitioned regression, the estimator of α_{m_i} in Equation 16.92 or 16.93 is given by

$$\hat{\alpha}_{m_i} = (p'_i b_h p_i)^- p'_i b_h y_i, \tag{16.94}$$

where $b_h \equiv I_T - h(h'h)^- h$, and $(\cdot)^-$ denotes any symmetric generalized inverse. The estimator of $m_i(x)$ is then given by

$$\hat{m}_i(x) = p^L(x)'\widehat{\alpha}_{m_i}. \tag{16.95}$$

Su and Jin (2010) establish the consistency and asymptotic normality of $\hat{m}_i(x)$ by passing $T \to \infty$.

16.6.2 Estimating the Homogenous Relationship

In practice, one may also be interested in estimating a restricted submodel of Equation 16.84:

$$y_{it} = m(x_{it}) + \gamma'_{1i} f_{1t} + e_{it}. \tag{16.96}$$

That is, $m_i(x) = m(x)$ for all i in model Equation 16.84. In the case where $\gamma_{1i} = 0$, Equation 16.96 can be regarded as a nonparametric extension of Bai's (2009) linear panel data model with multi-factor error structure or a simple extension of Huang's (2006) nonparametric panel data from his single-factor error structure to multiple-factor error structure.

If model Equation 16.96 is assumed to be correctly specified in conjunction with Equations 16.85 and 16.86, we can estimate $m(\cdot)$ by

$$\hat{m}(x) = p^L(x)'\tilde{\alpha}_m. \tag{16.97}$$

where $\tilde{\alpha}_m = (\sum_{i=1}^n p'_i b_h p_i)^- \sum_{i=1}^n p'_i b_h y_i$ and L is now allowed to depend on both n and T. The asymptotic normality of $\hat{m}(x)$ is also studied in Su and Jin (2010) by passing both n and T to infinity.

Clearly, besides the multifactor error structure, the key assumption that underlines the asymptotic analysis of Su and Jin (2010) is Equation 16.86 that specifies the relationship between the individual-specific regressor and the factors. The violation of such an assumption may invalidate their analysis. Therefore it is desirable to propose an alternative estimator without imposing such an assumption. By combining the series method with the principal component analysis, Su and Zhang (2010a) consider the estimation of homogenous relationship (m) in a simpler model

$$y_{it} = m(x_{it}) + \gamma_i' f_t + \varepsilon_{it}$$

where f_t is a $q \times 1$ vector of unobservable factors and γ_i's are factor loadings, m, ε_{it}, and x_{it} are as defined above. If $m(x_{it}) = x_{it}'\beta_0$ for a $p \times 1$ vector β_0, the model reduces to that of Bai (2009).

16.6.3 Specification Tests

Various specification tests can be conducted for the model in Equation 16.84. This includes tests for homogenous relationship ($m_i = m$ for all i) and tests for cross section independence or uncorrelatedness.

Jin and Su (2010) propose a nonparametric test for poolability in Equation 16.84. The null hypothesis is

$$H_0 : m_i(x) = m_j(x) \text{ a.e. on the joint support of } m_i \text{ and } m_j \text{ and for all } i, \\ j = 1, \ldots, n, \quad (16.98)$$

where a.e. is the abbreviation for almost everywhere. They propose a test statistic based on series estimation and the measure

$$\Gamma = \sum_{i=1}^{n-1} \sum_{j=i+1}^{n} \int (m_i(x) - m_j(x))^2 w(x) dx, \quad (16.99)$$

where $w(x)$ is a nonnegative weight function, and establish the asymptotic normality of their test under the null and a sequence of local alternatives. This extends and complements the work of Baltagi, Hidalgo, and Li (1996) who propose a kernel-based test for poolability in conventional panel data models.

Chen, Gao, and Li (2009) propose a kernel-based test for cross section *uncorrelatedness* in

$$y_{it} = m_i(x_{it}) + u_{it}, i = 1, \ldots, n, t = 1, \ldots, T, \quad (16.100)$$

where the error term satisfies $E(u_{it}) = 0$. They test whether $E(u_{it}u_{jt}) = 0$ for all $t \geq 1$ and $i \neq j$ by allowing both n and T to pass to the infinity. If n is fixed, then their test complements the test of conditional uncorrelatedness in Su and Ullah (2009). Su and Zhang (2010b) propose a test of cross section *independence* for the model in Equation 16.100. It is based on the comparison of the joint densities and the product of marginal densities and thus has extra power in detecting deviations from cross-section independence when compared with the test of Chen, Gao, and Li (2009).

16.7 Nonseparable Nonparametric Panel Data Models

In this section, we review papers on nonseparable nonparametric panel data models. We focus on two types of models. The first type is the partially separable nonparametric panel data model

$$y_{it} = m(x_{it}, \alpha_i) + u_{it}, i = 1, \ldots, n, t = 1, \ldots, T, \tag{16.101}$$

where x_{it} is a $p \times 1$ vector of explanatory variables, the scalar α_i is a parameter that represents unobserved individual heterogeneity, u_{it} is a scalar idiosyncratic error term, and m is an unknown smooth function. The second type is the fully nonseparable nonparametric panel data models

$$y_{it} = m(x_{it}, \alpha_i, u_{it}), i = 1, \ldots, n, t = 1, \ldots, T, \tag{16.102}$$

where x_{it}, α_i, and u_{it} are defined as above, and the structural function $m(x, \alpha, u)$ is unknown.

16.7.1 Partially Separable Nonparametric Panel Data Models

Evdokimov (2009) studies the identification and estimation of the structural function $m(x, \alpha)$ in Equation 16.101. For simplicity, we focus on the case where $T = 2$. Let $f_{A|B}(\cdot|b)$ and $\phi_{A|B}(\cdot|b)$ denote the conditional p.d.f. and conditional characteristic function (c.h.f.) of A given $B = b$, respectively. Let $x_{i,(-t)} = x_i \backslash x_{i,t}$ and $u_{i,(-t)} = u_i \backslash u_{i,t}$, where $x_i = (x_{i1}, x_{i2})$ and $u_i = (u_{i1}, u_{i2})$.

We first assume that (i) $\{x_i, \alpha_i, u_i\}$ is an i.i.d. random sample; (ii) $f_{u_{it}|x_{it},\alpha_i,x_{i,(-t)},u_{i,(-t)}}(u_t|x, \alpha, x_{(-t)}, u_{(-t)}) = f_{u_{it}|x_{it}}(u_t|x)$; (iii) $E(u_{it}|x_{it}) = 0$ a.s.; (iv) $\phi_{u_{it}|x_{it}}(u|x)$ does not vanish for all $u \in \mathbb{R}$, x on the support $\mathcal{X}$ of x_{it}, and $t \in \{1, 2\}$, (v) $E[|m(x_{it}, \alpha_i)||x_i]$ and $E[|u_{it}||x_{it}]$ are bounded a.s. for each t; (vi) the joint p.d.f. $f_{x_{i1},x_{i2}}(\cdot, \cdot)$ of x_{i1} and x_{i2} satisfies $f_{x_{i1},x_{i2}}(x, x) > 0$ for all $x \in \mathcal{X}$; (vii) $m(x, \alpha)$ is increasing in α for all x; $(viii)$ α_i and $x_i = (x_{i1}, x_{i2})$ are independent, and (ix) α_i has a uniform distribution on $[0, 1]$. Under these conditions, Evdokimov (2009) shows that

1. Under Assumptions (i)-(iv), the conditional distributions of $m(x, \alpha_i)$, u_{i1} and u_{i2} given $x_{i1} = x_{i2} = x$ is identified for all $x \in \mathcal{X}$.
2. Under (ii) and (iv), the c.h.f. of $m(x, \alpha_i)$ given $x_{it} = x$ is identified as $\phi_{m(x,\alpha_i)|x_{it}}(s|x) = \phi_{y_{it}|x_{it}}(s|x)/\phi_{u_{it}|x_{it}}(s|x)$.
3. By the equivalence of c.h.f. and conditional cumulative distribution function (c.d.f.), this implies that the conditional c.d.f. $F_{m(x,\alpha_i)|x_{it}}(\cdot|x)$ of $m(x, \alpha_i)$ given $x_{it} = x$ is identified by

$$F_{m(x,\alpha_i)|x_{it}}(w|x) = \frac{1}{2} - \lim_{A\to\infty} \int_{-A}^{A} \frac{e^{-\mathbf{i}sw}}{2\pi\mathbf{i}s} \phi_{m(x,\alpha_i)|x_{it}}(s|x)ds$$
$$\text{for all } (w, x) \in \mathbb{R} \times \mathcal{X}$$

at the continuity of the c.d.f. in w, where $\mathbf{i} = \sqrt{-1}$.

4. $m(x, \alpha)$ is then identified by $m(x, \alpha) = F^{-1}_{m(x,\alpha_i)|x_{it}}(\alpha|x)$ for all $x \in \mathcal{X}$ and $\alpha \in (0, 1)$, where $F^{-1}_{m(x,\alpha_i)|x_{it}}(\cdot|x)$ is the inverse function of $F_{m(x,\alpha_i)|x_{it}}(\cdot|x)$.

The key in the proof of the above identification results lie in the first step. Consider the special case when u_{i1} and u_{i2} are identically and symmetrically distributed, conditional on $x_{i1} = x_{i2} = x$. Then the c.h.f. of $y_{i2} - y_{i1}$ given $x_{i1} = x_{i2} = x$ equals

$$\begin{aligned}\phi_{y_{i2}-y_{i1}}(s|x_{i1} = x_{i2} = x) &= E[\exp(\mathbf{i}s(y_{i2} - y_{i1}))|x_{i1} = x_{i2} = x]\\ &= E[\exp(\mathbf{i}s(u_{i2} - u_{i1}))|x_{i1} = x_{i2} = x]\\ &= \phi_U(s|x)\phi_U(-s|x) = \phi_U(s|x)^2\end{aligned}$$

where $\phi_U(s|x)$ denotes the c.h.f. of U_{it} given $x_{it} = x$. Consequently, $\phi_U(s|x)$ is identified because $\phi_{y_{i2}-y_{i1}}(s|x_{i1} = x_{i2} = x)$ can be identified from the observed data.

When Assumption $(viii)$ is violated, Evdokimov (2009) shows that the structural equation in a correlated random effects model can also be identified. The key assumption in this case is the normalization condition: there exists $\bar{x} \in \mathcal{X}$ such that $m(\bar{x}, \alpha) = \alpha$ for all α. Similar conditions are imposed in early literature on nonseparable nonparametric models; see Matzkin (2003) and Altonji and Matzkin (2005).

Based on the identification results, Evdokimov (2009) considers consistent estimation of the structural function $m(x, \alpha)$ which boils down to the estimation of c.d.f. and conditional quantile functions. Nevertheless, he needs to estimate $\phi_{m(x,\alpha_i)|x_{it}}(s|x)$ by a conditional deconvolution approach which yields extremely slow convergence rates. In particular, if the idiosyncratic error term is normally distributed, the conditional deconvolution estimator converges to its truth at the logarithm rate. Besides, no distributional theory has yet been established so far for such an estimator, and no dynamic lagged dependent variable is allowed to be a regressor in the structural equation.

16.7.2 Fully Nonseparable Nonparametric Panel Data Models

For the fully nonseparable nonparametric panel data model in Equation 16.102, Altonji and Matzkin (2005), Bester and Hansen (2007), and Hoderlein and White (2009) study conditions for identification and estimation of the structural functional itself or the local average derivatives. Here we focus on the two estimators of Altonji and Matzkin (2005) and remark on other estimators.

Both estimators of Altonji and Matzkin (2005) involve nonseparable unobservable terms and endogenous regressors, and both are based on a conditional density restriction

$$f(\alpha, u|x', z') = f(\alpha, u|x'', z'') \tag{16.103}$$

for specific values (x', z') and (x'', z'') of the vector of conditioning variables (x_{it}, z_{it}). Here $f(\cdot, \cdot|x, z)$ denotes the conditional p.d.f. of (α_i, u_{it}) given

$(x_{it}, z_{it}) = (x, z)$. Similarly, $f(\cdot, \cdot|x)$ denotes the conditional p.d.f. of (α_i, u_{it}) given $x_{it} = x$.

16.7.2.1 Local Average Response (LAR) Estimator

The local average response (LAR) estimator is based on the identification of average marginal effects by assuming the existence of a control variable (CV) z_{it} that is sufficient for x_{it} in the distribution of unobservables.

Let $\varepsilon_{it} = (\alpha_i, u_{it})$. Then Equation 16.102 can be written as $y_{it} = m(x_{it}, \varepsilon_{it})$. When $m(x, \varepsilon)$ is differentiable in x, [2] we can define the local average response (LAR) $\beta(x)$ as

$$\beta(x) = \int m_x(x, \varepsilon) f(\varepsilon|x) d\varepsilon \tag{16.104}$$

where here and below the use of function arguments as subscripts to functions denotes partial derivatives. Under the conditional independence assumption that

$$f(\varepsilon|x, z) = f(\varepsilon|z), \tag{16.105}$$

$\beta(x)$ can be identified as follows

$$\begin{aligned}\beta(x) &= \int m_x(x, \varepsilon) f(\varepsilon|x, z) f(z|x) dz d\varepsilon \\ &= \int E[m_x(x, \varepsilon_{it})|x, z] f(z|x) dz \\ &= \int E_x[y_{it}|x, z] f(z|x) dz,\end{aligned} \tag{16.106}$$

where $f(z|x)$ denotes the conditional p.d.f. of z_{it} given x_{it}. Equation 16.106 forms the basis of Altonji and Matzkin's LAR estimator.

Let $\hat{E}_x[y_{it}|x, z]$ and $\hat{f}(z|x)$ denote kernel estimators of $E_x[y_{it}|x, z]$ and $f(z|x)$, respectively. In principle, one could estimate $\beta(x)$ by

$$\hat{\beta}(x) = \int \hat{E}_x[y_{it}|x, z] \hat{f}(z|x) dz.$$

But this estimator is not easy to analyze because it involves a random denominator problem. Noting that

$$\begin{aligned}\beta(x) &= \int E_x[y_{it}|x, z] f(z|x) dz \\ &= \frac{\int y f_x(y, x) dy}{f(x)} - \int \frac{f_x(x, z) \int y f(y, x, z) dy}{f(x, z) f(x)} dz,\end{aligned}$$

[2] The LAR $\beta(x)$ can also be defined if m is not differentiable as in the binary response case. See Altonji and Matzkin (2005).

where $f(x)$, $f(x, z)$, $f(y, x, z)$ denote the p.d.f.'s of x_{it}, (x_{it}, z_{it}), and (y_{it}, x_{it}, z_{it}), respectively, we can estimate $\beta(x)$ by

$$\hat{\beta}(x) = \frac{\int y\hat{f}_x(y, x)dy}{\hat{f}(x)} - \int \frac{\tau(\hat{f}(x, z), b)\hat{f}_x(x, z)\int y\hat{f}(y, x, z)dy}{\hat{f}(x)}dz$$

where $\hat{f}(x)$, $\hat{f}(x, z)$, $\hat{f}_x(x, z)$, $\hat{f}(y, x, z)$ are kernel estimates of $f(x)$, $f(x, z)$, $f_x(x, z)$, and $f(y, x, z)$, respectively, and τ is a trimming function defined by

$$\tau(s, b) = \begin{cases} 1/s & \text{if } s \geq 2b \\ \left[\frac{49(s-b)^3}{b^4} - \frac{76(s-b)^4}{b^5} + \frac{31(s-b)^5}{b^6}\right]/8 & \text{if } b \leq s < 2b \\ 0 & \text{if } s < b \end{cases}.$$

Altonji and Matzkin (2005) establish the asymptotic normality of $\hat{\beta}(x)$ first for the case of $T = 1$. If $T > 1$, then one can proceed by first observing an estimator of $\beta(x)$ for each $t = 1, \ldots, T$, and averaging the T estimators to obtain the final estimator of $\beta(x)$. It is well known from standard asymptotic analysis in the kernel literature, these T estimators are asymptotically independent because the covariance between each two of them is of smaller magnitude than the individual variances.

When the dimensional of x_{it} is high, the rate of convergence of $\beta(x)$ can be undesirably slow. As an alternative, one can consider some weighted average measure of the nonparametric LAR estimator to increase the precision of the estimator. For example, one can consider estimating

$$\overline{\beta} = \int \beta(x)w(x)dx \qquad (16.107)$$

for some prescribed weight function w. As usual, the estimator of $\overline{\beta}$ will have the regular $\sqrt{n}$-rate of convergence.

Clearly, the key assumption underlying the above analysis is the conditional independence assumption in Equation 16.105. This requires Equation 16.103 holds for all values of (α, u), (x, x'), and (z', z'') such that $z' = z''$. The LAR estimator is based upon the (derivative of) conditional expectation $E(y|x, z)$. Because of (16.105), holding z constant also holds the distribution of the unobservable term (ε) constant. Then one can undo the effect of conditioning on z by integrating $E_x(y|x, z)$ over an estimate of the distribution of z given x.

Bester and Hansen (2007) consider identification and estimation of average marginal effects in a correlated random coefficients models. Instead of assuming the existence of the known CV vector z_{it}, they assume the existence of a set of sufficient statistics for x_{it} in the distribution of individual heterogeneity, which is not known but takes on some index form. To be concrete, Bester and Hansen assume that

$$F(\alpha_i|\mathbf{x}_i) = F(\alpha_i|h_1(\mathbf{x}_{i,1}), \ldots, h_p(\mathbf{x}_{i,p})) \qquad (16.108)$$

for some unknown real-valued functions $h_s(\mathbf{x}_{i,s})$, $s = 1, \ldots, p$, where for example, $F(\alpha_i|\mathbf{x}_i)$ denotes the conditional c.d.f. of α_i given $\mathbf{x}_i = (x_{i1}, \ldots, x_{iT})'$, a $T \times p$ matrix, and $\mathbf{x}_{i,s}$ denotes the sth column in $\mathbf{x}_i$. This assumption is neither more or less general than the conditional independence assumption Equation 16.105 of Altonji and Matzkin (2005): the set of sufficient statistics is unknown but restricted so that there is one sufficient statistic for each covariate in x_{it} for the restriction in Equation 16.108; the CV z_{it} has to be unknown but may include interactions of covariates in Equation 16.105. In addition, neither Bester and Hansen's (2007) nor Altonji and Matzkin's (2005) LAR approach identifies the structural function itself.

16.7.2.2 Structural Function and Distribution (SFD) Estimator

To define the structural function and distribution (SFD) estimator, we impose the following assumptions: (i) There exists a real valued function $g(\varepsilon)$ such that $y_{it} = m(x_{it}, e_{it})$[3] for $e_{it} = g(\varepsilon_{it})$; (ii) $m(x, e)$ is strictly increasing in e for all x; (iii) there exists some value $\bar{x}$ of x, $m(\bar{x}, e) = e$ for all e; (iv) for any value $\tilde{x}$ of x there exist values $\tilde{z}$ and $\tilde{z}'$ of z such that $f(e|\tilde{x}, \tilde{z}) = f(e|\bar{x}, \tilde{z}')$, where $f(e|x, z)$ denotes the conditional p.d.f. of e_{it} given (x_{it}, z_{it}); (v) for all (x, z), $f(e|x, z)$ is strictly positive everywhere.

Clearly, (i) indicates that the effect of the vector ε_{it} can be aggregated by a scalar-valued unobservable random term $e_{it} = g(\varepsilon_{it})$. (ii) assumes monotonicity in the unobservable and (iii) is a normalization restriction. (iv) can be satisfied under some exchangeability conditions. (v) and (ii) guarantee that the conditional c.d.f. $F(\cdot|x, z)$ of y_{it} given $(x_{it}, z_{it}) = (x, z)$ is strictly increasing so that $m(x, e)$ can be identified via

$$m(x, e) = F^{-1}(F(e|\bar{x}, z')|x, z). \tag{16.109}$$

To see this, noticing that for any value x there exist values z and z' such that for any value e, we have

$$\begin{aligned}
P(e_{it} \le e|x, z) &= P(e_{it} \le e|\bar{x}, z') \text{ [by } (iv)] \\
&\Longrightarrow \\
P(m(x, e_{it}) \le m(x, e)|x, z) &= P(m(\bar{x}, e_{it}) \le m(\bar{x}, e)|\bar{x}, z') \text{ [by } (ii)] \text{ or} \\
P(y_{it} \le m(x, e)|x, z) &= P(y_{it} \le m(\bar{x}, e)|\bar{x}, z') \text{ or} \\
F(m(x, e)|x, z) &= F(m(\bar{x}, e)|\bar{x}, z').
\end{aligned}$$

The last line implies Equation 16.109 by (ii) and (v). Let $F_{e_{it}|x_{it}}(\cdot|x)$ and $F_{y_{it}|x_{it}}(\cdot|x)$ denote the conditional c.d.f. of e_{it} and y_{it} given $x_{it} = x$, respectively. Then under (ii), $F_{e_{it}|x_{it}}$ is identified via

$$F_{e_{it}|x_{it}}(e|x) = F_{y_{it}|x_{it}}(m(x, e)|x). \tag{16.110}$$

---

[3] We keep using the notation m in $m(x, e)$ but keep in mind that this is different from the original structrual function $m(x, \alpha, u)$.

Given the above identification results, we can obtain estimators of $m(x, e)$ and $F_{e_{it}|x_{it}}(e|x)$ straightforwardly via the kernel method. Both estimators involve the kernel estimates of $F(\cdot|x, z)$ and its inverse function (conditional quantile function) $F^{-1}(\cdot|x, z)$. The latter also involves the estimation of $F_{y_{it}|x_{it}}(\cdot|x)$. Altonji and Matzkin (2005) formally establish the asymptotic normality of either estimator.

It is worth mentioning that neither the LAR nor the SFD estimator deals with dynamics in the model. The LAR estimator can be used to estimate the marginal effects of x_{it} on y_{it} in a censored regression model but neither can be used to study the effects on a latent dependent variable. The SFD estimator estimates some structural function but it is different from the original one.

16.7.2.3 Nonparametric Identification and Estimation without Monotonicity

Hoderlein and White (2009) consider the general class of nonseparable panel models of the form

$$y_{it} = m(x_{it}, z_{it}, \alpha_i, u_{it}), i = 1, \ldots, n, t = 1, \ldots, T, \tag{16.111}$$

where z_{it} is a $q \times 1$ vector of observed variables, x_{it}, α_i, and u_{it} are defined as before. Their interest centers on the effect of x_{it} on y_{it} by controlling the influence of all other variables, whether observed like z_{it} or unobserved like α_i and u_{it}.

Without assuming that $m(x, z, \alpha, u)$ is monotonic in α or u, the structural function m itself and its derivatives are not identified, but certain of its conditional expectations and their derivatives are. Like early estimators in the nonseparable panel literature, Hoderlein and White's estimator does not allow for lagged dependent variables either. In addition, they can only identify effects for the subpopulation for which $x_{i1} - x_{i2} = 0$ and $z_{i1} - z_{i2} = 0$ in the case of $T = 2$.

16.7.3 Testing of Monotonicity in Nonseparable Nonparametric Panel Data Models

Despite the wide use of monotonicity of the structural function in individual heterogeneity (e.g., Matzkin 2003; Altonji and Matzkin 2005; Imbens and Newey 2009; Evdokimov 2009; among others), Hoderlein and Mammen (2007, 2009) argue that such an assumption may not be fully justified in economics, say, when the individual effects represent the unobserved heterogeneity in preferences or technologies. Moreover, as Hoderlein, Su, and White (2010) demonstrate, some key identification results fail when monotonicity is violated. This motivates them to consider tests of monotonicity in nonseparable nonparametric panel data models. Under some strict exogeneity conditions, they propose two tests for monotonicity of unobservables in panel nonseparable nonparametric panel data models. The first works under some ideal situation where the unobservables vary across i but not t dimension (t may not be time index). The second works in large dimensional panel where both n and T approach ∞ and both time-invariant and time-varying unobservables are present.

Consider first the case where the unobservables vary across individuals but not "time":

$$y_{it} = m(x_{it}, \alpha_i), i = 1, \ldots, n, t = 1, \ldots, T,$$

where α_i is i.i.d. uniformly distributed on $[0, 1]$, and (x_{it}, α_i) is identically distributed across i. Under the null hypothesis that $m(x, \cdot)$ is strictly increasing for any x, we have

$$\alpha_i = F_t(y_{it}|x_{it}) \text{ a.s. for all } (i, t)$$

where $F_t(\cdot|x)$ is the conditional c.d.f. of y_{it} given $x_{it} = x$. If we further assume that α_i is independent of x_{it} (x_{it} is exogenous), then we can show that this conditional c.d.f. is time-invariant, that is, F_t should not depend on t and can be abbreviated as F. Thus, we can write the null hypothesis as

$$H_0 : F(y_{it} \mid x_{it}) = F(y_{is} \mid x_{is}) \text{ a.s. for all } (t, s). \tag{16.112}$$

Significantly, exogeneity and the time invariance of α_i jointly ensure that F_t is time invariant. When exogeneity or monotonicity fails, we generally have the alternative

$$H_1 : P[F_t(y_{it} \mid x_{it}) = F_s(y_{is} \mid x_{is})] < 1 \text{ for some } t \neq s.$$

Let $\hat{F}_t$ be suitable estimator of F_t. We can consider the following test statistic

$$D_n \equiv \sum_{t=1}^{T-1} \sum_{s=t+1}^{T} \sum_{i=1}^{n} (\hat{F}_t(y_{it} \mid x_{it}) - \hat{F}_s(y_{is} \mid x_{is}))^2.$$

Hoderlein, Su, and White (2010) obtain the estimate $\hat{F}_t(y|x)$ by the local polynomial method and demonstrate after correct centering, $h^{p/2} D_n$ is asymptotically normality distributed under the null and diverges to infinity under the alternative, where h is the bandwidth parameter used in the local polynomial estimation.

Now consider the nonseparable structure of the form

$$y_{it} = m(x_{it}, \alpha_i, u_{it}), i = 1, \ldots, n, t = 1, \ldots, T,$$

α_i is i.i.d. uniformly distributed on $[0, 1]$, and $(x_{it}, \alpha_i, u_{it})$ is i.i.d. across i, and identically distributed across t. Define some nonnegative weight functions $w_\tau(x)$ on the support of x_{it}, $\tau = 1, \ldots, T$. Assuming that $(x_{it}, u_{it}) \perp \alpha_i$, we have

$$\tilde{Y}_{\tau,i} = E[y_{it} w_\tau(x_{it})|\alpha_i] = \int m(x, \alpha_i, u) w_\tau(x) dF(x, u) \equiv \bar{m}_\tau(\alpha_i),$$

where $F(x, u)$ denotes the c.d.f. of (x_{it}, u_{it}). Clearly, $\bar{m}_\tau(\cdot)$ is also monotonic under the null hypothesis that $m(x, \cdot, u)$ is monotone for all (x, u). Furthermore, α_i can be identified as

$$\alpha_i = \bar{m}_\tau^{-1}(\bar{Y}_{\tau,i}) = \bar{F}_\tau(\bar{Y}_{\tau,i})$$

where $\bar{F}_\tau$ is the c.d.f. of $\bar{Y}_{\tau,i}$. As a result, we can test the monotonicity by testing the following null hypothesis

$$\bar{H}_0 : \bar{F}_\tau(\bar{Y}_{\tau,i}) = \bar{F}_\varsigma(\bar{Y}_{\varsigma,i}) \text{ a.s. for all } (\tau, \varsigma). \tag{16.113}$$

The test statistic is

$$\hat{D}_{nT} \equiv \sum_{\tau=1}^{\mathcal{T}-1} \sum_{\varsigma=\tau+1}^{\mathcal{T}} \sum_{i=1}^{n} (\hat{F}_{n,T,\tau}(\bar{Y}_{T,\tau,i}) - \hat{F}_{n,T,\varsigma}(\bar{Y}_{T,\varsigma,i}))^2.$$

where for $\tau = 1, \ldots, \mathcal{T}$, $\hat{F}_{n,T,\tau}(y) = n^{-1} \sum_{i=1}^{n} \mathbf{1}\{\bar{Y}_{T,\tau,i} \leq y\}$, $\bar{Y}_{T,\tau,i} = T^{-1} \sum_{t=1}^{T} y_{it} w_\tau(x_{it})$ is a consistent estimate of $\bar{Y}_{\tau i}$ under weak conditions, and $\mathbf{1}\{\cdot\}$ is the usual indicator function. Under some regularity conditions, Hoderlein, Su, and White (2010) show that limit distribution of $\hat{D}_{nT}$ is given by weighted chi-squares under the null.

16.8 Concluding Remarks

In this chapter, we survey some of the recent developments on NP and SP panel data models. Due to space limitation, we omit some of the important areas in this literature. This includes NP and SP limited dependent variable models (see Ai and Li 2008), and NP and SP panel models with spatial dependence. It is worth mentioning that the latter area is underdeveloped in econometrics. Other areas that seem promising to us include NP or SP panel data models that impose some curvature restrictions (e.g., monotonicity, concavity, homogeneity) or require less restrictions (e.g., exogeneity, separability, monotonicity). In the nonseparable nonparametric models, no estimator has been proposed to deal with dynamic panel data models. Obviously, this is an interesting yet challenging research topic.

16.9 Acknowledgment

The first author gratefully acknowledges the financial support from the NSFC under the grant numbers 70501001 and 70601001. The second author acknowledges the financial support from the academic senate, UCR.

References

Ahmad, I., S. Leelahanon, and Q. Li. 2005. Efficient estimation of semiparametric varying coefficient models. *Annals of Statistics* 33:258–283.

Ai, C., and X. Chen. 2003. Efficient estimation of models with conditional moment restrictions containing unknown functions. *Econometrica* 71:1795–1843.

Ai, C., and Q. Li. 2008. Semi-parametric and non-parametric methods in panel data models. In *The Econometrics of Panel Data: Fundamentals and Recent Developments in Theory and Practice*, 3rd ed. L. Mátyás and P. Sevestre (eds). Berlin: Springer. pp. 451–478.

Altonji, J. G., and R. L. Matzkin. 2005. Cross section and panel data estimators for nonseparable models with endogenous regressors. *Econometrica* 73: 1053-1102.

Arellano, M. 2003. *Panel Data Econometrics*. Oxford, U.K.: Oxford University Press.

Bai, J. 2009. Panel data models with interactive fixed effects. *Econometrica* 77:1229–1279.

Baltagi, B. H. 2008. *Econometric Analysis of Panel Data*, 4th ed. West Sussex, U.K.: John Wiley & Sons.

Baltagi, B. H., J. Hidalgo, and Q. Li. 1996. A nonparametric test for poolability using panel data. *Journal of Econometrics* 75:345–367.

Baltagi, B. H., and Q. Li. 2002. On instrumental variable estimation of semiparametric dynamic panel data models. *Economics Letters* 76:1–9.

Baltagi, B. H., and Q. Li. 2000. Efficient instrumental variable estimation of semiparametric dynamic panel data models. Working paper, Department of Economics, Texas A&M University, College Station.

Baltagi, B. H., and D. Li. 2002. Series estimation of partially linear panel data models with fixed effects. *Annals of Economic and Finance* 3:103–116.

Bester, A., and C. Hansen. 2007. Identification of marginal effects in a nonparametric correlated random effects model. Working paper, Department of Economics, University of Chicago, Chicago, IL.

Cai, Z., and H. Xiong. 2006. Efficient estimation of partially varying-coefficient instrumental variables models. Working paper, WISE, Xiamen University, Fujian, China.

Cai, Z., and Q. Li. 2008. Nonparametric estimation of varying coefficient dynamic panel data models. *Econometric Theory* 24:1321–1342.

Cai, Z., M. Das, H. Xiong, and Z. Wu. 2006. Functional coefficient instrumental variables models. *Journal of Econometrics* 133:207–241.

Chen, J., J. Gao, and D. Li. 2009. A new diagnostic test for cross-section independence in nonparametric panel data models. Working paper 2009–16, Department of Economics, University of Adelaide, Adelaide, SA.

Das, M. 2005. Instrumental variables estimation of nonparametric models with discrete endogenous variables. *Journal of Econometrics* 124:335-361.

Evdokimov, K. 2009. Identification and estimation of a nonparametric panel data model with unobserved heterogeneity. Working paper, Department of Economics, Yale University, New Haven, CT.

González, A., T. Teräsvirta, and D. van Dijk. 2005. Panel smooth transition regression models. Working paper, Stockholm School of Economics, Stockholm, Sweden.

Hansen, B. E. 1999. Threshold effects in non-dynamic panels: Estimation, testing and inference. *Journal of Econometrics* 93:345–368.

Henderson, D. J., and A. Ullah. 2005. A nonparametric random effects estimator. *Economics Letters* 88:403–407.

Henderson, D. J., and A. Ullah. 2008. Nonparametric estimation in a one-way error component model: A Monte Carlo analysis. Working paper, Department of Economics, University of California, Riverside.

Henderson, D. J., R. J. Carroll, and Q. Li. 2008. Nonparametric estimation and testing of fixed effects panel data models. *Journal of Econometrics* 144:257–275.

Hoderlein, S., and E. Mammen. 2007. Identification of marginal effects in nonseparable models without monotonicity. *Econometrica* 75:1513–1518.

Hoderlein, S., and E. Mammen. 2009. Identification and estimation of local average derivatives in non-separable models without monotonicity. *Econometrics Journal* 12:1–25.

Hoderlein, S., and H. White. 2009. Nonparametric identification in nonseparable panel data models with generalized fixed effects. *Working paper*, Department of Economics, Brown University, Providence, RI.

Hoderlein, S., L. Su, and H. White. 2010. Nonparametric identification in nonseparable panel data models with generalized fixed effects. Working paper, Department of Economics, Brown University, Providence, RI.

Hsiao, C. 2003. *Analysis of Panel Data*, 2nd ed. Cambridge, U.K.: Cambridge University Press.

Hu, Z., N. Wang, and R. J. Carroll. 2004. Profile-kernel versus backfitting in the partially linear models for longitudinal/clustered data. *Biometrika* 91:251–262.

Huang, X., 2006. Nonparametric estimation in large panel with cross-section dependence. Working paper, Department of Economics & Finance, Kennesaw State University, Kennesaw, GA.

Imbens, G. W., and W. K. Newey. 2009. Identification and estimation of triangular simultaneous equations models without additivity. *Econometrica* 77: 1481–1512.

Jin, S., and L. Su. 2010. A nonparametric poolability test for panel data models with cross section dependence. Working paper, School of Economics, Singapore Management University, Singapore.

Kniesner, T., and Q. Li. 2002. Semiparametric panel data models with heterogeneous dynamic adjustment: Theoretical consideration and an application to labor supply. *Empirical Economics* 27:131–148.

Lee, Y. 2008. Nonparametric estimation of dynamic panel models with fixed effects. Working paper, Department of Economics, University of Michigan, Ann Arbor.

Lee, Y., and D. Mukherjee. 2008. New nonparametric estimation of the marginal effects in fixed-effects panel models: An application on the environmental Kuznets curve. Working paper, Department of Economics, University of Michigan, Ann Arbor.

Li, Q. 2000. Efficient estimation of additive partially linear models. *International Economic Review* 41:1073–1092.

Li, Q., and T. Stengos. 1996. Semiparametric estimation of partially linear panel data models. *Journal of Econometrics* 71:289–397.

Li, Q., and C. Hsiao. 1998. Testing serial correlation in semiparametric panel data models. *Journal of Econometrics* 87:207–237.

Li, Q., and A. Ullah. 1998. Estimating partially linear panel data models with one-way error components. *Econometric Reviews* 17:145–166.

Li, Q., and J. S. Racine. 2007. *Nonparametric Econometrics: Theory and Practice*. Princeton, NJ: Princeton University Press.

Lin, X., and R. J. Carroll. 2000. Nonparametric function estimation for clustered data when the predictor is measured without/with error. *Journal of the American Statistical Association* 95:520–534.

Lin, X., and R. J. Carroll. 2001a. Semiparametric regression for clustered data. *Biometrika* 88:1179–1185.

Lin, X., and R. J. Carroll. 2001b. Semiparametric regression for clustered data using generalized estimating equations. *Journal of the American Statistical Association* 96:1045–1055.

Linton, O., and J. P. Nielsen. 1995. A kernel method of estimating structured nonparametric regression based on marginal integration. *Biometrika* 82:93–101.

Martins-Filho, C., and F. Yao. 2009. Nonparametric regression estimation with general parametric error covariance. *Journal of Multivariate Analysis* 100:309–333.

Matzkin, R. L. 2003. Nonparametric estimation of nonadditive random functions. *Econometrica* 71:1339–1375.

Mundra, K. 2005. Nonparametric slope estimation for fixed-effect panel data. Working paper, Department of Economics, San Diego State University, San Diego, CA.

Mukherjee, D. 2002. Nonparametric and semiparametric generalized panel data analysis of convergence and growth. *Dissertaion*. University of California, Riverside.

Mukherjee, D., and A. Ullah. 2003. Semiparmaetric analysis of generalized panel data: an application unpublished manuscript. University of California, Riverside.

Newey, W. K. 1990. Efficient instrumental variables estimation of nonlinear models. *Econometrica* 58:809–837.

Pagan, A., and A. Ullah. 1999. *Nonparametric Econometrics*. Cambridge, U.K.: Cambridge University Press.

Pesaran, M. H. 2006. Estimation and inference in large heterogenous panels with multifactor error. *Econometrica* 74:967–1012.

Ruckstuhl, A. F., A. H. Welsh, and R. J. Carroll. 2000. Nonparametric function estimation of the relationship between two repeatedly measured variables. *Statistica Sinica* 10:51–71.

Su, L., and A. Ullah. 2006. Profile likelihood estimation of partially linear panel data models with fixed effects. *Economics Letters* 92:75–81.

Su, L., and A. Ullah. 2007. More efficient estimation of nonparametric panel data models with random effects. *Economics Letters* 96:375–380.

Su, L., and A. Ullah. 2009. Testing conditional uncorrelatedness. *Journal of Business and Economic Statistics* 27:18–29.

Su, L., and S. Jin. 2010. Sieve estimation of panel data models with cross section dependence. Working paper. School of Economics, Singapore Management University, Singapore.

Su, L., and Y. Zhang. 2010a. Semiparametric panel data model with interactive effects. Working paper, School of Economics, Singapore Management University, Singapore.

Su, L., and Y. Zhang. 2010b. Testing for cross section dependence in nonparametric panel data models. Working paper, School of Economics, Singapore Management University, Singapore.

Su, L., Y. Chen, and A. Ullah. 2009. Functional coefficient estimation with both categorical and continuous data. *Advances in Econometrics* 25:131–167.

Su, L., A. Ullah, and Y. Wang. 2010. A note on nonparametric regression estimation with general parametric error covariance. Working paper, School of Economics, Singapore Management University, Singapore.

Sun, Y., R. J. Carroll, and D. Li. 2009. Semiparametric estimation of fixed effects panel data varying coefficient models. *Advances in Econometrics* 25:101–130.

Tran, K. C., and E. G. Tsionas. 2010. Local GMM estimation of semiparametric panel data with smooth coefficient models. *Econometric Reviews* 29:39–61.

Ullah, A., and N. Roy. 1998. Nonparametric and semiparametric econometrics of panel data. In *Handbook of Applied Economic Statistics*. A. Ullah and D. E. A. Giles (eds). New York: Marcel Dekker, pp. 579–604.

Index

C

F

G

N

V

W

Z

Milton Keynes UK
Ingram Content Group UK Ltd.
UKHW030902141024
449569UK00025B/1265

9 781138 113664